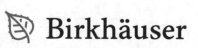

Marcelo R. Ebert • Michael Reissig

Methods for Partial Differential Equations

Qualitative Properties of Solutions, Phase Space Analysis, Semilinear Models

 Birkhäuser

Marcelo R. Ebert
University of São Paulo
Department of Computing and Mathematics
Ribeirão Preto, São Paulo, Brazil

Michael Reissig
TU Bergakademie Freiberg
Institute of Applied Analysis
Freiberg, Germany

ISBN 978-3-030-09772-1   ISBN 978-3-319-66456-9   (eBook)
https://doi.org/10.1007/978-3-319-66456-9

Mathematics Subject Classification (2010): 35-01, 35-02, 35A01, 35A02, 35A10, 35B30, 35B33, 35B40, 35B44, 35E20, 35F20, 35F35, 35J05, 35J10, 35J99, 35K05, 35K99, 35L05, 35L45, 35L71, 35L99

Printed on acid-free paper

This book is published under the trade name Birkhäuser, www.birkhauser-science.com
The registered company is Springer International Publishing AG
The registered company address is: Gewerbestrasse 11, 6330 Cham, Switzerland

Dedicated to our wives Helena and Steffi

Preface

It is the aim of this book to motivate young mathematicians at the postgraduate level to start studying different topics of the theory of partial differential equations. The book is addressed to Master and PhD students with interest in this theory. After attending introductory courses on PDE's, usually at the end of undergraduate studies, the backgrounds of those students may vary widely. Sometimes the theory on PDE's is reduced to Fourier's method and integral transformations only.

The main purpose of this book is to amplify an advanced course on PDE's during graduate studies. The book provides an overview on different topics of the theory of partial differential equations. As it is announced in the title, the explanation of qualitative properties of solutions of linear models, the introduction to phase space analysis on the one hand and modern methods how to treat semilinear models on the other hand, form the core of the book. The authors have chosen these subtitles because a deeper knowledge of these topics is an important base from which to apply or develop the theory of PDE's.

Usually, introductory courses on PDE's contain only very few properties of solutions to basic PDE's. We will illustrate qualitative properties of solutions of model equations showing fundamental differences between properties of solutions to Laplace, heat, wave or Schrödinger equations. A deep knowledge of such properties helps in understanding the applicability of different methods or techniques to treat more complex models.

Phase space analysis is widely applied in a lot of branches of the theory of PDE's. Only a very careful description of Fourier transform of functions or even distributions together with decomposition techniques of the phase space allows for attacking a lot of problems from the theory of partial differential equations. Without having tools from WKB analysis, the theory of pseudo-differential or para-differential operators, a lot of models can not be treated in an optimal way. For this reason the authors included chapters on basics of phase space analysis.

The move from linear models to nonlinear ones is a big challenge. Even if one adds the most simple nonlinearities $|u|^p$ or $\pm|u|^{p-1}u$ as a nonlinear right-hand side in a linear model it causes a lot of new difficulties. Sometimes such a term determines new trends in the theory. The authors show by semilinear heat, wave or Schrödinger models the big influence of these nonlinearities on methods, trends and expected results. A deep knowledge of how to treat semilinear models simplifies the move to understand quasi- or nonlinear models as well.

The authors have years of experience giving courses on PDE's at undergraduate and graduate level and supervising PhD students. The book contains their combined teaching and supervising expertise. The courses were held at various universities throughout the world.

In 2009, the first author gave a course about the Cauchy problem for partial differential operators within the 27th Brazilian Mathematics Colloquium at Instituto Nacional de Matemática Pura e Aplicada (IMPA) in Rio de Janeiro. Upon the invitation of Prof. Sandra Lucente, the second author stayed at the University of Bari in Italy during November 2013, and gave a series of lectures on partial differential equations within the "Project Messaggeri della Conoscenza 2012, ID424", supported by the government of Italy. The audience was composed of Master and PhD students not only from University of Bari. This comprehensive course was only one of a series of nine courses on "Basics of Partial Differential Equations" the second author gave during the period 2007–2016 at the Shanghai Jiao-Tong University, Kazakh National University Al Faraby Alma-Aty in Kazakhstan, the Eurasian National University L.N.Gumileva Astana in Kazakhstan, the Hanoi National University of Education and the Hanoi University of Science and Technology in Vietnam. The second author since 2002 gives the course "Partial Differential Equations 1/2" at the Faculty for Mathematics and Computer Science of Technical University Bergakademie Freiberg, and here he supervised 15 PhD students, several of them from abroad. The first author spent the period July 2014–July 2015, his sabbatical year, in Freiberg. During this time he gave advanced seminars on "Partial Differential Equations" for Master students. From all activities arose the joint idea to write this book.

It will be our great pleasure should this book stimulate young mathematicians to become familiar with the beautiful theory of partial differential equations.

Ribeirão Preto, São Paulo, Brazil   Marcelo R. Ebert
Freiberg, Germany   Michael Reissig
June 2017

Acknowledgements

The idea to write this book arose during the stay of the first author (July 2014–July 2015) at the Institute of Applied Analysis at TU Bergakademie Freiberg. The stay of the first author was supported by Fundação de Amparo à Pesquisa do Estado de São Paulo (FAPESP), grant 2013/20297-8. The book was completed within the DFG project RE 961/21-1 and FAPESP Grant 2015/16038-2.

The authors thank Vladimir Georgiev, Sandra Lucente, Winfried Sickel, Mitsuru Sugimoto, Hiroyuki Takamura and Karen Yagdjian for fruitful discussions on the content of some parts of this book. Moreover, the authors thank former PhD students Abdelhamid Mohammed Djaouti, Christian Jäh, Wanderley Nunes do Nascimento, Alessandro Palmieri and Maximilian Reich for reading some of the chapters.

Finally, we thank the staff from Birkhäuser publishing house, in particular, Sarah Goob and Dr. Thomas Hempfling, for the fruitful co-operation in preparing the final version of this book.

Contents

Part I

Chapter 1
Introduction

The present book is addressed to Master's and PhD students. The authors assume that readers have already attended basic courses on analysis, functional analysis ODE theory and a basic undergraduate course on PDE's. This book is our attempt to strike a balance between self-contained presentations of several aspects of PDE's and sketches of other aspects, in some cases we restrict ourselves to list results only. We hope that this approach motivates the reader to study the cited references and to complete in this way the non self-contained presentations. In following this way the reader becomes familiar with using mathscinet database, other textbooks or monographs to study several topics of the theory of PDE's, and develops skills useful in working independently.

One of the aims of the book is to familiarize advanced students with research. In order to reach this goal, some research projects for beginners are proposed in Chap. 23. Such projects are a good opportunity to apply mathematical methods and concepts which are presented in the book. Solving the proposed exercises at the end of each chapter, reading not only chapters of this monograph, but also studying the cited references and dealing with some of the research projects will help the reader become more involved in research activities. The authors used this philosophy in teaching their PhD and Master's students to think independently.

The book consists of 24 main chapters. Chapter 23 contains 13 research projects of different levels. All chapters have a short introduction. Most of the chapters contain concluding remarks and exercises. Here the authors collect on the one hand some knowledge which is not considered basic. On the other hand, recent trends, concepts and results, as well as results of the authors are listed. Some general remarks about the overall organization of the text are in order. The book is organized into five parts as we now describe.

Part I is considered as an introductory part of the book. Here the authors repeat or refresh some basic knowledge and mathematical prerequisites including a detailed discussion of classification of partial differential equations and systems of partial differential equations as well, classification of domains in which a process takes

© Springer International Publishing AG 2018

M.R. Ebert, M. Reissig, *Methods for Partial Differential Equations*,

https://doi.org/10.1007/978-3-319-66456-9_1

place, of notions of solutions and additional conditions as initial or boundary conditions to the solutions. Moreover, two of the most fundamental results in the theory of partial differential equations are given, the Cauchy-Kovalevskaja theorem and Holmgren's uniqueness theorem. In general, textbooks or monographs contain only the classical versions. The authors present here abstract versions as well in scales of Banach spaces and explain applications to fluid dynamics. Finally, the method of characteristics is introduced in detail. This method is applied in studying general quasilinear partial differential equations of first order as, for example, convection or transport equations. One of the most interesting applications of the method of characteristics is the study of Burger's equation. Notions as blow up, geometrical blow up, or life span of solutions to Burger's equation with and without mass term are introduced and discussed.

Part II is focused on qualitative properties of solutions to basic partial differential equations. Usual properties of solutions to elliptic, parabolic or hyperbolic equations we explain with the aid of the Laplace equation, heat equation or wave equation. Here the reader learns different features of each theory. A good knowledge of such properties helps develop some feeling for applicability of methods or tools within each theory. The reader should know some of the properties from basic courses on PDE's, but the authors explain, for example, such properties as hypo-ellipticity or local solvability, smoothing effects, finite or infinite speed of propagation of perturbations, existence of a domain of dependence, existence of forward or backward wave fronts and propagation of singularities, too. Moreover, relations to potential theory are explained. In this way the reader will understand how to apply potential theory to treat boundary value problems or mixed problems. Finally, the notion of energy of solutions, a very effective tool for the treatment of non-stationary or evolution models, will be discussed. The authors introduce energies for different models. Sometimes the choice of a suitable energy seems to be a miracle, sometimes one has different choices. All these issues are discussed. The reader can even find in addition to the description of the behavior of total energies the behavior of local energies as well, an issue one can not find generally in textbooks or monographs.

Part III is on phase space analysis, an efficient tool as was already pointed out in the preface. This part makes up Chaps. 12, 13, 14, 15, and 16. We illustrate the method for the treatment of the heat equation, Schrödinger equation, wave equation and plate equation. The main concern is to show how phase space analysis and interpolation techniques can be used to prove $L^p - L^q$ estimates on and away from the conjugate line. Such estimates are an essential tool for the study of nonlinear models. The reader learns how terms of lower order (mass or dissipation) or additional regularity of the data may influence expected results. Phase space analysis allows for proving the so-called diffusion phenomenon, a very interesting relation between classical damped wave models and heat models from the point of view of decay estimates. Finally, the method of stationary phase and applications is Covered. This method is used to derive $L^p - L^q$ estimates of solutions to dispersive equations (applications to wave and Schrödinger equations are presented). Key lemmas of this method are Littman-type lemmas. Two such lemmas are proved. In general, it is not easy to find such a proof in detail in the literature.

In Part IV we mainly discuss the treatment of semilinear models. More or less we focus on nonlinearities of power type on the right-hand side. Here we distinguish between source and absorbing power nonlinearities. In opposition to source power nonlinearities, one can "absorb" absorbing power nonlinearities in the definition of an energy. This implies benefits in the treatment of these models. The main goal here is to determine critical exponents. Two very famous critical exponents, the Fujita exponent and the Strauss exponent come into play. Depending on concrete models, these critical exponents divide the range of admissible powers in classes which allow for proving quite different qualitative properties of solutions, for example, the stability of the zero solution or blow up behavior of local (in time) solutions. One can also describe quite different asymptotic profiles of solutions (close to the Gauss kernel or close to self-similar solutions). To prove blow up results the authors introduce two methods, the test function method and the average method coupled with Kato type lemmas. Chapter 22 is a short introduction to linear hyperbolic systems. Here we have mainly in mind two classes of systems, symmetric hyperbolic and strictly hyperbolic ones.

In the last part (Chap. 24) we gather some background material as "Basics of Fourier transformation", some aspects of the "Theory of Fourier multipliers", some "Function spaces", "Some tools from distribution theory", and "Useful inequalities". There is no temptation to present these sections in a self-contained form. The readers are encouraged to study these topics in more detail by using related literature.

We hope this text will serve as an inviting source for further study by Master's and PhD students with interest in the theory of partial differential equations and as a useful future reference. The authors encourage students to solve the exercises proposed at the end of the chapters. Then it is the right moment to deal with one or two research projects from Chap. 23.

Chapter 2
Partial Differential Equations in Models

We begin with a discussion of various demands on mathematical modeling. We explain how to model technical processes as convection, diffusion, waves, or hydrodynamics. For this reason we introduce partial differential equations as Laplace equation, heat equation, wave equation or Schrödinger equation that play a central role in applications. These models are treated in later chapters.

2.1 A General Conservation Law

A general conservation law is a mathematical description of the following observations we often meet in nature or techniques.

Let G be an arbitrary domain in $\mathbb{R}^n$ and let u be a quantity in G. The amount of the quantity u in G is given by $\int_G u\,dx$. Let us denote by ϕ the flux vector of the quantity u. Then

$$q := -\int_{\partial G} \phi \cdot \mathbf{n}\, d\sigma,$$

is the amount of the quantity per unit time flowing across the boundary ∂G of G with exterior unit normal $\mathbf{n}$.

Then the rate of change of the amount of quantity u in G with respect to the time $t > 0$, that is,

$$d_t \int_G u\,dx = \int_G \partial_t u\,dx,$$

is equal to the flow q of the quantity u through the boundary ∂G of the domain G and the integral on the density f of sinks and sources of u inside of G.

M.R. Ebert, M. Reissig, *Methods for Partial Differential Equations*,
https://doi.org/10.1007/978-3-319-66456-9_2

A mathematical description of this conservation law is given in the form

$$\int_G \partial_t u \, dx = - \int_{\partial G} \phi \cdot \mathbf{n} \, d\sigma + \int_G f \, dx.$$

Thanks to the divergence theorem we may write

$$\int_G \left(\partial_t u + \text{div} \, \phi - f \right) dx = 0.$$

Under the assumption of smoothness of the integrands and the fact that G can be arbitrarily chosen we conclude the general conservation law

$$\partial_t u + \text{div} \, \phi = f(t, x), \quad x \in G, \quad t > 0.$$

2.2 Transport or Convection

In the case of linear convection the relation between the flow ϕ and the quantity u is given in homogeneous media by $\phi = \mathbf{c} u$. Here $\mathbf{c} = (c_1, \cdots, c_n)$ is a constant vector. So, every component ϕ_k of the flow ϕ is proportional to the quantity u. In this way we obtain the equation of linear *transport or convection*

$$\partial_t u + \sum_{k=1}^{n} c_k \, \partial_{x_k} u = f(t, x).$$

In the case of nonlinear convection the relation between the flow ϕ and the quantity u is given in homogeneous media by $\phi = \phi(u)$. So, every component of the flow depends nonlinearly on the quantity u. We obtain the equation of nonlinear *transport or convection*

$$\partial_t u + \sum_{k=1}^{n} \partial_{x_k} \phi_k(u) = 0$$

if we assume no sources or sinks inside of the domain G.

Remark 2.2.1 The partial differential equation

$$\partial_t u + \text{div} \, \phi(u) = 0$$

is called Burger's equation (see Chap. 7).

In non-homogeneous media we have the relation $\phi = \mathbf{c}(x)u$. Here it is allowed that the vector $\mathbf{c}$ may depend on the points x of the domain G. We get the equation of linear convection in non-homogeneous media as

$$\partial_t u + \sum_{k=1}^{n} c_k(x)\, \partial_{x_k} u + \Big(\sum_{k=1}^{n} \partial_{x_k} c_k(x) \Big) u = 0,$$

if we again assume no sources or sinks inside of the domain G. The last equation can be written in the form

$$\partial_t u + \mathbf{c} \cdot \nabla u + u \operatorname{div} c = 0.$$

We are interested in "solutions" of the linear transport equation in homogeneous media without any sources or sinks.

For this reason we choose the ansatz for solutions

$$u(t,x) = F(\mathbf{p} \cdot x - t) = F\Big(\sum_{k=1}^{n} p_k x_k - t \Big),$$

where F is a real differentiable function on $\mathbb{R}$ and $\mathbf{p} := (p_1, \cdots, p_n)$ is a given vector in $\mathbb{R}^n$. Then we get

$$\partial_t u + \sum_{k=1}^{n} c_k\, \partial_{x_k} u = \Big(-1 + \sum_{k=1}^{n} c_k p_k \Big) F'(\mathbf{p} \cdot x - t) = 0.$$

Consequently, $u(t,x) = F(\mathbf{p} \cdot x - t)$ is a (classical) solution of

$$\partial_t u + \mathbf{c} \cdot \nabla u = 0$$

for any differentiable function F if $\mathbf{c} \cdot \mathbf{p} = 1$. Such solutions are called *plane waves or traveling waves* for some direction $\mathbf{p} \in \mathbb{R}^n$ with velocity $\frac{1}{|\mathbf{p}|}$ and with profile F.

2.3 Diffusion

Let us come back to the general conservation law

$$\partial_t u + \operatorname{div} \phi = f(t,x).$$

Now we assume $\phi = -D\, \nabla u$. The constant D is called *coefficient of diffusion*. Then we obtain

$$\partial_t u - D\Delta u = f(t,x),$$

where $\Delta := \sum_{k=1}^{n} \partial_{x_k}^2$ is the Laplace operator. This partial differential equation is called *diffusion equation* (see Chap. 9). Diffusion of particles into media is described by the more general partial differential equation (among other things we use Fick's law of diffusion)

$$\rho(x)\partial_t u \; - \; \text{div}\left(D(x)\,\nabla u\right) \; + q(x)u = f(t,x),$$

where

- $\rho = \rho(x)$ is the porosity, $D = D(x)$ the coefficient of diffusion, and $q = q(x)$ the absorption
- $u = u(t,x)$ is the density of a particle in the point x for the time t.

The diffusion equation is also used to describe heat conduction. For this reason the equation is also called heat equation. Due to Fourier's law the heat flux is proportional to the gradient of temperature, that is, $\phi = -K\,\nabla T$. Here $T = T(t,x)$ describes the temperature in a point x for the time t. In this way we obtain

$$\partial_t T \; + \; \text{div}\,\phi = \partial_t T \; - \; K\,\text{div}\,\nabla T = \partial_t T \; - K\Delta T = f(t,x).$$

The thermal conductivity K depends in non-homogeneous media on x or, if big differences of temperature may appear, it depends on T also. Due to Fourier's law we have $\phi = -K(x,T)\,\nabla T$. Moreover, we conclude the nonlinear heat equation

$$\partial_t T \; - \; \text{div}\left(K(x,T)\,\nabla T\right) = f(t,x).$$

2.4 Stationary Models

Stationary (time independent) models are of special interest. If we observe, in general, a nonstationary process over a long time interval, then the solution might converge to a solution of the corresponding stationary model. Therefore, it is reasonable to suppose that sources or sinks are independent of time. The stationary models to the ones of Sect. 2.3 are

$$-D\Delta u = f(x), \quad -K\Delta T = f(x),$$

$$-\text{div}\left(D(x)\,\nabla u\right) + q(x)u = f(x) \quad \text{and}$$

$$-\text{div}\left(K(x,T)\,\nabla T\right) = f(x).$$

The partial differential equation $\Delta u = f(x)$ is called Poisson equation (see Chap. 8). In the case we have no sources or sinks, it is the Laplace equation $\Delta u = 0$. The Poisson equation is the basic equation of *potential theory*. Solutions of the Laplace equation describe potentials (gravitational potentials, single- or double-layer potentials, see Sect. 8.4.2) outside of sources or sinks.

2.5 Waves in Acoustics

We are interested in the propagation of sound waves in a tube with a circular cross-section. The tube is filled with gas and we suppose the state variables are constant along the cross-section, consequently depending on the length variable x and time variable t, only. The state variables are the mass density $\rho = \rho(t, x)$, the velocity $v = v(t, x)$ and the pressure $p = p(t, x)$. To derive the system of partial differential equations of acoustics we recall the general conservation law $\partial_t u + \partial_x \phi = f$ of Sect. 2.1. Taking into consideration the *mass conservation,* we choose $u = \rho$ and $\phi = \rho v$, the so-called *mass flow.* We assume no sources or sinks. The mass conservation is then described as

$$\partial_t \rho + \partial_x(\rho v) = 0.$$

This is the so-called *continuity equation.* If a given fluid is supposed to be incompressible, that is, the density ρ is supposed to be constant, then the continuity equation simplifies to $\partial_x v = 0$.

Now, let us devote consideration to the *conservation of momentum.* We choose the *density of momentum $u = \rho v$* and the *flow of momentum $\phi = (\rho v)v$.* Moreover, there is a source f. This source appears due to the gradient of the pressure along the axis of the length variable. It holds $f = -\partial_x p$. In this way we get the conservation of momentum

$$\partial_t(\rho v) + \partial_x(\rho v^2) = -\partial_x p.$$

Besides these two conservation laws, there is a third equation, *a constitutive equation,* connecting pressure and density by $p = F(\rho)$. It is reasonable to assume $F'(\rho) > 0$. So, pressure increases with increasing density. A usual example is the constitutive equation $p = k\rho^\gamma$ with constants $k > 0$ and $\gamma > 1$. To derive the model of acoustics we neglect thermal effects. So, the temperature is supposed to be constant and the *conservation of energy* can be omitted in the final model.

The theory of propagation of sound studies small perturbation of a homogeneous gas. Let us assume the gas is in an ideal state, that is, $\rho = \rho_0$, $v = 0$, $p_0 = F(\rho_0)$, where ρ_0 is a positive constant. There appears a small perturbation at one end of the tube. This perturbation is described by $\rho = \rho_0 + \tilde{\rho}(t, x)$, $v = \tilde{v}(t, x)$, $\tilde{\rho}$ and $\tilde{v}$ are small perturbations. The constitutive equation implies

$$p = F(\rho) = F(\rho_0 + \tilde{\rho}) = F(\rho_0) + F'(\rho_0)\tilde{\rho} + \frac{F''(\rho_0)}{2!}\tilde{\rho}^2 + \cdots$$

$$= p_0 + c^2\tilde{\rho} + \cdots,$$

where $c = \sqrt{F'(\rho_0)}$ is the speed of sound. Terms with $\tilde{\rho}^2$ are negligible with respect to terms with $\tilde{\rho}$, only. Together with the conservation of mass we obtain

$$\partial_t\left(\rho_0 + \tilde{\rho}\right) + \partial_x\big((\rho_0 + \tilde{\rho})\tilde{v}\big) = 0.$$

Setting the continuity equation into the conservation of momentum implies

$$\partial_t(\rho v) + \partial_x(\rho v^2) = v\,\partial_t\rho + \rho\,\partial_t v + v^2\,\partial_x\rho + \rho\,2v\partial_x v$$
$$= -v^2\,\partial_x\rho - \rho\,v\partial_x v + \rho\,\partial_t v + v^2\partial_x\rho + 2\rho\,v\partial_x v$$
$$= \rho\,\partial_t v + \rho\,v\partial_x v = -\partial_x p.$$

Summarizing, we conclude the relation

$$\rho\,\partial_t v + \rho\,v\,\partial_x v + \partial_x p = 0.$$

Setting the terms for ρ, v and p into the last relation yields

$$(\rho_0 + \tilde{\rho})\,\partial_t\tilde{v} + (\rho_0 + \tilde{\rho})\tilde{v}\,\partial_x\,\tilde{v} + \partial_x(p_0 + c^2\tilde{\rho} + \cdots) = 0.$$

The conservation of mass gives

$$\partial_t\tilde{\rho} + \rho_0\,\partial_x\tilde{v} + \partial_x\,(\tilde{\rho}\,\tilde{v}) = \partial_t\tilde{\rho} + \rho_0\,\partial_x\tilde{v} + \underbrace{\tilde{\rho}\,\partial_x\,\tilde{v} + \tilde{v}\,\partial_x\,\tilde{\rho}}\ = 0.$$

(these terms are small
because $\tilde{\rho}$ and $\tilde{v}$ are small)

Taking account of the conclusion of conservation of mass and momentum we get

$$\rho_0\,\partial_t\tilde{v} + c^2\,\partial_x\tilde{\rho} + \begin{array}{c}\text{small and therefore}\\ \text{negligible terms}\end{array} = 0.$$

There appears at least one of the terms $\tilde{v}$ or $\tilde{\rho}$ in the negligible terms. In this way we get the linear approximation

$$\partial_t\tilde{\rho} + \rho_0\,\partial_x\tilde{v} = 0, \qquad\qquad \rho_0\,\partial_t\tilde{v} + c^2\,\partial_x\tilde{\rho} = 0,$$

$$\downarrow \qquad\qquad\qquad\qquad\qquad \downarrow$$

differentiation according to t $\qquad\qquad$ differentiation according to x

$$\partial_t^2\tilde{\rho} + \rho_0\,\partial_{tx}^2\tilde{v} = 0, \qquad\qquad \rho_0\,\partial_{xt}^2\tilde{v} + c^2\,\partial_x^2\tilde{\rho} = 0.$$

Let us assume $\partial_{tx}^2\tilde{v} = \partial_{xt}^2\tilde{v}$. Then we verify the equation

$$\partial_t^2\tilde{\rho} - c^2\,\partial_x^2\tilde{\rho} = 0.$$

Analogously we derive

$$\partial_t^2\tilde{v} - c^2\,\partial_x^2\tilde{v} = 0.$$

So, we derived the wave equation $\partial_t^2 u - c^2 \partial_x^2 u = 0$, or in higher dimensions $\partial_t^2 u - c^2 \Delta u = 0$. We will study properties of solutions in Chap. 10.

Remark 2.5.1 The wave equation models, for example, transversal vibrations of an elastic string or of a membrane. If we are interested in the model of vibration of a string in an oil, then there appears additionally a damping term $k \, \partial_t u$, where k is a positive constant. Then we arrive at the classical damped wave equation

$$\partial_t^2 u - c^2 \Delta u + k \, \partial_t u = 0$$

(see Sects. 11.3 and 14.2). Moreover, we find wave equations in the description of small displacements of an elastic bar. The model is

$$\rho(x)S(x)\partial_t^2 u - \partial_x\big(E(x)S(x)\partial_x u\big) = F(t,x),$$

where $S(x)$ is the area of cross-section of the bar and $E(x)$ denotes Young's modulus.

2.6 Quantum Mechanics

Let us consider a particle with mass m that moves along the real axis by a force $F = F(x)$. Due to Newton's second law the position $x = x(t)$ is described by $m d_t^2 x = F(x)$. This relation does not hold any more in the atomic scale when taking into consideration Heisenberg's uncertainty principle. The state of a particle is explained by the aid of a so-called wave function $\psi = \psi(t,x)$. Here we restrict ourselves to the one-dimensional case. The function $|\psi(t,x)|^2$ is the probability density. Consequently, $\int_a^b |\psi(t,x)|^2 dx$ describes the probability of a particle being for a fixed time t in the interval $[a,b]$. The wave function ψ is a solution of the Schrödinger equation

$$i\tilde{h}\,\partial_t\psi = -\frac{\tilde{h}^2}{2m}\,\partial_x^2\psi + V(x)\psi$$

for $t > 0$ and $x \in \mathbb{R}$, where $V = V(x)$ is the potential energy, m the mass and $\tilde{h} = \frac{h}{2\pi}$ Planck's constant. The equation

$$i\,\partial_t\psi = -\Delta\psi$$

is called free Schrödinger equation in all dimensions (see Chap. 13).

2.7 Gas- and Hydrodynamics

Now we are interested in the description of a flow of an ideal fluid with vanishing viscosity. Let us restrict ourselves to the three-dimensional case. Let $v = (v_1, v_2, v_3)(t, x, y, z)$ be the velocity, $\rho = \rho(t, x, y, z)$ the density, $p = p(t, x, y, z)$ the pressure, $f = f(t, x, y, z)$ a source and $G(t, x, y, z) = (G_1(t, x, y, z), G_2(t, x, y, z), G_3(t, x, y, z))$ the weight. The basic equations of gas- or hydrodynamics are consequences of the conservation of mass and momentum. We obtain the system of partial differential equations

$$\partial_t \rho + \operatorname{div}(\rho v) = f \quad \text{(conservation of mass)},$$

$$\partial_t v + (v, \nabla)v + \frac{1}{\rho} \nabla p = G \quad \text{(conservation of momentum)},$$

where

$$(v, \nabla)v = \big((v_1, v_2, v_3) \cdot (\partial_x, \partial_y, \partial_z)\big)(v_1, v_2, v_3)$$

$$= \big(v_1 \partial_x v_1 + v_2 \partial_y v_1 + v_3 \partial_z v_1, v_1 \partial_x v_2 + v_2 \partial_y v_2 + v_3 \partial_z v_2,$$

$$v_1 \partial_x v_3 + v_2 \partial_y v_3 + v_3 \partial_z v_3\big).$$

Additionally, we have a constitutive equation $\Phi(p, \rho) = 0$ (cf. with Sect. 2.5). We devote more to the mathematical treatment of the system of gas- and hydrodynamics in Chap. 22.

2.8 Concluding Remarks

The Navier-Stokes equations are the equations of motion for a Newtonian fluid. If the fluid is supposed to be incompressible, that is, the density is supposed to be constant, the Navier-Stokes equations read as follows:

$$\rho \, \partial_t \mathbf{v} + \rho(\mathbf{v}, \nabla)\mathbf{v} - \mu \Delta_x \mathbf{v} + \nabla p = G, \quad \operatorname{div} \mathbf{v} = 0.$$

Here the viscosity μ is supposed to be constant. The first system of partial differential equations follows by the conservation of momentum, the second equation follows by the conservation of mass. If the velocity of the flow is small ($|\mathbf{v}|$ is small) or if the gradient of the velocity is small (($|\partial_x \mathbf{v}|, |\partial_y \mathbf{v}|, |\partial_z \mathbf{v}|$) is small, so we avoid any turbulence), then we may neglect the term $\rho(\mathbf{v}, \nabla)\mathbf{v}$. In this way we get the *Stokes equations*

$$\rho \, \partial_t \mathbf{v} - \mu \, \Delta_x \mathbf{v} + \nabla p = G, \quad \operatorname{div} \mathbf{v} = 0.$$

In the case of non-Newtonian fluids (slurry or floor screed) the viscosity is no longer constant. Without any elastic behavior the equations of motion are

$$\rho \, \partial_t \mathbf{v} + \rho(\mathbf{v}, \nabla)\mathbf{v} - \operatorname{div}\big(2\mu\big(\sqrt{2\operatorname{tr} D^2}\big)D\big) + \nabla p = G, \quad \operatorname{div} \mathbf{v} = 0,$$

where the viscosity $\mu = \mu\big(\sqrt{2 \operatorname{tr} D^2}\big)$ depends on $\sqrt{2 \operatorname{tr} D^2}$.

Here $D = \frac{1}{2}(L + L^T)$ denotes the symmetric part of the *velocity gradient tensor L*, where

$$L = \begin{pmatrix} \frac{\partial u_1}{\partial x} & \frac{\partial u_1}{\partial y} & \frac{\partial u_1}{\partial z} \\ \frac{\partial u_2}{\partial x} & \frac{\partial u_2}{\partial y} & \frac{\partial u_2}{\partial z} \\ \frac{\partial u_3}{\partial x} & \frac{\partial u_3}{\partial y} & \frac{\partial u_3}{\partial z} \end{pmatrix}, \quad L^T \quad \text{is the transposed matrix to } L$$

and $\operatorname{tr} D^2$ is the trace of D^2.

If the non-Newtonian fluid has an elastic behavior, the fluid has a *memory for former strains*, then, in the case of rheological elementary materials (rheology studies deformations and flow rating of matters), we have the model

$$\rho \, \partial_t \mathbf{v} + \rho(\mathbf{v}, \nabla)\mathbf{v} - \operatorname{div}\left(2\int_{-\infty}^{t} G(t-\tau)D(\tau, x, y, z)d\tau\right) + \nabla p = G,$$

$$\operatorname{div} \mathbf{v} = 0,$$

where $G = G(t)$ is the so-called *relaxation function for isotropic fluids*.

We will not present methods on how to treat the Stokes system or the system of Navier-Stokes equations. The interested reader can find comprehensive explanations to these research topics in the monographs [58, 203] or in the introductory papers [57, 218].

Exercises Relating to the Considerations of Chap. 2

Exercise 1 Try to find solutions to the wave equation $\partial_t^2 u - c^2 \Delta u = 0$ in the form of plane waves or traveling waves.

Exercise 2 Try to find solutions to the classical damped wave equation $\partial_t^2 u - c^2 \Delta u + k\partial_t u = 0$ in the form of plane waves or traveling waves.

Chapter 3
Basics for Partial Differential Equations

This chapter is devoted to mathematical prerequisites, including a detailed discussion of classification of partial differential equations and systems of partial differential equations, as well as classification of domains in which a process takes place, of notions of solutions and additional conditions as initial or boundary conditions to the solutions.

3.1 Classification of Linear Partial Differential Equations of Kovalevskian Type

Consider the linear partial differential equation of *Kovalevskian type*

$$D_t^m u + \sum_{k+|\alpha|\leq m,\, k\neq m} a_{k,\alpha}(t,x) D_x^\alpha D_t^k u = f(t,x),$$

where $D_t = -i\partial_t$ and $D_{x_k} = -i\partial_{x_k}, k = 1,\cdots,n, i^2 = -1$. Here $\alpha = (\alpha_1,\cdots,\alpha_n) \in \mathbb{N}^n$ is a multi-index and $m \in \mathbb{N}$. We introduce the notions *principal part of the given linear partial differential operator*, it is the linear partial differential operator

$$D_t^m + \sum_{k+|\alpha|=m,\, k\neq m} a_{k,\alpha}(t,x) D_x^\alpha D_t^k,$$

and *part of lower order terms*, it is the linear partial differential operator

$$\sum_{k+|\alpha|<m} a_{k,\alpha}(t,x) D_x^\alpha D_t^k.$$

© Springer International Publishing AG 2018
M.R. Ebert, M. Reissig, *Methods for Partial Differential Equations*,
https://doi.org/10.1007/978-3-319-66456-9_3

To decide which type a given partial differential equation of Kovalevskian type has, we study the *principal symbol* of the *principal part*. We replace in the principal part D_t by τ, D_{x_k} by ξ_k. Consequently, we replace $D_x^\alpha D_t^k = D_{x_1}^{\alpha_1} D_{x_2}^{\alpha_2} \cdots D_{x_n}^{\alpha_n} D_t^k$ by $\xi_1^{\alpha_1} \xi_2^{\alpha_2} \cdots \xi_x^{\alpha_n} \tau^k =: \xi^\alpha \tau^k$. In this way the principal symbol is given by

$$\tau^m + \sum_{k+|\alpha|=m,\, k\neq m} a_{k,\alpha}(t,x)\xi^\alpha \tau^k.$$

Definition 3.1 Consider the above differential operator of Kovalevskian type where the coefficients of the principal part are assumed to be real in a domain $G \subset \mathbb{R}^{n+1}$. Then the operator is called

- *elliptic in a point* $(t_0, x_0) \in G$ if the characteristic equation

$$\tau^m + \sum_{k+|\alpha|=m,\, k\neq m} a_{k,\alpha}(t_0,x_0)\xi^\alpha \tau^k = 0$$

  has for $\xi \neq 0$ no real roots $\tau_1 = \tau_1(t_0, x_0, \xi), \cdots, \tau_m = \tau_m(t_0, x_0, \xi)$
- *elliptic in a domain G* if the operator is elliptic in every point $(t_0, x_0) \in G$
- *hyperbolic in a point* $(t_0, x_0) \in G$ if the characteristic equation

$$\tau^m + \sum_{k+|\alpha|=m,\, k\neq m} a_{k,\alpha}(t_0,x_0)\xi^\alpha \tau^k = 0$$

  has for $\xi \neq 0$ only real roots $\tau_1 = \tau_1(t_0, x_0, \xi), \cdots, \tau_m = \tau_m(t_0, x_0, \xi)$
- *strictly hyperbolic in a point* $(t_0, x_0) \in G$ if the characteristic equation

$$\tau^m + \sum_{k+|\alpha|=m,\, k\neq m} a_{k,\alpha}(t_0,x_0)\xi^\alpha \tau^k = 0$$

  has for $\xi \neq 0$ only real and pairwise distinct roots $\tau_1 = \tau_1(t_0, x_0, \xi), \cdots, \tau_m = \tau_m(t_0, x_0, \xi)$
- *hyperbolic (strictly hyperbolic) in a domain G* if the operator is hyperbolic (strictly hyperbolic) in every point $(t_0, x_0) \in G$.

Now we are interested in which type does the *nonstationary plate operator* $\partial_t^2 + \Delta^2$ have?

At a first glance we are not able to give an answer because this operator is not of Kovalevskian type, so it does not fit into the class of operators from Definition 3.1. This operator is *not hyperbolic*, but it is a so-called *2-evolution operator*.

Let us introduce the notion *p-evolution operator*. For this reason we consider the non-Kovalevskian (if $p > 1$) linear partial differential equation

$$D_t^m u + \sum_{j=1}^m A_j(t,x,D_x)D_t^{m-j}u = f(t,x),$$

where $A_j = A_j(t, x, D_x) = \sum_{k=0}^{jp} A_{j,k}(t, x, D_x)$ are linear partial differential operators of order jp for a fixed integer $p \geq 1$, and $A_{j,k} = A_{j,k}(t, x, D_x)$ are linear partial differential operators of order k. The *principal part* of this linear partial differential operator *in the sense of Petrovsky* is defined by

$$D_t^m + \sum_{j=1}^{m} A_{j,jp}(t, x, D_x) D_t^{m-j}.$$

Definition 3.2   The given linear partial differential operator

$$D_t^m + \sum_{j=1}^{m} A_j(t, x, D_x) D_t^{m-j}$$

is called a *p-evolution operator* if the principal symbol in the sense of Petrovsky

$$\tau^m + \sum_{j=1}^{m} A_{j,jp}(t, x, \xi) \tau^{m-j}$$

has only real and distinct roots $\tau_1 = \tau_1(t, x, \xi), \cdots, \tau_m = \tau_m(t, x, \xi)$ for all points (t, x) from the domain of definition of coefficients and for all $\xi \neq 0$.

Remark 3.1.1   The set of 1-evolution operators coincides with the set of strictly hyperbolic operators. The p-evolution operators with $p \geq 2$ represent generalizations of Schrödinger operators.

Remark 3.1.2   One of the later goals is to study Cauchy problems for p-evolution equations. Taking account of the Lax-Mizohata theorem for the principal symbol [140], the assumption that the characteristic roots are real in Definition 3.2 is necessary for proving well-posedness for the Cauchy problem.

3.2   Classification of Linear Partial Differential Equations of Second Order

Let us consider the general linear partial differential equation of second order

$$\sum_{j,k=1}^{n} a_{jk}(x) \partial_{x_j x_k}^2 u + \sum_{k=1}^{n} b_k(x) \partial_{x_k} u + c(x)u = f(x)$$

with real and continuous coefficients in a domain $G \subset \mathbb{R}^n$. Let u be a twice continuously differentiable solution. Then we may assume the matrix of coefficients $(a_{jk}(x))_{j,k=1}^{n}$ to be symmetric in G. Consequently,

- all eigenvalues $\lambda_1, \cdots, \lambda_n$ are real
- the numbers of positive, negative and vanishing eigenvalues remain invariant under a regular coordinate transformation.

These properties allow for the following definition.

Definition 3.3 Let

$$\sum_{j,k=1}^{n} a_{jk}(x)\partial^2_{x_j x_k} u + \sum_{k=1}^{n} b_k(x)\partial_{x_k} u + c(x)u = f(x)$$

be a given linear partial differential equation with real and continuous coefficients and right-hand side in a domain $G \subset \mathbb{R}^n$. The matrix $(a_{jk}(x))_{j,k=1}^{n}$ is supposed to be symmetric in G. Let $x_0 \in G$.

- If $\lambda_1, \cdots, \lambda_n$ are non-vanishing and have the same sign in x_0, then the differential equation is called *elliptic in x_0*.
- If $\lambda_1, \cdots, \lambda_n$ are non-vanishing and if all eigenvalues, except exactly one, have the same sign in x_0, then the differential equation is called *hyperbolic in x_0*.
- If $\lambda_1, \cdots, \lambda_n$ are non-vanishing and if at least two are positive and at least two are negative in x_0, then the differential equation is called *ultrahyperbolic in x_0*.
- If one eigenvalue is vanishing in x_0, then the differential equation is called *parabolic in x_0*.
- If exactly one eigenvalue is vanishing in x_0 and if the other eigenvalues have the same sign in x_0, then the differential equation is called *parabolic of normal type in x_0*.

Example 3.2.1 Consider in $\mathbb{R}^2$ the partial differential equation

$$\partial_t^2 u - t^2 \partial_x^2 u = 0.$$

The characteristic roots $\tau_{1,2}$ are given by $\tau_1(t, x, \xi) = t\xi$ and $\tau_2(t, x, \xi) = -t\xi$. Due to Definition 3.1, this partial differential equation is hyperbolic in $\mathbb{R}^2$ and simultaneously strictly hyperbolic away from the line $\{(t, x) \in \mathbb{R}^2 : t = 0\}$. On the other hand, the eigenvalues in Definition 3.3 are given by $\lambda_1(t, x) \equiv 1$ and $\lambda_2(t, x) = -t^2$. Hence, with respect to Definition 3.3 this partial differential equation is hyperbolic away from the line $\{(t, x) \in \mathbb{R}^2 : t = 0\}$. On the line $t = 0$ this partial differential equation is parabolic of normal type. For this reason, this line is called a parabolic line, too.

Example 3.2.2 Let

$$\sum_{j,k=1}^{n} a_{jk}(x_1, \cdots, x_n) \, \partial^2_{x_j x_k}$$

be a linear differential operator with real and continuous coefficients and symmetric matrix $(a_{jk}(x))_{j,k=1}^n$ in a domain $G \subset \mathbb{R}^n$. If its symbol satisfies the estimate

$$\sum_{j,k=1}^n a_{jk}(x_1, \cdots, x_n)\, \xi_j \xi_k \geq C|\xi|^2,$$

that is, the quadratic form on the left-hand side is positive definite, then:

1. the differential equation

$$\sum_{j,k=1}^n a_{jk}(x)\partial^2_{x_j x_k} u + \sum_{k=1}^n b_k(x)\partial_{x_k} u + c(x)u = f(x)$$

   is elliptic in G, if f is defined in G;

2. the differential equation

$$\partial^2_t u - \sum_{j,k=1}^n a_{jk}(x_1, \cdots, x_n)\, \partial^2_{x_j x_k} u + \sum_{k=1}^n b_k(x)\partial_{x_k} u + c(x)u = f(t, x)$$

   is strictly hyperbolic in every cylinder $(0, T) \times G$, $T > 0$, if f is defined in $(0, T) \times G$;

3. the differential equation

$$\partial_t u - \sum_{j,k=1}^n a_{jk}(x_1, \cdots, x_n)\, \partial^2_{x_j x_k} u + \sum_{k=1}^n b_k(x)\partial_{x_k} u + c(x)u = f(t, x)$$

   is parabolic of normal type in every cylinder $(0, T) \times G$, $T > 0$, if f is defined in $(0, T) \times G$.

3.3   Classification of Linear Systems of Partial Differential Equations

In this section, we firstly study linear systems of m first order partial differential equations in m unknowns and two independent variables of the form

$$\partial_t u_k + \sum_{j=1}^m \left(a_{kj}(t, x)\partial_x u_j + b_{kj}(t, x)u_j \right) = f_k(t, x), \ k = 1, \cdots, m.$$

In matrix notation this system takes the form $(U = (u_1, \cdots, u_m)^T$, $F = (f_1, \cdots, f_m)^T)$

$$\partial_t U + A(t, x)\partial_x U + B(t, x)U = F(t, x),$$

where we denote

$$A = A(t, x) := (a_{kj}(t, x))_{k,j=1}^m \text{ and } B = B(t, x) := (b_{kj}(t, x))_{k,j=1}^m.$$

Just as in the case of a single partial differential equation, it turns out that most of the properties of solutions depend on the *principal part* $\partial_t U + A(t, x)\partial_x U$ of this system. Since the principal part is completely characterized by the matrix $\tau I + A\xi$ (∂_t is replaced by τ and ∂_x is replaced by ξ), this matrix plays a fundamental role in the study of these systems. There are two important classes of systems of the above form defined by properties of the matrix A.

Definition 3.4 Let us consider the above system of partial differential equations, where the entries of the matrix A are assumed to be real and continuous in a domain $G \subset \mathbb{R}^2$. Then the system is called

- *elliptic in a point* $(t_0, x_0) \in G$ if the matrix $A(t_0, x_0)$ has no real eigenvalues $\lambda_1 = \lambda_1(t_0, x_0), \cdots, \lambda_m = \lambda_m(t_0, x_0)$
- *elliptic in a domain G* if the system is elliptic in every point $(t_0, x_0) \in G$
- *hyperbolic in a point* $(t_0, x_0) \in G$ if the matrix $A(t_0, x_0)$ has real eigenvalues $\lambda_1 = \lambda_1(t_0, x_0), \cdots, \lambda_m = \lambda_m(t_0, x_0)$ and a full set of right eigenvectors
- *strictly hyperbolic in a point* $(t_0, x_0) \in G$ if the matrix $A(t_0, x_0)$ has distinct real eigenvalues $\lambda_1 = \lambda_1(t_0, x_0), \cdots, \lambda_m = \lambda_m(t_0, x_0)$
- *hyperbolic (strictly hyperbolic) in a domain G* if the operator is hyperbolic (strictly hyperbolic) in every point $(t_0, x_0) \in G$.

Now we would like to present a classification for linear systems of partial differential equations of first order having the form

$$\partial_t U + \sum_{k=1}^n A_k(t, x)\partial_{x_k} U + A_0(t, x)U = F(t, x) \quad \text{in} \quad [0, \infty) \times \mathbb{R}^n,$$

where A_k, $k = 0, 1, \cdots n$, are continuous $m \times m$ matrices, subject to the Cauchy condition

$$U(0, x) = U_0(x).$$

For the following we introduce the notation

$$A(t, x, \xi) := \sum_{k=1}^n \xi_k A_k(t, x) \quad \text{for} \quad t \geq 0, \ (x, \xi) \in \mathbb{R}^{2n}.$$

Just as in the case of scalar partial differential equations, it turns out that most of the properties of solutions depend on the "principal part" of the differential operator. So, the matrix $A(t, x, \xi)$ plays a fundamental role in the study of these systems. There are two important classes of systems of the above form defined by properties of the matrix A.

Definition 3.5 Let us consider the above differential system where the entries of the $m \times m$ matrix A_k are assumed to be real and continuous in a domain $G \subset \mathbb{R}^{1+n}$. Then the system is called

- *elliptic in a point* $(t_0, x_0) \in G$ if the matrix $A(t_0, x_0, \xi)$ has no real eigenvalues $\lambda_1 = \lambda_1(t_0, x_0, \xi), \cdots, \lambda_m = \lambda_m(t_0, x_0, \xi)$ for all $\xi \in \mathbb{R}^n \setminus \{0\}$
- *elliptic in a domain* G if the system is elliptic in every point $(t_0, x_0) \in G$
- *hyperbolic* in a point $(t_0, x_0) \in G$ if the matrix $A(t_0, x_0, \xi)$ has m real eigenvalues $\lambda_1(t_0, x_0, \xi) \leq \lambda_2(t_0, x_0, \xi) \leq \cdots \leq \lambda_m(t_0, x_0, \xi)$ and a full set of right eigenvectors
- *hyperbolic in a domain* G if the system is hyperbolic in every point $(t_0, x_0) \in G$.

There are two important special cases.

Definition 3.6 We say that the above system is *strictly hyperbolic* if for each $x, \xi \in \mathbb{R}^n$, $\xi \neq 0$, and each $t \geq 0$, the matrix $A(t, x, \xi)$ has m distinct real eigenvalues:

$$\lambda_1(t, x, \xi) < \lambda_2(t, x, \xi) < \cdots < \lambda_m(t, x, \xi).$$

We say that the above system is a *symmetric hyperbolic system* if all $m \times m$ matrices $A_k(t, x)$ are symmetric for $k = 1, \cdots, m$.

Example 3.3.1 In the complex function theory holomorphic functions $w = w(z)$, which are defined in a given domain $G \subset \mathbb{C}$, are classical solutions of $\partial_{\bar{z}} w = \frac{1}{2}(\partial_x + i\partial_y)w = 0$. If we introduce $w(z) =: u(x, y) + iv(x, y)$, then this elliptic equation is equivalent to the Cauchy-Riemann system

$$\partial_x U + A \, \partial_y U = 0,$$

where $U(x, y) = (u(x, y), v(x, y))^T$ and the matrix

$$A = \begin{pmatrix} 0 & -1 \\ 1 & 0 \end{pmatrix}.$$

The Cauchy-Riemann system is an *elliptic* system in the domain $G \subset \mathbb{R}^2$.

Example 3.3.2 The study of long gravity waves on the surface of a fluid in a channel leads to the linear system

$$\partial_t u + g \, \partial_x \rho = 0,$$
$$\partial_t \rho + \frac{S_0(x)}{b} \, \partial_x u + \frac{S_0'(x)}{b} \, u = 0,$$

where b and g are positive constants and $S_0 = S_0(x)$ is a given differentiable positive function representing the equilibrium cross-sectional area of the fluid in the channel (for more details see [120]). This system is *strictly hyperbolic* in the domain of the definition of the coefficients in the (t, x)-plane.

Example 3.3.3 In electrical engineering the modeling of transmission lines leads to the system

$$L\partial_t I + \partial_x E + RI = 0,$$
$$C\partial_t E + \partial_x I + GE = 0,$$
$$I(0, x) = I_0(x), \quad E(0, x) = E_0(x), \quad x \in \mathbb{R}.$$

Here C denotes the capacitance to ground per unit length, G is the conductance to ground per unit length, L is the inductance per unit length and R the resistance per unit length. The unknowns $I = I(t, x)$ and $E = E(t, x)$ are, respectively, the current and potential at point x of the line at time t. This system is strictly hyperbolic in the whole (t, x)-plane (see [231] and for its solvability Example 22.4.1).

Example 3.3.4 The equations governing the electromagnetic field in $\mathbb{R}^3$ are given by

$$\partial_t B + \nabla \times E = 0, \qquad \partial_t E - \nabla \times B = 0,$$

where $B := (B_1, B_2, B_3)$ and $E := (E_1, E_2, E_3)$ denote the magnetic and electric fields, respectively. This system of partial differential equations forms a *symmetric hyperbolic system* for $U := (B_1, B_2, B_3, E_1, E_2, E_3)^T$.

The class of symmetric hyperbolic systems can be enlarged to *general symmetric hyperbolic systems* with continuous coefficients of the form

$$A_0(t, x)\partial_t U + \sum_{k=1}^{n} A_k(t, x)\partial_{x_k} U + B(t, x)U = F(t, x),$$

where the real matrix A_0 is supposed to be *positive definite* uniformly with respect to (t, x) from the domain of definition of the coefficients, and the matrices A_k are *real and symmetric*.

In a lot of cases we are able to transform a single linear partial differential equation of higher order to a linear system of partial differential equations.

For example, every linear hyperbolic equation of second order with smooth coefficients and smooth right-hand side can be transformed into a symmetric hyperbolic system. Let us consider the special class of linear hyperbolic equations of second order

$$\partial_t^2 u - \sum_{j,k=1}^{n} a_{j,k}(t, x)\partial_{x_j x_k}^2 u + \sum_{j=1}^{n} b_j(t, x)\partial_{x_j} u + c(t, x)\partial_t u + d(t, x)u$$
$$= f(t, x),$$

where all coefficients are real-valued and $(a_{j,k}(t,x))_{j,k=1}^n$ is a *symmetric positive definite $n \times n$ matrix* (we assume smoothness of solutions) uniformly with respect to t and x. Let $u_1 := \partial_{x_1} u, \cdots, u_n := \partial_{x_n} u$, $u_{n+1} := \partial_t u$, $u_{n+2} := u$. We then obtain the following system of partial differential equations for the $N := n + 2$ functions $u_1, \cdots, u_{n+2}$:

$$\sum_{k=1}^n a_{j,k}(t,x)\partial_t u_k - \sum_{k=1}^n a_{j,k}(t,x)\partial_{x_k} u_{n+1} = 0, \quad k = 1, \cdots, n,$$

$$\partial_t u_{n+1} - \sum_{j,k=1}^n a_{j,k}(t,x)\partial_{x_k} u_j + \sum_{j=1}^n b_j(t,x)u_j + c(t,x)u_{n+1} + d(t,x)u_{n+2}$$

$$= f(t,x),$$

$$\partial_t u_{n+2} - u_{n+1} = 0.$$

Setting $U = (u_1, \cdots, u_{n+2})^T$ and $F = (0, \cdots, 0, f, 0)^T$, this system is equivalent to a symmetric hyperbolic system

$$A_0(t,x)\partial_t U + \sum_{k=1}^n A_k(t,x)\partial_{x_k} U + B(t,x)U = F(t,x),$$

where

$$A_0 := \begin{pmatrix} a_{1,1} & \dots & a_{1,n} & 0 & 0 \\ \vdots & & \vdots & \vdots & \vdots \\ a_{n,1} & \dots & a_{n,n} & 0 & 0 \\ 0 & \dots & 0 & 1 & 0 \\ 0 & \dots & 0 & 0 & 1 \end{pmatrix}, \quad B := \begin{pmatrix} 0 & \dots & 0 & 0 & 0 \\ \vdots & & \vdots & \vdots & \vdots \\ 0 & \dots & 0 & 0 & 0 \\ b_1 & \dots & b_n & c & d \\ 0 & \dots & 0 & -1 & 0 \end{pmatrix},$$

$$A_k := \begin{pmatrix} 0 & \dots & 0 & -a_{1,k} & 0 \\ \vdots & & \vdots & \vdots & \vdots \\ 0 & \dots & 0 & -a_{n,k} & 0 \\ -a_{1,k} & \dots & -a_{n,k} & 0 & 0 \\ 0 & \dots & 0 & 0 & 1 \end{pmatrix} \quad \text{for} \quad k = 1, \cdots, n.$$

The matrix A_0 is positive definite. The matrices A_k are symmetric.

Remark 3.3.1 There are other approaches to reduce scalar hyperbolic equations of higher order to systems of first order. Consider

$$D_t^m u + \sum_{k+|\alpha| \le m, k \ne m} a_{k,\alpha}(t,x)D_x^\alpha D_t^k u = f(t,x).$$

If we introduce $U = \left(\langle D_x \rangle^{m-1} u, \langle D_x \rangle^{m-2} D_t u, \cdots, \langle D_x \rangle D_t^{m-2} u, D_t^{m-1} u \right)^T$ and $F = (0, \cdots, 0, f)^T$, then we get the system of first order

$$D_t U - A(t, x, D_x)U = F(t, x),$$

where

$$
-A(t,x,D_x) := \begin{pmatrix}
0 & & & -\langle D_x \rangle & & \cdots \\
\cdot & & & 0 & & \\
\vdots & & & \vdots & & \\
0 & & & 0 & & \\
\displaystyle\sum_{|\alpha| \le m} a_{0,\alpha}(t,x) D_x^\alpha \langle D_x \rangle^{1-m} & & \displaystyle\sum_{|\alpha| \le m-1} a_{1,\alpha}(t,x) D_x^\alpha \langle D_x \rangle^{2-m} & \cdots \\
\\
0 & & & 0 \\
\vdots & & & \vdots \\
-\langle D_x \rangle & & & 0 \\
0 & & & -\langle D_x \rangle \\
\displaystyle\sum_{|\alpha| \le 2} a_{m-2,\alpha}(t,x) D_x^\alpha \langle D_x \rangle^{-1} & & \displaystyle\sum_{|\alpha| \le 1} a_{m-1,\alpha}(t,x) D_x^\alpha
\end{pmatrix}.
$$

The matrix A has the so-called *Sylvester structure*. Here we used the pseudodifferential operator $\langle D_x \rangle$.
There is a difference between both systems of first order.
The second system is a *pseudodifferential system of first order*. After the reduction to the Sylvester structure, the *theory of pseudodifferential operators* should be applied.

3.4   Classification of Domains and Statement of Problems

Models in nature and technique are often described by partial differential equations or systems of partial differential equations. In general, we need additional conditions describing the behavior of solutions of such models. First we shall distinguish between   *stationary processes* and *nonstationary processes*. Moreover, we shall distinguish between different domains in which a process takes place.
On the one hand we have *interior domains* (e.g. transversal vibration of a string which is fixed in two points, bending of a plate, or heat conduction in a body), *exterior domains* (e.g. potential flows around a cylinder), *whole space* $\mathbb{R}^n$ (e.g. propagation of electro-magnetic waves, potential of a mass point, potential of a single-layer or a double-layer), *wave guides* (e.g. propagation of sound in an infinite tube), wave guides are domains $\{(x, y) \in \mathbb{R}^{n+m} : (x, y) \in \mathbb{R}^n \times G, \, G \subset \mathbb{R}^m\}$, where G is an interior domain, the *exterior to wave guides* (e.g. diffraction of electromagnetic

waves around an infinite cylinder), or, finally, the *half-space* (e.g. reflection of waves on a large plane obstacle).

If we have a stationary (nonstationary) process, then the process takes place in a domain G (in a cylinder $(0, T) \times G$), where G is one of the above described domains. We are only interested in observing the forward (in time) process. If we observe the process for a long time, we set $T = \infty$. Depending on a given domain, we have to pose different additional conditions.

3.4.1   Stationary Processes

We pose *boundary conditions* on the boundary ∂G of the domain G. If we have an interior domain, then the usual boundary conditions are those of first, second, or third kind. These are called boundary conditions of *Dirichlet, Neumann* or *Robin*-type, respectively.

If we study boundary value problems for solutions to the elliptic equation

$$\sum_{k,j=1}^{n} a_{jk}(x)\partial^2_{x_k x_j}u + \sum_{k=1}^{n} b_k(x)\partial_{x_k}u + c(x)u = f(x)$$

in an interior domain G, then we state, instead of the Neumann boundary condition, *the co-normal boundary condition*

$$\sum_{j,k=1}^{n} a_{jk}(x)\partial_{x_j}u \cos(\mathbf{n}, \mathbf{e}_k)\Big|_{\partial G} = g(x).$$

This boundary condition is related to the structure of the given elliptic operator. If we consider the special case

$$\Delta u + \sum_{k=1}^{n} b_k(x)\partial_{x_k}u + c(x)u = f(x),$$

then the coefficients $a_{jk}(x)$ are equal to the Kronecker symbol δ_{jk}. Hence, the co-normal derivative is equal to

$$\sum_{j=1}^{n} a_{jj}(x)\partial_{x_j}u \cos(\mathbf{n}, \mathbf{e}_j)\Big|_{\partial G} = \sum_{j=1}^{n} \partial_{x_j}u \cos(\mathbf{n}, \mathbf{e}_j)\Big|_{\partial G} = \partial_{\mathbf{n}}u\Big|_{\partial G} = g(x),$$

and this is the classical Neumann condition.

If we study boundary value problems in exterior domains, then decay conditions (conditions for the solution if $|x|$ tends to infinity) might select the physical one among all solutions. Thus, the problem

$$\Delta u = 0, \quad u\big|_{\partial G} = g(x), \quad u(x) = O\left(\frac{1}{|x|}\right) \text{ for } x \to \infty,$$

in $G = \{x \in \mathbb{R}^3 : |x| > 1\}$ is of interest in the potential theory. Both conditions determine the solution uniquely.

If we are interested in boundary value problems for solutions to the *Helmholtz equation* $\Delta u + k^2 u = 0, k^2 > 0$, then spectral properties of the Helmholtz operator require an additional condition for the boundary and the decay condition to determine the solution uniquely.

Example 3.4.1 Let us study the boundary value problem

$$\Delta u + (2\pi)^2 u = 0, \quad u\big|_{\partial G} = 0, \quad u(x) = O\left(\frac{1}{|x|}\right) \text{ for } x \to \infty$$

in the exterior domain $G = \{x \in \mathbb{R}^3 : |x| > 1\}$. Then, the family of functions $\{u_C = u_C(x), C \in \mathbb{R}\}$ with $u_C(x) = -C\sin(2\pi|x|)/(4\pi|x|)$ is a family of radial solutions to this Dirichlet boundary value problem. Sommerfeld proposed *Sommerfeld's radiation condition*,

     either $r\partial_r u - i2\pi r u \to 0$ or $r\partial_r u + i2\pi r u \to 0$ for $r = |x| \to \infty$,

     and for solutions to the general Helmholtz equation,

     either $r\partial_r u - ikr u \to 0$ or $r\partial_r u + ikr u \to 0$ for $r = |x| \to \infty$

to select the solution $u \equiv 0$.

3.4.2 Nonstationary Processes

We state *boundary conditions* on the lateral surface $(0, T) \times \partial G$ of the cylinder $(0, T) \times G$ and initial conditions on the bottom $\{t = 0\} \times G$. The number m of initial conditions corresponds to the order m of partial derivatives with respect to t in

$$\partial_t^m u = \cdots .$$

In general, it is permissible to state the initial conditions $u(0, x) = u_0(x), \partial_t u(0, x) = u_1(x), \cdots, \partial_t^{m-1} u(0, x) = u_{m-1}(x)$ on $\{t = 0\} \times G$. The number of boundary conditions on the lateral surface is half of the order of the elliptic part if we consider $\partial_t^m u - P(t, x, D_x)u = f(t, x)$, where $P = P(t, x, D_x)$ is supposed to be an elliptic operator in $(0, T) \times G$. Problems consisting of partial differential equations, initial conditions and boundary conditions are called *initial boundary value problems* or *mixed problems*. Finally, we have to pay attention to so-called *compatibility*

conditions between initial conditions and boundary conditions on the boundary of
the bottom $\{t = 0\} \times \partial G$.

Now let us look at to *initial value problems,* also called *Cauchy problems,* in the
strip $(0, T) \times \mathbb{R}^n$ or in the half-space $(0, \infty) \times \mathbb{R}^n$. Consider the model linear partial
differential equation of p-evolution type

$$D_t^m u + \sum_{j=1}^{m} \sum_{|\alpha| \le pj} a_{j,\alpha}(t, x) D_x^\alpha D_t^{m-j} u = f(t, x).$$

Then, the Cauchy problem means that for the solution to this equation we pose m
initial conditions or *Cauchy conditions* on the hyperplane $\{(t, x) \in \{t = 0\} \times \mathbb{R}^n\}$ in
the following form:

$$(D_t^r u)(0, x) = u_r(x) \text{ for } r = 0, 1, \cdots, m - 2, m - 1.$$

A usual question is whether or not the *Cauchy problem is well-posed.* Here well-
posedness means *the existence of a solution, the uniqueness of the solution and the
continuous dependence of the solution on the data* $u_0, u_1, \cdots u_{m-2}, u_{m-1}$*, the right-
hand side* $f(t, x)$*, and the coefficients* $a_{k,\alpha}(t, x)$. However, all these desired properties
depend heavily on the choice of suitable function spaces.

Definition 3.7 The Cauchy problem

$$D_t^m u + \sum_{j=1}^{m} \sum_{|\alpha| \le pj} a_{j,\alpha}(t, x) D_x^\alpha D_t^{m-j} u = f(t, x),$$

$$(D_t^r u)(0, x) = u_r(x) \text{ for } r = 0, 1, \cdots, m - 2, m - 1$$

is *well-posed* if we can fix function spaces $A_0, A_1, \cdots, A_{m-2}, A_{m-1}$ for the
data $u_0, u_1, \cdots, u_{m-2}, u_{m-1}$, B and M for the right-hand side $f = f(t, x)$,
$B_0, B_1, \cdots, B_{m-2}, B_{m-1}$ and $M_0, M_1, \cdots, M_{m-2}, M_{m-1}$ for the solution u in such
a way that for given data and right-hand side

$$u_0 \in A_0, \ u_1 \in A_1, \cdots, u_{m-2} \in A_{m-2}, \ u_{m-1} \in A_{m-1}, \quad f \in M([0, T], B),$$

there exists a uniquely determined solution

$$u \in M_0([0, T], B_0) \cap M_1([0, T], B_1)$$

$$\cap \cdots \cap M_{m-2}([0, T], B_{m-2}) \cap M_{m-1}([0, T], B_{m-1}).$$

Moreover, the solution is required to depend continuously on the data and on the
right-hand side, that is, if we introduce small perturbations of the data and the right-
hand side according to the topologies of $A_0, A_1, \cdots, A_{m-2}, A_{m-1}$ and of $M([0, T], B)$,
then we get only a small perturbation of the solution with respect to the topology of

the space of solutions $M_0([0, T], B_0) \cap M_1([0, T], B_1) \cap \cdots \cap M_{m-2}([0, T], B_{m-2}) \cap M_{m-1}([0, T], B_{m-1})$.

But what are the usual function spaces to prove well-posedness?

The choice of the function spaces depends on the one hand on the regularity of coefficients of the partial differential equation and on the other hand on the partial differential equation itself. Let us suppose that the coefficients are smooth enough, thus having no "bad influence" on the choice of the function spaces. The usual spaces of functions or distributions to prove well-posedness are (for the definition of spaces of functions or distributions see Sect. 24.3)

- in the case $p = 1$: $A_k = H^{s-k}$, $B_k = H^{s-k}$, $B = H^{s-m+1}$, $M_k = C^k$, $M = C$ or $A_k = C^{mk}$, $B_k = C^{nk}$, $B = C^{nm}$, $M_k = C^{rk}$, $M = C^{rm}$, where $k = 0, \cdots, m - 1$
- in the case $p > 1$: $A_k = H^{s-pk}$, $B_k = H^{s-pk}$, $B = H^{s-p(m-1)}$, $M_k = C^k$, $M = C$.

Let us introduce some examples.

Example 3.4.2

1. For strictly hyperbolic Cauchy problems $(p = 1)$

$$D_t^m u + \sum_{k+|\alpha| \leq m, k \neq m} a_{k,\alpha}(t, x) D_x^\alpha D_t^k u = f(t, x),$$

$$(D_t^r u)(0, x) = u_r(x) \text{ for } r = 0, 1, \cdots, m - 2, m - 1,$$

one can prove well-posedness for solutions

$$u \in C([0, T], H^{m-1}(\mathbb{R}^n)) \cap C^1([0, T], H^{m-2}(\mathbb{R}^n))$$

$$\cap \cdots \cap C^{m-2}([0, T], H^1(\mathbb{R}^n)) \cap C^{m-1}([0, T], L^2(\mathbb{R}^n))$$

for given data and right-hand side

$$u_0 \in H^{m-1}(\mathbb{R}^n), \ u_1 \in H^{m-2}(\mathbb{R}^n), \cdots, u_{m-2} \in H^1(\mathbb{R}^n), \ u_{m-1} \in L^2(\mathbb{R}^n),$$

$$f \in C([0, T], L^2(\mathbb{R}^n))$$

if the coefficients are smooth enough.

2. For strictly hyperbolic Cauchy problems $(p = 1)$

$$D_t^m u + \sum_{k+|\alpha| \leq m, k \neq m} a_{k,\alpha}(t, x) D_x^\alpha D_t^k u = f(t, x),$$

$$(D_t^r u)(0, x) = u_r(x) \text{ for } r = 0, 1, \cdots, m - 2, m - 1,$$

one can prove well-posedness for solutions

$$u \in C^{r_0}([0,T], C^{n_0}(\mathbb{R}^n)) \cap C^{r_1}([0,T], C^{n_1}(\mathbb{R}^n))$$

$$\cap C^{r_{m-2}}([0,T], C^{n_{m-2}}(\mathbb{R}^n)) \cdots \cap \cap C^{r_{m-1}}([0,T], C^{n_{m-1}}(\mathbb{R}^n))$$

for given data and right-hand side

$$u_0 \in C^{m_0}(\mathbb{R}^n), \ u_1 \in C^{m_1}(\mathbb{R}^n), \cdots, u_{m-2} \in C^{m_{m-2}}(\mathbb{R}^n), \ u_{m-1} \in C^{m_{m-1}}(\mathbb{R}^n),$$

$$f \in C^{r_m}([0,T], C^{n_m}(\mathbb{R}^n))$$

if the coefficients are smooth enough.

3. For the Cauchy problem to p-evolution equations ($p \geq 1$, $p \in \mathbb{N}$)

$$D_t^m u + \sum_{j=1}^{m} \sum_{|\alpha| \leq pj,} a_{j,\alpha}(t,x) D_x^\alpha D_t^{m-j} u = f(t,x),$$

$$(D_t^r u)(0,x) = u_r(x) \ \text{ for } r = 0, 1, \cdots, m-2, m-1,$$

one can prove well-posedness for solutions

$$u \in C([0,T], H^{p(m-1)}(\mathbb{R}^n)) \cap C^1([0,T], H^{p(m-2)}(\mathbb{R}^n))$$

$$\cap \cdots \cap C^{m-2}([0,T], H^p(\mathbb{R}^n)) \cap C^{m-1}([0,T], L^2(\mathbb{R}^n))$$

for given data and right-hand side

$$u_0 \in H^{p(m-1)}(\mathbb{R}^n), \ u_1 \in H^{p(m-2)}(\mathbb{R}^n), \cdots, u_{m-2} \in H^p(\mathbb{R}^n),$$

$$u_{m-1} \in L^2(\mathbb{R}^n), f \in C([0,T], L^2(\mathbb{R}^n))$$

if the coefficients are smooth enough.

In general, it is impossible to choose for the Cauchy problem to the p-evolution equation with $p > 1$ the function spaces $A_k = C^{m_k}(\mathbb{R}^n)$, $B_k = C^{n_k}(\mathbb{R}^n)$, $B = C^{n_m}(\mathbb{R}^n)$, $M_k = C^{r_k}(\mathbb{R}^n)$, $M = C^{r_m}(\mathbb{R}^n)$. The reason is that in this case the solutions do not possess the property of *finite speed of propagation of perturbations* (compare with Sects. 10.1.2 and 10.1.3 of Chap. 10). So, we are not able to apply localization techniques to reduce this case to the case explained in the third statement of Example 3.4.2.

Remark 3.4.1 Without new difficulties, we can consider for p-evolution equations instead of *forward Cauchy problems* (only $t \geq 0$ is of interest) *backward Cauchy problems* (only $t \leq 0$ is of interest). This is different from parabolic Cauchy problems which, in general, can be only studied in one time direction (see Sect. 9.3.1).

3.5   Classification of Solutions

In this section we explain different notions of solutions for partial differential equations or systems of partial differential equations. Consider the linear partial differential equation (not necessarily of Kovalevskian type)

$$L(t, x, D_t, D_x)u := D_t^m u + \sum_{k+|\alpha| \le r,\, k<m} a_{k,\alpha}(t, x)D_x^\alpha D_t^k u = f(t, x).$$

Classical solutions are all functions satisfying this partial differential equation in the *classical sense*, that is, taking a function, forming all partial derivatives appearing in the partial differential equation and setting these into the partial differential equation gives an identity. The notion of *classical solutions* is, however, too restrictive in general. If some of the coefficients, or the right-hand side, or the initial, or the boundary data are not smooth enough, then we can not, in general, expect classical solutions. Even a non-smooth boundary (with corners, cusps and so on) of a given domain has an influence on regularity properties of the solution.

Definition 3.8 (Notion of Sobolev Solutions)   Let $G \subset \mathbb{R}^{n+1}$ be a domain and let $u \in L^1_{loc}(G)$ be a given function. Then, u is called a *Sobolev solution* of $L(t, x, D_t, D_x)u = f(t, x)$ if for all test functions $\phi \in C_0^{\max(r,m)}(G)$ the following integral identity holds:

$$\int_G u(t,x)L^*(t,x,D_t,D_x)\phi(t,x)\,d(t,x) = \int_G f(t,x)\phi(t,x)\,d(t,x),$$

where $L^*(t, x, D_t, D_x)$ denotes the adjoint operator to $L(t, x, D_t, D_x)$. Here we assume that all coefficients $a_{k,\alpha}$ belong to $C^{\max(r,m)}(G)$.

Definition 3.9 (Notion of Sobolev Solutions with Suitable Regularity)   Let $G \subset \mathbb{R}^{n+1}$ be a domain and let $u \in W_p^m(G) \subset L^1_{loc}(G)$ be a given function. Then, u is called a *Sobolev solution* from $W_p^m(G)$ of $L(t, x, D_t, D_x)u = f(t, x)$ if for all test functions $\phi \in C_0^{\max(r,m)}(G)$ the following integral identity holds:

$$\int_G u(t,x)L^*(t,x,D_t,D_x)\phi(t,x)\,d(t,x) = \int_G f(t,x)\phi(t,x)\,d(t,x),$$

where $L^*(t, x, D_t, D_x)$ denotes the adjoint operator to $L(t, x, D_t, D_x)$. Here we assume that all coefficients $a_{k,\alpha}$ belong to $C^{\max(r,m)}(G)$.

For the notion of Sobolev solution compare with Remarks 24.4.1 and 24.4.2. Sometimes one is interested in solutions of $Lu = f$ which can not be Sobolev solutions, that is, for some reasons these solutions do not belong to $L^1_{loc}(G)$. One possibility would be to introduce distributional solutions (cf. with Definition 24.29 and Remark 24.4.1).

**Definition 3.10 (Notion of Distributional Solutions or Solutions in the Distribu-
tional Sense)** Let $G \subset \mathbb{R}^{n+1}$ be a domain and let $u \in D'(G)$ be a distribution.
Then, u is called *a distributional solution* of $L(t, x, D_t, D_x)u = f(t, x)$ if for all test
functions $\phi \in C_0^\infty(G)$ the following identity holds:

$$u(t, x)\big(L^*(t, x, D_t, D_x)\phi(t, x)\big) = f(t, x)\big(\phi(t, x)\big),$$

where $u(\phi)$ denotes the action of the distribution u on the test function ϕ. Here we
assume that all coefficients $a_{k,\alpha}$ belong to $C^\infty(G)$.

Sometimes one is interested in Sobolev solutions having additional properties.
Consider, for example, the wave equation $u_{tt} - \Delta u = 0$. Then, one can ask for
solutions having an energy $E(u)(t)$ (cf. with the energy $E_W(u)(t)$ of Sect. 11.1), that
is, for almost all $t \in (0, T)$ we have $u(t, \cdot) \in H^1(\mathbb{R}^n)$ and $u_t(t, \cdot) \in L^2(\mathbb{R}^n)$. Such
solutions are called *energy solutions*.

For strictly hyperbolic Cauchy problems

$$D_t^m u + \sum_{k+|\alpha|\leq m, k\neq m} a_{k,\alpha}(t, x)D_x^\alpha D_t^k u = f(t, x),$$

$$(D_t^r u)(0, x) = u_r(x) \quad \text{for } r = 0, 1, \cdots, m-2, m-1,$$

one can study the existence of energy solutions

$$u \in C\big([0, T], H^{m-1}(\mathbb{R}^n)\big) \cap C^1\big([0, T], H^{m-2}(\mathbb{R}^n)\big)$$

$$\cap \cdots \cap C^{m-2}\big([0, T], H^1(\mathbb{R}^n)\big) \cap C^{m-1}\big([0, T], L^2(\mathbb{R}^n)\big).$$

We will not discuss in detail how boundary or initial conditions need to be
understood in a *weak sense*. We only mention that one has to use the notion of
traces. In the case of energy solutions for hyperbolic Cauchy problems we have a
regular behavior with respect to the time variable t. Thus, Cauchy conditions on the
hyperplane $t = 0$ are understood in the *classical sense, that is, the restriction* of
$D_t^k u(t, x)$ on $t = 0$ exists in a given function space.

Finally, we want to remark that different notions of solutions can be transferred
directly to systems of partial differential equations.

Definition 3.11 (Notion of Sobolev Solutions for Systems) Let $G \subset \mathbb{R}^{n+1}$ be a
domain and let $U = (u_1, \cdots, u_n)^T \in (W_p^m(G))^n \subset (L_{loc}^1(G))^n$ be a given vector
function. Then, U is called a *Sobolev solution* from $(W_p^m(G))^n$ of

$$A_0(t, x)\partial_t U + \sum_{k=1}^n A_k(t, x)\partial_{x_k} U + B(t, x)U = F(t, x)$$

with real and continuously differentiable matrices A_0 and A_k, with a real and
continuous matrix B, and with a real, integrable vector F if for all test vector

functions $\Phi = (\phi_1, \cdots, \phi_n)^T \in (C_0^1(G))^n$ the following integral identity holds:

$$\int_G U(t,x) \cdot \big(-\partial_t (A_0^T \Phi) - \partial_{x_k} (A_k^T \Phi) + B^T \Phi \big)(t,x)\, d(t,x)$$

$$= \int_G F(t,x) \cdot \Phi(t,x)\, d(t,x),$$

where A^T denotes the transposed matrix of A.

Exercises Relating to the Considerations of Chap. 3

Exercise 1 Find elliptic, hyperbolic and strictly hyperbolic operators. What type has the *stationary plate operator* Δ^2? Is the heat operator $\partial_t - \Delta$ elliptic or hyperbolic or something else? Find an operator with constant coefficients which is hyperbolic but not strictly hyperbolic. Is there an elliptic operator of order 3?

Exercise 2 Show that the classical Schrödinger operators $\frac{1}{i}\partial_t \pm \Delta$ and the non-stationary plate operator $\partial_t^2 + \Delta^2$ are 2-evolution operators. Is the heat operator $\partial_t - \Delta$ a 2-evolution operator? No, it is a parabolic operator. Compare this with the classification of linear operators of second order by using the principal symbol.

Exercise 3 Find one example of an ultrahyperbolic partial differential equation that is not hyperbolic. Find one example of a parabolic partial differential equation which is not of normal type.

Exercise 4 How many boundary conditions can we pose for solutions to $\Delta^m u = 0$ in an interior domain?

Exercise 5 Find in $\mathbb{R}^n \setminus \{0\}$ a classical radial solution $u = u(r)$ of the Laplace equation $\Delta u = 0$ (cf. with Example 24.4.4 from Chap. 24).

Exercise 6 Study the boundary value problem of Example 3.4.1. Use the Laplace operator in polar-coordinates

$$\Delta u = \frac{1}{r^2}\partial_r\big(r^2\partial_r u\big) + \frac{1}{r^2 \sin\theta}\partial_\theta\big(\sin\theta\,\partial_\theta u\big) + \frac{1}{r^2 \sin^2\theta}\partial_\phi^2 u.$$

How do we interpret the *radiation condition* in the case $k = 0$ in the Helmholtz equation?

Exercise 7 What kind of additional conditions to solutions may we pose for solutions to the nonstationary plate $\partial_t^2 u + \Delta^2 u = 0$ in the *interior domain* $\{x \in \mathbb{R}^3 : |x| < 1\}$ or in the *exterior domain* $\{x \in \mathbb{R}^3 : |x| > 1\}$?

Exercise 8 Explain the compatibility conditions for the model of a vibrating string which is fixed in two points $x = 0$ and $x = 1$. Explain compatibility conditions for the model of a potential inside of a rectangular

$$R = \{(x,y) \in \mathbb{R}^2 : (x,y) \in [0,a] \times [0,b]\}.$$

Exercise 9 In which time directions are we able to study the Cauchy problems

$$u_t - \Delta u = f(t, x), \; u(0, x) = u_0(x), \quad u_t + \Delta u = f(t, x), \; u(0, x) = u_0(x)?$$

Exercise 10 Look for a classical solution u in the form $u(t, x) = t^{-n/2} h\left(\frac{|x|}{\sqrt{t}}\right)$ for the heat equation $u_t - \Delta u = 0$ for $t > 0$ and $x \in \mathbb{R}^n$. Here h is a suitable chosen function (cf. with Example 24.4.5 from Chap. 24).

Exercise 11 What is the definition of a Sobolev solution of the Poisson equation $\Delta u = f$? Determine all Sobolev solutions of $d_t u = f$, where $f(t) = 0$ if $t \le 0$ and $f(t) = 1$ for $t > 0$.

Exercise 12 Find all distributional solutions of $d_t u = \delta_0$ in $\mathbb{R}^1$ and of $-\Delta u = \delta_0$ in $\mathbb{R}^n$, where δ_0 denotes *Dirac's delta distribution* at the origin.

Exercise 13 Suppose that we have an energy solution

$$u \in C\big([0, T], H^s(\mathbb{R}^n)\big) \cap C^1\big([0, T], H^{s-1}(\mathbb{R}^n)\big)$$

of the Cauchy problem for the wave equation with $s \ge 1$. For which s do we have a classical solution?

Exercise 14 Show that any classical solution of $L(t, x, D_t, D_x)u = 0$ is a Sobolev solution. Show that any Sobolev solution of $L(t, x, D_t, D_x)u = 0$ is a distributional solution. Here $L = L(t, x, D_t, D_x)$ is a linear partial differential operator with smooth coefficients.

Chapter 4
The Cauchy-Kovalevskaja Theorem

The classical Cauchy-Kovalevskaja theorem is one of the fundamental results in the theory of partial differential equations. This theorem makes two assertions. On the one hand it yields the local existence of analytic solutions to a large class of Cauchy problems and on the other hand it yields the uniqueness of this solution in the class of analytic functions. This chapter deals not only with the Cauchy-Kovalevskaja theorem in its classical form but in its abstract form in scales of Banach spaces as well. Some applications in the theory of Hele-Shaw flows complete this chapter. These applications serve as an interesting field for verifying the importance of the tool of an abstract form of the Cauchy-Kovalevskaja theorem.

4.1 Classical Version

The Cauchy-Kovalevskaja theorem is one of the oldest results in the theory of partial differential equations. We explain this result for the linear Cauchy problem

$$\partial_t^m u + \sum_{k+|\alpha|\leq m,\, k\neq m} a_{k,\alpha}(t,x)\partial_x^\alpha \partial_t^k u = f(t,x),$$

$$\left(\partial_t^r u\right)(0,x) = \varphi_r(x) \text{ for } r = 0,1,\cdots,m-2,m-1,$$

where the coefficients $a_{k,\alpha} = a_{k,\alpha}(t,x)$, the right-hand side $f = f(t,x)$, and the data $\varphi_r = \varphi_r(x)$, $r = 0,\cdots,m-1$, are real and defined in a cylinder $[-T,T] \times G$ containing as an interior point the origin in $\mathbb{R}^{n+1}$. It is important that the operator is given as an operator of *Kovalevskian type*, that is, an operator of the form

$$\partial_t^m + \sum_{k+|\alpha|\leq m,\, k\neq m} a_{k,\alpha}(t,x)\partial_x^\alpha \partial_t^k.$$

M.R. Ebert, M. Reissig, *Methods for Partial Differential Equations*,
https://doi.org/10.1007/978-3-319-66456-9_4

Example 4.1.1 The operators $\partial_t^2 + \Delta$ or $\partial_t^2 - \Delta$ are of Kovalevskian type. The operators $\partial_t \pm \Delta$ or $\partial_t^2 \pm \Delta^2$ are not of Kovalevskian type. Here Δ denotes the Laplace operator in $\mathbb{R}^n$.

Theorem 4.1.1 *Let us assume that the coefficients $a_{k,\alpha}$ and the right-hand side f are analytic in a neighborhood U of the origin in $\mathbb{R}^{n+1}$. Moreover, we suppose that the data $\varphi_0, \varphi_1, \cdots, \varphi_{m-2}, \varphi_{m-1}$ are analytic in $U \cap \{t = 0\}$. Then, there exists a neighborhood $W \subset U$ of the origin and a unique analytic solution $u = u(t,x)$ of the Cauchy problem*

$$\partial_t^m u + \sum_{k+|\alpha|\leq m,\, k\neq m} a_{k,\alpha}(t,x)\partial_x^\alpha \partial_t^k u = f(t,x),$$

$$(\partial_t^r u)(0,x) = \varphi_r(x) \text{ for } r = 0, 1, \cdots, m-2, m-1,$$

in W.

Remark 4.1.1 This theorem is independent of the type of the differential operator. What is only important is that the operator is of Kovalevskian type.

The proof of the Cauchy-Kovalevskaja theorem is based on the assumption of analyticity, which allows local representations of the coefficients, the right-hand side, and the data into a power series. Let us briefly sketch the proof for the 1d case in x only. We consider the Cauchy problem

$$\partial_t^m u + \sum_{k+l\leq m,\, k\neq m} a_{k,l}(t,x)\partial_x^l \partial_t^k u = f(t,x),$$

$$(\partial_t^r u)(0,x) = \varphi_r(x) \text{ for } r = 0, 1, \cdots, m-2, m-1.$$

The assumption of analyticity allows us to represent the coefficients, the right-hand side, and the data in a neighborhood of the origin in $\mathbb{R}^2$ in form of a power series. So we have

$$a_{k,l}(t,x) = \sum_{p,q=0}^\infty a_{k,l,p,q}t^p x^q, \quad f(t,x) = \sum_{p,q=0}^\infty f_{p,q}t^p x^q,$$

$$\varphi_r(x) = \sum_{q=0}^\infty \varphi_{r,q}x^q \text{ for } r = 0, 1, \cdots, m-2, m-1.$$

All these power series' converge in a small neighborhood $U_1 \subset U$ of the origin in $\mathbb{R}^2$. For the solution u, we make the ansatz $u(t,x) = \sum_{p,q=0}^\infty u_{p,q}t^p x^q$. Now we have to determine the unknown coefficients $u_{p,q}$ by the given coefficients $a_{k,l,p,q}$, $f_{p,q}$, and $\varphi_{r,q}$. This procedure works as follows:

1. The coefficients $u_{r,q}$ for $r = 0, \cdots, m - 1$ are determined by the coefficients $\varphi_{r,q}$ from the initial conditions by forming in the ansatz for the solution $(\partial_t^r u)(t = 0, x)$.

2. Plugging all power series' into the differential equation then, by the method of comparison of coefficients, we can determine the coefficients $u_{p,q}$ for $p \geq m$ step by step. First we determine $u_{m,q}$ by the already determined coefficients $u_{r,q}$, $r = 0, \cdots, m - 1$ and known coefficients $a_{k,l,p,q}$, $f_{p,q}$. Then, we determine $u_{m+1,q}$ by the already determined coefficients $u_{r,q}$, $r = 0, \cdots, m$, and known coefficients $a_{k,l,p,q}$, $f_{p,q}$, then, $u_{m+2,q}$, then $u_{m+3,q}$, and so on. The uniform convergence of the power series on compact subsets of U_1 allows to interchange the order of differentiation and summation, that is, a term by term differentiation of all power series is allowed.

3. After the determination of all coefficients $u_{p,q}$, we have to show that the power series $\sum\limits_{p,q=0}^{\infty} u_{p,q} t^p x^q$ is converging in a small neighborhood W of the origin. This is done by applying the *method of majorants*.

It is impossible to generalize the Cauchy-Kovalevskaja theorem to non-analytic classes. If we are interested in the Cauchy problem

$$\partial_t^2 u + \partial_x^2 u = 0, \ u(0, x) = \varphi(x), \ \partial_t u(0, x) = \psi(x)$$

with data $\varphi, \psi \in C^\infty(\mathbb{R}^1)$, then the property *of interior analytic regularity of harmonic functions* (see Theorem 8.2.4 of Sect. 8.2.3) shows that there exists, in general, *no classical solution*. But, if φ, ψ are analytic in a neighborhood of the origin, then there exists a unique analytic solution of this Cauchy problem in a neighborhood of the origin.

An important contribution to the Cauchy-Kovalevskaja theory has been given by Hans Lewy (see [121]).

Lewy's Example There exists a non-analytic infinitely differentiable function $F = F(t, x, y)$ such that the differential equation

$$\partial_x u + i \partial_y u - 2i(x + iy)\partial_t u = F(t, x, y)$$

has no Sobolev solution in any neighborhood of any point $(t_0, x_0, y_0) \in \mathbb{R}^3$. Take into consideration the fact that the coefficients are complex in Lewy's example. Lewy's example tells us that we can not expect any Cauchy-Kovalevskaja type theorem for C^∞ spaces. One can even show that the operator

$$\partial_x + i \partial_y - 2i(x + iy)\partial_t$$

is not locally solvable in any neighborhood of any point $(t_0, x_0, y_0) \in \mathbb{R}^3$ (see Sect. 8.3.2).

4.2 Abstract Version

There exists an abstract version of the linear Cauchy-Kovalevskaja theorem (see, for example, [206] or [211]). To formulate this theorem, we introduce a so-called scale of Banach spaces $\{B_s, \|\cdot\|_s\}_s$ with parameter $s \in [0, 1]$. This scale has the following properties:

- $B_s \subset B_{s'}$ for $0 \le s' \le s \le 1$,
- $\|u\|_{s'} \le \|u\|_s$ for all $u \in B_s$ and for $0 \le s' \le s \le 1$.

There exist operators which are unbounded on all function spaces B_s but which are bounded on the scale of Banach spaces $\{B_s\}_{s \in [0,1]}$, that is, which are bounded as mappings from B_s to $B_{s'}$ with $s' < s$. Let A be a linear operator acting in $\{B_s\}_{s \in [0,1]}$ in the following way:

- the operator A maps B_s into $B_{s'}$ for all $0 \le s' < s \le 1$
- the norm of the operator $A \in L(B_s \to B_{s'})$ can be estimated by

$$\frac{a}{s - s'}, \text{ that is, } \|Au\|_{s'} \le \frac{a}{s - s'}\|u\|_s \text{ for } 0 \le s' < s \le 1,$$

where the constant a is independent of s and s'.

Then, the norm of A in $L(B_s \to B_{s'})$ is defined by

$$\|A\|_{L(B_s \to B_{s'})} = \sup_{u \in B_s} (s - s')\|Au\|_{s'}\|u\|_s^{-1}.$$

Finally, we assume that the family $\{A(t)\}$, $t \in [0, T]$, of linear abstract operators from $L(B_s \to B_{s'})$ depends continuously on t, that is, to every $\varepsilon > 0$, $t_0 \in [0, T]$, $0 \le s' < s \le 1$ there exists a constant $\delta = \delta(\varepsilon, t_0, s, s')$ such that

$$\|A(t) - A(t_0)\|_{L(B_s \to B_{s'})} \le \varepsilon \text{ if } |t - t_0| < \delta.$$

Theorem 4.2.1 (Abstract Cauchy-Kovalevskaja Theorem)
Let us consider the abstract Cauchy problem

$$d_t u = A(t)u + f(t), \quad u(0) = u_0$$

under the following assumptions:

- $u_0 \in B_1$
- $f \in C([0, T], B_1)$

- *the family $\{A(t)\}$, $t \in [0, T]$, of linear abstract operators from $L(B_s \to B_{s'})$, $0 \le s' < s \le 1$, satisfies the above introduced conditions.*

Then, there exists a unique solution $u \in C([0, T_s), B_s)$ for $s \in [0, 1)$, where $T_s = \min\{T; K(1 - s)\}$ and with a suitable positive constant K.

Proof The proof is based on ideas for treating nonlinear ordinary differential equations.

Existence of a Solution We transform the abstract Cauchy problem to an equivalent abstract integral equation

$$u(t) = u_0 + \int_0^t f(\tau)d\tau + \int_0^t A(\tau)u(\tau)d\tau.$$

We apply the method of successive approximation to get for $k \ge 1$ the iterates

$$u^{(k+1)}(t) = u_0 + \int_0^t f(\tau)d\tau + \int_0^t A(\tau)u^{(k)}(\tau)d\tau, \quad u^{(0)}(t) = u_0 + \int_0^t f(\tau)d\tau.$$

All iterates $u^{(k)}$ belong to $C([0, T], B_s)$ for $s \in [0, 1)$. Let

$$v_k(t) := u^{(k)}(t) - u^{(k-1)}(t) \text{ for } k \ge 1, \ v_0(t) := u^{(0)}(t).$$

We show the uniform convergence of $\sum_{k=0}^{\infty} v_k(t)$ in B_s for all $t \in [0, T_s - \varepsilon]$, where $\varepsilon < T_s$ is an arbitrary small positive number. We may then conclude that $u(t) = \lim_{k \to \infty} u^{(k)}(t)$ is a solution of the abstract integral equation. Using

$$v_{k+1}(t) = \int_0^t A(\tau)v_k(\tau)d\tau \text{ for } k = 0, 1, \cdots$$

we show the estimates

$$\|v_k(t)\|_s \le M\left(\frac{aet}{1-s}\right)^k \text{ for } t \in [0, T] \text{ and } s \in [0, 1),$$

where e is the Euler's number. The constant M is defined by $M := \|u_0\|_1 + \int_0^T \|f(t)\|_1 dt$. The desired estimate is true for $k = 0$ due to the definition of $v_0(t)$. Let us assume that the estimate is valid for $k = p$. Then, we show its validity for $k = p + 1$, too. Using the definition of $v_{p+1}(t)$ and the assumptions for the family of operators $\{A(t)\}_{t \in [0,T]}$, we obtain

$$\|v_{p+1}(t)\|_{s'} \le \int_0^t \|A(\tau)v_p(\tau)\|_{s'}d\tau \le \frac{a}{s-s'}\int_0^t \|v_p(\tau)\|_s d\tau.$$

Taking account of the desired estimate for $v_p(t)$, it follows

$$\|v_{p+1}(t)\|_{s'} \leq \frac{a}{s-s'} M\left(\frac{ae}{1-s}\right)^p \frac{t^{p+1}}{p+1}.$$

Now let us choose a suitable $s' < s$, namely, $p(s - s') = 1 - s$. Then, $p(1 - s') = (p + 1)(1 - s)$. The last inequality implies

$$\|v_{p+1}(t)\|_{s'} \leq Me^p\left(\frac{a}{1-s}\right)^{p+1} \frac{p}{p+1} t^{p+1} = Me^p\left(\frac{at}{1-s'}\right)^{p+1}\left(\frac{p+1}{p}\right)^p$$

$$\leq M\left(\frac{aet}{1-s'}\right)^{p+1},$$

which is the desired estimate for v_{p+1}. It is clear that the series $\sum_{k=0}^{\infty} \|v_k(t)\|_s$ is converging for $t \in [0, \frac{1-s}{ae}) \cap [0, T]$. Hence, we have the existence of a solution $u \in C([0, T_s), B_s)$ for $s \in [0, 1)$ and $T_s = \min\{T; K(1 - s)\}$, where $K = \frac{1}{ae}$.

Uniqueness of the Solution Let us assume the existence of two different solutions, $u_1, u_2 \in C([0, T'), B_s)$. Then, the difference $u = u(t)$ of both solutions belongs to $C([0, T'), B_s)$, too and satisfies

$$u(t) = \int_0^t A(\tau)u(\tau)d\tau.$$

As in the proof for existence of a solution we can show

$$\|u(t)\|_{s'} \leq M(\varepsilon)\left(\frac{aet}{s-s'}\right)^k \quad \text{for } k \geq 0 \text{ and } s' \in [0, s),$$

where $M(\varepsilon)$ is taken from the estimate $\|u(t)\|_s \leq M(\varepsilon)$ for all $t \in [0, T' - \varepsilon]$. If we let $k \to \infty$ in the inequality for $\|u(t)\|_{s'}$, then it leads, together with the continuity in t, to $u(t) \equiv 0$ in the interval $t \in [0, \min\{T' - \varepsilon; \frac{s-s'}{ae}\}]$. If $t_1 = \frac{s-s'}{ae} < T' - \varepsilon$, then we repeat our approach for the abstract integral equation

$$u(t) = \int_{t_1}^t A(\tau)u(\tau)d\tau.$$

A finite number of such steps gives the uniqueness of solutions in the evolution space $C([0, T' - \varepsilon], B_s)$. The arbitrary choice of ε implies the uniqueness of solutions in $C([0, T'), B_s)$. This completes the proof.

Example 4.2.1 Consider the Cauchy problem for the first order system of complex partial differential equations

$$\partial_t U - \sum_{k=1}^n A_k(t, z)\partial_{z_k} U - B(t, z)U = F(t, z), \quad U(0, z) = U_0(z),$$

where $z = (z_1, z_2, \cdots, z_{n-1}, z_n)$ and the matrix functions A_k and B are holomorphic in the poly-cylinder $K = K_1 \times K_2 \times \cdots \times K_{n-1} \times K_n$ and continuous on its closure $\overline{K}$, where K_j denotes the unit disk with respect to the variable z_j. Finally, the matrices A_k and B and the vector function F are continuous with respect to t, that is, $A_k, B, F \in C([0, T], H(K) \cap C(\overline{K}))$. With a small positive constant a_0 let us introduce the family of poly-cylinders $K_s = K_{1,s} \times K_{2,s} \times \cdots \times K_{n-1,s} \times K_{n,s}$, where $K_{j,s}$ denotes the disk around the origin with radius $1 - a_0(1 - s)$ for $s \in [0, 1]$ with respect to the variable z_j. Now we are able to introduce $B_s := H(K_s) \cap C(\overline{K}_s)$, that is the space of functions which are holomorphic in the poly-cylinder K_s and continuous on the closure $\overline{K}_s$. Introducing the family of operators

$$A(t) := \sum_{k=1}^{n} A_k(t, z)\partial_{z_k} + B(t, z)$$

and applying Cauchy's integral formula, one can show that

$$\|A(t)\|_{L(B_s \to B_{s'})} \leq \frac{a}{s - s'}$$

with a positive constant a (see Exercise 2 of this chapter). Consequently, all assumptions of Theorem 4.2.1 are satisfied and one has a unique solution $u \in C((-T_s, T_s), B_s)$ for $s \in [0, 1)$ with $T_s = \min(T, M(1 - s))$ and with a suitable positive constant M. All constants are independent of s and s'.

We may transform the above complex system of first order to a real system of first order. Thus, in general, one can expect for type independent real systems of first order with analytic coefficients only solutions possessing a *conical evolution*. Here *conical evolution* means the larger the domain of existence K_s, the smaller the life span $(-T_s, T_s)$.

The difference between the statements of Theorems 4.1.1 and 4.2.1 is that in Theorem 4.1.1 we suppose analyticity of the coefficients in t, where as in Theorem 4.2.1 we suppose only continuity of the coefficients in t. What property does the solution from Theorem 4.2.1 have with respect to t?

4.3 Concluding Remarks

4.3.1 Generalizations of the Classical Cauchy-Kovalevskaja Theorem

The Cauchy-Kovalevskaja theorem holds for the Cauchy problem for linear systems of the form

$$\partial_t u + \sum_{j=1}^{n} A_j(t, x)\partial_{x_j} u + A_0(t, x)u = f(t, x), \quad u(0, x) = \varphi(x).$$

The coefficients A_j and A_0 are $m \times m$ matrices with entries which are analytic in the spatial variables and thanks to the abstract Cauchy-Kovalevskaja Theorem 4.2.1 only continuous with respect to the time variable.

We have also a Cauchy-Kovalevskaja theorem for nonlinear Cauchy problems which are of Kovalevskian type (see [117]). A complete proof of this general result based on the comparison of coefficients method can be found in [27] or [163].

4.3.2 Generalizations of the Abstract Cauchy-Kovalevskaja Theorem

There exists also an abstract Cauchy-Kovalevskaja theorem for nonlinear Cauchy problems of the form

$$d_t u = A(t, u), \quad u(t = 0) = u_0.$$

The reader can find a first proposal for such a theorem in [148] under special assumptions to the right-hand side $A(t, u)$. In [149] the author proved the following more general abstract form of the nonlinear Cauchy-Kovalevskaja theorem:

Theorem 4.3.1 *Consider in a scale of Banach spaces $\{B_s, \| \cdot \|_s\}_{s \in [0,1]}$ the abstract Cauchy problem*

$$d_t u = A(t, u), \quad u(t = 0) = 0$$

under the following assumptions:

- *The mapping $(t, u) \to A(t, u)$ is a continuous mapping of*

$$\{t : t \in [0, T]\} \times \{u \in B_s : \|u\|_s < R\} \ \ into \ \ B_{s'}$$

  for some positive T and R and for all $0 \le s' < s \le 1$.
- *The above mapping is Lipschitz continuous in u, that is,*

$$\|A(t, u) - A(t, v)\|_{s'} \le \frac{C}{s - s'}\|u - v\|_s,$$

  where the constant C depends only on T and R.
- *The function $A(t, 0)$ is a continuous function of $t \in [0, T]$ with values in B_s for every $s \in [0, 1)$. Moreover, it satisfies the estimate*

$$\|A(t, 0)\|_s \le \frac{K}{1 - s}$$

  with a fixed constant $K = K(T)$ which is independent of s.

Then, there is a uniquely determined solution

$$u \in C^1\big((-a(1-s), a(1-s)), B_s\big), \ s \in [0, 1), \ \|u\|_s < R,$$

with a suitable positive constant a.
An alternative proof of this result was given in [211].

4.3.3 Applications of the Abstract Cauchy-Kovalevskaja Theorem

We have a lot of applications of Theorem 4.3.1. The operator $A(t, u)$ can be linear, nonlinear or even non-local as well in a scale of Banach spaces. It can be applied to real or complex models. Such abstract versions are very useful in studying, among other things, problems of fluid dynamics (for example Hele-Shaw flows), transient wave-body interaction, filtration problems, Korteweg-de Vries equation and Boussinesq equation of water surface waves or motion of the oil contour (see, for example, [81, 166, 170, 173, 175] and references therein).
Let us explain one of these applications of [166].

4.3.3.1 The Model

Our starting point is the problem of finding the velocity vector field $v = (v_x, v_y)$ of a two-dimensional flow driven by a single sink at the origin with continuous intensity $q_0 = q_0(t)$, that is, div $v = -q_0(t)\delta_0$. Here, and in the following δ_0 is the two-dimensional Dirac distribution centered at the origin (see Example 24.4.2). Moreover, we suppose that the vector field is irrotational, meaning, rot $v = 0$. Both conditions together imply the existence of a potential function p satisfying $v = -\text{grad}\,p$, $\Delta p = q_0(t)\delta_0$. If $\Gamma(t)$ denotes the smooth moving boundary of the simply connected domain $\Omega(t)$ which is occupied by liquid, then the kinematic condition is written in the form $\frac{d\Gamma(t)}{dt} = v$. This condition explains the fact that a particle of the fluid on the free boundary stays on it during the whole period of the motion. If $p = 0$ on $\Gamma(t)$, then $p < 0$ in $\Omega(t)$. Consequently, $\partial_n p > 0$ on $\Gamma(t)$, where **n** denotes the outward normal to the boundary $\Gamma(t)$. Thus a nonlinear kinetic undercooling condition (dynamic condition) for the suction problem should be written in the form

$$p + F(\partial_n p) = 0 \quad \text{on } \Gamma(t)$$

under the constraint $\partial_n p \geq 0$. The function $F = F(v)$ stays for a class of nonlinear functions containing as a typical example $F(s) = s^\beta$, $\beta > 1$. For this reason we suppose the following assumptions for F:

- F is defined on $[0, \infty)$ and $F(0) = 0$
- F is strictly monotonic increasing on $[0, \infty)$

- $\lim\limits_{s \to \infty} F(s) = \infty$
- F is analytic on $(0, \infty)$.

4.3.3.2   Mathematical Formulation

Now we formulate the problem under consideration for Hele-Shaw flows with a nonlinear kinetic undercooling effect on the boundary starting from a similar formulation of the problem via the parametrization

$$\Gamma(t) := \big\{\mathbf{w}(t, x, y) : \ (t, x, y) \in I \times \partial U\big\},$$

where $I = (-\varepsilon, \varepsilon)$ and $\partial U := \{(x, y) \in \mathbb{R}^2 : \ |(x, y)| = 1\}$.

The problem NKUR (nonlinear kinetic undercooling regularization) is introduced by several conditions which we explain as follows:

- *the parametrization condition*:

$$\mathbf{w}(t, x, y) \in \Gamma(t) \ \text{ for all } \ (t, x, y) \in I \times \partial U$$

- *the analyticity condition*:

$$\mathbf{w}(t, \cdot) : \ \partial U \to \Gamma(t)$$

  is an analytic diffeomorphism for $t \in I$

- *the equation of motion*:

$$\partial_t \mathbf{w}(t, \cdot) = \nabla G_{\Omega(t)}(\mathbf{w}(t, x, y)) \ \text{ for all } \ (t, x, y) \in I \times \partial U,$$

  where $G_{\Omega(t)}$ is the solution of the nonlinear Robin problem

$$\Delta G = q_0(t)\delta_0 \ \text{ in } \ \Omega(t), \ \ G + F(\partial_\mathbf{n} G) = 0 \ \ \text{ on } \ \Gamma(t)$$

  with $\partial_\mathbf{n}$ taken along the outward normal to $\Gamma(t)$

- *the initial condition*:

$$\Gamma(0) := \{\mathbf{w}(0, x, y) = \mathbf{w}_0(x, y) : \ (x, y) \in \partial U\}.$$

A model based on parameterizations of $\Gamma(t)$ was firstly proposed for the two-dimensional case with Dirichlet boundary condition $p = 0$ on $\Gamma(t)$ in [71].

4.3.3.3 The Result

Now we should define a suitable scale of Banach spaces of analytic functions $\{B_s, \|\cdot\|_s\}_{s\in[0,1]}$. The main work consists in showing that our problem NKUR can be interpreted as a special abstract Cauchy problem in the scale of Banach spaces $\{B_s, \|\cdot\|_s\}_{s\in[0,1]}$ which satisfies the assumptions of Theorem 4.3.1. Then, the application of Theorem 4.3.1 yields the following result.

Theorem 4.3.2 *Let the initial fluid region be a simply connected domain containing the origin such that the boundary of the domain is parameterized by an analytic diffeomorphism. Then, for a small period of time (backward and forward) there exists a unique classical solution of the problem NKUR. The fluid region remains simply connected and has an analytic boundary. If $q_0 = q_0(t)$ is continuous in t, then the solution is continuously differentiable in t on the life span interval (backward and forward in time).*

Exercises Relating to the Considerations of Chap. 4

Exercise 1 Apply the classical Cauchy-Kovalevskaja theorem to the Cauchy problems

$$\partial_t u + 2\partial_x u = 6x^2, \quad u(0, x) = x^3,$$

$$\partial_t u + t\partial_x u = x, \quad u(0, x) = x^2.$$

Find the power series representation of the solutions in a neighborhood of the origin in $\mathbb{R}^2$. Use the methods of Chap. 6 to find the exact solutions.

Exercise 2 Show that the Cauchy-Kovalevskaja theorem does, in general, not hold for the Cauchy problem for non-Kovalevskian equations. For this reason study for $t > 0$ the Cauchy problem

$$\partial_t u = \partial_x^2 u, \quad u(0, x) = \frac{1}{1-x} \text{ for } |x| < 1.$$

Exercise 3 Show the scale type estimate for $\{B_s, \|\cdot\|_s\}_{s\in[0,1]}$ of Example 4.2.1, that is,

$$\|Au\|_{s'} \leq \frac{a}{s-s'}\|u\|_s \text{ for } A(t) := \sum_{k=1}^{n} A_k(t, z)\partial_{z_k}.$$

To solve this exercise it is sufficient to understand that the operator of complex differentiation d_z is continuous as a mapping from B_s into $B_{s'}$ for $0 \leq s' < s \leq 1$ with the norm estimate $\|d_z\|_{L(B_s \to B_{s'})} \leq \frac{M}{s-s'}$. Here B_s denotes the space of functions which are holomorphic in K_s and continuous on its closure $\overline{K}_s$. By K_s we denote the disk around the origin with radius $1 - a_0(1-s)$ for $s \in [0, 1]$ and $a_0 \in (0, 1)$.

Exercise 4 Show that the Cauchy problem

$$\partial_t u = 0, \quad (t, x) \in \mathbb{R}^+ \times \mathbb{R}^1, \quad u(0, x^2) = \varphi(x), \quad x \in \mathbb{R}^1,$$

where φ is an odd non-vanishing analytic function, has no analytic solution at any neighborhood of the origin. Is this observation a contradiction to the Cauchy-Kovalevskaja Theorem 4.1.1?

The answer to this question is related to the notion of a non-characteristic initial manifold, see Definition 6.2.

Chapter 5
Holmgren's Uniqueness Theorem

Holmgren's uniqueness theorem is one of the fundamental results in the theory of partial differential equations. It is related to the Cauchy-Kovalevskaja theorem. Theorem 4.1.1 implies a uniqueness result in the class of analytic solutions to a large class of Cauchy problems for partial differential equations. This uniqueness assertion still allows for the possibility that there may exist other classical or even distributional solutions which are not necessarily analytic. The classical theorem of Holmgren states that this can not happen in the set of classical solutions. As in Chap. 4, we explain the classical version and the abstract version in scales of Banach spaces as well.

5.1 Classical Version

Holmgren was interested in the Cauchy-Kovalevskaja theorem. Due to this theorem we know that in the class of analytic solutions we have a unique solution. Holmgren posed the very reasonable question asking if the Cauchy problem from Theorem 4.1.1 *has non-analytic solutions* (see [82]).

Theorem 5.1.1 (Classical Theorem of Holmgren)
In the set of m times differentiable solutions, that is, in the set of classical solutions, the analytic solution from Theorem 4.1.1 is the only one.

Proof Let us assume in Theorem 4.1.1 the existence of two classical solutions u_1 and u_2 in a neighborhood W of the origin. Then, the difference $u := u_1 - u_2$ is a classical solution of the homogeneous Cauchy problem

$$\partial_t^m u + \sum_{k+|\alpha|\leq m,\, k\neq m} a_{k,\alpha}(t,x)\partial_x^\alpha \partial_t^k u = 0 \text{ in } W,$$

$$\left(\partial_t^r u\right)(0,x) = 0 \quad \text{for all } x \in W \cap \{t = 0\} \text{ and } r = 0, 1, \cdots, m-2, m-1.$$

© Springer International Publishing AG 2018
M.R. Ebert, M. Reissig, *Methods for Partial Differential Equations*,
https://doi.org/10.1007/978-3-319-66456-9_5

Our goal then is to show that $u \equiv 0$ in a (possible smaller) neighborhood $W_0 \subset W$ of the origin. We prove it only on the set $\{(t, x) \in W_0 : t \geq 0\}$. The same proof gives the vanishing behavior of u in $\{(t, x) \in W_0 : t \leq 0\}$. We take u for $t \geq 0$. The homogeneous Cauchy conditions allow for setting $u \equiv 0$ for $t \leq 0$. Consequently, we obtain a classical solution $u \in C^m(\overline{W_0})$ with support in $\{(t, x) \in W_0 : t \geq 0\}$. We divide the proof into several steps.

Step 1. Application of Holmgren transformation

Introduce the change of variables (Holmgren transformation) $v(s, y) = u(t, x)$ with $y = x$ and $s = t + |x|^2$. This change of variables makes a bijective mapping of (a possible smaller) W_0 onto a neighborhood V_0 of the origin in the (s, y)-space. Let us fix ε sufficiently small. We introduce the set $M_\varepsilon := \{(s, y) \in V_0 : |y|^2 \leq s \leq \varepsilon\}$. By taking into account that $t = 0$ is transformed onto $\{(s, y) : s = |y|^2\}$, it follows that $v \in C^m(\overline{V_0})$.

Step 2. The new Cauchy problem

After the change of variables, our new Cauchy problem has the same form with respect to the variables (s, y) as the starting one, the coefficients are analytic, too. But there is a big difference. We know that $v(s, y) \equiv 0$ outside M_ε and $v(s, \cdot)$ has compact support for all $s \in [0, \varepsilon]$. This allows us to restrict our further considerations (we use again (t, x) instead of (s, y)) to the following Cauchy problem:

$$Lv := \partial_t^m v + \sum_{k+|\alpha| \leq m,\, k \neq m} b_{k,\alpha}(t, x) \partial_x^\alpha \partial_t^k v = 0 \text{ in } V_0,$$

$$\operatorname{supp} v \subset M_\varepsilon,$$

where the coefficients $b_{k,\alpha} = b_{k,\alpha}(t, x)$ are analytic in V_0 with $M_\varepsilon \subset V_0$.

Step 3. Influence of the adjoint operator

Let us recall the adjoint operator L^* (cf. with Remarks 24.4.1 and 24.4.2) which is defined by

$$L^* v = (-1)^m \partial_t^m v + \sum_{k+|\alpha| \leq m,\, k \neq m} (-1)^{|\alpha|+k} \partial_x^\alpha \partial_t^k \big(b_{k,\alpha}(t, x) v\big).$$

If $w \in C^m(M_\varepsilon)$ is a solution of $L^* w = 0$, then

$$\int_{M_\varepsilon} \big(v L^* w - w L v\big)\, d(x, t) = 0.$$

Furthermore, if $(\partial_t^r w)(\varepsilon, x) = 0$ for $r = 0, 1, \cdots, m - 2$, then after integration by parts we have

$$\int_{M_\varepsilon} \big(v L^* w - w L v\big)\, d(x, t) = \int_{K_\varepsilon} (-1)^m v(\varepsilon, x) \big(\partial_t^{m-1} w\big)(\varepsilon, x)\, dx = 0,$$

where $K_\varepsilon = \operatorname{supp} v(\varepsilon, x)$ is a compact subset.

Step 4.   Application of Weierstraß approximation theorem

Now, we prescribe besides the Cauchy conditions $(\partial_t^r w)(\varepsilon, x) = 0$ for $r = 0, 1, \cdots, m - 2$ the Cauchy condition $(\partial_t^{m-1} w)(\varepsilon, x) = P(x)$, where $P = P(x)$ is an arbitrarily chosen polynomial. One can then apply the Cauchy-Kovalevskaja theorem as now the data are prescribed on $t = \varepsilon$. In this way we obtain for all P an analytic solution in M_ε for all small ε, where ε is independent of P. Here, if the initial data P in the Cauchy condition $(\partial_t^{m-1} w)(\varepsilon, x) = P(x)$ is a polynomial, then the life span in time of the existence of analytic solutions is independent of P (see page 249 in [139]). So, we may conclude

$$\int_{K_\varepsilon} v(\varepsilon, x) P(x) \, dx = 0,$$

for an arbitrarily chosen polynomial $P(x)$. Due to the Weierstraß approximation theorem (polynomials are dense in the set of continuous functions with compact support) we may conclude $v(\varepsilon, x) = 0$. This we may conclude for all small $\varepsilon > 0$. Hence, $v \equiv 0$ in M_ε and we obtain $u(t, x) \equiv 0$ for all $(t, x) \in W_0$. The proof is completed.

The Holmgren theorem remains not valid for non-Kovalevskian equations. Let us consider, for example, the Cauchy problem

$$\partial_t u - \partial_x^2 u = 0, \ u(0, x) = 0.$$

Then, there exists the solution $u \equiv 0$. Let us try to construct a second solution. For this reason we choose the non-analytic function $\psi(t) = \exp(-t^{-2})$ for $t > 0$ and $\psi(t) = 0$ for $t \leq 0$. Then

$$u = u(t, x) = \sum_{k=0}^{\infty} \frac{1}{(2k)!} x^{2k} \psi^{(k)}(t)$$

is a second (formal) solution which is infinitely times differentiable, so a classical solution which is non-analytic.

The statement of the classical theorem of Holmgren can be used to conclude the following *uniqueness result* for Cauchy problems with non-analytic data and right-hand side.

Corollary 5.1.1 *Let us consider the Cauchy problem*

$$\partial_t^m u + \sum_{k+|\alpha| \leq m, \, k \neq m} a_{k,\alpha}(t, x) \partial_x^\alpha \partial_t^k u = f(t, x),$$

$$(\partial_t^r u)(0, x) = u_r(x) \ for \ r = 0, 1, \cdots, m - 2, m - 1,$$

where the real coefficients $a_{k,\alpha} = a_{k,\alpha}(t,x)$ are analytic in a neighborhood of the origin in $\mathbb{R}^{n+1}$. Then, there exists at most one uniquely determined m times differentiable solution in a neighborhood of the origin even if the smooth data and smooth right-hand side are supposed to be non-analytic.

5.2 Abstract Version

Section 4.2 is devoted to an abstract Cauchy-Kovalevskaja Theorem 4.2.1 in a scale of Banach spaces $\{B_s, \|\cdot\|_s\}_{s\in[0,1]}$. This theorem is applied to abstract Cauchy problems

$$d_t u = A(t)u + f(t), \quad u(0) = u_0,$$

where $\{A(t)\}_{t\in[0,T]}$ is a family of operators acting in this scale as follows:

$$\|A(t)u\|_{s'} \leq \frac{a}{s-s'}\|u\|_s \text{ for } 0 \leq s' < s \leq 1, u \in B_s.$$

In this section we will deal with the question as to whether an abstract Holmgren theorem does exist (see [206]).

To answer this question we assume an important assumption:

The scale of Banach spaces $\{B_s, \|\cdot\|_s\}_{s\in[0,1]}$ is dense into itself, which means that every space B_s is dense in $B_{s'}$ for $0 \leq s' \leq s \leq 1$.

Lemma 5.2.1 *Let us assume that a given scale of Banach spaces $\{B_s, \|\cdot\|_s\}_{s\in[0,1]}$ is dense into itself. Then, the family of Banach spaces $\{F_s, \|\cdot\|'_s\}_{s\in[0,1]}$, where $F_s := B'_{1-s}$, forms the so-called dual scale of Banach spaces to the given scale $\{B_s, \|\cdot\|_s\}_{s\in[0,1]}$.*

Proof The assumption of density implies that the dual operator of the natural injection of B_s into $B_{s'}$ is an injective continuous linear mapping of the dual $B'_{s'}$ of $B_{s'}$ into the dual B'_s of B_s. We shall refer to the latter mapping as the natural injection of $B'_{s'}$ into B'_s. Since the dual operator preserves the norm of the natural injection, we conclude immediately the statement that $\{F_s, \|\cdot\|'_s\}_{s\in[0,1]}$ is a scale of Banach spaces.

Now let us explain mapping properties of the family $\{A(t)'\}_{t\in[0,T]}$ of dual operators to the given family $\{A(t)\}_{t\in[0,T]}$ in the dual scale $\{F_s, \|\cdot\|'_s\}_{s\in[0,1]}$.

Lemma 5.2.2 *Let $\{A(t)\}_{t\in[0,T]}$ be a family of operators belonging to $L(B_s \to B_{s'})$ and mapping in a dense into itself scale of Banach spaces $\{B_s, \|\cdot\|_s\}_{s\in[0,1]}$ as follows:*

$$\|A(t)u\|_{s'} \leq \frac{a}{s-s'}\|u\|_s \text{ for } 0 \leq s' < s \leq 1, u \in B_s.$$

Then, the family of dual operators $\{A(t)'\}_{t\in[0,T]}$ *maps in the dual scale* $\{F_s, \|\cdot\|_s'\}_{s\in[0,1]}$ *as follows:*

$$\|A(t)'v\|_{s'}' \leq \frac{a}{s-s'}\|v\|_s' \text{ for } 0 \leq s' < s \leq 1, \ v \in F_s.$$

Proof We use the identity $A(t)'v(u) = v(A(t)u)$ for all $v \in F_s$ and $u \in B_{1-s'}$. Then, $A(t)u \in B_{1-s}$ and the functional $v(A(t)u)$ is well-defined on B_{1-s}. Consequently, $A(t)'v$ belongs to $F_{s'}$. Moreover,

$$|A(t)'v(u)| = |v(A(t)u)| \leq \|v\|_s'\|A(t)u\|_{1-s} \leq \|v\|_s'\frac{a}{s-s'}\|u\|_{1-s'}.$$

The last inequality implies together with

$$|A(t)'v(u)| \leq \|A(t)'v\|_{s'}'\|u\|_{1-s'} \text{ the estimate } \|A(t)'v\|_{s'}' \leq \frac{a}{s-s'}\|v\|_s'.$$

This we wanted to prove.

With these preparations in hand we are able to formulate and prove an abstract Holmgren theorem.

Theorem 5.2.1 (Abstract Holmgren Theorem)
Let us consider the abstract Cauchy problem

$$d_t u = A(t)u + f(t), \ u(0) = u_0,$$

under the assumptions of Theorem 4.2.1. Moreover, we assume that the scale of Banach spaces $\{B_s, \|\cdot\|_s\}_{s\in[0,1]}$ *is dense into itself. Then, there exists at most one solution* $v \in C([0,T], F_s)$ *for* $s \in (0,1]$ *of the abstract Cauchy problem*

$$d_t v = A(t)'v, \ v(0) = 0.$$

Here, $\{F_s, \|\cdot\|_s'\}_{s\in[0,1]}$ *denotes the dual scale to the given scale* $\{B_s, \|\cdot\|_s\}_{s\in[0,1]}$.

Proof The statement follows immediately after application of Lemma 5.2.2 and Theorem 4.2.1.

5.3 Concluding Remarks

5.3.1 Classical Holmgren Theorem

The statement of Theorem 5.1.1 can be generalized to the set of distributional solutions (see, for example, [85] or [206]), cf. with Definition 24.29 and Remark 24.4.1.

There exists a Holmgren type theorem for the Cauchy problem for the nonlinear partial differential equation of first order

$$\partial_t u = F(t, x, u, \partial_x u), \quad u(0, x) = u_0(x),$$

where F is a holomorphic function in its arguments, the variables (t, x) are taken from some set in $\mathbb{R}^{n+1}$ (see [133]).

In general there exists no Holmgren type result for Cauchy problems to nonlinear equations of higher order or to nonlinear systems (see [134]). Let us now explain the basic idea of [134].

It is well-known that a uniqueness result does, in general, not hold for linear Cauchy problems of Kovalevskian type with C^∞ coefficients. The proof of non-uniqueness results involves the construction (see, for instance, [86]) of a nontrivial classical solution u to the homogeneous partial differential equation of first order

$$\partial_t u + a(t, x)\partial_x u = 0,$$

together with the construction of a coefficient a of the equation. Such a nontrivial solution satisfies

$$u(t, x) = 0 \quad \text{for } t < 0 \text{ and } 0 \in \text{supp } u.$$

In [134] the author added an extra variable such that the vector (u, a) can be seen as a solution of a system or to eliminate the coefficient a so to reveal u as a solution of an higher order nonlinear equation. Let us now explain the idea for systems. For $s \in \mathbb{R}^1$ we define

$$v(t, x, s) = \chi(s)u(t - s, x) \text{ and } w(t, x, s) = a(t - s, x),$$

where $\chi \in C^\infty(\mathbb{R}^1)$ vanishes for $s \leq 0$ and $\chi(s) \neq 0$ for $s > 0$. Then,

$$v(t, x) = 0 \quad \text{for } t < 0 \text{ and } 0 \in \text{supp } v.$$

Therefore, $U = (0, w)$ and $V = (v, w)$ are different solutions of the following nonlinear system of first order with analytic coefficients:

$$\partial_t v + w \partial_x v = 0, \quad \partial_t w - \partial_s w = 0.$$

5.3.2   Abstract Holmgren Theorem

If we really want to apply the abstract Holmgren Theorem 5.2.1 to a given Cauchy problem we should at first find a suitable scale of Banach spaces. Then, we should show that a family of operators $\{A(t)\}_{t \in [0,T]}$ is acting in this scale as required in

Theorem 4.2.1. We then need the density into itself property. This step can cause difficulties. Then, we are able to apply the abstract Holmgren theorem. Finally, one has to understand which functions or distributions belong to the scale of dual spaces. In many examples this step is not trivial. The research project in Sect. 23.1 is devoted to the application of this abstract theory to one concrete example.

Exercises Relating to the Considerations of Chap. 5

Exercise 1 Study the Gevrey property of the nontrivial non-analytic solution u to

$$\partial_t u - \partial_x^2 u = 0, \quad u(0, x) = 0$$

of Sect. 5.1. Show, that u is really a classical solution.

Exercise 2 Show, that $u(t, x) = 0$ for $t \geq 0$ and $u(t, x) = e^{1/t}$ for $t < 0$ is a nontrivial classical solution to the Cauchy problem

$$\partial_x u = 0, \quad (t, x) \in \mathbb{R}^+ \times \mathbb{R}^1, \quad u(0, x) = 0, \quad x \in \mathbb{R}^1.$$

Is this observation a contradiction to Holmgren's uniqueness Theorem 5.1.1?
The answer to this question is related to the notion of a non-characteristic initial manifold, see Definition 6.2.

Chapter 6
Method of Characteristics

The method of characteristics is applied in studying general quasilinear partial differential equations of first order such as, for example, convection or transport equations. It is shown how the notion of characteristics allows for reducing the considerations to those for nonlinear systems of ordinary differential equations. An application to the continuity equation describing mass conservation completes this chapter.

6.1 Quasilinear Partial Differential Equations of First Order

Let us consider the following quasilinear partial differential equation of first order:

$$\sum_{k=1}^{n} a_k(x, u)\partial_{x_k} u = b(x, u).$$

The real coefficients $a_k = a_k(x, u) = a_k(x_1, \cdots, x_n, u)$ and the real right-hand side $b = b(x, u) = b(x_1, \cdots, x_n, u)$ are continuously differentiable in a domain G_0 of the $(n+1)$-dimensional space $\mathbb{R}^{n+1}_{x,u}$ with $(x, u) = (x_1, x_2, \cdots, x_n, u)$. We try to find real solutions $u = u(x)$. Here, one of the variables is allowed to be the time variable. If we are interested in the model

$$\partial_t u + \sum_{k=1}^{n} a_k(t, x, u)\partial_{x_k} u = b(t, x, u),$$

then it is hyperbolic and it is allowed to prescribe an additional initial condition or Cauchy condition.

© Springer International Publishing AG 2018

M.R. Ebert, M. Reissig, *Methods for Partial Differential Equations*,
https://doi.org/10.1007/978-3-319-66456-9_6

Remark 6.1.1 Transport or convection processes are special cases of the above model (see Sect. 2.2). A typical example is the mass conservation law

$$\partial_t \rho(t,x) + \text{div}\left(\rho(t,x)\, v(t,x)\right) = 0, \quad \rho(0,x) = \rho_0(x).$$

By $x = (x_1, \cdots, x_n)$ we denote the vector of spatial variables in $\mathbb{R}^n$ by t the time variable, by $v = v(t,x)$ *the velocity field* and by $\rho = \rho(t,x)$ *the density*. Finally, $\rho_0 = \rho_0(x)$ denotes the density for $t = 0$. The velocity field is supposed to be known.

We write the conservation of mass in the form

$$\partial_t \rho + \sum_{k=1}^{n} v_k(t,x)\partial_{x_k}\rho + \rho\,\text{div}\, v(t,x) = 0.$$

This linear model for the density ρ is a special case of our given quasilinear model.

6.2   The Notion of Characteristics: Relation to Systems of Ordinary Differential Equations

In every point $(x_1, \cdots, x_n, u) \in G_0$ we construct the so-called *characteristic vector* $(a_1(x,u), \cdots, a_n(x,u), b(x,u))$. The direction of this vector is called *characteristic direction*. The given quasilinear partial differential equation can be written in the form

$$\sum_{k=1}^{n} a_k(x,u)\partial_{x_k}u - b(x,u) = 0.$$

Hence, the normal to the surface given in $\mathbb{R}^{n+1}_{x,u}$ by the equation

$$u(x_1, \cdots, x_n) - u = 0$$

(form the gradient with respect to x and u) is *orthogonal to the characteristic direction*. But this implies that the characteristic vector lies in the plane tangent to the surface. We obtain a so-called *field of directions in G_0*, which is formed by all characteristic directions in all points of G_0 (compare with the field of directions from the course "Ordinary differential equations").

Our goal is to construct curves in this field of directions having the property that in every point of the curve the tangent vector coincides with the characteristic vector.

A curve in $\mathbb{R}^{n+1}_{x,u}$ is given by the parameter representation

$$\{(x_1, x_2, \cdots, x_n, u) \in \mathbb{R}^{n+1}$$

$$: x_1 = x_1(\tau), x_2 = x_2(\tau), \cdots, x_n = x_n(\tau), u = u(\tau), \ \tau \in I\},$$

where the parameter τ is taken from an open interval I in $\mathbb{R}^1$. The vector

$$\left(\frac{dx_1}{d\tau}(\tau_0), \cdots, \frac{dx_n}{d\tau}(\tau_0), \frac{du}{d\tau}(\tau_0)\right)$$

gives the direction of the tangent vector in the point $(x_1(\tau_0), \cdots, x_n(\tau_0), u(\tau_0))$, where τ_0 is from the parameter interval I for τ. Due to our goal, the tangent vector should coincide with the characteristic vector in the point $(x_1(\tau_0), \cdots, x_n(\tau_0), u(\tau_0))$. For this reason, the following system of ordinary differential equations should be satisfied:

$$\frac{dx_k}{d\tau} = a_k(x_1(\tau), \cdots, x_n(\tau), u(\tau)), \; k = 1, \cdots, n,$$

$$\frac{du}{d\tau} = b(x_1(\tau), \cdots, x_n(\tau), u(\tau)).$$

Let $(x_{01}, \cdots, x_{0n}, u_0) \in G_0$ be a given point. Let us find a solution of this nonlinear system of ordinary differential equations satisfying for $\tau = 0$ the Cauchy condition $x_k(0) = x_{0k}$, $k = 1, \cdots, n$, and $u(0) = u_0$. Then, such a solution is a curve passing through $(x_{01}, \cdots, x_{0n}, u_0)$ and having the property that in every point the tangent vector coincides with the characteristic vector.

Definition 6.1 Every curve having these properties is called a characteristic.

Remark 6.2.1 Let us assume that the given partial differential equation of first order is linear, that is,

$$\sum_{k=1}^{n} a_k(x)\partial_{x_k} u = b(x)u.$$

In general, to describe the characteristics we get nevertheless a nonlinear system of ordinary differential equations. Thus we lose the linear structure of our Cauchy problem in the case of non-constant coefficients a_k.

6.3 Influence of the Initial Condition

We are not only interested in the construction of solutions to a given quasilinear partial differential equation of first order

$$\sum_{k=1}^{n} a_k(x, u)\partial_{x_k} u = b(x, u),$$

but want to pose a *Cauchy condition* to the solutions.

How do we prescribe such a Cauchy condition? We will not only construct a characteristic through a point, but through all points of a $(n-1)$-*dimensional smooth surface* $M_0 \subset G_0$. A parametrization of such a surface is given by $n+1$ real functions

$$x_1^0 = x_1^0(t_1, \cdots, t_{n-1}), \cdots, x_n^0 = x_n^0(t_1, \cdots, t_{n-1}), \ u^0 = u^0(t_1, \cdots, t_{n-1})$$

with parameters $(t_1, \cdots, t_{n-1})$ from a compact closure $\mathcal{B} \subset \mathbb{R}^{n-1}$ of a domain of parameters. Moreover, for every point $(t_1, \cdots, t_{n-1})$ from $\mathcal{B}$ the corresponding point

$$\left(x_1^0(t_1, \cdots, t_{n-1}), \cdots, x_n^0(t_1, \cdots, t_{n-1}), u^0(t_1, \cdots, t_{n-1}) \right)$$

must belong to M_0.

Let us determine the characteristic for every point from M_0. Then, this set of characteristics depends on τ and the $n-1$ parameters $t_1, \cdots, t_{n-1}$ as well. So we have characteristics

$$x_k = x_k(\tau, t_1, \cdots, t_{n-1}), \ k = 1, \cdots, n, \ u = u(\tau, t_1, \cdots, t_{n-1}).$$

These characteristics define a n-dimensional surface in $\mathbb{R}^{n+1}$. Due to our principle of construction, every characteristic vector in every point of this n-dimensional surface lies in the tangential plane to this point. The vector $\left(\partial_{x_1} u, \cdots, \partial_{x_n} u, -1 \right)$ to the surface $u(x_1, \cdots, x_n) - u = 0$ is orthogonal to the tangential plane. This follows from

$$\sum_{k=1}^{n} a_k(x, u) \partial_{x_k} u - b(x, u) = 0.$$

Summary   We have shown that all functions

$$x_k = x_k(\tau, t_1, \cdots, t_{n-1}), \ k = 1, \cdots, n, \ u = u(\tau, t_1, \cdots, t_{n-1})$$

define a n-dimensional surface in $\mathbb{R}^{n+1}$ which is a solution to our given quasilinear partial differential equation,

$$\sum_{k=1}^{n} a_k(x, u) \partial_{x_k} u = b(x, u),$$

satisfying a Cauchy condition on a $(n-1)$-dimensional surface $M_0 \subset G_0$. We may conclude this from

$$\frac{du}{d\tau} = \sum_{k=1}^{n} \frac{\partial u}{\partial x_k} \frac{dx_k}{d\tau} = \sum_{k=1}^{n} a_k(x, u) \partial_{x_k} u = b(x, u).$$

The reader expects that this surface, the solution, respectively, can be represented in the form $u = u(x_1, \cdots, x_n)$. Let us return to this problem in the next section.

6.4 Application of the Inverse Function Theorem

We are interested in under which conditions representations $\tau = \tau(x_1, \cdots, x_n)$, $t_l = t_l(x_1, \cdots, x_n)$, $l = 1, \cdots, n-1$, can be derived from the representations $x_k = x_k(\tau, t_1, \cdots, t_{n-1})$, $k = 1, \cdots, n$. Using such representations

$u = u(\tau, t_1, \cdots, t_{n-1})$ implies immediately

$u = u(x_1, \cdots, x_n) = u\big(\tau(x_1, \cdots, x_n), t_1(x_1, \cdots, x_n), \cdots, t_{n-1}(x_1, \cdots, x_n)\big)$, which is

what we wanted to have. If any two different characteristics appearing in different points of the surface M_0 do not intersect, then such representations exist. But, in general this property does not hold. For this reason we restrict ourselves to a locally (in τ) defined solution. The question of the existence of local solutions in τ is connected with the behavior of the *Jacobi matrix*

$$\frac{\partial(x_1, \cdots, x_n)}{\partial(\tau, t_1, \cdots, t_{n-1})} =: J_x(\tau, t_1, \cdots, t_{n-1}).$$

For $\tau = 0$ we get

$$J_x(0, t_1, \cdots, t_{n-1}) = \begin{pmatrix} a_1(x_k^{(0)}(t_1, \cdots, t_{n-1}), u^0(t_1, \cdots, t_{n-1})) & \frac{\partial x_1^0}{\partial t_1} & \cdots & \frac{\partial x_1^0}{\partial t_{n-1}} \\ a_n(x_k^{(0)}(t_1, \cdots, t_{n-1}), u^0(t_1, \cdots, t_{n-1})) & \frac{\partial x_n^0}{\partial t_1} & \cdots & \frac{\partial x_n^0}{\partial t_{n-1}} \end{pmatrix}.$$

Here we assume that the functions $x_k^0 = x_k^0(t_1, \cdots, t_{n-1})$ depend continuously differentiable on $t_1, \cdots, t_{n-1}$, that is, we choose a sufficiently smooth parametrization of the initial surface. Let us assume, that $J_x(0, t_1, \cdots, t_{n-1})$ *is a regular matrix*. Then, the continuous dependence of J_x with respect to τ implies that $J_x(\tau, t_1, \cdots, t_{n-1})$ *remains regular for small* $|\tau|$. The application of the inverse function theorem gives a locally defined solution for small values of $|\tau|$.

Geometric Explanation of the Condition "$J_x(0, t_1, \cdots, t_{n-1})$ is Regular in $\tau = 0$"

- The surface (or initial manifold) M_0 has the parameter representation

$$x_1^0 = x_1^0(t_1, \cdots, t_{n-1}), \cdots, x_n^0 = x_n^0(t_1, \cdots, t_{n-1}), \ u^0 = u^0(t_1, \cdots, t_{n-1}).$$

Then, the tangential plane at every point

$$x^0 = x^0(t_1, \cdots, t_{n-1}), \ (t_1, \cdots, t_{n-1}) \in \mathcal{B},$$

is $(n-1)$-dimensional in the space $\mathbb{R}_x^n$. Consequently, one has to choose a suitable smooth parametrization of M_0.

- The projection of the characteristic vector

$$\left(a_1\left(x_k^{(0)}(t_1,\cdots,t_{n-1}),u^0(t_1,\cdots,t_{n-1})\right),\cdots,a_n\left(x_k^{(0)}(t_1,\cdots,t_{n-1}),\right.\right.$$
$$\left.\left.u^0(t_1,\cdots,t_{n-1})\right),b\left(x_k^{(0)}(t_1,\cdots,t_{n-1}),u^0(t_1,\cdots,t_{n-1})\right)\right)$$

in $\mathbb{R}_{x,u}^{n+1}$ on $\{u=0\}$, is the vector

$$\left(a_1\left(x_k^{(0)}(t_1,\cdots,t_{n-1}),u^0(t_1,\cdots,t_{n-1})\right),\cdots,a_n\left(x_k^{(0)}(t_1,\cdots,t_{n-1}),\right.\right.$$
$$\left.\left.u^0(t_1,\cdots,t_{n-1})\right),0\right)$$

and can be associated with the vector

$$\left(a_1\left(x_k^{(0)}(t_1,\cdots,t_{n-1}),u^0(t_1,\cdots,t_{n-1})\right),\cdots,a_n\left(x_k^{(0)}(t_1,\cdots,t_{n-1}),\right.\right.$$
$$\left.\left.u^0(t_1,\cdots,t_{n-1})\right)\right)$$

in $\mathbb{R}_x^n$, which does not belong to the tangential plane to every point

$$x^0=x^0(t_1,\cdots,t_{n-1}),\quad (t_1,\cdots,t_{n-1})\in\mathcal{B}.$$

Definition 6.2 If the surface (or initial manifold) M_0 satisfies the last two conditions, then it is called *non-characteristic*.

The notion of non-characteristic surfaces can be generalized to partial differential operators of higher order. For this reason we choose now $(n-1)$-dimensional surfaces $M_0\subset\mathbb{R}_x^n$ which are defined by $\varphi(x)=0$ with $\nabla\varphi\neq 0$, where φ is a real function.

Let us first explain this for linear partial differential operators of order m having the representation

$$P(x,D_x)=\sum_{|\alpha|\leq m}a_\alpha(x)D_x^\alpha,\quad x\in\mathbb{R}^n.$$

Now, the principal symbol of P is given by

$$P_m(x,\xi)=\sum_{|\alpha|=m}a_\alpha(x)\xi^\alpha,\quad \xi\in\mathbb{R}^n.$$

Definition 6.3 If a $(n-1)$-dimensional surface $M_0\subset\mathbb{R}_x^n$ is defined by $\varphi(x)=0$ ($\nabla\varphi\neq 0$) and satisfies $P_m(x,\nabla\varphi(x))\neq 0$ for all $x\in M_0$, then we say that M_0 is a non-characteristic surface. If $P_m(x,\nabla\varphi(x))=0$ for some $x_0\in M_0$, then we say that M_0 is a characteristic surface at $x=x_0$.

Remark 6.4.1 The property of the surface M_0 to be non-characteristic with respect to the above introduced operator $P(x, D_x)$ is invariant under a regular change of variables. In this way, if our analysis is restricted to a neighborhood of a point on M_0, then we may assume, without loss of generality, that this neighborhood coincides with a neighborhood in the hyperplane $\{x \in \mathbb{R}^n : x_1 = 0\}$.

Definition 6.4 Let $M_0 \subset \mathbb{R}^n_x$ be a $(n-1)$-dimensional surface. If M_0 is a non-characteristic surface with respect to $P(x, D_x)$, then we say that the Cauchy problem

$$P(x, D_x)u = \sum_{|\alpha| \leq m} a_\alpha(x) D_x^\alpha u = f(x),$$

$$\partial_n^r u \, |_{M_0} = u_r(x) \quad \text{for} \quad r = 0, 1, \cdots, m-2, m-1,$$

is non-characteristic. Here, ∂_n denotes the normal derivative on M_0.

Example 6.4.1 Consider the Cauchy problem

$$D_t^m u + \sum_{k+|\alpha| \leq m, \, k \neq m} a_{k,\alpha}(t, x) D_x^\alpha D_t^k u = f(t, x),$$

$$(D_t^r u)(0, x) = \varphi_r(x) \quad \text{for} \quad r = 0, 1, \cdots, m-2, m-1,$$

with Cauchy conditions on the set $M_0 := \{(t, x) \in \mathbb{R}^{n+1} : t = 0\}$. Due to Definitions 6.3 and 6.4 this Cauchy problem is non-characteristic. Consequently, the statements of Theorems 4.1.1 and 5.1.1 hold for this class of non-characteristic Cauchy problems.

Remark 6.4.2 Thanks to Remark 6.4.1, if the initial surface M_0 is supposed to be analytic, then the statements of Theorems 4.1.1 and 5.1.1 are still valid for non-characteristic Cauchy problems. Indeed, if we look for solutions which are defined in a neighborhood of some point $x_0 \in M_0$, then a suitable regular and analytic change of variables maps x_0 to the origin and a neighborhood of x_0 in M_0 onto a neighborhood of the origin in $\{x \in \mathbb{R}^n : x_1 = 0\}$.

We will discuss a special characteristic Cauchy problem in Sect. 6.6.2.

For nonlinear partial differential operators the definition should also depend on the solution and it's derivatives prescribed on the initial surface M_0. If for the quasilinear partial differential equation of first order

$$\sum_{k=1}^n a_k(x, u) \partial_{x_k} u = b(x, u)$$

with prescribed initial data $u = u^0$ on a surface M_0 (see Definition 6.1 in the case that M_0 has a parameter representation), the surface M_0 is given by $\varphi(x) = 0 \, (\nabla \varphi \neq 0)$. Then, the non-characteristic condition can be written as follows:

$$\left(a_1(x, u^0), \cdots, a_n(x, u^0)\right) \cdot \nabla \varphi(x) \neq 0 \quad \text{for all} \quad x \in M_0.$$

6.5   Summary

Summarizing the considerations of the previous sections leads to the following main
result:

Theorem 6.5.1 *Let us assume that a given $(n - 1)$-dimensional surface (initial
manifold) M_0 is non-characteristic in $\mathbb{R}^{n+1}_{x,u}$ and can be described by*

$$M_0 = \Big\{(x^0, u^0) = (x^0_1, \cdots, x^0_n, u^0) \in \mathbb{R}^{n+1}_{x,u} : x^0_1 = x^0_1(t_1, \cdots, t_{n-1}),$$

$$\cdots, x^0_n = x^0_n(t_1, \cdots, t_{n-1}), u^0 = u^0(t_1, \cdots, t_{n-1})\Big\},$$

*where $(t_1, \cdots, t_{n-1})$ belongs to a compact parameter set $\mathcal{B}$ in $\mathbb{R}^{n-1}$. Then, there
exists a neighborhood of the initial manifold M_0 with the following property:
To a given continuously differentiable data $u^0 = u^0(t_1, \cdots, t_{n-1})$ there exists a
uniquely determined classical solution $u = u(x_1, \cdots, x_n)$ of the quasilinear partial
differential equation of first order*

$$\sum_{k=1}^{n} a_k(x_1, \cdots, x_n, u)\partial_{x_k} u = b(x_1, \cdots, x_n, u).$$

*The unique solution is the solution of the system of in general, nonlinear ordinary
differential equations*

$$\frac{dx_k}{d\tau} = a_k(x_1, \cdots, x_n, u), \quad \frac{du}{d\tau} = b(x_1, \cdots, x_n, u)$$

with the initial values

$$x_k(0, t_1, \cdots, t_{n-1}) = x^0_k(t_1, \cdots, t_{n-1}), k = 1, \cdots, n,$$

$$u(0, t_1, \cdots, t_{n-1}) = u^0(t_1, \cdots, t_{n-1}).$$

*One has to solve the equations $x_k = x_k(\tau, t_1, \cdots, t_{n-1})$ with respect to $\tau, t_1, \cdots, t_{n-1}$.
The solutions $\tau = \tau(x_1, \cdots, x_n), t_l = t_l(x_1, \cdots, x_n), l = 1, \cdots, n - 1$, are substi-
tuted in $u(\tau, t_1, \cdots, t_{n-1})$. Finally, the solution $u = u(x_1, \cdots, x_n)$ is continuously
differentiable in a neighborhood of the initial manifold if we assume that the
functions a_k, b, the data, and the initial manifold are continuously differentiable.*

Remark 6.5.1 We omitted the proof of the uniqueness. To conclude this property
the initial manifold is supposed to be non-characteristic. Moreover, we need the
uniqueness of the characteristics which follows from standard arguments of the
theory of ordinary differential equations.

6.6 Examples

6.6.1 Continuity Equation

Let us consider the Cauchy problem for the mass conservation

$$\partial_t \rho(t, x) + \text{div} \left(\rho(t, x) \, v(t, x) \right) = 0, \quad \rho(0, x) = \rho_0(x), \quad x = (x_1, x_2, x_3) \in \mathbb{R}^3.$$

We write the mass conservation in the form

$$\partial_t \rho + v_1(t, x) \partial_{x_1} \rho + v_2(t, x) \partial_{x_2} \rho + v_3(t, x) \partial_{x_3} \rho + \rho \, \text{div} \, v(t, x) = 0.$$

Then we obtain the system of ordinary differential equations

$$\frac{dt}{d\tau} = 1, \quad \frac{dx_1}{d\tau} = v_1(t, x), \quad \frac{dx_2}{d\tau} = v_2(t, x), \quad \frac{dx_3}{d\tau} = v_3(t, x),$$

$$t(0) = 0, \quad x_1(0) = t_1, \quad x_2(0) = t_2, \quad x_3(0) = t_3,$$

$$\frac{d\rho}{d\tau} = -(\text{div} \, v(t, x))\rho, \quad \rho(0) = \rho_0(t_1, t_2, t_3).$$

The initial manifold is isomorphic to $\mathbb{R}^3$. It is *non-characteristic*. Here we use

$$J_x(0, t_1, t_2, t_3) = \begin{pmatrix} 1 & 0 \ 0 \ 0 \\ v_1(t_1, t_2, t_3, 0) & 1 \ 0 \ 0 \\ v_2(t_1, t_2, t_3, 0) & 0 \ 1 \ 0 \\ v_3(t_1, t_2, t_3, 0) & 0 \ 0 \ 1 \end{pmatrix}.$$

The first column vector is linear independent to a subspace formed by the second up to the fourth column vector. Let us put the velocity field $v_1(t, x) = x_1$, $v_2(t, x) = v_3(t, x) = 0$ and the initial condition for the density $\rho_0(x) = x_1^3$. Then we obtain

$$\frac{dt}{d\tau} = 1, \quad \frac{dx_1}{d\tau} = x_1, \quad \frac{d\rho}{d\tau} = -\rho,$$

$$t(0) = 0, \quad x_1(0) = t_1, \quad \rho(0) = t_1^3,$$

$$t = \tau, \quad x_1(t) = t_1 e^t, \quad \rho(t) = t_1^3 e^{-t}, \rho(t, x) = x_1^3 e^{-4t}.$$

6.6.2 An Example of a Characteristic Cauchy Problem

Let us discuss a typical example of a characteristic Cauchy problem.

Consider the following Cauchy problem for a quasilinear partial differential equation of first order:

$$a_1(x, y, u)\partial_x u + a_2(x, y, u)\partial_y u = b(x, y, u), \quad u(\sigma(t), \gamma(t)) = f(t), \quad t \in I.$$

Here I is an interval in $\mathbb{R}^1$. Suppose that the Jacobi matrix

$$J_x(0, t) = \begin{pmatrix} a_1(\sigma(t), \gamma(t), f(t)) & \sigma'(t) \\ a_2(\sigma(t), \gamma(t), f(t)) & \gamma'(t) \end{pmatrix}$$

vanishes at $t = t_0 \in I$, i.e.,

$$a_1(\sigma(t_0), \gamma(t_0), f(t_0))\gamma'(t_0) - a_2(\sigma(t_0), \gamma(t_0), f(t_0))\sigma'(t_0) = 0.$$

The Cauchy problem is then a characteristic one. Let us assume that a_1 and a_2 do not vanish. We may conclude that

$$\frac{\sigma'(t_0)}{a_1(\sigma(t_0), \gamma(t_0), f(t_0))} = \frac{\gamma'(t_0)}{a_2(\sigma(t_0), \gamma(t_0), f(t_0))} = \mu$$

with a real constant μ. By using the initial data we obtain

$$f'(t) = \frac{d}{dt}u(\sigma(t), \gamma(t)) = \sigma'(t)\partial_x u(\sigma(t), \gamma(t)) + \gamma'(t)\partial_y u(\sigma(t), \gamma(t)).$$

Finally, by using the partial differential equation at $t = t_0$, we derive the compatibility condition

$$f'(t_0) = \mu b(\sigma(t_0), \gamma(t_0), f(t_0)).$$

Therefore, if $f'(t_0) \neq \mu b(\sigma(t_0), \gamma(t_0), f(t_0))$, then a solution of the initial value problem does not exist in any neighborhood of $(\sigma(t_0), \gamma(t_0))$.

Remark 6.6.1 If the initial manifold M_0 is characteristic, then there is, in general, no solution to the Cauchy problem. But, if there exists a solution, then there are, in general, infinitely many solutions for the Cauchy problem (see Exercise 3 below).

6.7   Concluding Remarks

We complete this chapter with some remarks concerning generalizations of the Cauchy-Kovalevskaja and Holmgren theorem for characteristic Cauchy problems. To find additional conditions in order to have existence and uniqueness of solutions for characteristic Cauchy problems has been of interest. After one of the pioneering papers [73], in [5] the authors introduced the class of partial differential operators of so-called Fuchsian type and proved for this class some generalizations of the classical Cauchy-Kovalevskaja theorem and the Holmgren uniqueness theorem. In [127] the author considered characteristic Cauchy problems for some essentially non-Fuchsian partial differential operators whose principal parts are essentially of Fuchsian type. Considering functions that are of class C^∞ with respect to the variable t and holomorphic with respect to x, the author proved some theorems that are similar to the Cauchy-Kovalevskaja theorem and the Holmgren uniqueness theorem.

Exercises Relating to the Considerations of Chap. 6

Exercise 1 We recall from the course "Ordinary differential equations" that the Theorems of Peano and Picard-Lindelöf yield the existence (Peano) and uniqueness (Picard-Lindelöf) of locally defined solutions in an open interval around $\tau = 0$ of the Cauchy problem for nonlinear systems of ordinary differential equations. But, under which assumptions do we get a globally defined solution in τ?

Exercise 2 Find the globally (in time) classical solution to the following Cauchy problem for the transport equation:

$$\partial_t u + b \cdot \nabla u = f(t, x), \quad u(0, x) = g(x),$$

where b is a real constant vector, $f \in C(\mathbb{R}^{n+1})$ and $g \in C^1(\mathbb{R}^n)$.

Exercise 3 Consider the Cauchy problem

$$\partial_x u = 2xy, \quad u(x, x^2) = f(x).$$

Decide, if there exists or does not exist a locally defined classical solution in the following cases:

1. $f(x) \equiv 1$,
2. $f(x) = x$,
3. f is an odd function on $\mathbb{R}^1$.

Can we expect uniqueness of solutions if the existence of at least one solution is verified?

Exercise 4 Solve the Cauchy problem

$$x\partial_x u - y\partial_y u = u^2, \quad u(x, 1) = 1$$

and find the interval of existence of a classical solution with respect to y.

Exercise 5 Solve the Cauchy problem

$$x\partial_x u + y\partial_y u = u + 1, \quad u(x, x^2) = x^2.$$

Conclude that, although the equation is linear, we have no classical solution on $\mathbb{R}^2$. What happens if we change the initial condition to $u(x, x) = x^2$?

Exercise 6 Solve the Cauchy problem

$$\partial_t u + u \partial_x u = 0, \quad u(x, 0) = f(x).$$

Show that if $f'(x) \leq \alpha$, $\alpha > 0$, then after a finite time the classical solution does not exist any more (see Chap. 7 for a generalization of this model).

Exercise 7 Let $P(D_x)$ be a homogeneous partial differential operator with constant coefficients in $\mathbb{R}^n$ and let $N \neq 0$ be a real vector. Show that the plane $H_N = \{x \in \mathbb{R}^n : x \cdot N = 0\}$ is characteristic with respect to $P(D_x)$ if and only if there exists a solution u of the equation $P(D_x)u = 0$ with $u \in C^\infty(\mathbb{R}^n)$ and $\operatorname{supp} u = \{x \in \mathbb{R}^n : x \cdot N \geq 0\}$.

Chapter 7
Burgers' Equation

In the previous chapter the method of characteristics was introduced. This method allows among other things, a precise description of the life span of solutions to Cauchy problems for quasilinear partial differential equations of first order. We will now elaborate such an application by means of Burgers' equation. Johannes (Jan) Martinus Burgers (1895–1981) was a Dutch physicist. He is credited with being the father of Burgers' equation. Notions as blow up, geometrical blow up or life span of solutions to Burgers' equation with and without mass term are introduced and discussed.

7.1 Classical Burgers' Equation

Now let us apply the method of characteristics to the treatment of an important model equation, the so-called *Burgers' equation*

$$u_t + a(u)u_x = 0.$$

Here $a = a(u)$ is a real function. Consequently, the partial differential equation is hyperbolic. Moreover, we prescribe the Cauchy condition $u(0, x) = f(x)$.
We assume that the functions $a = a(s)$ and $f = f(s)$ belong to $C^\infty(\mathbb{R}^1)$, that is, they are infinitely times differentiable on $\mathbb{R}^1$.
Let us consider the forward Cauchy problem. We are interested in finding infinitely differentiable solutions in the largest possible strip $[0, T) \times \mathbb{R}^1$.

We shall apply the method of characteristics to the forward Cauchy problem $u_t + a(u)u_x = 0$, $u(0, x) = f(x)$. So, we obtain the characteristic system

$$\frac{dt}{d\tau} = 1, \quad \frac{dx}{d\tau} = a(u), \quad \frac{du}{d\tau} = 0,$$

$$x(0, s) = s, \quad u(0, s) = f(s).$$

© Springer International Publishing AG 2018
M.R. Ebert, M. Reissig, *Methods for Partial Differential Equations*,
https://doi.org/10.1007/978-3-319-66456-9_7

We may set $\tau = t$. Let $u = u(t, x)$ be a given continuously differentiable solution on the strip $[0, T) \times \mathbb{R}^1$. We introduce the characteristic Γ_s by

$$\Gamma_s := \{x(t) : \frac{dx}{dt} = a(u), \ x(0, s) = s\}.$$

Then we have, along Γ_s, the identity

$$\frac{du}{dt} = \frac{d}{dt}u(t, x) = u_t + \frac{dx}{dt}u_x = u_t + a(u)u_x = 0.$$

It follows that $u(t, s) = u(0, s) = f(s)$ is constant along Γ_s. Hence,

$$\frac{dx}{dt} = a(f(s)) \quad \text{or} \quad x(t, s) = s + a(f(s))t.$$

Summary Let $u = u(t, x)$ be a given continuously differentiable solution. Then this solution is implicitly defined by $u = f(x - a(u)t)$. This follows from

$$u(t, s) = u(0, s) = f(s) = f(x(t, s) - a(f(s))t).$$

What do we conclude from this implicit definition?

Case 1:   We assume $a'(f(s))f'(s) \geq 0$ for all $s \in \mathbb{R}^1$.

Then we have $\frac{dx}{ds} = 1 + a'(f(s))f'(s)t \geq 1$ for all $s \in \mathbb{R}^1$ and $t \geq 0$.
So, two different characteristics Γ_{s_1} and Γ_{s_2}, $s_1 \neq s_2$, never intersect each other. This follows from the fact that for all times $t_0 > 0$ the position $x = x(t_0, s)$ is a strictly monotonous increasing function with respect to the parameter s.

Summary In the first case the solution $u = u(t, x)$ exists globally in t, that is on the upper half plane $(0, \infty) \times \mathbb{R}^1$.

Case 2:   Let us assume $\kappa := \sup_{s \in \mathbb{R}^1}(-a'(f(s))f'(s)) > 0$. Then $\frac{dx}{ds} = 1 + a'(f(s))f'(s)t > 0$ for all $s \in \mathbb{R}^1$ and $t \in [0, T)$, $T := \frac{1}{\kappa}$. Thus, two different characteristics Γ_{s_1} and Γ_{s_2}, $s_1 \neq s_2$, never intersect each other for all $t \in [0, T)$. Due to our assumption, there exist two different intersecting characteristics Γ_{s_1} and Γ_{s_2}, $s_1 \neq s_2$, for $t = T$.

Summary The solution exists only on the strip $(0, T) \times \mathbb{R}^1$.
If two characteristics Γ_{s_1} and Γ_{s_2}, $s_1 \neq s_2$, intersect each other, then we have, in general, two different values in the intersection point. This follows from the fact that the solution is constant along characteristics. One value $f(s_1)$ is transported along Γ_{s_1}, a second value $f(s_2)$ is transported along Γ_{s_2}. For this reason the solution fails to exist in the intersection point. But, *the solution remains bounded* along the two different characteristics up to the point of intersection.
But what about the behavior of the derivative $u_x = u_x(t, x)$ for $t \to T - 0$?

To answer this question we introduce $w := u_x$. Then, we have along the characteristic Γ_s the equation

$$\frac{dw}{dt} = \frac{\partial^2 u}{\partial x \partial t} + \frac{dx}{dt} \frac{\partial^2 u}{\partial x^2} = u_{xt} + a(u)u_{xx}$$

$$= \partial_x(-a(u)u_x) + a(u)u_{xx} = -a'(u)u_x^2 = -a'(f(s))w^2.$$

Consequently, u_x satisfies along Γ_s the semilinear ordinary differential equation

$$\frac{dw}{dt} = -a'(f(s))w^2.$$

For $t = 0$ it holds that $u_x(0, s) = w(0, s) = f'(s)$. Applying the method of separation of variables implies

$$u_x = w = \frac{f'(s)}{1 + a'(f(s))f'(s)t}.$$

Under the assumptions of the second case the modulus $|u_x(t, x)|$ of the partial derivative u_x tends to infinity on some Γ_s for $t \to T - 0$. Such a blow up behavior is called *geometrical blow up*: solutions themselves remain bounded if t tends to the point of intersection (to the *life span time* $t=T$) along two different characteristics, but derivatives are unbounded in a certain sense along these two different characteristics.

Example 7.1.1 Let us consider the Cauchy problem

$$u_t + uu_x = 0, \quad u(0, x) = -x.$$

The characteristics Γ_s are given by $x(t, s) = s(1 - t)$. So, all characteristics Γ_s intersect each other at the point $(t_0, x_0) = (1, 0)$. The life span time T is equal to 1. Consequently, the solution exists only on the strip $(0, 1) \times \mathbb{R}^1$. Indeed, the solution has the representation $u(t, x) = \frac{x}{t-1}$ for all $0 \le t < 1$. We conclude that not only $u_x(t, x(t, s)) = \frac{1}{t-1}$ blows up at $t = 1$ but also the solution $u = u(t, x)$ itself. It is clear that $u(t, x(t, s)) = -s$, i.e., the solution remains bounded along arbitrarily chosen intersecting characteristics Γ_s.

Summary

Case 1: The *life span* $T = \infty$, the solution exists globally (in t).

Case 2: The *life span* $T < \infty$, the derivative u_x has a blow up for $t = T$ (geometrical blow up) along two different intersecting characteristics while the solution u itself has no blow up along these intersecting characteristics, it remains bounded approaching the blow up time. This follows from the implicit definition of the solution $u = f(x - a(u)t)$.

Case 3: If $\inf_{s\in\mathbb{R}^1}(a'(f(s))f'(s)) = -\infty$, then there is no classical solution in any strip $(0, T) \times \mathbb{R}^1$ with $T > 0$ chosen arbitrarily small.

Example 7.1.2 Let us consider the Cauchy problem

$$u_t - uu_x = 0, \quad u(0, x) = 1 + x^2.$$

By using the method of characteristics we derive the solution

$$u(t, x) = \frac{2 + 2x^2}{1 - 2tx + \sqrt{1 - 4tx - 4t^2}}.$$

Therefore, there is no classical solution in any strip $(0, T) \times \mathbb{R}^1$, $T > 0$, as already summarized in Case 3.

Let us turn again to the Cauchy problem

$$u_t + a(u)u_x = 0, \quad u(0, x) = f(x).$$

But now we assume only a *smooth data f* in $C_0^\infty(\mathbb{R}^1)$, that is, the data has compact support and is infinitely times differentiable.

Definition 7.1 Burgers' equation $u_t + a(u)u_x = 0$ is called *genuinely nonlinear* if $a'(0) \neq 0$ ($a(u) \approx a(0) + a'(0)u + \cdots$, the nonlinear term uu_x really appears).

Let us assume that Burgers' equation is genuinely nonlinear and the nontrivial data has compact support and is infinitely times differentiable. Although we have a smooth data, the condition $a'(f(s))f'(s) \geq 0$ *for all* $s \in \mathbb{R}^1$ from Case 1, is never satisfied. We will not prove this result, but leave it as an exercise for the reader.

Corollary 7.1.1 *Assume that Burgers' equation* $u_t + a(u)u_x = 0$ *with* $a'(0) \neq 0$ *is genuinely nonlinear. Moreover, the data has compact support and is infinitely times differentiable. Then, classical solutions have, in general, a geometrical blow up at some time* $T > 0$. *The time* $T = T(f)$ *depends on the data f.*

If the data $u(0, x) \equiv 0$, then the solution $u = u(t, x) \equiv 0$ exists globally in $t \in (0, \infty)$. Intuitively we expect that a compactly supported infinitely differentiable data $u(0, x) = \varepsilon f(x)$ leads to a solution in an interval $(0, T_\varepsilon)$ in t, where the *life span time* T_ε tends to infinity for $\varepsilon \to +0$.

Let us choose the data $u(0, x) = \varepsilon f(x)$ with $f = f(x) \in C_0^\infty(\mathbb{R}^1)$.

Mathematicians are interested in the following question:

How does the life span time $T_\varepsilon = T(\varepsilon)$ *behave for* $\varepsilon \to +0$?

We get

$$\lim_{\varepsilon \to 0} \varepsilon T(\varepsilon) = \left(\sup_{s\in\mathbb{R}^1}(-a'(0)f'(s)) \right)^{-1} > 0.$$

It follows that $T_\varepsilon = O(\frac{1}{\varepsilon})$ for $\varepsilon \to +0$. Hence, T_ε tends to ∞.

If Burgers' equation is *not genuinely nonlinear*, that is $a'(0) = 0$, then classical solutions to smooth compactly supported data do not necessarily blow up. Choosing a coefficient $a(u)$ which is 0 in a neighborhood of 0, we do not necessarily have a blow up behavior of the solution. If we assume for not genuinely nonlinear Burgers' equations $a''(0) \neq 0$, then we may conclude

$$\lim_{\varepsilon \to 0} \varepsilon^2 T(\varepsilon) = \sup_{s \in \mathbb{R}^1} \left(-a''(0)f(s)f'(s) \right)^{-1} > 0, \quad T_\varepsilon = O\left(\frac{1}{\varepsilon^2}\right).$$

7.2 Other Models Related to Burgers' Equation

Let us study some *modified Burgers' equations*. First we address to the following Cauchy problem for a special Burgers' equation with a mass term:

$$u_t + uu_x + u = 0, \quad u(0,x) = f(x), \quad f \in C_0^\infty(\mathbb{R}^1).$$

The mass term u should have an improving influence on the life span time of classical solutions. Applying the method of characteristics leads to the system of ordinary differential equations

$$\frac{dx}{dt} = u, \ x(0,s) = s, \ \frac{du}{dt} = -u,$$

$$u(t,s) = f(s)e^{-t}, \ x(t,s) = s + f(s)\left(1 - e^{-t}\right).$$

Consequently, if $f'(s) > -1$ for all $s \in \mathbb{R}^1$, then $\frac{dx}{ds} > 0$ for all $t > 0$. Thus the solution exists globally. We feel the improving influence of the mass term. There exist global (in time) solutions for classes of nontrivial compactly supported smooth data.

Let us suppose that a geometrical blow up appears. Then, we get for the blow up time

$$T = \log\left(\min_{s \in \mathbb{R}^1} \frac{f'(s)}{1 + f'(s)} \right).$$

Let us explain this formula. The derivative u_x satisfies along the characteristic Γ_s the equation

$$\frac{du_x}{dt} = u_{xt} + \frac{dx}{dt} u_{xx} = u_{xt} + uu_{xx} = -uu_{xx} - u_x^2 - u_x + uu_{xx};$$

$$\frac{du_x}{dt} = -u_x^2 - u_x \ \text{with the Cauchy condition} \ u_x(0,x) = f'(x).$$

Separation of variables yields

$$u_x(t, s) = \frac{f'(s)e^{-t}}{1 + f'(s)(1 - e^{-t})}.$$

In consequence, the solution possesses a geometrical blow up with the above introduced blow up time.

Finally, we are interested in the model problem

$$u_t + u_x = u_x^2, \quad u(0, x) = f(x), \quad f \in C_0^\infty(\mathbb{R}^1).$$

The function $w := u_x$ solves the quasilinear Cauchy problem

$$w_t + (1 - 2w)w_x = 0, \quad w(0, x) = f'(x).$$

We have for a nontrivial $f \in C_0^\infty(\mathbb{R}^1)$ the blow up time $T = \left(\sup_{s \in \mathbb{R}^1} 2f''(s) \right)^{-1}$. The solution u and the partial derivatives $u_x = w$ and u_t remain bounded along two different intersecting characteristics. Only second order derivatives blow up along these characteristics for $t \to T-0$. Thus, it may happen that a higher order derivative of a solution blows up in a certain sense. Let us summarize the blow up definitions in the following remark.

Remark 7.2.1 A solution of some differential equation has *a blow up* behavior in $t = T$ if the solution tends pointwise to infinity or some norm $\|u(t, \cdot)\|_B$ of the solution tends to infinity for $t \to T-0$. If a solution remains bounded for $t \to T-0$, but some partial derivative tends pointwise to infinity or the norm of some partial derivative of the solution tends to infinity for $t \to T - 0$, then the solution has a *geometrical blow up*. The *life span time* of the solution is T.

7.3   Concluding Remarks

The method of characteristics shows that, in general, there do not exist global (in time) smooth solutions to the Cauchy problem for Burgers' equation. If $u \in C^1(G)$, then Burgers' equation can be written in $G \subset \mathbb{R}^2$ in divergence form as a class of conservation laws

$$u_t + \partial_x f(u) = 0.$$

Then it is clear to say that a function in $L^1_{loc}(G)$ is a weak (Sobolev) solution. But, as one can verify in Exercise 5 below, the uniqueness property may fail if we look for weak solutions to the Cauchy problem. Hence, one has to include additional criteria to ensure uniqueness. Such a kind of criterion is, for example, the entropy condition.

Several authors are devoted to this topic for the Cauchy problem to the general class of conservation laws

$$u_t + \partial_x f(u) = 0, \quad u(0, x) = \varphi(x).$$

For further discussions we refer to [87] and the references therein.

Exercises Relating to the Considerations of Chap. 7

Exercise 1   Consider the Cauchy problem $u_t + a(u)u_x = 0$, $u(x, 0) = f(x)$ with data $f \in C_0^\infty(\mathbb{R}^1)$ and $f(x) = 0$ for $|x| > r$. Is it true that $u(t, x) = 0$ for $x < -r + a(0)t$ and for $x > r + a(0)t$?

Exercise 2   Assume that Burgers' equation is genuinely nonlinear and the nontrivial data has compact support and is infinitely times differentiable. Show that the solution has a geometrical blow up.

Exercise 3   Prove the formula for the life span time T_ε if we assume for not genuinely nonlinear Burgers' equations the condition $a''(0) \neq 0$.

Exercise 4   Compare the results for Burgers' equation with and without mass term. Why does the mass term have a positive influence on the existence of classical solutions? What happens with this influence if we study Burgers' equation with a negative mass term, for example, the equation $u_t + uu_x - u = 0$?

Exercise 5   Consider the Cauchy problem for Burgers' equation in the form

$$u_t + \frac{1}{2}\partial_x u^2 = 0, \quad u(0, x) = \varphi(x),$$

where $\varphi(x) = 0$ for $x \leq 0$ and $\varphi(x) = 1$ for $x > 0$. Show that $u_1(t, x) = 0$ for $x \leq \frac{t}{2}$ and $u_1(t, x) = 1$ for $x > \frac{t}{2}$ is a weak solution. Verify that $u_2(t, x) = 1$ for $x \geq t$, $u_2(t, x) = \frac{x}{t}$ for $0 \leq x < t$ and $u_2(t, x) = 0$ for $x < 0$ is another weak solution.

Part II

Chapter 8
Laplace Equation—Properties of Solutions—Starting Point of Elliptic Theory

Check for updates

There exists comprehensive literature on the theory of elliptic partial differential equations. One of the simplest elliptic partial differential equations is the Laplace equation. By means of this equation we explain usual properties of solutions. Here we have in mind maximum-minimum principle or regularity properties of classical solutions. On the other hand we explain properties as hypoellipticity or local solvability, too. Both properties are valid even for larger classes than elliptic equations. Moreover, a boundary integral representation for solutions of the Laplace equation shows the connection to potential theory. Boundary value problems of potential theory of first, second and third kind are introduced and relations to the theory of integral equations are described.

8.1 Poisson Integral Formula

8.1.1 How Does Potential Theory Come into Play?

The notion of fundamental solution or elementary potential is known from the course "Theoretical Physics" (cf. with Definition 24.30 and Example 24.4.4). Let Δ be the Laplace operator. Then, a practical relevant *fundamental solution* is

$$H_2(y, x) = -\frac{1}{2\pi} \ln \frac{1}{|y - x|} \quad \text{in } \mathbb{R}^2$$

$$\text{and } H_n(y, x) = -\frac{1}{(n - 2)\sigma_n} \frac{1}{|y - x|^{n-2}} \quad \text{in } \mathbb{R}^n \text{ for } n \geq 3,$$

where σ_n denotes the n-dimensional measure of the unit sphere.

Let a point mass of amount 1 or a point charge of amount 1 be located in the point $x \in \mathbb{R}^n$. The practical relevant fundamental solution or elementary potential

© Springer International Publishing AG 2018
M.R. Ebert, M. Reissig, *Methods for Partial Differential Equations*,
https://doi.org/10.1007/978-3-319-66456-9_8

is the potential made by this mass or charge. A straight forward calculation implies $\Delta_y H_2(y, x) = \Delta_y H_n(y, x) = 0$ for $y \neq x$.

Let $G \subset \mathbb{R}^n$ be a bounded domain with smooth boundary ∂G. Let $u \in C^2(\overline{G})$ be *an arbitrary given function.* Moreover, let x be a point in G and $\mathcal{K}_\varepsilon(x)$ the closed ball around x with radius ε. We suppose that $\mathcal{K}_\varepsilon(x)$ belongs to G. Let us introduce the domain $G_\varepsilon := G \setminus \mathcal{K}_\varepsilon(x)$. The functions $u = u(y)$ and $H_n = H_n(y, x)$ belong as functions in y to the function space $C^2(\overline{G_\varepsilon})$. Let us now apply the second Green's formula. Then we get

$$\int_{G_\varepsilon} \big(H_n(y, x) \Delta u(y) - u(y) \Delta_y H_n(y, x) \big) \, dy$$

$$= \int_{\partial G_\varepsilon} \Big(\frac{\partial u(y)}{\partial \mathbf{n}_y} H_n(y, x) - u(y) \frac{\partial H_n(y, x)}{\partial \mathbf{n}_y} \Big) \, d\sigma_y,$$

where $\mathbf{n}_y$ denotes the outer normal in $y \in \partial G_\varepsilon$ to G_ε. Using $\Delta_y H_n(y, x) = 0$ in G_ε and the fact that the fundamental solution has in x a weak singularity (the exponent $n - 2$ is smaller than the dimension n of G_ε) and, consequently, an integrable singularity, we may conclude

$$\lim_{\varepsilon \to 0} \int_{G_\varepsilon} \big(H_n(y, x) \Delta u(y) - u(y) \Delta H_n(y, x) \big) \, dy = \int_G H_n(y, x) \Delta u(y) \, dy.$$

The boundary ∂G_ε consists of two parts, ∂G and $\partial \mathcal{K}_\varepsilon(x)$. Let us turn to

$$\lim_{\varepsilon \to 0} \int_{\partial \mathcal{K}_\varepsilon(x)} \Big(H_n(y, x) \frac{\partial u(y)}{\partial \mathbf{n}_y} - u(y) \frac{\partial H_n(y, x)}{\partial \mathbf{n}_y} \Big) \, d\sigma_y$$

$$= \lim_{\varepsilon \to 0} \int_{\partial \mathcal{K}_\varepsilon(x)} H_n(y, x) \frac{\partial u(y)}{\partial \mathbf{n}_y} \, d\sigma_y + \lim_{\varepsilon \to 0} \int_{\partial \mathcal{K}_\varepsilon(x)} \big(u(x) - u(y) \big) \frac{\partial H_n(y, x)}{\partial \mathbf{n}_y} \, d\sigma_y$$

$$- u(x) \lim_{\varepsilon \to 0} \int_{\partial \mathcal{K}_\varepsilon(x)} \frac{\partial H_n(y, x)}{\partial \mathbf{n}_y} \, d\sigma_y.$$

We treat the terms on the right-hand side separately.

Term 1:   It holds $\lim_{\varepsilon \to 0} \int_{\partial \mathcal{K}_\varepsilon(x)} H_n(y, x) \frac{\partial u(y)}{\partial \mathbf{n}_y} \, d\sigma_y = 0$ after using $H_n(y, x) = O(\varepsilon^{-(n-2)})$ on $\partial \mathcal{K}_\varepsilon(x)$ and $d\sigma_y = O(\varepsilon^{n-1})$.

Term 2:   It holds $\lim_{\varepsilon \to 0} \int_{\partial \mathcal{K}_\varepsilon(x)} \big(u(x) - u(y) \big) \frac{\partial H_n(y, x)}{\partial \mathbf{n}_y} \, d\sigma_y = 0$ after using $|u(x) - u(y)| \leq L|x - y| = O(\varepsilon)$, $\big| \frac{\partial H_n(y, x)}{\partial \mathbf{n}_y} \big| = O(\varepsilon^{-(n-1)})$ on $\partial \mathcal{K}_\varepsilon(x)$ and $d\sigma_y = O(\varepsilon^{n-1})$.

Term 3:   It holds $\lim_{\varepsilon \to 0} \int_{\partial \mathcal{K}_\varepsilon(x)} \frac{\partial H_n(y, x)}{\partial \mathbf{n}_y} \, d\sigma_y = -1$.

Lemma 8.1.1 *It holds with the above introduced notations*

$$\lim_{\varepsilon \to 0} \int_{\partial K_\varepsilon(x)} \frac{\partial H_n(y, x)}{\partial \mathbf{n}_y} \, d\sigma_y = -1.$$

Proof We have

$$\lim_{\varepsilon \to 0} \int_{\partial K_\varepsilon(x)} \frac{\partial H_n(y, x)}{\partial \mathbf{n}_y} \, d\sigma_y = -\lim_{\varepsilon \to 0} \int_{\partial K_\varepsilon(x)} \frac{\partial}{\partial \mathbf{n}_y} \frac{1}{(n-2)\sigma_n |y - x|^{n-2}} \, d\sigma_y.$$

The exterior normal vector is opposite the radial vector. For this reason

$$\frac{\partial}{\partial \mathbf{n}_y} \frac{1}{(n-2)\sigma_n |y - x|^{n-2}} = -\frac{1}{(n-2)\sigma_n} \frac{d}{dr} \frac{1}{r^{n-2}} (r = \varepsilon) = \frac{1}{\varepsilon^{n-1}\sigma_n}.$$

So, we may conclude

$$\lim_{\varepsilon \to 0} \int_{\partial K_\varepsilon(x)} \frac{\partial H_n(y, x)}{\partial \mathbf{n}_y} \, d\sigma_y = -\lim_{\varepsilon \to 0} \frac{1}{\varepsilon^{n-1}\sigma_n} \int_{\partial K_\varepsilon(x)} 1 \, d\sigma_y = -1.$$

This we wanted to prove.

Summary If $u \in C^2(\overline{G})$ is an arbitrary given function and $G \subset \mathbb{R}^n$ a bounded domain with smooth boundary ∂G, then we have the following representation for u :

$$u(x)$$

$$= \int_G H_n(y, x) \Delta u(y) \, dy - \int_{\partial G} \frac{\partial u(y)}{\partial \mathbf{n}_y} H_n(y, x) \, d\sigma_y + \int_{\partial G} u(y) \frac{\partial H_n(y, x)}{\partial \mathbf{n}_y} \, d\sigma_y.$$

$$\text{volume potential} \qquad \text{single-layer potential} \qquad \text{double-layer potential}$$

Here we used the notions

volume potential (with the density $\mu = \mu(y)$): $\displaystyle \int_G \mu(y) \frac{1}{|x - y|^{n-2}} \, d\sigma_y,$

single-layer potential (with the density $\mu = \mu(y)$): $\displaystyle \int_{\partial G} \mu(y) \frac{1}{|x - y|^{n-2}} \, d\sigma_y,$

double-layer potential (with the density $\mu = \mu(y)$):

$$\int_{\partial G} \mu(y) \frac{\partial}{\partial \mathbf{n}_y} \frac{1}{|x - y|^{n-2}} \, d\sigma_y.$$

Consequently, every twice continuously differentiable function u on $\overline{G}$ is the sum of a *volume potential with the density* $-\frac{1}{(n-2)\sigma_n} \Delta u(y)$, *a single-layer potential with* the density $\frac{1}{(n-2)\sigma_n} \frac{\partial u}{\partial \mathbf{n}_y}$ and a double-layer potential with the density $-\frac{1}{(n-2)\sigma_n}u$ for $n \geq 3$. A corresponding representation exists in the case $n = 2$.

If the given function $u = u(y)$ is *harmonic in G*, then the *volume potential vanishes*. Here we recall the following definition.

Definition 8.1 Let G be a domain in $\mathbb{R}^n$. Then, every twice continuously differentiable solution of the Laplace equation

$$\Delta u = \sum_{k=1}^{n} \partial_{x_k}^2 u = 0$$

in G is called a *harmonic function*.

Remark 8.1.1 There exist other possible ways to define harmonic functions. In Sect. 8.2.1 we will introduce the so-called mean value property. Let G be a domain in $\mathbb{R}^n$. Let $x_0 \in G$ be an arbitrary given point and $K_R(x_0)$ be an arbitrary closed ball around the point x_0 with radius R which belongs to G. Then, a function u is called *harmonic* in G if it satisfies the mean value property for all x_0 and $K_R(x_0)$ in the sense of Sect. 8.2.1. Both definitions are equivalent. It holds the following statement:

"If a continuous function u satisfies the mean value property in a given domain $G \subset \mathbb{R}^n$, then it is a twice continuously differentiable solution of $\Delta u = 0$ in G and vice versa."

Theorem 8.1.1 *Let $u \in C^2(\overline{G})$ be a harmonic function in a bounded domain G with smooth boundary ∂G. Then, u is the sum of a single-layer potential with the density* $\frac{1}{(n-2)\sigma_n} \frac{\partial u}{\partial \mathbf{n}_y}$ *for $n \geq 3$ $\left(\frac{1}{2\pi} \frac{\partial u}{\partial \mathbf{n}_y}\right.$ for $\left. n = 2\right)$ and a double-layer potential with the density* $-\frac{1}{(n-2)\sigma_n} u$ *for $n \geq 3$ $\left(-\frac{1}{2\pi} u\right.$ *for $n = 2$*).
What kind of information do we have from this theorem?

If we know a harmonic function on the closure $\overline{G}$ of a bounded domain G, then some knowledge of this function on ∂G implies immediately the harmonic function in the domain G itself. The value $u(x_0)$ of u in x_0, $x_0 \in G$, can be determined by the boundary behavior of $\frac{\partial u}{\partial \mathbf{n}}$ and u on ∂G. This hints at the opportunity to study *boundary value problems* for harmonic functions (see Sect. 8.4).

8.1.2 Green's Function and Poisson Integral Formula

Opinion of a Physicist It is impossible to reconstruct a potential u in a bounded domain G by arbitrary given data for u and $\frac{\partial u}{\partial \mathbf{n}}$ on ∂G. To reconstruct the potential it is permissible to prescribe only data for u on ∂G.
Is this opinion a contradiction to the statement of Theorem 8.1.1?
No! In Theorem 8.1.1 we only claim a representation of a given harmonic function on $\overline{G}$ by the known behavior of u and $\frac{\partial u}{\partial \mathbf{n}}$ on ∂G. It was not our goal to reconstruct harmonic functions.

How can we verify the opinion of the physicist? The main idea is the use of *Green's function* for the Laplace operator Δ with respect to a given bounded domain $G \subset \mathbb{R}^n$. Green's function $G_n = G_n(y, x)$ has the following properties:

- $G_n(y, x) = H_n(y, x) + h_n(y, x)$, where $h_n = h_n(y, x)$ is a solution to the Laplace equation $\Delta_y h_n(y, x) = 0$ in G for all $x \in G$
- $G_n(y, x) = 0$ for $y \in \partial G$ and $x \in G$.

If we use Green's function $G_n(y, x)$ in the representation from Theorem 8.1.1 instead of the fundamental solution $H_n(y, x)$, then the single-layer potential vanishes. In this way we obtain for a given harmonic function u with a suitable behavior on the boundary ∂G the representation

$$u(x) = \int_{\partial G} u(y) \, \frac{\partial G_n(y, x)}{\partial \mathbf{n}_y} \, d\sigma_y.$$

This formula is called *Poisson integral formula* and verifies the opinion of the physicist that a potential in a domain G can be determined by Dirichlet data on the boundary ∂G.

Theorem 8.1.2 *Let u be a twice continuously differentiable solution of the Poisson equation $\Delta u = f$ in a bounded domain G with smooth boundary ∂G, where the right-hand side f is supposed to be continuous on $\overline{G}$. Then this solution has the following representation*:

$$u(x) = \int_G G_n(y, x) f(y) \, dy + \int_{\partial G} u(y) \, \frac{\partial G_n(y, x)}{\partial \mathbf{n}_y} \, d\sigma_y.$$

Here $G_n = G_n(y, x)$ is Green's function for the Laplace operator Δ with respect to the given domain $G \subset \mathbb{R}^n$.

For the construction of Green's function we have to solve the *Dirichlet problem* $\Delta_y h_n(y, x) = 0$ in G, $h_n(y, x) = -H_n(y, x)$ on ∂G. We are able to determine explicitly Green's function for the Laplace operator Δ in balls. This leads to the *Poisson integral formula.*

Theorem 8.1.3 *Let $K_R(x_0)$ be the ball with radius R around $x_0 \in \mathbb{R}^n$. Let g be a given continuous function on $\partial K_R(x_0)$. Then*

$$u(x) = \frac{1}{R\sigma_n} \int_{\partial K_R(x_0)} g(y) \, \frac{R^2 - r_{x_0 x}^2}{r_{yx}^n} \, d\sigma_y \quad \text{for} \quad n \geq 2$$

defines a harmonic function in the interior of $K_R(x_0)$. Here r_{yx} denotes the distance of a point x to a point y. If y_0 is an arbitrarily chosen point on the boundary $\partial K_R(x_0)$, then u has in y_0 the limit $g(y_0)$. Consequently, u is continuous on $\overline{K_R(x_0)}$ and a solution of the Dirichlet problem

$$\Delta u = 0 \quad \text{in} \quad K_R(x_0), \quad u|_{\partial K_R(x_0)} = g.$$

Proof The proof is divided into several steps.

Step 1:   A solution of the Laplace equation

First we verify if $u = u(x)$ is really a classical solution of the Laplace equation and a harmonic function, respectively. We apply the Laplace operator to both sides of the Poisson integral formula. For a fixed $x \in K_R(x_0)$ the integrand has no singularity (y belongs to $\partial K_R(x_0)$). This allows changing the order of integration and the action of the Laplace operator. It holds

$$\Delta_x \frac{R^2 - r_{x_0 x}^2}{r_{yx}^n} = 0$$

for all $x \in K_R(x_0)$ and all $y \in \partial K_R(x_0)$. Hence, u is a harmonic function.

Step 2:   Constant boundary data

Let $g(y) = c$ for $y \in \partial K_R(x_0)$. Then it holds

$$u(x) = \frac{1}{R\sigma_n} \int_{\partial K_R(x_0)} c \, \frac{R^2 - r_{x_0 x}^2}{r_{yx}^n} \, d\sigma_y = c,$$

that is, the harmonic function u is even constant in $K_R(x_0)$. We shall prove this statement in the following.

(a) Let x and x' be two different points in $K_R(x_0)$ having the same distance to the center x_0. There exists a rotation around the center which transfers x in x'. If $y \in \partial K_R(x_0)$ is an arbitrary point, then we denote by $y' \in \partial K_R(x_0)$ the image of y by carrying out this rotation. Due to our assumption we have $g(y) = g(y') = c$. A rotation does not change any distance. Hence, $r_{x_0 x} = r_{x_0 x'}$ and $r_{yx} = r_{y'x'}$. Denoting by $d\sigma_y'$ the image of the surface element $d\sigma_y$ after rotation around the center x_0 we obtain

$$u(x) = \frac{1}{R\sigma_n} \int_{\partial K_R(x_0)} c \, \frac{R^2 - r_{x_0 x}^2}{r_{yx}^n} \, d\sigma_y$$

$$= \frac{1}{R\sigma_n} \int_{\partial K_R(x_0)} c \, \frac{R^2 - r_{x_0 x'}^2}{r_{y'x'}^n} \, d\sigma_{y'} = u(x').$$

Taking into consideration that the distances of x and x' to x_0 coincide, the harmonic function depends only on the polar distance r if the data g is constant on $\partial K_R(x_0)$.

(b) The Laplace operator in spherical harmonics and the dependence of u only on the polar distance r imply that $u = u(r)$ solves the ordinary differential equation

$$\frac{d^2 u}{dr^2} + \frac{n-1}{r} \frac{du}{dr} = 0.$$

The method of separation of variables leads for $n \geq 3$ to the solution $u(r) = -\frac{c_1}{r^{n-2}} + c_2$ with real constants c_1 and c_2. For $n = 2$ we get $u(r) = -c_1 \log r + c_2$. The harmonic function $u = u(x)$ is finite in the center $x = x_0$. So, $c_1 = 0$, and $u(x) = c_2$ in $K_R(x_0)$. Taking account of $g(y) = c$, we get by the Poisson integral formula for the center x_0:

$$u(x_0) = \frac{1}{R^{n-1}\sigma_n} \int_{\partial K_R(x_0)} c \, d\sigma_y = c.$$

Consequently, $u(x) \equiv c$ in $K_R(x_0)$.

Step 3: Verification of boundary behavior

Let x_1 be an arbitrarily chosen point on $\partial K_R(x_0)$. We shall prove that u has the limit $g(x_1)$ in x_1. After proving this limit behavior, the harmonic function u is continuous on the closure of the ball $K_R(x_0)$ after defining u in the boundary points of $K_R(x_0)$ by $g(x_1)$. The results of the second step imply

$$g(x_1) = \frac{1}{R\sigma_n} \int_{\partial K_R(x_0)} g(x_1) \frac{R^2 - r_{x_0 x}^2}{r_{yx}^n} \, d\sigma_y.$$

We form the difference with the Poisson integral formula and get

$$u(x) - g(x_1) = \frac{1}{R\sigma_n} \int_{\partial K_R(x_0)} (g(y) - g(x_1)) \frac{R^2 - r_{x_0 x}^2}{r_{yx}^n} \, d\sigma_y.$$

So, we may estimate

$$|u(x) - g(x_1)| \leq \frac{1}{R\sigma_n} \int_{\partial K_R(x_0)} |g(y) - g(x_1)| \frac{R^2 - r_{x_0 x}^2}{r_{yx}^n} \, d\sigma_y.$$

The integration over the sphere is divided over two sets, S_1 and S_2. Here S_1 denotes the part of the sphere lying inside of a small ball $K_{\delta_1}(x_1)$ around x_1, the remaining part is denoted by S_2.

(a) Let us study the integral over S_1. It holds

$$\frac{1}{R\sigma_n} \int_{S_1} |g(y) - g(x_1)| \frac{R^2 - r_{x_0 x}^2}{r_{yx}^n} \, d\sigma_y$$

$$\leq \sup_{S_1} |g(y) - g(x_1)| \frac{1}{R\sigma_n} \int_{\partial K_R(x_0)} 1 \frac{R^2 - r_{x_0 x}^2}{r_{yx}^n} \, d\sigma_y$$

$$= \sup_{S_1} |g(y) - g(x_1)|$$

due to the results of the second step.

(b) Let us turn to the integral over S_2. We choose only points $x \in K_R(x_0)$ having a distance to x_1 smaller than δ_2. Choosing now $\delta_2 \leq \frac{1}{2}\delta_1$ we have $r_{yx} \geq \delta_1 - \delta_2 \geq \frac{1}{2}\delta_1$. Hence, $\frac{1}{r_{yx}} \leq \frac{2}{\delta_1}$ for all $y \in S_2$. For $x \in U_{\delta_2}(x_1)$ we use the estimate

$$R^2 - r_{x_0x}^2 \leq R^2 - (R - \delta_2)^2 = 2R\delta_2 - \delta_2^2 < 2R\delta_2.$$

This estimate leads together with

$$\sup_{S_2} |g(y) - g(x_1)| \leq 2 \sup_{\partial K_R(x_0)} |g|$$

and

$$\int_{S_2} d\sigma < \int_{\partial K_R(x_0)} d\sigma = \sigma_n R^{n-1}$$

to the estimate

$$\frac{1}{R\sigma_n} \int_{S_2} |g(y) - g(x_1)| \frac{R^2 - r_{x_0x}^2}{r_{yx}^n} \, d\sigma_y$$

$$\leq \frac{2}{R\sigma_n} \sup_{\partial K_R(x_0)} |g(x)| 2R\delta_2 \left(\frac{2}{\delta_1}\right)^n \sigma_n R^{n-1}$$

$$= 2^{n+2} R^{n-1} \frac{\delta_2}{\delta_1^n} \sup_{\partial K_R(x_0)} |g(x)|.$$

Let us take $\varepsilon > 0$ as sufficiently small. The continuity of g in the point x_1 allows the choice of $\delta_1 = \delta_1(\varepsilon)$ so small that $\sup_{S_1} |g(y) - g(x_1)| < \frac{\varepsilon}{2}$ is valid. After this choice we choose $\delta_2 = \delta_2(\delta_1)$ so small that

$$2^{n+2} R^{n-1} \frac{\delta_2}{\delta_1^n} \sup_{\partial K_R(x_0)} |g(x)| < \frac{\varepsilon}{2}.$$

Summarizing we conclude $|u(x) - g(x_1)| < \varepsilon$ for all $x \in U_{\delta_2}(x_1)$ belonging to $K_R(x_0)$, that is, $\lim_{y \to x_1} u(y) = g(x_1)$. This completes the proof.

Remark 8.1.2 The statements of Theorem 8.1.3 yield an existence result and a representation of one solution for the Dirichlet problem for harmonic functions in balls of $\mathbb{R}^n$, $n \geq 2$. Up to now we have no statement on the uniqueness of classical solutions.

Remark 8.1.3 The second Green's formula

$$\int_G (v \, \Delta u - u \Delta v) \, dy = \int_{\partial G} \left(v \frac{\partial u}{\partial \mathbf{n}} - u \frac{\partial v}{\partial \mathbf{n}}\right) d\sigma_y$$

allows for deriving a necessary condition for the solvability of the Neumann problem

$$\Delta u = f \text{ in } G, \quad \frac{\partial u}{\partial \mathbf{n}} = g \text{ on } \partial G$$

in bounded domains with smooth boundary. The Neumann problem is in general ill-posed (cf. with Sects. 8.4.1.2 and 8.4.1.5). Setting $v \equiv 1$ implies

$$\int_G f \, dy = \int_{\partial G} g \, d\sigma_y.$$

This means that the integral $\int_G f \, dy$ over all sources in the domain G is equal to the flow $\int_{\partial G} g \, d\sigma_y$ through the closed surface ∂G.

8.2 Properties of Harmonic Functions

8.2.1 Mean Value Property

Setting $x = x_0$ in the Poisson integral formula of Theorem 8.1.3 leads to

$$u(x_0) = \frac{1}{\sigma_n R^{n-1}} \int_{\partial K_R(x_0)} u(y) d\sigma_y.$$

This is the so-called *mean value property in balls*.

Theorem 8.2.1 *Let $u = u(x)$ be a given harmonic function in the ball $K_R(x_0)$, which is continuous on its closure $\overline{K_R(x_0)}$. Then, the value $u(x_0)$ of u in the center x_0 is equal to the mean over all values of u on the surface $\partial K_R(x_0)$ of the ball $K_R(x_0)$.*

8.2.2 Maximum-Minimum Principle

The mean value property is an essential tool for proving the next statement.

Theorem 8.2.2 (Maximum Principle)
Let u be a given harmonic function in a bounded domain G with smooth boundary ∂G. Let u be continuous on $\overline{G}$. Moreover, we suppose that u is not a constant function in G. Under these assumptions, the harmonic function u takes its maximum on the boundary ∂G. This means there is no point $x_0 \in G$ and no neighborhood $U(x_0) \subset G$ such that $u(x_0) \geq u(x)$ for all $x \in U(x_0)$.
If u is harmonic, then $-u$ is harmonic, too. So, we have the maximum principle for $-u$, but this gives a minimum principle for u itself.

Theorem 8.2.3 (Minimum Principle)

Let u be a given harmonic function in a bounded domain G with smooth boundary ∂G. Let u be continuous on $\overline{G}$. Moreover, we suppose that u is not a constant function in G. Under these assumptions, the harmonic function u takes its minimum on the boundary ∂G. This means there is no point $x_0 \in G$ and no neighborhood $U(x_0) \subset G$ such that $u(x_0) \leq u(x)$ for all $x \in U(x_0)$.

8.2.3 Regularity of Harmonic Functions

Let u be a harmonic function in a bounded domain G and continuous on its closure $\overline{G}$. Using the Poisson integral formula we obtain

$$\partial_x^\alpha u(x) = \partial_x^\alpha \int_{\partial G} u(y) \, \frac{\partial G_n(y, x)}{\partial \mathbf{n}_y} \, d\sigma_y = \int_{\partial G} u(y) \, \frac{\partial}{\partial \mathbf{n}_y} \, \partial_x^\alpha G_n(y, x) \, d\sigma_y.$$

Here we use properties of Green's function $G_n = G_n(y, x)$. On the one hand G_n is infinitely differentiable in x. On the other hand G_n is even *analytic in x*, that is, G_n can be represented locally in x by its Taylor series. These properties in x are transferred to the behavior of solutions in x. Because $x \in G$ and $y \in \partial G$ the order of integration and differentiation is allowed to change. After a formal change of the limiting processes of differentiation and integration, we take account of the uniform convergence of the integral with respect to x in a small neighborhood of a given point $x_0 \in G$. These explanations lead us to expect the following result.

Theorem 8.2.4 *Let u be a harmonic function in a given domain G. Then, u is analytic in x, that is, for a given $x_0 \in G$ the function u can be represented in a neighborhood $U(x_0)$ of x_0 by its Taylor series*

$$u(x) = \sum_{|\alpha|=0}^{\infty} \frac{\partial_x^\alpha u(x_0)}{\alpha!} (x - x_0)^\alpha \quad \text{for all} \ \ x \in U(x_0).$$

Proof Let $x_0 \in G$ and $K_R(x_0)$ be a ball such that its closure is contained in G. Then we apply Theorem 8.1.3 and follow the above explanations for the general Poisson formula.

The statement of Theorem 8.2.4 can be used to determine *zero sets* of harmonic functions.

- If a harmonic function u vanishes in a subdomain $G_0 \subset G$, then u vanishes in G.
- From the theory of *holomorphic functions* we know that a holomorphic function $f = f(z)$ vanishes in a domain $G \subset \mathbb{C}$ if f vanishes on a set $M \subset G$ having an accumulation point in the domain G. This statement does not hold any more for harmonic functions. Let us choose the harmonic function $u = u(x, y) = x^2 - y^2$. This function vanishes on the set $\{(x, y) : |x| = |y|\}$. The point $(0, 0)$ is an

accumulation point of this set, but u does not vanish in any neighborhood of $(0,0)$.

8.2.4 Weyl's Lemma and Interior Regularity

Due to Theorem 8.2.4 we know that every harmonic function u in a domain G is even analytic in G.

These assertions still allow for the possibility that there may exist Sobolev or even distributional solutions of $\Delta u = 0$, that is, the equation $\Delta u(\phi) = u(\Delta \phi) = 0$ is valid for a distribution $u \in D'(G)$ and arbitrary test functions $\phi \in C_0^\infty(G)$ (cf. with Definitions 3.10 and 24.29). Weyl's lemma states that this can not happen.

Theorem 8.2.5 (Weyl's Lemma)
Every distributional solution $u \in D'(G)$ of $\Delta u = 0$ is a classical solution of $\Delta u = 0$, that is, the solution is twice continuously differentiable and, consequently, due to Theorem 8.2.4 even analytic in G.

Summary We have only "beautiful solutions" (analytic solutions) of $\Delta u = 0$. We shall prove Theorem 8.2.5 in a special case only.

Theorem 8.2.6 *Let u be a continuous Sobolev solution of $\Delta u = 0$. Then u is a classical solution, that is, there exist continuous partial derivatives up to the second order. Consequently, u is even analytic in G.*

Proof We prove the statement in the two-dimensional case $n = 2$. For $n > 2$ the proof works in a similar way. Let u be a continuous Sobolev solution of $\Delta u = 0$. Then, the relation $\int_G u \, \Delta \phi \, dx = 0$ is satisfied for all test functions $\phi \in C_0^2(G)$ (cf. with Definition 3.9).

Step 1: Choice of a special test function

Let us choose an arbitrary point $x_0 \in G$, but after this choice it is fixed. Then, we can find a closed disk $\overline{K_R(x_0)}$ around x_0 with a possibly small radius R which belongs to G. Let r be the distance of a point x to x_0. Then we choose

$$\phi(x) = \begin{cases} (R^2 - r^2)^3 & \text{if } r \le R, \\ 0 & \text{otherwise.} \end{cases}$$

So, $\phi \in C_0^2(G)$ and $\Delta \phi = -12(R^4 - 4R^2 r^2 + 3r^4)$. Using the Definition 3.9 for a continuous Sobolev solution of $\Delta u = 0$ we get

$$\int_{K_R(x_0)} u(x)\left(R^4 - 4R^2 r^2 + 3r^4\right) dx = 0.$$

Step 2:   Properties of an auxiliary function

We define the auxiliary function $F(R) = \int_{K_R(x_0)} f(x,R)\, dx$, where the function $f(x,R)$ is continuously differentiable in R.

Lemma 8.2.1 *It holds*

$$\frac{dF}{dR}(R) = \int_{K_R(x_0)} \frac{\partial f}{\partial R}(x,R)\, dx + \int_{\partial K_R(x_0)} f(x,R)\, ds.$$

Proof For the proof we only take into account

$$F(R) = \int_0^R \left(\int_0^{2\pi} f(r\cos\varphi, r\sin\varphi, R) d\varphi \right) r\, dr.$$

Hence,

$$\frac{dF}{dR}(R) = \int_0^R \left(\int_0^{2\pi} \frac{\partial}{\partial R} f(r\cos\varphi, r\sin\varphi, R) d\varphi \right) r\, dr$$

$$+ \int_0^{2\pi} f(R\cos\varphi, R\sin\varphi, R) R\, d\varphi$$

$$= \int_{K_R(x_0)} \frac{\partial f}{\partial R}(x,R)\, dx + \int_{\partial K_R(x_0)} f(x,R)\, ds.$$

This we wanted to prove.
Using the identity of Lemma 8.2.1 gives

$$\frac{d}{dR} \int_{K_R(x_0)} u(x)\left(R^4 - 4R^2 r^2 + 3r^4\right) dx = \int_{K_R(x_0)} u(x)(4R^3 - 8Rr^2)\, dx = 0.$$

Here we have taken account of the vanishing behavior of the boundary integral. Division by $4R$ and differentiation with respect to R yield

$$2R \int_{K_R(x_0)} u(x)\, dx - R^2 \int_{\partial K_R(x_0)} u(x)\, ds = 0,$$

$$\frac{1}{2\pi R} \int_{\partial K_R(x_0)} u(x)\, ds = \frac{1}{\pi R^2} \int_{K_R(x_0)} u(x)\, dx, \quad \text{respectively.}$$

Step 3:   Relation to the mean value property (see Sect. 8.2.1)

After differentiation of the last identity with respect to R we obtain

$$\frac{d}{dR} \left(\frac{1}{2\pi R} \int_{\partial K_R(x_0)} u(x)\, ds \right)$$

$$= \frac{-2}{\pi R^3} \int_{K_R(x_0)} u(x)\, dx + \frac{1}{\pi R^2} \int_{\partial K_R(x_0)} u(x)\, ds = 0.$$

Hence, the mean value of u on $\partial K_R(x_0)$ is independent of R. Using the continuity of u this mean value is equal to $u(x_0)$. Summarizing, we have shown the *mean value property*

$$u(x_0) = \frac{1}{2\pi R} \int_{\partial K_R(x_0)} u(x)\, ds.$$

In Sect. 8.2.1 we discussed the mean value property. It holds some converse statement:
If a continuous function u satisfies the mean value property in a given domain $G \subset \mathbb{R}^n$, then it is a twice continuously differentiable solution of $\Delta u = 0$ in G (cf. with Remark 8.1.1).

From this it follows together with Theorem 8.2.4 the desired statement. In this way the proof is completed.

8.3 Other Properties of Elliptic Operators or Elliptic Equations

In the previous sections we explained properties of solutions to the Laplace equation. Some of these properties as interior regularity or maximum-minimum principle are also valid for solutions of large classes of elliptic equations.

In the following we will explain some more typical properties for solutions to linear elliptic equations. For this reason let us consider $P(x, D_x)u = f$, where

$$P(x, D_x) = \sum_{|\alpha| \leq m} a_\alpha(x) D_x^\alpha$$

is a linear elliptic operator in a domain $G \subset \mathbb{R}^n$ (see Definition 3.1). The right-hand side f is defined in G, too. As a special example we have in mind the Poisson equation $\Delta u = f$. Weyl's lemma (Theorem 8.2.5) tells us that if $f \equiv 0$, then we have only analytic solutions of the Laplace equation $\Delta u = 0$. In the next two sections we explain some results which describe how a behavior of the right-hand side f influences properties of solutions to $P(x, D_x)u = f$.

8.3.1 Hypoellipticity

Let $G \subset \mathbb{R}^n$ be a given domain and $P(x, D_x) = \sum_{|\alpha| \leq m} a_\alpha(x) D_x^\alpha$ be a linear partial differential operator. The coefficients are supposed to be analytic in G. Now let us consider the inhomogeneous linear partial differential equation $P(x, D_x)u = f$. Let us assume that $f \in C^\infty(G)$.

Can we exclude the existence of non-smooth solutions $u \in D'(G)$? This is the problem of hypoellipticity or interior regularity. Let us be a bit more precise.

Definition 8.2 We say that the given linear partial differential operator

$$P(x, D_x) = \sum_{|\alpha| \leq m} a_\alpha(x) D_x^\alpha$$

with analytic coefficients in a domain G is hypoelliptic in G if for any subdomain $G_0 \subset G$ and for all distributional solutions $u \in D'(G)$ the following property holds: If $f = P(x, D_x)u \in C^\infty(G_0)$, then we also have $u \in C^\infty(G_0)$.

Elliptic Operators are Hypoelliptic, But the Converse 'Is Not True If $P(D_x)$ is a linear differential operator with constant coefficients, thanks to the well-known Malgrange-Ehrenpreis' Theorem, there exists a fundamental solution E, that is a distributional solution of $P(D_x)E = \delta_0$ (cf. with Sect. 8.1.1 and Definition 24.30). Moreover, if some fundamental solution E belongs to $C^\infty(\mathbb{R}^n \setminus \{0\})$, then every fundamental solution has the same regularity and $P(D_x)$ is hypoelliptic. In this way we understand that the Laplace operator Δ, the Cauchy-Riemann operator $\partial_{\bar{z}}$ and the heat operator $\partial_t - \Delta$ are hypoelliptic.

The theory of hypoelliptic operators is a modern topic of research. The following example of Mizohata operators shows, that this theory needs deep understanding for applying the results in other fields of Mathematics.

Example 8.3.1 Let us consider the Mizohata operator $\partial_x + ix^h \partial_y$ in a domain $G \subset \mathbb{R}^2$ containing the origin, where $h \geq 1$ is a fixed integer. If h is an even number, then the operator is hypoelliptic (see [138] or [130]). If h is an odd number, then the operator is not hypoelliptic. The Mizohata operators themselves are not elliptic operators. Why?

8.3.2 Local Solvability

Let $G \subset \mathbb{R}^n$ be a given domain and $P(x, D_x) = \sum_{|\alpha| \leq m} a_\alpha(x) D_x^\alpha$ be a linear partial differential operator. The coefficients are supposed to be analytic in G. Now let us consider the inhomogeneous linear partial differential equation $P(x, D_x)u = f$. We shall content ourselves with finding only one solution. This problem may turn out to be very difficult in general. Much depends on the space of functions or distributions where the right-hand side is chosen and to the space of functions or distributions the solution u is allowed to belong.

Definition 8.3 We say that the given linear partial differential operator

$$P(x, D_x) = \sum_{|\alpha| \leq m} a_\alpha(x) D_x^\alpha$$

with analytic coefficients in a domain G is solvable in G if for any $f \in C_0^\infty(G)$ there exists a distributional solution $u \in D'(G)$ of $P(x, D_x)u = f$.

We say that $P(x, D_x)$ is locally solvable at $x_0 \in G$ if it is solvable in a neighborhood of x_0.

Elliptic Operators Are Solvable, But the Converse Is Not True If $P(D_x)$ is a linear differential operator with constant coefficients, then $P(D_x)$ is solvable in $\mathbb{R}^n$ and, consequently, locally solvable at any point $x_0 \in \mathbb{R}^n$. Thus the classical heat, wave and Schrödinger operators are locally solvable.

Example 8.3.2 Let us consider the Mizohata operator $\partial_x + ix^h \partial_y$ in $\mathbb{R}^2$, where $h \geq 1$ is a fixed integer. If h is an even number, then the operator is solvable. If h is an odd number, then the operator is not locally solvable at any point $(0, y)$, $y \in \mathbb{R}^1$ (see [138] or [130]).

If we have a Kovalevskian type operator (cf. with Sect. 3.1) with analytic coefficients, then these operators are locally solvable at any point of a domain G. This follows from the Cauchy-Kovalevskaja theorem. Lewy's example (see Sect. 4.1) is not locally solvable. There exist hyperbolic operators $\partial_t^2 - a(t)\partial_x^2$ with a suitably high oscillating coefficient $a = a(t)$ in a neighborhood of $t = 0$ which are not locally solvable at any point $(x, t = 0) \in \mathbb{R}^2$ (see [49]).

8.4 Boundary Value Problems of Potential Theory

In this section we address boundary value problems of potential theory. First, we propose some boundary value problems. Then, we explain the approach to use single or double-layer potentials as representations of solutions (cf. with Sect. 8.1.1). Finally, we shall point out how functional analytic tools are used to treat the corresponding integral equations.

8.4.1 Basic Boundary Value Problems of Potential Theory

8.4.1.1 Interior Dirichlet Problem

Let $G \subset \mathbb{R}^n$ be an interior domain, that is, a bounded domain. We suppose that the boundary ∂G is smooth. Find a harmonic function u in G with prescribed continuous data g for u on ∂G.

Mathematical Formulation

$$\Delta u = 0 \text{ in } G, \quad u = g \text{ on } \partial G.$$

Let us restrict ourselves to balls $K_R(x_0) \subset \mathbb{R}^n$. Then the statement of Theorem 8.1.3 implies the existence of a solution. Together with the statements of Theorems 8.2.2 and 8.2.3 we conclude the uniqueness of solutions and continuous dependence on the data as well in the space of continuous solutions on the closure $\overline{K_R(x_0)}$. Consequently, the interior Dirichlet problem is well-posed.

Remark 8.4.1 We are able to study the interior Dirichlet problem in more general domains G. Using Theorem 8.1.2 implies formally the solution

$$u(x) = \int_{\partial G} g(y) \, \frac{\partial G_n(y, x)}{\partial \mathbf{n}_y} \, d\sigma_y.$$

This is a harmonic function in G, but we have to clarify in which sense the boundary condition is really satisfied. This requires choosing suitable spaces for the data g which allow for verifying the boundary behavior as a restriction of the solution on ∂G or in the sense of traces. In this way one is able to prove well-posedness results.

8.4.1.2   Interior Neumann Problem

Let $G \subset \mathbb{R}^n$ be an interior domain with smooth boundary ∂G. Find a harmonic function u in G with prescribed continuous data g for the normal derivative (in the direction of the outer normal) $\partial_{\mathbf{n}} u$ on ∂G.

Mathematical Formulation

$$\Delta u = 0 \text{ in } G, \quad \partial_{\mathbf{n}} u = g \text{ on } \partial G.$$

We have no uniqueness of solutions. If u solves the interior Neumann problem, then $u + c$ is a solution, too. Consequently, we have ill-posedness. We learned in Remark 8.1.3 the necessary condition $\int_{\partial G} g \, d\sigma_y = 0$ for the existence of a solution. The sufficiency of this condition is explained in Sect. 8.4.3.

8.4.1.3   Interior Robin Problem

Let $G \subset \mathbb{R}^n$ be an interior domain with smooth boundary ∂G. Find a harmonic function u in G with prescribed continuous data g for $\partial_{\mathbf{n}} u + c(x)u$ (in the direction of the outer normal) on ∂G. Here $c = c(x)$ is supposed to be a continuous function on ∂G.

Mathematical Formulation

$$\Delta u = 0 \text{ in } G, \quad \partial_{\mathbf{n}} u + c(x)u = g \text{ on } \partial G.$$

To get well-posedness of this boundary value problem we assume $c(x) \geq 0$ on ∂G and $c(x) > 0$ on some subset of ∂G with positive Lebesgue measure. These conditions are important for showing that a maximum principle is valid. As pointed out in Remark 8.4.1, one has to discuss suitable regularity of $c = c(x)$ and of $g = g(x)$ on the boundary ∂G to prove well-posedness results.

8.4.1.4 Exterior Dirichlet Problem

Let $G \subset \mathbb{R}^n$ be an exterior domain, that is, the complement in $\mathbb{R}^n$ is the closure of an interior domain. We suppose the boundary ∂G to be smooth. Find a harmonic function u in G with prescribed continuous data g for u on ∂G. Moreover, the solution satisfies the decay condition $u(x) = O(|x|^{-(n-2)})$ for $|x| \to \infty$ and $n \geq 3$.

Mathematical Formulation

$$\Delta u = 0 \quad \text{in} \quad G, \quad u = g \quad \text{on} \quad \partial G,$$

$$\text{and} \quad u(x) = O(|x|^{-(n-2)}) \quad \text{for} \quad |x| \to \infty.$$

Let us restrict to the exterior of balls $K_R(x_0) \subset \mathbb{R}^n$. Then, a solution is given by the following Poisson integral formula for the exterior domain $\mathbb{R}^n \setminus \overline{K_R(x_0)}$:

$$u(x) = \frac{1}{R\sigma_n} \int_{\partial K_R(x_0)} g(y) \frac{r_{x_0 x}^2 - R^2}{r_{yx}^n} \, d\sigma_y$$

if the data is supposed to belong to $C(\partial K_R(x_0))$. This function is harmonic and satisfies the decay condition $u(x) = O(|x|^{-(n-2)})$ for $|x| \to \infty$ and $n \geq 3$. The uniqueness can be concluded by the following modified maximum principle:
Let u be a harmonic function in the exterior domain $\mathbb{R}^n \setminus \overline{K_R(x_0)}$. If the solution is continuous on the closure of this domain and if $u(x) \to 0$ for $|x| \to \infty$, then it holds

$$\sup_{x \in \mathbb{R}^n \setminus K_R(x_0)} |u(x)| \leq \sup_{x \in \partial(\mathbb{R}^n \setminus K_R(x_0))} |u(x)|.$$

Consequently, the exterior Dirichlet problem is well-posed. For more general exterior domains we recall Remark 8.4.1.

8.4.1.5 Exterior Neumann Problem

Let $G \subset \mathbb{R}^n$ be an exterior domain with smooth boundary ∂G. Find a harmonic function u in G with prescribed continuous data g for the normal derivative (in the direction of the outer normal) $\partial_n u$ on ∂G. Moreover, the solution satisfies the decay condition $u(x) = O(|x|^{-(n-2)})$ for $|x| \to \infty$ and $n \geq 3$.

Mathematical Formulation

$$\Delta u = 0 \text{ in } G, \quad \partial_n u = g \text{ on } \partial G,$$

$$\text{and } u(x) = O(|x|^{-(n-2)}) \text{ for } |x| \to \infty.$$

We do not have any trouble with constant solutions because the only constant solution satisfying the decay condition is the solution $u \equiv 0$. One can prove well-posedness results (see Sect. 8.4.3 and Remark 8.4.1).

8.4.1.6 Exterior Robin Problem

Let $G \subset \mathbb{R}^n$ be an exterior domain with smooth boundary ∂G. Find a harmonic function u in G with prescribed continuous data g for $\partial_n u + c(x)u$ (in the direction of the outer normal) on ∂G. Here $c = c(x)$ is supposed to be a continuous function on ∂G. Moreover, the solution satisfies the decay condition $u(x) = O(|x|^{-(n-2)})$ for $|x| \to \infty$ and $n \geq 3$.

Mathematical Formulation

$$\Delta u = 0 \text{ in } G, \quad \partial_n u + c(x)u = g \text{ on } \partial G,$$

$$\text{and } u(x) = O(|x|^{-(n-2)}) \text{ for } |x| \to \infty.$$

To get well-posedness of this boundary value problem we assume $c(x) \leq 0$ on ∂G and $c(x) < 0$ on some subset of ∂G with positive Lebesgue measure. Here we take into consideration that the direction of the outer normal for exterior domains is opposite of that for interior domains. These conditions are important for showing that a maximum principle is valid. As pointed out in Remark 8.4.1, one has to discuss suitable regularity of $c = c(x)$ and of $g = g(x)$ on the boundary ∂G to prove well-posedness results.

8.4.2 How to Use Potentials in Representations of Solutions?

8.4.2.1 Dirichlet Problems

Let us consider the interior or exterior Dirichlet problem for harmonic functions in a domain $G \subset \mathbb{R}^n$ with smooth boundary ∂G and $n \geq 3$. The key idea is to choose the ansatz for a solution as a double-layer potential, that is,

$$u(x) = \int_{\partial G} \mu(y) \, \partial_{n_y} \frac{1}{|x - y|^{n-2}} \, d\sigma_y.$$

Here we assume that the density is taken from a suitable function space on ∂G such that the integral really exists for all $x \in G$. By using the Poisson integral formula of Sect. 8.1.2, the function u is harmonic in G. What remains is to find the density function μ on ∂G. Therefore, we apply the Dirichlet condition

$$g(x_0) = \lim_{x \to x_0 \in \partial G} u(x) = \lim_{x \to x_0 \in \partial G} \int_{\partial G} \mu(y) \, \partial_{\mathbf{n}_y} \frac{1}{|x - y|^{n-2}} \, d\sigma_y,$$

where $x \in G$. In general, it is not allowed to form the limit under the integral sign. The main reason is jump relations for the double-layer potential after crossing the layer (see [135]). After some calculations one can show for all boundary points x_0 the relation

$$g(x_0) = \lim_{x \to x_0 \in \partial G} \int_{\partial G} \mu(y) \, \partial_{\mathbf{n}_y} \frac{1}{|x - y|^{n-2}} \, d\sigma_y$$

$$= \pm \frac{(n - 2)\sigma_n}{2} \mu(x_0) + \int_{\partial G} \mu(y) \partial_{\mathbf{n}_y} \frac{1}{|x_0 - y|^{n-2}} \, d\sigma_y.$$

The negative sign appears in the case of the interior, the positive sign in the case of the exterior Dirichlet problem. The last relation implies a Fredholm integral equation of second kind (see Sect. 8.4.3).

Remark 8.4.2 Solutions of the exterior Dirichlet problem as double-layer potentials satisfy the decay condition $u(x) = O(|x|^{-(n-1)})$ for $|x| \to \infty$. But we require only the decay behavior $u(x) = O(|x|^{-(n-2)})$ for $|x| \to \infty$. For this reason we can not expect that every solution of an exterior Dirichlet problem is given as a double-layer potential.

What do we conclude if we choose a single-layer potential as an ansatz for the solution to an interior or exterior Dirichlet problem? Is it a good idea to choose such an ansatz?

8.4.2.2 Neumann or Robin Problems

Let us consider the interior or exterior Neumann or Robin problem for harmonic functions in a domain $G \subset \mathbb{R}^n$ with smooth boundary ∂G and $n \geq 3$. The key idea is to choose the ansatz for a solution as a single-layer potential, that is,

$$u(x) = \int_{\partial G} \mu(y) \, \frac{1}{|x - y|^{n-2}} \, d\sigma_y.$$

Here we assume that the density is taken from a suitable function space on ∂G such that the integral really exists for all $x \in G$. The explanations of Sect. 8.1.1 imply that the function u is harmonic in G. What remains is to find the density function μ

on ∂G. Therefore, we apply at first the Neumann condition

$$g(x_0) = \lim_{x \to x_0 \in \partial G} \partial_{\mathbf{n}_x} u(x) = \lim_{x \to x_0 \in \partial G} \partial_{\mathbf{n}_x} \int_{\partial G} \mu(y) \, \frac{1}{|x - y|^{n-2}} \, d\sigma_y,$$

where $x \in G$. In general, it is not allowed to change the order of differentiation and integration and to form the limit under the integral sign. The main reason is jump relations of the normal derivative of the single-layer potential after crossing the layer (see [135]). After some calculations one can show for all boundary points x_0 the relation

$$g(x_0) = \lim_{x \to x_0 \in \partial G} \partial_{\mathbf{n}_x} \int_{\partial G} \mu(y) \, \frac{1}{|x - y|^{n-2}} \, d\sigma_y$$

$$= \pm \frac{(n-2)\sigma_n}{2} \mu(x_0) + \int_{\partial G} \mu(y) \left(\partial_{\mathbf{n}_x} \frac{1}{|x - y|^{n-2}} \right)(x = x_0) \, d\sigma_y.$$

The positive sign appears in the case of the interior, the negative sign in the case of the exterior Neumann problem. The last relation implies a Fredholm integral equation of second kind (see Sect. 8.4.3).

What happens if we choose a double-layer potential as an ansatz for the solution to an interior or exterior Neumann (or Robin) problem? Is it a good idea to choose such an ansatz?

We can choose a single-layer potential as ansatz for solutions to the interior and exterior Robin problem as well. In this way we obtain on the one hand

$$g(x_0) = \lim_{x \to x_0 \in \partial G} \left(\partial_{\mathbf{n}_x} + c(x) \right) u(x)$$

$$= \lim_{x \to x_0 \in \partial G} \left(\partial_{\mathbf{n}_x} + c(x) \right) \int_{\partial G} \mu(y) \, \frac{1}{|x - y|^{n-2}} \, d\sigma_y,$$

where $x \in G$, and on the other hand

$$g(x_0) = \lim_{x \to x_0 \in \partial G} \left(\partial_{\mathbf{n}_x} + c(x) \right) \int_{\partial G} \mu(y) \, \frac{1}{|x - y|^{n-2}} \, d\sigma_y$$

$$= \pm \frac{(n-2)\sigma_n}{2} \mu(x_0)$$

$$+ \int_{\partial G} \mu(y) \left(\left(\partial_{\mathbf{n}_x} \frac{1}{|x - y|^{n-2}} \right)(x = x_0) + \frac{c(x_0)}{|x_0 - y|^{n-2}} \right) d\sigma_y.$$

The positive sign appears in the case of the interior, the negative sign in the case of the exterior Robin problem. The last relation implies a Fredholm integral equation of second kind (see Sect. 8.4.3).

8.4.3 *Integral Equations of Potential Theory*

In the last section we derived the following integral equations of potential theory:

interior Dirichlet problem:

$$\mu(x) - \frac{2}{(n-2)\sigma_n} \int_{\partial G} \mu(y)\, \partial_{\mathbf{n}_y} \frac{1}{|x-y|^{n-2}}\, d\sigma_y = -\frac{2}{(n-2)\sigma_n} g(x);$$

exterior Dirichlet problem:

$$\mu(x) + \frac{2}{(n-2)\sigma_n} \int_{\partial G} \mu(y)\, \partial_{\mathbf{n}_y} \frac{1}{|x-y|^{n-2}}\, d\sigma_y = \frac{2}{(n-2)\sigma_n} g(x);$$

interior Neumann problem:

$$\mu(x) + \frac{2}{(n-2)\sigma_n} \int_{\partial G} \mu(y)\, \partial_{\mathbf{n}_x} \frac{1}{|x-y|^{n-2}}\, d\sigma_y = \frac{2}{(n-2)\sigma_n} g(x);$$

exterior Neumann problem:

$$\mu(x) - \frac{2}{(n-2)\sigma_n} \int_{\partial G} \mu(y)\, \partial_{\mathbf{n}_x} \frac{1}{|x-y|^{n-2}}\, d\sigma_y = -\frac{2}{(n-2)\sigma_n} g(x);$$

interior Robin problem:

$$\mu(x) + \frac{2}{(n-2)\sigma_n} \int_{\partial G} \mu(y)\left(\partial_{\mathbf{n}_x} \frac{1}{|x-y|^{n-2}} + \frac{c(x)}{|x-y|^{n-2}} \right) d\sigma_y$$
$$= \frac{2}{(n-2)\sigma_n} g(x);$$

exterior Robin problem:

$$\mu(x) - \frac{2}{(n-2)\sigma_n} \int_{\partial G} \mu(y)\left(\partial_{\mathbf{n}_x} \frac{1}{|x-y|^{n-2}} + \frac{c(x)}{|x-y|^{n-2}} \right) d\sigma_y$$
$$= -\frac{2}{(n-2)\sigma_n} g(x).$$

Consequently, the density μ is determined as a solution of an integral equation. All these integral equations "living on the boundary ∂G" are Fredholm integral equations of second kind. We can write them in the abstract form

$$\mu(x) - \lambda \int_{\partial G} K(x,y)\mu(y)\, d\sigma_y = f(x),$$

where λ is a complex parameter. To solve this integral equation we have to

1. Investigate properties of the integral operator

$$T : \mu \in B \rightarrow T\mu := \int_{\partial G} K(x, y)\mu(y)\, d\sigma_y,$$

   in particular, we should fix function spaces B which allow for proving that the integral operator $T \in L(B \rightarrow B)$ is a compact operator mapping B into itself.
2. Choose the right-hand side f from suitable function spaces related to B.
3. Understand the role of the parameter $\lambda = \pm \frac{2}{(n-2)\sigma_n}$ in the integral equation.

In this way we get the opportunity to apply Fredholm alternative from functional analysis.

One can, for example, show that a suitable choice of the function space B allows for proving, that $\lambda = \frac{2}{(n-2)\sigma_n}$ is a regular value for T, that is, it does not belong to the spectrum of T. This implies the well-posedness of the interior Dirichlet and exterior Neumann problem in suitable function spaces. Opposite to this, it turns out that $\lambda = -\frac{2}{(n-2)\sigma_n}$ is an eigenvalue of T. This implies ill-posedness of the exterior Dirichlet or interior Neumann problem (for the exterior Dirichlet problem take account of Remark 8.4.2).

If we apply Fredholm alternative to the integral equation for the interior Neumann problem, then we know that the integral equation is only solvable if the right-hand side f (or better g) is orthogonal to all solutions of the adjoint homogeneous integral equation

$$z(x) + \frac{2}{(n-2)\sigma_n} \int_{\partial G} z(y)\, \partial_{\mathbf{n}_y} \frac{1}{|x-y|^{n-2}}\, d\sigma_y = 0.$$

One can show that this equation has in suitable function spaces only constant solutions. Hence, $\int_{\partial G} g \cdot 1 \, d\sigma_x = 0$ is a sufficient condition for the solvability of the interior Neumann problem. The necessity is already explained in Remark 8.1.3.

Exercises Relating to the Considerations of Chap. 8

Exercise 1 What are harmonic functions in $\mathbb{R}^1$? Check all the above introduced properties.

Exercise 2 Prove the relation

$$\Delta_x \frac{R^2 - r_{x_0 x}^2}{r_{yx}^n} = 0$$

of the first step of the proof to Theorem 8.1.3.

Exercise 3 Write the Laplace operator in spherical harmonics. Compare with the second step of the proof to Theorem 8.1.3.

Exercise 4 Prove the following statement:
If a harmonic function u vanishes in a subdomain $G_0 \subset G$, then u vanishes in G.

Exercise 5 Study zero sets M_0 for harmonic functions. Try to understand possible structures of M_0 which imply that a harmonic function vanishing on $M_0 \subset G$ will vanish in all G. From the previous exercise we know that every subdomain G_0 of a given domain G is a zero set, but there exist, of course, "smaller zero sets" (cf. with [4]).

Exercise 6 Study the proof of the following statement:
If a function u satisfies the mean value property, then it is a twice continuously differentiable solution of $\Delta u = 0$ (see [135]).

Exercise 7 Is it reasonable to study the property of hypoellipticity for the Schrödinger equation or the wave equation?

Exercise 8 Verify the relation

$$\lim_{x \to x_0 \in \partial G} \int_{\partial G} \mu(y)\, \partial_{\mathbf{n}_y} \frac{1}{|x - y|^{n-2}}\, d\sigma_y$$

$$= \pm \frac{(n-2)\sigma_n}{2} \mu(x_0) + \int_{\partial G} \mu(y)\partial_{\mathbf{n}_y} \frac{1}{|x_0 - y|^{n-2}}\, d\sigma_y$$

for double-layer potentials after crossing the layer.

Exercise 9 Verify the relation

$$\lim_{x \to x_0 \in \partial G} \partial_{\mathbf{n}_x} \int_{\partial G} \mu(y)\, \frac{1}{|x - y|^{n-2}}\, d\sigma_y$$

$$= \pm \frac{(n-2)\sigma_n}{2} \mu(x_0) + \int_{\partial G} \mu(y)\partial_{\mathbf{n}_x} \left(\frac{1}{|x - y|^{n-2}} \right)(x = x_0)\, d\sigma_y$$

for single-layer potentials after crossing the layer.

Chapter 9
Heat Equation—Properties of Solutions—Starting Point of Parabolic Theory

There exists comprehensive literature on the theory of parabolic partial differential equations. One of the simplest parabolic partial differential equation is the heat equation. By means of this equation we explain qualitative properties of solutions as maximum-minimum principle, non-reversibility in time, infinite speed of propagation and smoothing effect. Moreover, we explain connections to thermal potential theory. Thermal potentials prepare the way for integral equations for densities in single- or double-layer potentials as solutions to mixed problems.

9.1 Potential Theory and Representation Formula

By $Z_T = (0, T) \times G$ we denote a space-time cylinder. Here $G \subset \mathbb{R}^n$ is supposed to be a bounded domain. We try to find solutions of the heat equation in this cylinder. By $S_T = (0, T) \times \partial G$ we denote the lateral surface of Z_T and by $B_T = S_T \cup (\{t = 0\} \times \overline{G})$ the so-called *parabolic boundary*.

Usually *different regularity* in x and t (one time derivative corresponds to two spatial derivatives) is assumed for classical solutions to the heat equation.

Definition 9.1 By $C^{2,1}(Z_T)$ we denote the space of functions which are in Z_T twice continuously differentiable in x and continuously differentiable in t. By $C(\overline{Z_T})$ we denote the space of continuous functions on the closure $\overline{Z_T}$ of Z_T.

Similar to the considerations in Sect. 8.1.1 one can define *thermal potentials* for solutions to the heat equation. To introduce these potentials we choose the following fundamental solution or elementary potential for $\partial_t - \Delta$ (cf. with Definition 24.30 and Example 24.4.5):

$$H_n(t, x) = \frac{\theta(t)}{(4\pi t)^{n/2}} \exp\left(-\frac{|x|^2}{4t}\right),$$

© Springer International Publishing AG 2018
M.R. Ebert, M. Reissig, *Methods for Partial Differential Equations*,
https://doi.org/10.1007/978-3-319-66456-9_9

here $\theta = \theta(t)$ denotes the Heaviside function. Similar to the considerations in Sect. 8.1.1 one can show the following identity for all functions u from $C^{2,1}(Z_T) \cap C^{1,0}(\overline{Z_T})$:

$$u(t,x) = \int_{Z_t} H_n(t-\tau, x-y)(u_\tau - \Delta_y u)\, d(\tau, y)$$

$$+ \int_{S_t} \left(H_n(t-\tau, x-y)\partial_{\mathbf{n}_y} u - u\, \partial_{\mathbf{n}_y} H_n(t-\tau, x-y) \right) d(\tau, \sigma_y)$$

$$+ \int_G H_n(t, x-y)u(0,y)\, dy.$$

In the following we shall use the notions

thermal volume potential (with the density $\mu = \mu(\tau, y)$):

$$\int_{Z_t} H_n(t-\tau, x-y)\mu(\tau, y)\, d(\tau, y),$$

thermal single-layer potential (with the density $\mu = \mu(\tau, y)$):

$$\int_{S_t} H_n(t-\tau, x-y)\mu(\tau, y)\, d(\tau, \sigma_y),$$

thermal double-layer potential (with the density $\mu = \mu(\tau, y)$):

$$\int_{S_t} \partial_{\mathbf{n}_y} H_n(t-\tau, x-y)\mu(\tau, y)\, d(\tau, \sigma_y).$$

Theorem 9.1.1 *Let $u \in C^{2,1}(Z_T) \cap C^{1,0}(\overline{Z_T})$ be a classical solution to $u_t - \Delta u = h(t,x)$ in Z_T, where the right-hand side h is supposed to belong to $C(\overline{Z_T})$. Then, u can be represented as a sum of a thermal volume potential*

$$\int_{Z_t} H_n(t-\tau, x-y)h(\tau, y)\, d(\tau, y),$$

a thermal single-layer potential with the density $\partial_{\mathbf{n}_y} u$, a thermal double-layer potential with the density $-u$ and the integral

$$\int_G H_n(t, x-y)u(0,y)\, dy$$

over the given domain G.

Remark 9.1.1 We can verify the meaning of the parabolic boundary by these integrals because the layer potentials are taken over by the lateral surface S_T, the domain integral is taken over by the lower base $\{t = 0\} \times G$. The union of both of these sets forms the parabolic boundary. The representation from Theorem 9.1.1 shows that mixed problems are appropriate for solutions to $u_t - \Delta u = h(t, x)$.

9.2 Maximum-Minimum Principle

Lemma 9.2.1 *Let us assume that $u \in C^{2,1}(Z_T)$ solves the differential inequality $u_t - \Delta u > 0$ in Z_T. Then, there does not exist any local minimum of u in Z_T. Moreover, there does not exist any local minimum in the set $\{t = T\} \times G$.*

Lemma 9.2.2 *Let $u \in C(\overline{Z_T}) \cap C^{2,1}(Z_T)$ be a function satisfying the differential inequality $u_t - \Delta u \geq 0$ in Z_T and $u \geq 0$ on the parabolic boundary B_T. It then holds that $u \geq 0$ on the closure $\overline{Z_T}$ of Z_T. Analogously, the differential inequality $u_t - \Delta u \leq 0$ in Z_T and $u \leq 0$ on B_T imply $u \leq 0$ on $\overline{Z_T}$.*

Proof After introducing the auxiliary function $w = u + \varepsilon t$, $\varepsilon > 0$, we obtain $w_t - \Delta w = u_t - \Delta u + \varepsilon > 0$ in Z_T. Using Lemma 9.2.1, the function w takes its minimum on B_T. Taking account of $u \geq 0$ on B_T implies $w = u + \varepsilon t \geq 0$ on $\overline{Z_T}$. Passing to the limit $\varepsilon \to 0$ leads to $u \geq 0$ on $\overline{Z_T}$. The second statement follows from the first one after applying it to $-u$.

Theorem 9.2.1 (Maximum-Minimum Principle for Classical Solutions to the Heat Equation) *Let $u \in C(\overline{Z_T}) \cap C^{2,1}(Z_T)$ be a classical solution of $u_t - \Delta u = 0$ in Z_T. Then it holds*

$$\min_{B_T} u(t, x) \leq u(t, x) \leq \max_{B_T} u(t, x).$$

Consequently, maximum and minimum are taken on the parabolic boundary.
The *maximum-minimum principle* is useful for proving *uniqueness and a priori estimates* for solutions of the interior Dirichlet problem for the heat equation.

Theorem 9.2.2 *Let us turn to the Dirichlet problem*

$$u_t - \Delta u = h(t, x) \text{ in } Z_T, \quad u(t, x) = g(t, x) \text{ on } S_T,$$

$$u(0, x) = \varphi(x) \text{ on } \{t = 0\} \times \overline{G},$$

where $h \in C(\overline{Z_T})$, $g \in C(\overline{S_T})$ and $\varphi \in C(\overline{G})$. Moreover, we suppose the compatibility condition $g(0, x) = \varphi(x)$ for $x \in \partial G$.

1. *Then, the Dirichlet problem has at most one classical solution $u \in C(\overline{Z_T}) \cap C^{2,1}(Z_T)$.*
2. *If there exists a classical solution $u \in C(\overline{Z_T}) \cap C^{2,1}(Z_T)$ and if $|h(t, x)| \leq N, |g(t, x)| \leq M$ and $|\varphi(x)| \leq M$ on the corresponding domains of definition for h, g and φ, then it follows $|u(t, x)| \leq M + NT$ on $\overline{Z_T}$.*

Proof The first statement follows from Theorem 9.2.1. The functions $M + Nt \mp u(t, x)$ satisfy the first pair of inequalities from Lemma 9.2.2. Hence, $M + Nt \geq \pm u(t, x)$. So, it follows $|u(t, x)| \leq M + NT$ on $\overline{Z_T}$.

Corollary 9.2.1 *If there exists a classical solution of the Dirichlet problem, then this solution depends continuously on the data with respect to the norms given by Theorem 9.2.2.*

The maximum-minimum principle is also used to prove the *uniqueness of bounded solutions for the exterior Dirichlet problem.* Let us explain.

Let G be an exterior domain. Let us define the cylinder Z_T and its lateral surface S_T as before. Let us assume the existence of two *bounded classical solutions.* Then, the difference $w = u_1 - u_2$ satisfies an exterior Dirichlet problem with homogeneous data. From the boundedness of u_1 and u_2 there exists a constant M with $M \geq |u_1(t, x)| + |u_2(t, x)| \geq |w(t, x)|$. Let us restrict ourselves to the 3d case. Choosing R large enough guarantees that the interior domain $G_1 := \mathbb{R}^3 \setminus \overline{G}$ belongs to $K_R(0)$. Here $K_R(0)$ denotes the closed ball around the origin with radius R. Then

$$W = W(t, x) := \frac{6M}{R^2} \left(\frac{|x|^2}{6} + t \right)$$

is a classical solution of $W_t - \Delta W = 0$. Let us introduce the family of cylinders $Z_{R,T} = \{(t, x) : 0 < t < T, \ x \in K_R(0) \cap G\}$. Then $v := W \mp w$ is a solution to $v_t - \Delta v = 0$ in $Z_{R,T}$. Moreover, $v \geq 0$ on $\{t = 0\} \times \overline{G}$ and on S_T. For $|x| = R$ and $t \geq 0$ it holds $|w| \leq M \leq W$. Hence, $v \geq 0$ on the parabolic boundary of $Z_{R,T}$ and

$$|w(t, x)| \leq \frac{6M}{R^2} \left(\frac{|x|^2}{6} + t \right) \quad \text{on } \overline{Z_{R,T}}.$$

Let us fix a point (t, x). Let the radius R tend to infinity. Then, $w(t, x) = 0$. Hence, $w \equiv 0$ in Z_T, which is what we wanted to prove.

One can prove uniqueness also for classical solutions to other mixed problems (see Exercises 2 and 3 below).

Remark 9.2.1 In [147] the maximum principle for the heat equation was extended to second order parabolic equations of the form

$$Lu = L(t, x, \partial_t, \nabla)u$$

$$= \sum_{j,k=1}^{n} a_{jk}(t, x) \partial^2_{x_j x_k} u + \sum_{j=1}^{n} b_j(t, x) \partial_{x_j} u + c(t, x) u - \partial_t u = 0$$

$$\text{in } Z_T = (0, T) \times G,$$

with sufficiently regular coefficients in Z_T, where $G \subset \mathbb{R}^n$ is supposed to be a bounded domain. Moreover, $c(t, x) \leq 0$ in Z_T and

$$\sum_{j,k=1}^{n} a_{jk}(t, x) \xi_j \xi_k \geq C |\xi|^2 \quad \text{in } Z_T \text{ for all } \xi \in \mathbb{R}^n$$

with a nonnegative constant C. Later, in [54] a maximum principle (Theorem 9.2.3) was proved for weakly sub-parabolic solutions to $Lu = 0$ which are defined as follows:

Definition 9.2 A bounded measurable function u in Z_T is called *weakly subparabolic* if for any compact subset $K \subset Z_T$ (the set K is chosen as the closure of a domain) with piecewise smooth boundary

$$\int_K u(t,x) L^*(t,x,\partial_t,\nabla) v(t,x)\, d(t,x) \geq 0$$

for any function v satisfying the properties

1. $v \geq 0$ in Z_T,
2. $v, \partial_{x_j} v, \partial^2_{x_j x_k} v, \partial_t v$ are continuous in K and vanish on the boundary ∂K of K.

Here L^* is the adjoint operator to L (see Remark 24.4.1).

Before stating the result we introduce some kind of necessary regularity to the solutions.

Definition 9.3 For any point $P = (t_0, x_0)$ in Z_T we denote by $C(P)$ the set of all points (t,x) in Z_T such that there exists a differentiable curve connecting (t_0, x_0) to (t,x) and along which the t-coordinate is non-increasing. A function u is said to be *continuous from below* at a point $P = (t_0, x_0)$ if u is continuous as a function defined in $C(P)$.

Theorem 9.2.3 *Let u be weakly sub-parabolic in Z_T. If u takes its essential supremum M (in Z_T) at a point $P = (t_0, x_0)$ at which u is continuous from below and if $M \geq 0$, then $u = M$ almost everywhere in $C(P)$.*

9.3   Qualitative Properties of Solutions of the Cauchy Problem for the Heat Equation

Let us discuss the Cauchy problem

$$\partial_t u - \partial_x^2 u = 0, \quad u(0,x) = \varphi(x).$$

We assume that the data $\varphi = \varphi(x)$ is nonnegative, continuous and has compact support on an interval $[a,b]$. Then, a classical solution of this Cauchy problem is given in the form

$$u(t,x) = \frac{1}{2\sqrt{\pi t}} \int_a^b e^{-\frac{(x-y)^2}{4t}} \varphi(y)\, dy$$

for $(t,x) \in (0,\infty) \times \mathbb{R}^1$.

9.3.1 Non-reversibility in Time

Consider the Cauchy problem $\partial_t u - \partial_x^2 u = 0$, $u(0, x) = \varphi(x)$ with data from a suitable function space. We are only interested in the *forward Cauchy problem*, that is, the Cauchy problem for $t > 0$. We are able to prove well-posedness results for the forward Cauchy problem.

But, in general, we are not able to prove some well-posedness results for the *backward parabolic Cauchy problem*, that is, the data is given for $t = 0$ and we solve the Cauchy problem for $t < 0$. In some sense we would like to *reconstruct the solution for $t = 0$ from its behavior in the past $t < 0$*. But this backward parabolic Cauchy problem is, in general, ill-posed.

The main reason is that for $t > 0$ the kernel function $\exp\left(-\frac{(x-y)^2}{4t}\right)$ remains uniformly bounded with respect to $(t, x) \in (0, \infty) \times \mathbb{R}^1$. This implies together with data φ from a suitable function space the existence of the integral for all $(t, x) \in (0, \infty) \times \mathbb{R}^1$. We get a uniform estimate up to $t = 0$. This implies $\lim_{t \to +0} u(t, x) = \varphi(x)$.

On the other hand, in the case $t < 0$ the kernel function is unbounded for $t \to -0$ it increases faster than any power t^{-n} for every fixed x. It is clear that the representation gives a solution of the heat equation, but it does, in general, not satisfy the given Cauchy condition.

Remark 9.3.1 If we consider instead the backward parabolic Cauchy problem

$$\partial_t u + \partial_x^2 u = 0, \quad u(0, x) = \varphi(x),$$

then the solvability behavior is converse. We are able to derive general well-posedness results for $t < 0$, but not for $t > 0$. For this reason the operator $\partial_t + \Delta$ is called *a backward parabolic operator* in comparison to $\partial_t - \Delta$ which is called *a (forward) parabolic operator*.

9.3.2 Infinite Speed of Propagation

The considerations of the previous subsection allow us to explain another important property, the so-called *infinite speed of propagation of perturbations*. Let us assume that the nonnegative data $\varphi \neq 0$ is small in the norm of $C[a, b]$. Thus, the data is a small perturbation of the zero data. Nevertheless, we feel this perturbation for any positive time t in any point $x \in \mathbb{R}^1$. This follows from the fact that

$$\frac{1}{2\sqrt{\pi t}} \int_a^b e^{-\frac{(x-y)^2}{4t}} \varphi(y) \, dy > 0 \quad \text{for any } (t, x) \in (0, \infty) \times \mathbb{R}^1.$$

Hence this small perturbation propagates with an infinite speed.

Remark 9.3.2 This property does not allow for proving a result for C^∞ well-posedness for the forward Cauchy problem. The main reason is that the infinite speed of propagation of perturbations does not allow for having a domain of dependence (see Sect. 10.1.3 for wave equations). Thus, we are not able to reduce the study of C^∞ well-posedness to that for H^∞ well-posedness (see also Exercise 1 of Chap. 5).

9.3.3   Smoothing Effect

Let us turn to the Cauchy problem

$$u_t - \Delta u = 0, \quad u(0,x) = \varphi(x) \text{ in } \mathbb{R}^n \text{ for } t > 0.$$

Let us assume low regularity for the data, let us say, $\varphi \in L^p(\mathbb{R}^n)$ for $p \in [1,\infty]$. Then the solution is very regular, namely, C^∞ with respect to t and x in the set $(0,\infty) \times \mathbb{R}^n$. This effect is called *smoothing effect*. The smoothing effect follows from properties of the kernel function in the representation of solution

$$u(t,x) = \frac{1}{(4\pi t)^{\frac{n}{2}}} \int_{\mathbb{R}^n} e^{-\frac{|x-y|^2}{4t}} \varphi(y)\,dy.$$

This kernel function is C^∞ in any neighborhood of any point $(t_0, x_0) \in (0,\infty) \times \mathbb{R}^n$. Consequently, this behavior is transferred to the solution.

In the treatment of nonlinear models one is interested in a *global smoothing behavior*. That is, we expect some behavior of $u(t, \cdot)$ in some function spaces of smoother functions than the given data. The next lemma is a key lemma in these considerations.

Lemma 9.3.1 *Let us suppose $\varphi \in L^p(\mathbb{R}^n)$ for $p \in [1,\infty]$. Then, the solution u to the above Cauchy problem satisfies for $t > 0$ the following inequalities:*

$$\|u(t,\cdot)\|_{L^p} \le \|\varphi\|_{L^p}, \quad \|\partial_t^k \partial_x^\alpha u(t,\cdot)\|_{L^q} \le C t^{-\frac{n}{2r} - \frac{|\alpha|}{2} - k} \|\varphi\|_{L^p},$$

where $\frac{1}{p} = \frac{1}{q} + \frac{1}{r}$, $1 \le r, q \le \infty$. The constant $C = C(p, q, r, |\alpha|, k)$ is a nonnegative constant.

The proof to this lemma is given in Sect. 12.1. The statements of this lemma imply the smoothing effect. If the data φ belongs to $L^p(\mathbb{R}^n)$, then the solution belongs to all spaces $C^k((0,\infty), W^{m,q}(\mathbb{R}^n))$ for all $k, m \in \mathbb{N}_0$ and all p, q satisfying $q \ge p$. But we have, of course, no smoother regularity up to $t = 0$. We can only prove $u \in C([0,\infty), L^p(\mathbb{R}^n))$ if the data is supposed to belong to $L^p(\mathbb{R}^n)$ (see Sect. 12.1).

9.3.4 Uniqueness of Classical Solutions to the Cauchy Problem

Let us discuss the Cauchy problem

$$\partial_t u - \Delta u = 0, \quad u(0, x) = \varphi(x) \text{ in } \mathbb{R}^n \text{ for } t > 0.$$

We assume that the data $\varphi = \varphi(x)$ is continuous and has compact support in $\mathbb{R}^n$. Then, a classical solution of this Cauchy problem is given by the convolution of a suitable fundamental solution (cf. with Example 24.4.5) with the initial data, that is,

$$u(t, x) = \frac{1}{(4\pi t)^{\frac{n}{2}}} \int_{\mathbb{R}^n} e^{-\frac{|x-y|^2}{4t}} \varphi(y)\, dy.$$

In Sect. 5.1 one can find with a suitable function $\psi = \psi(t)$ a classical non-analytic solution

$$u(t, x) = \sum_{k=0}^{\infty} \frac{1}{(2k)!} x^{2k} \psi^{(k)}(t)$$

to the Cauchy problem

$$\partial_t u - \partial_x^2 u = 0, \quad u(0, x) = 0.$$

After some considerations, one can show that this solution can not be bounded by Ce^{ax^2} with positive constants a and C (see [100]). As one can expect from this example, in order to derive a maximum principle and the uniqueness of classical solutions to the Cauchy problem for the heat equation we have to impose some control on the behavior of solutions for large $|x|$.

Theorem 9.3.1 (Maximum Principle for Classical Solutions to the Cauchy Problem)

Let

$$u \in C\big([0, T] \times \mathbb{R}^n\big) \cap C^{2,1}\big((0, T) \times \mathbb{R}^n\big)$$

be a classical solution to the Cauchy problem

$$\partial_t u - \Delta u = 0, \quad u(0, x) = \varphi(x) \text{ in } \mathbb{R}^n \text{ for } t > 0,$$

which satisfies the growth estimate

$$u(t, x) \leq Ce^{a|x|^2} \text{ for } (t, x) \in [0, T] \times \mathbb{R}^n,$$

where a and C are positive constants. Then it holds

$$\max_{[0,T] \times \mathbb{R}^n} u(t, x) = \max_{\mathbb{R}^n} \varphi.$$

Proof A proof can be found in [100]. It uses, among other things, Theorem 9.2.1. As a corollary of the maximum principle we have the uniqueness of classical solutions for the Cauchy problem.

Corollary 9.3.1 (Uniqueness)
Let

$$u \in C\big([0, T] \times \mathbb{R}^n\big) \cap C^{2,1}\big((0, T) \times \mathbb{R}^n\big)$$

be a classical solution of the Cauchy problem

$$\partial_t u - \Delta u = 0, \quad u(0, x) = 0 \ \text{in} \ \mathbb{R}^n \ \text{for} \ t > 0,$$

which satisfies the growth estimate

$$u(t, x) \le C e^{a|x|^2} \ \text{for} \ (t, x) \in [0, T] \times \mathbb{R}^n,$$

where a and C are positive constants. Then, u is identically zero.
This corollary shows that for continuous initial data with compact support in $\mathbb{R}^n$ the solution given by the convolution represents the only bounded solution to the Cauchy problem for the heat equation.

9.4 Mixed Problems for the Heat Equation

First, we propose some mixed problems for the heat equation. We restrict ourselves to mixed problems with Dirichlet or Neumann boundary condition. Then, we explain the approach of using heat potentials in representations of solutions. We derive the corresponding integral equations. Finally, we point out how to use functional analytic tools to treat the corresponding integral equations.

9.4.1 Basic Mixed Problems

9.4.1.1 Interior Domain and Dirichlet Boundary Condition

Let $G \subset \mathbb{R}^n$ be an interior domain. We suppose that the boundary ∂G is sufficiently smooth. Find a solution u of the heat equation in $(0, \infty) \times G$ with prescribed Dirichlet data $g = g(t, x)$ for u on $(0, \infty) \times \partial G$ and prescribed initial data $\varphi = \varphi(x)$ for u on $\{t = 0\} \times G$. The compatibility condition between boundary and initial data is satisfied on $\{t = 0\} \times \partial G$.

Mathematical Formulation

$$\partial_t u - \Delta u = h(t, x) \quad \text{in} \ (0, \infty) \times G,$$

$$u(t, x) = g(t, x) \ \text{on} \ (0, \infty) \times \partial G, \quad u(0, x) = \varphi(x) \ \text{for} \ x \in G,$$

$$\text{compatibility condition} \ g(0, x) = \varphi(x) \ \text{on} \ \{t = 0\} \times \partial G.$$

9.4.1.2   Interior Domain and Neumann Boundary Condition

Let $G \subset \mathbb{R}^n$ be an interior domain. We suppose that the boundary ∂G is sufficiently smooth. Find a solution u of the heat equation in $(0, \infty) \times G$ with prescribed Neumann data $g = g(t, x)$ for u on $(0, \infty) \times \partial G$ and prescribed initial data $\varphi = \varphi(x)$ for u on $\{t = 0\} \times G$. The compatibility condition between boundary and initial data is satisfied on $\{t = 0\} \times \partial G$.

Mathematical Formulation

$$\partial_t u - \Delta u = h(t, x) \quad \text{in} \ (0, \infty) \times G,$$

$$\partial_\mathbf{n} u(t, x) = g(t, x) \ \text{on} \ (0, \infty) \times \partial G, \quad u(0, x) = \varphi(x) \ \text{for} \ x \in G,$$

$$\text{compatibility condition} \ g(0, x) = \partial_\mathbf{n} \varphi(x) \ \text{on} \ \{t = 0\} \times \partial G.$$

9.4.1.3   Exterior Domain and Dirichlet Boundary Condition

Let $G \subset \mathbb{R}^n$ be an exterior domain. We suppose that the boundary ∂G is sufficiently smooth. Find a solution u of the heat equation in $(0, \infty) \times G$ with prescribed Dirichlet data $g = g(t, x)$ for u on $(0, \infty) \times \partial G$ and prescribed initial data $\varphi = \varphi(x)$ for u on $\{t = 0\} \times G$. The compatibility condition between boundary and initial data is satisfied on $\{t = 0\} \times \partial G$.

Mathematical Formulation

$$\partial_t u - \Delta u = h(t, x) \quad \text{in} \ (0, \infty) \times G,$$

$$u(t, x) = g(t, x) \ \text{on} \ (0, \infty) \times \partial G, \quad u(0, x) = \varphi(x) \ \text{for} \ x \in G,$$

$$\text{compatibility condition} \ g(0, x) = \varphi(x) \ \text{on} \ \{t = 0\} \times \partial G.$$

9.4.1.4   Exterior Domain and Neumann Boundary Condition

Let $G \subset \mathbb{R}^n$ be an exterior domain. We suppose that the boundary ∂G is sufficiently smooth. Find a solution u of the heat equation in $(0, \infty) \times G$ with prescribed Neumann data $g = g(t, x)$ for u on $(0, \infty) \times \partial G$ and prescribed initial data $\varphi = \varphi(x)$

for u on $\{t = 0\} \times G$. The compatibility condition between boundary and initial data is satisfied on $\{t = 0\} \times \partial G$.

Mathematical Formulation

$$\partial_t u - \Delta u = h(t, x) \quad \text{in} \ (0, \infty) \times G,$$

$$\partial_\mathbf{n} u(t, x) = g(t, x) \ \text{on} \ (0, \infty) \times \partial G, \ \ u(0, x) = \varphi(x) \ \text{for} \ x \in G,$$

$$\text{compatibility condition} \ \ g(0, x) = \partial_\mathbf{n} \varphi(x) \ \text{on} \ \{t = 0\} \times \partial G.$$

9.4.2 How to Use Thermal Potentials in Representations of Solutions?

We introduced already thermal potentials in Sect. 9.1. In the following we will use the thermal single-layer potential with density $\mu = \mu(\tau, y)$

$$\int_{S_t} H_n(t - \tau, x - y) \mu(\tau, y) \, d(\tau, \sigma_y),$$

and the thermal double-layer potential with density $\mu = \mu(\tau, y)$

$$\int_{S_t} \partial_{\mathbf{n}_y} H_n(t - \tau, x - y) \mu(\tau, y) \, d(\tau, \sigma_y)$$

to treat the above formulated mixed problems.

9.4.2.1 Mixed Problems with Dirichlet Boundary Condition

To treat mixed problems with Dirichlet boundary condition we use thermal double-layer potentials with unknown density. As in Sect. 8.4.2.1, jump relations for thermal double-layer potentials after crossing $(0, \infty) \times \partial G$ are used. After some calculations (see [135]) one can show for all $(t_0, x_0) \in (0, \infty) \times \partial G$ the relation

$$\lim_{x \to x_0 \in \partial G} \int_{S_{t_0}} \partial_{\mathbf{n}_y} H_n(t_0 - \tau, x - y) \mu(\tau, y) \, d(\tau, \sigma_y)$$

$$= \pm \frac{1}{2} \mu(t_0, x_0) + \int_{S_{t_0}} \partial_{\mathbf{n}_y} H_n(t_0 - \tau, x_0 - y) \mu(\tau, y) \, d(\tau, \sigma_y).$$

The negative sign appears in the case of an interior, the positive sign in the case of an exterior domain.

9.4.2.2 Mixed Problems with Neumann Boundary Condition

To treat mixed problems with Neumann boundary condition we use thermal single-layer potentials with unknown density. As in Sect. 8.4.2.2, jump relations for the normal derivative of thermal single-layer potentials after crossing $(0, \infty) \times \partial G$ are used. After some calculations (see [135]) one can show for all $(t_0, x_0) \in (0, \infty) \times \partial G$ the relation

$$\lim_{x \to x_0 \in \partial G} \partial_{\mathbf{n}_x} \int_{S_{t_0}} H_n(t_0 - \tau, x - y) \mu(\tau, y) \, d(\tau, \sigma_y)$$

$$= \pm \frac{1}{2} \mu(t_0, x_0) + \int_{S_{t_0}} \left(\partial_{\mathbf{n}_x} H_n(t_0 - \tau, x - y) \right)(x = x_0) \mu(\tau, y) \, d(\tau, \sigma_y).$$

The positive sign appears in the case of an interior, the negative sign in the case of an exterior domain.

9.4.3 Integral Equations of Mixed Problems for the Heat Equation

We write the solution u in the form $u = w + v$. The function $w = w(t, x) \in C^{2,1}(Z_T) \cap C(\overline{Z_T})$, $T > 0$, which is defined for all $t \in (0, T)$ by

$$w(t, x) = \int_{Z_t} H_n(t - \tau, x - y) h(\tau, y) \, d(\tau, y) + \int_G H_n(t, x - y) \varphi(y) \, dy$$

and satisfies

$$\partial_t w - \Delta w = h(t, x) \text{ in } Z_T, \quad w(0, x) = \varphi(x) \text{ on } \{t = 0\} \times \overline{G}.$$

Here we assume that the right-hand side h and the data φ are continuous and bounded on $\overline{Z_T}$ and $\overline{G}$, respectively. So, it remains to study the above mixed problems for $\partial_t v - \Delta v = 0$ with a homogeneous initial condition and corresponding boundary condition.

If we choose the ansatz of solutions in the form of thermal single or double-layer potentials with density from suitable function spaces, then these thermal potentials solve the homogeneous heat equation with homogeneous initial condition. Thus, it remains only to determine the density of thermal potentials in such a way that the boundary condition is satisfied. Taking into consideration the jump conditions for the normal derivative of the thermal single-layer potential or for the thermal double-layer potential, we obtain the following integral equations where w is defined as above:

interior domain with Dirichlet condition:

$$g(t, x_0) - w(t, x_0) = \lim_{x \to x_0 \in \partial G} v(t, x)$$

$$= \lim_{x \to x_0 \in \partial G} \int_{S_t} \partial_{\mathbf{n}_y} H_n(t - \tau, x - y) \mu(\tau, y) \, d(\tau, \sigma_y)$$

$$= -\frac{1}{2} \mu(t, x_0) + \int_{S_t} \partial_{\mathbf{n}_y} H_n(t - \tau, x_0 - y) \mu(\tau, y) \, d(\tau, \sigma_y);$$

exterior domain with Dirichlet condition:

$$g(t, x_0) - w(t, x_0) = \lim_{x \to x_0 \in \partial G} v(t, x)$$

$$= \lim_{x \to x_0 \in \partial G} \int_{S_t} \partial_{\mathbf{n}_y} H_n(t - \tau, x - y) \mu(\tau, y) \, d(\tau, \sigma_y)$$

$$= \frac{1}{2} \mu(t, x_0) + \int_{S_t} \partial_{\mathbf{n}_y} H_n(t - \tau, x_0 - y) \mu(\tau, y) \, d(\tau, \sigma_y);$$

interior domain with Neumann condition:

$$g(t, x_0) - (\partial_{\mathbf{n}_x} w)(t, x_0) = \lim_{x \to x_0 \in \partial G} \partial_{\mathbf{n}_x} v(t, x)$$

$$= \lim_{x \to x_0 \in \partial G} \partial_{\mathbf{n}_x} \int_{S_t} H_n(t - \tau, x - y) \mu(\tau, y) \, d(\tau, \sigma_y)$$

$$= \frac{1}{2} \mu(t, x_0) + \int_{S_t} \left(\partial_{\mathbf{n}_x} H_n(t - \tau, x - y) \right)(x = x_0) \mu(\tau, y) \, d(\tau, \sigma_y);$$

exterior domain with Neumann condition:

$$g(t, x_0) - (\partial_{\mathbf{n}_x} w)(t, x_0) = \lim_{x \to x_0 \in \partial G} \partial_{\mathbf{n}_x} v(t, x)$$

$$= \lim_{x \to x_0 \in \partial G} \partial_{\mathbf{n}_x} \int_{S_t} H_n(t - \tau, x - y) \mu(\tau, y) \, d(\tau, \sigma_y)$$

$$= -\frac{1}{2} \mu(t, x_0) + \int_{S_t} \left(\partial_{\mathbf{n}_x} H_n(t - \tau, x - y) \right)(x = x_0) \mu(\tau, y) \, d(\tau, \sigma_y).$$

There is a difference in the integral equations of potential theory from Sect. 8.4.3. Now, the integral equations "living on the lateral surface S_t for all $t > 0$" are Volterra integral equations of second kind (the domain of integration S_t depends on t). They have the abstract form

$$\mu(t, x) - \lambda \int_{S_t} K(t - \tau, x - y) \mu(\tau, y) \, d(\tau, \sigma_y) = f(t, x),$$

where λ is a complex parameter. To solve this integral equation we may apply the principle of successive approximation. Thus, we define for all $p \geq 0$

$$\mu_0(t, x) = f(t, x), \quad \mu_{p+1}(t, x) = \lambda \int_{S_t} K(t - \tau, x - y) \mu_p(\tau, y) \, d(\tau, \sigma_y) + f(t, x).$$

These are the main steps in studying mixed problems for the heat equation with Dirichlet or Neumann boundary condition. We omit more in detail discussions about regularity of solutions. But, we can expect well-posedness results for all these mixed problems if we investigate the following problems.

1. Study properties of the family of integral operators

$$T_t : \mu \in B_t \to T_t \mu := \int_{S_t} K(t - \tau, x - y) \mu(\tau, y) \, d(\tau, \sigma_y),$$

   in particular, we should fix spaces B_t which allow to prove that $T_t \in L(B_t \to B_t)$ are bounded operators for all $t > 0$.
2. Choose the right-hand side f, the data g, respectively, from suitable function spaces related to B_t.
3. Discuss the regularity of w to verify the necessary regularity of f.

The main concern of the research project in Sect. 23.2 is to solve step by step the interior Robin problem for the heat equation. Following all the steps one can solve in the same way the other mixed problems taking account of the specific character of any of these problems.

Exercises Relating to the Considerations of Chap. 9

Exercise 1 Prove the statement of Lemma 9.2.1.

Exercise 2 Let us suppose the usual assumptions for h, g and φ (cf. with Sects. 9.1 and 9.2). Show that there exists at most one solution $u \in C^{2,1}(Z_T) \cap C^{1,0}(\overline{Z}_T)$ of the Neumann problem

$$u_t - \Delta u = h(t, x) \text{ in } Z_T, \quad \partial_\mathbf{n} u(t, x) = g(t, x) \text{ on } S_T,$$

$$u(0, x) = \varphi(x) \text{ on } \{t = 0\} \times \overline{G}.$$

Hint   Let w be the difference of two classical solutions. Define the energy functional $I(t) = \frac{1}{2} \int_G w(t, x)^2 \, dx$ and show that $I'(t) \leq 0$.

Exercise 3 Let us consider the Dirichlet problem

$$u_t - \Delta u + b \cdot \nabla u + cu = h(t, x) \text{ in } Z_T, \quad u(t, x) = g(t, x) \text{ on } S_T,$$

$$u(0, x) = \varphi(x) \text{ on } \{t = 0\} \times \overline{G},$$

where $c \geq 0$ on $\overline{Z}_T$ and b is a constant vector. The usual assumptions for h, g and φ are supposed to be satisfied (cf. with Sects. 9.1 and 9.2). Show, that there exists at most one classical solution $u \in C^{2,1}(Z_T) \cap C(\overline{Z}_T)$.

Exercise 4 Prove that

$$u(t, x) = \frac{1}{2\sqrt{\pi t}} \int_a^b e^{-\frac{(x-y)^2}{4t}} \varphi(y)\, dy$$

is a classical solution to the Cauchy problem

$$\partial_t u - \partial_x^2 u = 0, \quad u(0, x) = \varphi(x).$$

To which function space is the data supposed to belong?

Exercise 5 Prove the statements of Lemma 9.3.1.

Exercise 6 Verify the relation

$$\lim_{x \to x_0 \in \partial G} \int_{S_t} \partial_{n_y} H_n(t - \tau, x - y) \mu(\tau, y)\, d(\tau, \sigma_y)$$

$$= \pm \frac{1}{2} \mu(t, x_0) + \int_{S_t} \partial_{n_y} H_n(t - \tau, x_0 - y) \mu(\tau, y)\, d(\tau, \sigma_y)$$

for thermal double-layer potentials after crossing the layer.

Exercise 7 Verify the relation

$$\lim_{x \to x_0 \in \partial G} \partial_{n_x} \int_{S_t} H_n(t - \tau, x - y) \mu(\tau, y)\, d(\tau, \sigma_y)$$

$$= \pm \frac{1}{2} \mu(t, x_0) + \int_{S_t} \big(\partial_{n_x} H_n(t - \tau, x - y)\big)(x = x_0) \mu(\tau, y)\, d(\tau, \sigma_y)$$

for thermal single-layer potentials after crossing the layer.

Chapter 10
Wave Equation—Properties of Solutions—Starting Point of Hyperbolic Theory

There exists comprehensive literature on the theory of hyperbolic partial differential equations. One of the simplest hyperbolic partial differential equations is the free wave equation. First, we introduce d'Alembert's representation in 1d and derive usual properties of solutions as finite speed of propagation of perturbations, existence of a domain of dependence, existence of forward or backward wave fronts and propagation of singularities. There has been a long way to get representation of solutions in higher dimensions, too. The emphasis is on two and three spatial dimensions in the form of Kirchhoff's representation in three dimensions and by using the method of descent in two dimensions, too. Representations in higher-dimensional cases are only sketched. Some comments on hyperbolic potential theory and the theory of mixed problems complete this chapter.

10.1 d'Alembert's Representation in $\mathbb{R}^1$

We turn to the Cauchy problem

$$u_{tt} - u_{xx} = 0, \quad u(0, x) = \varphi(x), \quad u_t(0, x) = \psi(x).$$

Replacing t by $-t$ we see that this model is *time reversible*. In the following we restrict ourselves to the forward Cauchy problem, that is, to $t > 0$. A change of variables $\xi = x - t$, $\eta = x + t$ (motivated by the *notion of characteristics* if we use $\partial_t^2 - \partial_x^2 = (\partial_t - \partial_x)(\partial_t + \partial_x)$, see Definition 6.1 in Sect. 6.2) leads to $-4u_{\xi\eta} = 0$. This partial differential equation has the classical solution $u = u(\xi, \eta) = u_1(\xi) + u_2(\eta)$ with arbitrarily given twice continuously differentiable functions u_1 and u_2. The backward transformation gives $u = u(t, x) = u_1(x - t) + u_2(x + t)$. The general solution u is a linear superposition of two waves, the wave $u_1(x - t)$ is moving with velocity 1 to the right-hand side. The wave $u_2(x + t)$ is moving with velocity 1 to

M.R. Ebert, M. Reissig, *Methods for Partial Differential Equations*, https://doi.org/10.1007/978-3-319-66456-9_10

the left-hand side. Both solutions are called *traveling wave solutions* (see Sect. 2.2). Using both Cauchy conditions we obtain

$$u(0, x) = \varphi(x) = u_1(x) + u_2(x), \quad u_t(0, x) = \psi(x) = -u_1'(x) + u_2'(x).$$

Integration of the second equation yields

$$-u_1(x) + u_2(x) = \int_{x_0}^{x} \psi(r)dr,$$

where x_0 is an arbitrary real number. Hence,

$$u_1(x) = \frac{1}{2}\varphi(x) - \frac{1}{2}\int_{x_0}^{x} \psi(r)dr, \quad u_2(x) = \frac{1}{2}\varphi(x) + \frac{1}{2}\int_{x_0}^{x} \psi(r)dr.$$

Summarizing, we derived the so-called *d'Alembert's representation of solution in 1d*

$$u(t, x) = \frac{1}{2}\big(\varphi(x - t) + \varphi(x + t)\big) + \frac{1}{2}\int_{x-t}^{x+t} \psi(r)dr.$$

From d'Alembert's representation formula we conclude remarkable properties for solutions of wave equations which in most cases are typical properties for solutions of *hyperbolic partial differential equations,* too. The free wave equation is one representative of this class.

In the following sections we explain what kind of properties we have in mind.

10.1.1   Regularity of Solutions

Let us consider the Cauchy problem

$$u_{tt} - u_{xx} = 0, \quad u(0, x) = \varphi(x), \quad u_t(0, x) = \psi(x)$$

with data $\varphi \in C^k(\mathbb{R}^1)$ and $\psi \in C^{k-1}(\mathbb{R}^1)$. To understand the statement of the following theorem and its proof we recommend the reader review the Definitions 24.16, 24.25 and 24.26.

Theorem 10.1.1 *This Cauchy problem possesses one and only one classical solution $u \in C^k([0, \infty) \times \mathbb{R}^1)$, $k \geq 2$. The solution depends continuously on the data, that is, if we change φ and ψ a bit with respect to the topologies of $C^k(\mathbb{R}^1)$ and $C^{k-1}(\mathbb{R}^1)$, then the solution u changes a bit with respect to the topology of $C^k([0, \infty) \times \mathbb{R}^1)$.*

Proof The existence of a solution is given by d'Alembert's representation formula. The uniqueness follows from the fact that the general solution of $u_{tt} - u_{xx} = 0$ is given by the formula $u(t, x) = u_1(x - t) + u_2(x + t)$. The continuous dependence of the solution on the data is concluded from the representation formula.

10.1.2   Finite Speed of Propagation of Perturbations

Let us turn to the Cauchy problem with data $\varphi \in C^2(\mathbb{R}^1)$ and $\psi \in C^1(\mathbb{R}^1)$. We perturb these data by the aid of data $\varphi_s \in C^2(\mathbb{R}^1)$ and $\psi_s \in C^1(\mathbb{R}^1)$ supported on the interval $[a, b]$. We are interested in the propagation of these perturbations. The superposition principle allows for restricting ourselves to studying the Cauchy problem

$$u_{tt} - u_{xx} = 0, \quad u(0, x) = \varphi_s(x), \quad u_t(0, x) = \psi_s(x)$$

with $\varphi_s = \psi_s = 0$ outside of $[a, b]$. We get the solution

$$u_s(t, x) = \frac{1}{2}\big(\varphi_s(x - t) + \varphi_s(x + t)\big) + \frac{1}{2}\int_{x-t}^{x+t} \psi_s(r)dr.$$

When do we feel the perturbation in a given point $x_0 \in \mathbb{R}^1$ lying outside of $[a, b]$? For small times t we have of course $u(t, x_0) = 0$ in x_0.
But we feel the perturbation after the finite time $T = \text{dist}(x_0, [a, b])$. This property is called *finite speed of propagation of perturbations or existence of a forward wave front*.

10.1.3   Domain of Dependence

Which information on the data has an influence on the solution in a given point (t_0, x_0)?
To determine the solution $u(t_0, x_0)$ in the point (t_0, x_0) we need the data φ in the points $x_0 - t_0$ and $x_0 + t_0$ and the data ψ on the interval $[x_0 - t_0, x_0 + t_0]$. The interval $[x_0 - t_0, x_0 + t_0]$ is called *domain of dependence* for the solution u in the point (t_0, x_0).

10.1.4   Huygens' Principle

The *Huygens' principle* describes the existence of a *backward wave front*, that is, the property that in a point $x_0 \in \mathbb{R}^1$ the solution vanishes after the time $T(x_0)$ if we are interested in the propagation of perturbations located in an interval $[a, b]$. In general, we can not expect the existence of a *backward wave front* having in mind that the *domain of dependence* for the solution u in the point (t_0, x_0) is the interval $[x_0 - t_0, x_0 + t_0]$. If we choose the data ψ such that $\int_a^b \psi(s)ds = 0$, then the solution u in (t_0, x_0) is determined by the values of φ in $x_0 - t_0$ and $x_0 + t_0$. Consequently, after time $T = \max\{x_0 - a, b - x_0\}$ we have $u = 0$ in x_0. Summarizing the *Huygens' principle* holds under the assumption $\int_a^b \psi(s)ds = 0$.

10.2   Wave Models with Sources or Sinks

Consider the Cauchy problem

$$u_{tt} - u_{xx} = F(t, x), \quad u(0, x) = \varphi(x), \quad u_t(0, x) = \psi(x).$$

We suppose that the source or sink F is integrable. Let us say $F \in L^1_{loc}([0, \infty) \times \mathbb{R}^1)$ (cf. with Definition 24.20). Thus we are interested in non-classical solutions. For the solution u we choose the ansatz $u = v + w$, where v and w are solutions of the Cauchy problems (here we take account of the linearity of the homogeneous wave equation)

$$v_{tt} - v_{xx} = F(t, x), \quad v(0, x) = 0, \quad v_t(0, x) = 0,$$
$$w_{tt} - w_{xx} = 0, \quad w(0, x) = \varphi(x), \quad w_t(0, x) = \psi(x).$$

The Cauchy problem for w is studied in Sect. 10.1. So, we restrict ourselves to the study of the Cauchy problem for v. Straight forward calculations (cf. with Exercise 3) imply the representation of solution

$$v(t, x) = \frac{1}{2} \int_0^t \left(\int_{x-(t-s)}^{x+(t-s)} F(s, y) \, dy \right) ds.$$

In which domain of dependence do the values of F determine the solution v in a given point (t_0, x_0)?
Denoting the domain of dependence by $\Omega(t_0, x_0)$ we conclude

$$\Omega(t_0, x_0) = \left\{ (t, x) \in \mathbb{R}^2 : (t, x) \in [0, t_0) \times \{|x - x_0| \leq t_0 - t\} \right\}.$$

Summarizing the considerations of Sects. 10.1 and 10.2 we have a basic knowledge on properties of solutions to wave equations in 1d. What about a corresponding knowledge in higher dimensions? We devote the following sections to this question.

10.3   Kirchhoff's Representation in $\mathbb{R}^3$

As in the previous section our main concern is the treatment of the Cauchy problem

$$u_{tt} - \Delta u = 0, \quad u(0, x) = \varphi(x), \quad u_t(0, x) = \psi(x), \quad x \in \mathbb{R}^3,$$

but now in the 3d case. To find a representation of solution is more complicated than in the 1d case. A first observation tells us the following:

Lemma 10.3.1 *If $u_p = u_p(t, x)$ solves the Cauchy problem*

$$u_{tt} - \Delta u = 0, \quad u(0, x) = 0, \quad u_t(0, x) = p(x), \quad x \in \mathbb{R}^n,$$

where $p = p(x)$ is sufficiently smooth, then $\partial_t u_p =: v$ solves the Cauchy problem

$$v_{tt} - \Delta v = 0, \quad v(0, x) = p(x), \quad v_t(0, x) = 0.$$

From this lemma we may conclude the following statement.

Corollary 10.3.1 *The solution $u = u(t, x)$ of*

$$u_{tt} - \Delta u = 0, \quad u(0, x) = \varphi(x), \quad u_t(0, x) = \psi(x), \quad x \in \mathbb{R}^n,$$

has the representation $u(t, x) = u_\psi(t, x) + \partial_t u_\varphi(t, x)$, where the data φ and ψ are supposed to be smooth.

Thus, it is enough to derive a formula for $u_p = u_p(t, x)$ as follows. In a first step we derive a formula for u_p. In a second step we verify for u_p to be the desired solution to the Cauchy problem

$$u_{tt} - \Delta u = 0, \quad u(0, x) = 0, \quad u_t(0, x) = p(x), \quad x \in \mathbb{R}^3.$$

10.3.1 How Can the Reader Guess Kirchhoff's Formula?

Let us consider the auxiliary Cauchy problem

$$u_{tt} - \Delta u = 0, \quad u(0, x) = 0, \quad u_t(0, x) = \phi_\varepsilon(x), \quad x \in \mathbb{R}^3,$$

where

$$\phi_\varepsilon(x) = (4\pi\varepsilon)^{-\frac{3}{2}} \exp\left(-\frac{|x|^2}{4\varepsilon}\right), \quad \varepsilon > 0.$$

Here the reader might recall Example 24.4.2. It holds

$$\int_{\mathbb{R}^3} \phi_\varepsilon(x)dx = 1, \quad \lim_{\varepsilon \to +0} \phi_\varepsilon(x) = 0 \quad \text{for all } x \neq 0, \quad \text{and} \quad \lim_{\varepsilon \to +0} \phi_\varepsilon = \delta_0$$

in the sense of distributions. This data depends on the polar distance only, so it is *radially symmetric*. Then, we may expect the solution to be radially symmetric too, thus it depends only on $r = |x|$ and t.
Consequently, $u = u(t, r)$ satisfies the wave equation if and only if

$$u_{tt} - u_{rr} - \frac{2}{r}u_r = 0.$$

One can verify (cf. with Exercise 6) that this partial differential equation has the general classical solution

$$u(t, r) = \frac{u_1(r + t)}{r} + \frac{u_2(r - t)}{r},$$

with arbitrarily given twice differentiable functions u_1, u_2.
Using the Cauchy conditions then, one integration leads to

$$u_2(r) = \int_0^r -\frac{s}{2} \, (4\pi\varepsilon)^{-\frac{3}{2}} \exp\left(-\frac{s^2}{4\varepsilon}\right) ds$$

$$= \varepsilon \, (4\pi\varepsilon)^{-\frac{3}{2}} \exp\left(-\frac{r^2}{4\varepsilon}\right) - \varepsilon \, (4\pi\varepsilon)^{-\frac{3}{2}}.$$

It follows the representation of solution

$$u(t, x) = I_\varepsilon(t, r) - J_\varepsilon(t, r) :=$$

$$\frac{1}{4\pi r} \frac{1}{\sqrt{4\pi\varepsilon}} \exp\left(-\frac{(r - t)^2}{4\varepsilon}\right) - \frac{1}{4\pi r} \frac{1}{\sqrt{4\pi\varepsilon}} \exp\left(-\frac{(r + t)^2}{4\varepsilon}\right).$$

The data $p = p(y)$ is supposed to be continuous. Thus it is nearly constant in a small cube. Hence, the solution $u = u(t, x)$ to the data

$$p(y)\phi_\varepsilon(|x - y|)\Delta y,$$

where Δy describes a localization near y, is

$$u(t, x) = p(y)\big(I_\varepsilon(t, |x - y|) - J_\varepsilon(t, |x - y|)\big)\Delta y.$$

The superposition of all influences of localized data leads to

$$u(t, x) = \int_{\mathbb{R}^3} p(y)\big(I_\varepsilon(t, |x - y|) - J_\varepsilon(t, |x - y|)\big) \, dy.$$

The desired formula results from

$$u(t, x) = \lim_{\varepsilon \to +0} \int_{\mathbb{R}^3} p(y)\big(I_\varepsilon(t, |x - y|) - J_\varepsilon(t, |x - y|)\big) \, dy$$

$$= \lim_{\varepsilon \to +0} \int_{\mathbb{R}^3} p(y)I_\varepsilon(t, |x - y|) \, dy.$$

Introducing spherical harmonics we get with $y = x + \rho\,\omega$, ω is a unit vector in $\mathbb{R}^3$,

$$u(t,x) = \lim_{\varepsilon \to +0} \int_{\mathbb{R}^3} p(y) \frac{1}{4\pi |x-y|} \frac{1}{\sqrt{4\pi\varepsilon}} \exp\left(-\frac{(|x-y|-t)^2}{4\varepsilon}\right) dy$$

$$= \lim_{\varepsilon \to +0} \frac{1}{4\pi} \int_0^\infty \exp\left(-\frac{(\rho-t)^2}{4\varepsilon}\right) \frac{1}{\sqrt{4\pi\varepsilon}} \left(\frac{1}{\rho} \int_{|\omega|=1} p(x+\rho\omega)\rho^2 \, d\sigma_\omega\right) d\rho.$$

Finally, setting $\rho - t = 2\sqrt{\varepsilon z}$ and changing the order of integration it follows

$$u(t,x)$$

$$= \frac{1}{4\pi} \int_{|\omega|=1} \frac{1}{\sqrt{\pi}} \left(\lim_{\varepsilon \to +0} \int_{\frac{-t}{2\sqrt{\varepsilon}}}^\infty p\big(x + (t + 2\sqrt{\varepsilon z})\omega\big)(t + 2\sqrt{\varepsilon z}) \exp(-z^2) dz\right) d\sigma_\omega$$

$$= \frac{t}{4\pi} \int_{|\omega|=1} p(x+t\omega)\, d\sigma_\omega.$$

The element of the surface of a ball with radius t around the center x is $d\sigma_y = t^2 d\sigma_\omega$. Setting $x + t\omega = y$ we arrive at the equivalent representation

$$u(t,x) = \frac{1}{4\pi t} \int_{S_t(x)} p(y)\, d\sigma_y,$$

where $S_t(x)$ is the surface of a ball of radius t and center x.

Remark 10.3.1 The above heuristic considerations serve to derive a representation of solution to the Cauchy problem

$$u_{tt} - \Delta u = 0, \quad u(0,x) = 0, \quad u_t(0,x) = p(x), \quad x \in \mathbb{R}^3.$$

This formula was proposed by Kirchhoff in [113].

10.3.2   Verification of Kirchhoff's Formula

Theorem 10.3.1 *Let $p \in C^k(\mathbb{R}^3)$ with $k \geq 2$. Then a classical solution $u_p = u_p(t,x)$ of the Cauchy problem*

$$u_{tt} - \Delta u = 0, \quad u(0,x) = 0, \quad u_t(0,x) = p(x),$$

is given by Kirchhoff's formula

$$u_p(t,x) = \frac{1}{4\pi t} \int_{S_t(x)} p(y)\, d\sigma_y.$$

The solution belongs to $C^k\big([0,\infty) \times \mathbb{R}^3\big)$.

Proof We introduce $y = x + t\alpha$, $\alpha = (\alpha_1, \alpha_2, \alpha_3)$, where α is a unit vector in the direction $y - x$. Using $d\sigma_t = t^2 d\sigma_1$ gives

$$u_p(t, x) = \frac{t}{4\pi} \int_{S_1(0)} p(x + t\alpha)\, d\sigma_1.$$

Thus, we get $\lim_{t \to +0} u_p(t, x) = 0$. Differentiating with respect to t implies together with the supposed regularity for p the relation

$$\partial_t u_p(t, x) = \frac{1}{4\pi} \int_{S_1(0)} p(x + t\alpha)\, d\sigma_1 + \frac{t}{4\pi} \int_{S_1(0)} \nabla p(x + t\alpha) \cdot \alpha\, d\sigma_1.$$

The regularity of p allows for changing integration and passage to the limit $\lim_{t \to +0}$ in the last equation. Hence, it follows $\lim_{t \to +0} \partial_t u_p(t, x) = p(x)$. It remains to show that u_p solves the wave equation, that is, $(\partial_t^2 - \Delta)u_p(t, x) = 0$. We use the representation

$$\partial_t u_p(t, x) = \frac{1}{t} u_p(t, x) + \frac{1}{4\pi t} \int_{S_t(x)} \nabla p(y) \cdot \alpha\, d\sigma_y.$$

Now, using the fact that α is the exterior unit normal vector to $S_t(x)$ and applying the *Divergence Theorem* we obtain

$$\partial_t u_p(t, x) = \frac{1}{t} u_p(t, x) + \frac{1}{4\pi t} \int_{B(x,t)} \Delta p(y)\, dy.$$

Here $B(x, t) \subset \mathbb{R}^3$ denotes the ball around the center x with radius t. Differentiation with respect to t yields

$$\partial_t^2 u_p(t, x) = -\frac{1}{t^2} u_p(t, x) + \frac{1}{t} \partial_t u_p(t, x)$$
$$-\frac{1}{4\pi t^2} \int_{B(x,t)} \Delta p(y)\, dy + \frac{1}{4\pi t} \frac{\partial}{\partial t} \int_{B(x,t)} \Delta p(y)\, dy.$$

Setting into this equation the above relation for $\partial_t u_p$ we obtain

$$\partial_t^2 u_p(t, x) = \frac{1}{4\pi t} \partial_t \int_{B(x,t)} \Delta p(y)\, dy.$$

Taking account of

$$\partial_t \int_{B(x,t)} \Delta p(y)\, dy = \partial_t \int_0^t \int_{S_r(x)} \Delta p(x + r\alpha)\, d\sigma_r\, dr = \int_{S_t(x)} \Delta p(x + t\alpha)\, d\sigma_t$$

we derive

$$\partial_t^2 \, u_p(t, x) = \frac{1}{4\pi t} \int_{S_t(x)} \Delta p(y) \, d\sigma_t = \frac{t}{4\pi} \int_{S_1(0)} \Delta p(x + t\alpha) \, d\sigma_1.$$

Finally, the relation

$$\Delta u_p(t, x) = \frac{t}{4\pi} \int_{S_1(0)} \Delta p(x + t\alpha) \, d\sigma_1$$

completes the verification that $u_p = u_p(t, x)$ is really the solution of our Cauchy problem.

Corollary 10.3.2 *The Cauchy problem*

$$u_{tt} - \Delta u = 0, \quad u(0, x) = \varphi(x), \quad u_t(0, x) = \psi(x)$$

with data $\varphi \in C^k(\mathbb{R}^3)$ and $\psi \in C^{k-1}(\mathbb{R}^3)$, $k \geq 3$, has a classical solution $u \in C^{k-1}([0, \infty) \times \mathbb{R}^3)$. This solution has the representation

$$u(t, x) = \frac{1}{4\pi t} \int_{S_t(x)} \psi(y) \, d\sigma_y + \partial_t \left(\frac{1}{4\pi t} \int_{S_t(x)} \varphi(y) \, d\sigma_y \right).$$

Do we see differences between the statements of Theorem 10.1.1 and Corollary 10.3.2?

We have no uniqueness result in the statement of Corollary 10.3.2. Moreover, the solution from Corollary 10.3.2 belongs only to the function space $C^{k-1}([0, \infty) \times \mathbb{R}^3)$. Thus *we lose one order of regularity* in comparison with the data $\varphi \in C^k(\mathbb{R}^3)$.

But, using Kirchhoff's representation formula from Corollary 10.3.2 we may conclude the following qualitative properties of solutions:

1. finite speed of propagation of perturbations
2. existence of a domain of dependence
3. Huygens' principle is valid, which means, there exists not only a forward but also a backward wave front. *This is one of the main reasons that human beings are able to hear.*

10.4   Kirchhoff's Representation in $\mathbb{R}^2$

One can derive Kirchhoff's representation formula in the 2d case from the Kirchhoff's formula in the 3d case. Therefore, we shall apply the *method of descent*.

10.4.1   Method of Descent

Let us explain the method of descent. First, we turn to the 2d case. The data $\varphi(x) = \varphi(x_1, x_2)$ and $\psi(x) = \psi(x_1, x_2)$ are considered as data in $\mathbb{R}^3$, which are independent of x_3. The application of Theorem 10.3.1 gives

$$u_p(t, x) = \frac{1}{4\pi t} \int_{S_t(x_1, x_2, 0)} p(y) \, d\sigma_t(y) = \frac{1}{2\pi} \int_{B(x_1, x_2, t)} \frac{p(y)}{\sqrt{t^2 - |y - x|^2}} \, dy.$$

To get the last relation we transfer the surface integral to an integral over the set

$$\{y = (y_1, y_2) \in \mathbb{R}^2 : |y - x|^2 \leq t^2\}, \quad x = (x_1, x_2).$$

Therefore, we choose the parameter representation of the upper or lower half sphere in the following way:

$$\Phi_1(y_1, y_2) = y_1, \quad \Phi_2(y_1, y_2) = y_2, \quad y_3 = \Phi_3(y_1, y_2) = \pm(t^2 - |y - x|^2)^{1/2}.$$

For transferring the surface element we calculate the Gauß fundamentals E, G, F and obtain

$$\sqrt{EG - F^2} = \sqrt{1 + (\partial_{y_1} y_3)^2 + (\partial_{y_2} y_3)^2},$$

and, finally,

$$d\sigma_t(y) = \frac{2t \, dy}{\sqrt{t^2 - |y - x|^2}}.$$

Using Corollary 10.3.2 we have the following result.

Theorem 10.4.1 *Under the assumptions $\varphi \in C^3(\mathbb{R}^2)$ and $\psi \in C^2(\mathbb{R}^2)$ there exists a classical solution $u = u(t, x) \in C^2([0, \infty) \times \mathbb{R}^2)$ to the Cauchy problem*

$$u_{tt} - \Delta u = 0, \quad u(0, x) = \varphi(x), \quad u_t(0, x) = \psi(x), \quad x \in \mathbb{R}^2,$$

having the representation

$$u(t, x) = \frac{1}{2\pi} \int_{B(x, t)} \frac{\psi(y)}{\sqrt{t^2 - |y - x|^2}} \, dy + \frac{\partial}{\partial t} \left(\frac{1}{2\pi} \int_{B(x, t)} \frac{\varphi(y)}{\sqrt{t^2 - |y - x|^2}} \, dy \right).$$

Kirchhoff's representation formula from Theorem 10.4.1 yields the following qualitative properties of solutions:

1. finite speed of propagation of perturbations
2. existence of a domain of dependence

3. existence of a forward wave front, but we do not have any backward wave front
4. we have a loss of regularity of order 1 of the solution in comparison with the data φ.

Applying the method of descent again one can even derive d'Alembert's representation formula in the 1d case (see Sect. 10.1) by means of the representation formula for the 2d case. Therefore, we write

$$
\frac{1}{2\pi} \int_{B(x_1,0,t)} \frac{p(y_1)}{\sqrt{t^2 - (y_1 - x_1)^2 - y_2^2}} \, d(y_1, y_2)
$$

$$
= \frac{1}{2\pi} \int_{x_1-t}^{x_1+t} p(y_1) \left(\int_{-\sqrt{t^2-(y_1-x_1)^2}}^{\sqrt{t^2-(y_1-x_1)^2}} \frac{1}{\sqrt{t^2 - (y_1 - x_1)^2 - y_2^2}} \, dy_2 \right) dy_1
$$

$$
= \frac{1}{2} \int_{x_1-t}^{x_1+t} p(y_1) \, dy_1
$$

by using

$$
\int_{-a}^{a} \frac{1}{\sqrt{a^2 - y_2^2}} \, dy_2 = \pi \quad \text{for all } a > 0.
$$

This leads, together with Corollary 10.3.1, to d'Alembert's representation formula

$$
u(t, x) = \frac{1}{2} \left(\varphi(x - t) + \varphi(x + t) \right) + \frac{1}{2} \int_{x-t}^{x+t} \psi(r) dr.
$$

10.5 Representation Formulas in Higher Dimensions

In this section we shall expound explicit representations of classical solutions to the Cauchy problem for the wave equation in higher dimensions. The formulas are taken from [214]. Moreover, we draw conclusions for qualitative properties of classical solutions.

10.5.1 Odd Space Dimension

Let us consider the Cauchy problem

$$
u_{tt} - \Delta u = 0, \quad u(0, x) = \varphi(x), \quad u_t(0, x) = \psi(x), \quad x \in \mathbb{R}^{2n+1}, \quad n \geq 1.
$$

Theorem 10.5.1 *To given data* $\varphi \in C^k(\mathbb{R}^{2n+1})$ *and* $\psi \in C^{k-1}(\mathbb{R}^{2n+1})$ *with* $k \geq n+2$, $n \geq 1$, *there exists a classical solution* $u \in C^{k-n}([0,\infty) \times \mathbb{R}^{2n+1})$. *The solution has the representation*

$$u(t,x) = \sum_{j=0}^{n-1} \left((j+1)a_j t^j \partial_t^j + a_j t^{j+1} \partial_t^{j+1} \right) \frac{1}{\sigma_{2n+1}} \int_{|y|=1} \varphi(x+ty)\, d\sigma_y$$

$$+ \sum_{j=0}^{n-1} a_j t^{j+1} \partial_t^j \frac{1}{\sigma_{2n+1}} \int_{|y|=1} \psi(x+ty)\, d\sigma_y,$$

where $a_j = a_j(n)$ *are constants with* $a_{n-1} \neq 0$ *and where* σ_{2n+1} *denotes the measure of the unit sphere in* $\mathbb{R}^{2n+1}$.
What do we conclude from this representation of solution?
We may immediately conclude the following properties of solutions:

- The loss of regularity of solutions in comparison with the data φ is n.
- The solutions have the properties of finite speed of propagation of perturbations, of existence of a domain of dependence and of existence of a forward and of a backward wave front as well. Consequently, Huygens' principle is valid.

Example 10.5.1 If $n = 1$, then $a_0 = 1$, and we conclude Kirchhoff's representation formula in the 3d case:

$$u(t,x) = (1 + t\,\partial_t) \frac{1}{4\pi} \int_{|y|=1} \varphi(x+ty)\, d\sigma_y + \frac{t}{4\pi} \int_{|y|=1} \psi(x+ty)\, d\sigma_y.$$

10.5.2   Even Space Dimension

Let us consider the Cauchy problem

$$u_{tt} - \Delta u = 0, \quad u(0,x) = \varphi(x), \quad u_t(0,x) = \psi(x), \quad x \in \mathbb{R}^{2n}, \quad n \geq 1.$$

Theorem 10.5.2 *To given data* $\varphi \in C^k(\mathbb{R}^{2n})$ *and* $\psi \in C^{k-1}(\mathbb{R}^{2n})$ *with* $k \geq n+2$, $n \geq 1$, *there exists a classical solution* $u \in C^{k-n}([0,\infty) \times \mathbb{R}^{2n})$ *having the representation*

$$u(t,x) = \sum_{j=0}^{n-1} \left((j+1)b_j t^j \partial_t^j + b_j t^{j+1} \partial_t^{j+1} \right) G_\varphi(t,x)$$

$$+ \sum_{j=0}^{n-1} b_j t^{j+1} \partial_t^j G_\psi(t,x),$$

where $b_j = b_j(n)$ are constants with $b_{n-1} \neq 0$, σ_{2n} denotes the measure of the unit sphere in $\mathbb{R}^{2n}$ and

$$G_\phi(t, x) = \frac{2\Gamma(\frac{2n+1}{2})}{\sqrt{\pi}\Gamma(n)t^{2n-1}} \int_0^t \frac{r^{2n-1}}{\sigma_{2n}(t^2 - r^2)^{1/2}} \int_{|y|=1} \phi(x + ry)\, d\sigma_y\, dr.$$

What do we conclude from this representation of solution?
We may immediately conclude the following properties of solutions:

- The loss of regularity of solutions in comparison with the data φ is n.
- The properties of solutions as finite speed of propagation of perturbations, of existence of a domain of dependence and of existence of a forward wave front are fulfilled.
- But we do not have any backward wave front. Thus, Huygens' principle is not valid.

Example 10.5.2 For $n = 1$ we get Kirchhoff's representation formula in the 2d case:

$$u(t, x) = (b_0 + b_0 t\, \partial_t) \frac{2\Gamma(\frac{3}{2})}{\sqrt{\pi}\Gamma(1)t} \int_0^t \frac{r}{\sigma_2(t^2 - r^2)^{1/2}} \int_{|y|=1} \varphi(x + ry)\, d\sigma_y\, dr$$

$$+ b_0 \frac{2\Gamma(\frac{3}{2})}{\sqrt{\pi}\Gamma(1)} \int_0^t \frac{r}{\sigma_2(t^2 - r^2)^{1/2}} \int_{|y|=1} \psi(x + ry)\, d\sigma_y\, dr$$

$$= b_0 \frac{2\Gamma(\frac{3}{2})}{\sqrt{\pi}\Gamma(1)} \left(\partial_t \int_{B(t,x)} \frac{\varphi(y)}{\sqrt{t^2 - |y - x|^2}}\, dy + \int_{B(t,x)} \frac{\psi(y)}{\sqrt{t^2 - |y - x|^2}}\, dy \right)$$

after a suitable choice for the constant $b_0 \neq 0$.

10.6 Propagation of Singularities

In this section we will discuss an important property for solutions to wave (hyperbolic) equations, the so-called *propagation of singularities along characteristics*. Let us recall d'Alembert's representation of solution to the 1d free wave equation (see Sect. 10.1)

$$u(t, x) = \frac{1}{2}\big(\varphi(x - t) + \varphi(x + t)\big) + \frac{1}{2} \int_{x-t}^{x+t} \psi(r)\, dr.$$

We assume that the data φ has a *jump* in $x = x_0$. Then, there still will be a jump in the solution and the jump propagates along the characteristics $x - x_0 = t$ and $x - x_0 = -t$. If there is a jump in ψ, then we do not feel this jump in the solution, but in its first Sobolev derivatives $\partial_x u$ and $\partial_t u$. This observation can be generalized to higher-dimensional cases.

Thus, singularities in the data propagate along the lateral surface of the forward characteristic cone for $t > 0$ (or along the lateral surface of the backward characteristic cone for $t < 0$) (see Sect. 11.1).

If we have obstacles, for example, domains with boundaries (so boundary conditions become important), then singularities from the data for solutions of the wave equation will be reflected on the boundary. There exist special models where the study of propagation of singularities for solutions of mixed problems, and thus the property of *reflection of singularities* on the boundary, can be explained from the study of the propagation of singularities for solutions to Cauchy problems. Let us introduce two such models.

Model 1:   Vibrations of an infinite string with fixed end

The mathematical model is

$$u_{tt} - u_{xx} = 0, \quad u(0, x) = \varphi(x), \quad u_t(0, x) = \psi(x) \text{ for } (t, x) \in (0, \infty) \times (0, \infty),$$

$$\text{and } u(t, 0) = 0 \text{ for } t \geq 0.$$

The compatibility conditions imply $\varphi(0) = \psi(0) = 0$. Let $\tilde{\varphi}$ and $\tilde{\psi}$ be *odd continuations* of φ and ψ onto the negative part of the real axis. Instead we consider the Cauchy problem

$$v_{tt} - v_{xx} = 0, \quad v(0, x) = \tilde{\varphi}(x), \quad v_t(0, x) = \tilde{\psi}(x).$$

From d'Alembert's formula we get the solution

$$v(t, x) = \frac{1}{2} \left(\tilde{\varphi}(x - t) + \tilde{\varphi}(x + t) \right) + \frac{1}{2} \int_{x-t}^{x+t} \tilde{\psi}(s) \, ds.$$

Thus the boundary condition $v(t, 0) = 0$ is satisfied. The solution of our given mixed problem is

$$u(t, x) = \frac{1}{2} \left(\tilde{\varphi}(x - t) + \tilde{\varphi}(x + t) \right) + \frac{1}{2} \int_{x-t}^{x+t} \tilde{\psi}(s) \, ds$$

for $(t, x) \in (0, \infty) \times (0, \infty)$. Now let us suppose that φ has a jump in $x_0 > 0$. Then $\tilde{\varphi}$ has a jump in x_0 and $-x_0$. Thus, from the propagation picture we know that the jump propagates along the characteristics $x - x_0 = t$ (propagation of the jump into the direction of larger x), $x + x_0 = -t$ (virtual picture, because the string is only modeled for $x \geq 0$), $x - x_0 = -t$ (only real for $t \in [0, x_0]$) and $x + x_0 = t$ (only real for $t \in [x_0, \infty]$). The last two observations explain the *propagation and reflection of singularities* issuing from $x = x_0$.

Model 2:   *Vibrations of an infinite string with a free end*

The mathematical model is

$$u_{tt} - u_{xx} = 0, \quad u(0,x) = \varphi(x), \quad u_t(0,x) = \psi(x) \ \text{ for } \ (t,x) \in (0,\infty) \times (0,\infty),$$

$$\text{and } u_x(t,0) = 0 \ \text{ for } \ t \geq 0.$$

The compatibility conditions imply $\varphi'(0) = \psi'(0) = 0$. Let $\tilde{\varphi}$ and $\tilde{\psi}$ be *even continuations* of φ and ψ onto the negative part of the real axis. Instead we consider the Cauchy problem

$$v_{tt} - v_{xx} = 0, \quad v(0,x) = \tilde{\varphi}(x), \quad v_t(0,x) = \tilde{\psi}(x).$$

We then get the solution

$$u(t,x) = \frac{1}{2}\big(\tilde{\varphi}(x - t) + \tilde{\varphi}(x + t)\big) + \frac{1}{2}\int_{x-t}^{x+t} \tilde{\psi}(s)ds.$$

We verify that the boundary condition $u_x(t,0) = 0$ is satisfied for $t \geq 0$. We explain the propagation and reflection of singularities as for Model 1.

10.6.1   More About Propagation of Singularities

The above introduced propagation of singularities along characteristics is one possibility to describe this phenomenon. Mathematicians are often interested in more precise descriptions, so-called micro-local descriptions. Let us briefly sketch the main ideas of this concept.

This is based on the notion of *wave front set*. Let x_0 be a point in $\mathbb{R}^n$ and let $u = u(x)$ be a function which is C^∞ in a neighborhood $U_0(x_0)$ of x_0. Further, let us choose a *cut-off* function $\chi \in C_0^\infty(U_0(x_0))$ which is identical to 1 in a smaller neighborhood $U_1(x_0)$, $\overline{U_1(x_0)} \subset U_0(x_0)$, of x_0. If we form the Fourier transform of χu, then this is a Schwartz function (see Definition 24.1). So, we have, due to Theorem 24.1.1, the following estimate:

to each N there exists a constant C_N such that

$$|F(\chi u)(\xi)| \leq C_N \langle \xi \rangle^{-N} \ \text{ for all } \ \xi \in \mathbb{R}^n.$$

If a function is not smooth, then the last estimate fails. This is one way to motivate the following definitions for the notion *wave front set*. We use the notation $T^*G \setminus \{0\} := G \times (\mathbb{R}^n \setminus \{0\})$. In the following definition we use the notion of distributions. Basics of distribution theory are introduced in Sect. 24.4.

Definition 10.1

1. Let us consider a distribution $u \in D'(G)$. A point $(x_0, \xi_0) \in T^*G \setminus \{0\}$ does not belong to the *wave front set* WF u if there exist two functions ψ and χ with the following properties:

   - $\chi \in C_0^\infty(U_0(x_0))$, $\chi \equiv 1$ on a neighborhood $U_1(x_0)$, $\overline{U_1(x_0)} \subset U_0(x_0)$, of x_0
   - $\psi \in C^\infty(\mathbb{R}^n)$, $\psi \equiv 1$ on a conical neighborhood $V_1(\xi_0)$ of ξ_0, $\psi \equiv 0$ outside a larger conical neighborhood $V_0(\xi_0)$, $\overline{V_1(\xi_0)} \subset V_0(\xi_0)$, of ξ_0
   - $\psi(D)(\chi u) \in C^\infty(\mathbb{R}^n)$.

2. Let us consider a distribution $u \in D'(G)$. A point $(x_0, \xi_0) \in T^*G \setminus \{0\}$ does not belong to the *wave front set* WF u if there exists a function χ with the following properties:

   - $\chi \in C_0^\infty(U_0(x_0))$, $\chi \equiv 1$ on a neighborhood $U_1(x_0)$, $\overline{U_1(x_0)} \subset U_0(x_0)$, of x_0
   - to each N there exists a constant C_N such that $|F(\chi u)(\xi)| \leq C_N \langle \xi \rangle^{-N}$ in a conical neighborhood $V_0(\xi_0)$ of ξ_0.

The term $\psi(D)$ is a pseudodifferential operator and $\psi(D)(\chi u)$ is defined as follows:

$$\psi(D)(\chi u)(x) = F^{-1}\big(\psi(\xi)F(\chi u)(\xi)\big)(x).$$

Example 10.6.1 Let δ_{x_0} be the Dirac δ-distribution in a point $x_0 \in \mathbb{R}^n$ (cf. with Example 24.4.2). Then WF $\delta_{x_0} = \{x_0\} \times (\mathbb{R}^n \setminus \{0\})$. If $x \neq x_0$, then $\delta_{x_0} \equiv 0$ (in the sense of distributions) in a small neighborhood $U(x)$ of x, so any point $(x, \xi) \in T^*(\{x\})$ does not belong to the wave front set WF δ_{x_0}. By Example 24.1.2 we have

$$F(\delta_{x_0}) = \frac{e^{ix_0 \cdot \xi}}{(2\pi)^{\frac{n}{2}}}.$$

From Definition 10.1 it follows that all points (x_0, ξ), $\xi \neq 0$, belong to the wave front set WF δ_{x_0}.

To apply the concept of wave front set to solutions of wave equations we introduce an important result that explains the structure of the wave front set for *Fourier integral operators*.

Theorem 10.6.1 *Let A be the Fourier integral operator*

$$Au(x) := \int_{\mathbb{R}^n} e^{i\phi(x,\xi)} a(x, \xi) F(u)(\xi) d\xi,$$

where the amplitude $a = a(x, \xi)$ is from $S^m(G)$ and $u \in E'(G)$ is a distribution with compact support in G. Let us suppose that the set

$$\{(y, \xi) \in WF u : \text{ there exists } x \in G : y = \nabla_\xi \phi(x, \xi), \nabla_x \phi(x, \xi) = 0\}$$

is empty. Then it holds that

$$WF Au \subset \{(x, \xi) : \text{ there exists } \eta \in \mathbb{R}^n \setminus \{0\} \text{ such that}$$
$$\xi = \nabla_x \phi(x, \eta), (\nabla_\eta \phi(x, \eta), \eta) \in WF u\}.$$

This tool in hand, we are able to consider the *propagation of microlocal singularities* for the solutions to the Cauchy problem

$$u_{tt} - \Delta u = 0, \quad u(0, x) = \varphi(x), \quad u_t(0, x) = \psi(x).$$

In Sect. 14.1 we shall explain how the solution to this Cauchy problem is represented using Fourier multipliers.

Theorem 10.6.2 *The wave front set of $u(t, \cdot)$ is described for $t \neq 0$ as follows:*

$$WF u(t, \cdot) \subset \{(x \pm t\xi|\xi|^{-1}, \xi) : (x, \xi) \in WF \varphi \cup WF \psi\}.$$

This result explains that microlocal singularities of the solution are contained on the lateral surface of the characteristic cone with apex in the singularities of the data, that is, in those points having no small neighborhood where the data are C^∞. More precisely, if $(x_0, \xi_0) \in WF \varphi \cup WF \psi$, that is, in the direction ξ_0 the estimate from Definition 10.1 (second part) does not hold, then the wave front $WF u(t, \cdot)$ is contained in the set (with respect to x) of points on the lateral surface of the characteristic cone with apex in x_0 in the direction $\omega_0 := \xi_0|\xi_0|^{-1}$. With respect to ξ the ξ_0 direction is a bad one.

Proof From the representation of the solution to the wave equation by Fourier multipliers (see Sect. 14.1) we know we have to take account of phase functions $\phi_\pm(t, x, \xi) := x \cdot \xi \pm t|\xi|$. These phase functions satisfy the assumption from Theorem 10.6.1 because $\nabla_x \phi_\pm(t, x, \eta) = \eta = 0$ is excluded in the Definition 10.1 for $\xi = \eta$. Hence, with $\xi = \eta$ we get from $(x_0, \xi_0) \in WF \varphi \cup WF \psi$ immediately

$$\nabla_\xi \phi_\pm(t, x_0, \xi_0) = x_0 \pm t\xi_0|\xi_0|^{-1}.$$

This proves the statement.

10.7 Concluding Remarks

In the following sections we sketch how hyperbolic potential theory can be applied to treat mixed problems for the wave equation.

10.7.1 Derivation of Wave Layer Potentials

First we need fundamental solutions or elementary potentials to the operator $\partial_t^2 - \Delta$, these are distributional solutions of $(\partial_t^2 - \Delta)H_n = \delta_0$ in $\mathbb{R}^{n+1}$ (see Definition 24.30). The results depend on the dimension n as was shown for the representations of solutions and qualitative properties of solutions in Sects. 10.5.1 and 10.5.2.

For this reason, let us restrict in the following to the case $n = 3$. Our goal is to sketch the main ideas of hyperbolic potential theory (see [70]).

How can we obtain fundamental solutions for $n = 3$?

Here we recall the considerations in Sect. 10.3.1. There, we derived for small positive ε the representation of solutions $u_\varepsilon = u_\varepsilon(t, x)$ to the Cauchy problem

$$u_{tt} - \Delta u = 0, \quad u(0, x) = 0, \quad u_t(0, x) = \phi_\varepsilon(x)$$

in the form

$$u_\varepsilon(t, x) = \frac{4\pi\varepsilon}{4\pi r}\big(\phi_\varepsilon(t - r) - \phi_\varepsilon(t + r)\big), \quad r = |x|.$$

Let ε tend to $+0$. Taking account of

$$\lim_{\varepsilon \to +0} 4\pi\varepsilon\phi_\varepsilon(t \mp r) = \delta_{t \mp r}$$

we arrive at the fundamental solution

$$H_3(t, x) = \frac{1}{4\pi r}\delta_{t-r}, \quad r = |x|.$$

Here we take account of $\delta_{t+r} = 0$ because of $t + r > 0$. In this way we obtain the distributional solution to the Cauchy problem

$$u_{tt} - \Delta u = 0, \quad u(0, x) = 0, \quad u_t(0, x) = \delta_0.$$

As in Sect. 9.4.2, we need the distributional solution to the Cauchy problem

$$u_{tt} - \Delta u = 0, \quad u(0, x) = 0, \quad u_t(0, x) = \delta_y,$$

where $y \in \mathbb{R}^3$ is a parameter. The same approach implies

$$H_3(t, x - y) = \frac{1}{4\pi r}\delta_{t-r}, \quad r = |x - y|.$$

Now we are able to introduce wave potentials.

The *wave single-layer potential (with the density* $\mu = \mu(\tau, y)$) is formally defined by

$$\int_{S_t} H_3(t - \tau, x - y)\mu(\tau, y)\, d(\tau, \sigma_y) = \int_0^t \int_{\partial G} H_3(t - \tau, x - y)\mu(\tau, y)\, d\sigma_y\, d\tau.$$

The *wave double-layer potential (with the density* $\mu = \mu(\tau, y)$) is formally defined by

$$\int_{S_t} \partial_{\mathbf{n}_y} H_3(t - \tau, x - y)\mu(\tau, y)\, d(\tau, \sigma_y)$$

$$= \int_0^t \int_{\partial G} \partial_{\mathbf{n}_y} H_3(t - \tau, x - y)\mu(\tau, y)\, d\sigma_y\, d\tau.$$

Here S_t denotes the lateral surface of the space-time cylinder Z_t. Both wave layer potentials involve H_3 delta distributions. Consequently, we have to clarify the meaning of both potentials.

First we assume that the density is zero for $t < 0$. Proceeding formally, we may conclude after the change of variables $\eta := t - \tau - r$ for the wave single-layer potential

$$\int_0^t \int_{\partial G} H_3(t - \tau, x - y)\mu(\tau, y)\, d\sigma_y\, d\tau = \int_{\partial G} \int_0^t \frac{1}{4\pi r}\delta_{t-\tau-r}\mu(\tau, y)\, d\tau\, d\sigma_y$$

$$= \int_{\partial G} \int_{-\infty}^t \frac{1}{4\pi r}\delta_\eta \mu(t - r - \eta, y)\, d\eta\, d\sigma_y = \int_{\partial G} \frac{1}{4\pi r}\mu(t - r, y)\, d\sigma_y.$$

Similarly, we get for the wave double-layer potential

$$\int_0^t \int_{\partial G} \partial_{\mathbf{n}_y} H_3(t - \tau, x - y)\mu(\tau, y)\, d\sigma_y\, d\tau$$

$$= \int_{\partial G} \int_0^t \partial_{\mathbf{n}_y} \frac{1}{4\pi r}\delta_{t-\tau-r}\mu(\tau, y)\, d\tau\, d\sigma_y$$

$$= -\mathrm{div}_x \int_{\partial G} \int_{-\infty}^t \frac{1}{4\pi r}\delta_{t-\tau-r}\mu(\tau, y)\mathbf{n}_y\, d\tau\, d\sigma_y$$

$$= -\mathrm{div}_x \int_{\partial G} \int_{-\infty}^t \frac{1}{4\pi r}\delta_\eta \mu(t - r - \eta, y)\mathbf{n}_y\, d\eta\, d\sigma_y$$

$$= -\mathrm{div}_x \int_{\partial G} \frac{1}{4\pi r}\mu(t - r, y)\mathbf{n}_y\, d\sigma_y.$$

Summarizing, we derived the exact definitions for the wave single-layer potential (with density $\mu = \mu(\tau, y)$) in the form

$$\int_{\partial G} \frac{1}{4\pi r} \mu(t - r, y) \, d\sigma_y$$

and for the wave double-layer potential (with density $\mu = \mu(\tau, y)$) in the form

$$-\mathrm{div}_x \int_{\partial G} \frac{1}{4\pi r} \mu(t - r, y) \mathbf{n}_y \, d\sigma_y.$$

Here $\mathbf{n}_y$ describes the outer unit normal to ∂G.

10.7.2   Basic Mixed Problems for the Wave Equation

Interior Domain and Dirichlet Boundary Condition
Let $G \subset \mathbb{R}^n$ be an interior domain. We suppose that the boundary ∂G is sufficiently smooth. Find a solution u of the wave equation in $(0, \infty) \times G$ with prescribed Dirichlet data $g = g(t, x)$ for u on $(0, \infty) \times \partial G$ and prescribed initial data $\varphi = \varphi(x)$, $\psi = \psi(x)$ for u, u_t respectively, on $\{t = 0\} \times G$. The compatibility conditions between boundary and initial data are satisfied on $\{t = 0\} \times \partial G$.

Mathematical Formulation

$$u_{tt} - \Delta u = h(t, x) \quad \text{in} \quad (0, \infty) \times G, \quad u(t, x) = g(t, x) \quad \text{on} \quad (0, \infty) \times \partial G,$$

$$u(0, x) = \varphi(x), \quad u_t(0, x) = \psi(x) \quad \text{for } x \in G,$$

compatibility conditions: $g(0, x) = \varphi(x)$, $g_t(0, x) = \psi(x)$ on $\{t = 0\} \times \partial G$.

Interior Domain and Neumann Boundary Condition
Let $G \subset \mathbb{R}^n$ be an interior domain. We suppose that the boundary ∂G is sufficiently smooth. Find a solution u of the wave equation in $(0, \infty) \times G$ with prescribed Neumann data $g = g(t, x)$ for u on $(0, \infty) \times \partial G$ and prescribed initial data $\varphi = \varphi(x)$, $\psi = \psi(x)$ for u, u_t respectively, on $\{t = 0\} \times G$. The compatibility conditions between boundary and initial data are satisfied on $\{t = 0\} \times \partial G$.

Mathematical Formulation

$$u_{tt} - \Delta u = h(t, x) \quad \text{in} \quad (0, \infty) \times G, \qquad \partial_{\mathbf{n}} u(t, x) = g(t, x) \quad \text{on} \quad (0, \infty) \times \partial G,$$

$$u(0, x) = \varphi(x), \quad u_t(0, x) = \psi(x) \quad \text{for } x \in G,$$

compatibility conditions:

$$g(0, x) = \partial_{\mathbf{n}} \varphi(x), \quad g_t(0, x) = \partial_{\mathbf{n}} \psi(x) \quad \text{on} \quad \{t = 0\} \times \partial G.$$

Exterior Domain and Dirichlet Boundary Condition
Let $G \subset \mathbb{R}^n$ be an exterior domain. We suppose that the boundary ∂G is sufficiently smooth. Find a solution u of the wave equation in $(0, \infty) \times G$ with prescribed Dirichlet data $g = g(t, x)$ for u on $(0, \infty) \times \partial G$ and prescribed initial data $\varphi = \varphi(x)$, $\psi = \psi(x)$ for u, u_t respectively, on $\{t = 0\} \times G$. The compatibility conditions between boundary and initial data are satisfied on $\{t = 0\} \times \partial G$.

Mathematical Formulation

$$u_{tt} - \Delta u = h(t, x) \quad \text{in} \quad (0, \infty) \times G, \qquad u(t, x) = g(t, x) \quad \text{on} \quad (0, \infty) \times \partial G,$$

$$u(0, x) = \varphi(x), \quad u_t(0, x) = \psi(x) \quad \text{for } x \in G,$$

$$\text{compatibility conditions: } g(0, x) = \varphi(x), \quad g_t(0, x) = \psi(x) \quad \text{on } \{t = 0\} \times \partial G.$$

Exterior Domain and Neumann Boundary Condition
Let $G \subset \mathbb{R}^n$ be an exterior domain. We suppose that the boundary ∂G is sufficiently smooth. Find a solution u of the wave equation in $(0, \infty) \times G$ with prescribed Neumann data $g = g(t, x)$ for u on $(0, \infty) \times \partial G$ and prescribed initial data $\varphi = \varphi(x)$, $\psi = \psi(x)$ for u, u_t respectively, on $\{t = 0\} \times G$. The compatibility conditions between boundary and initial data are satisfied on $\{t = 0\} \times \partial G$.

Mathematical Formulation

$$u_{tt} - \Delta u = h(t, x) \quad \text{in} \quad (0, \infty) \times G, \qquad \partial_{\mathbf{n}} u(t, x) = g(t, x) \quad \text{on} \quad (0, \infty) \times \partial G,$$

$$u(0, x) = \varphi(x), \quad u_t(0, x) = \psi(x) \quad \text{for } x \in G,$$

$$\text{compatibility conditions:}$$

$$g(0, x) = \partial_{\mathbf{n}} \varphi(x), \quad g_t(0, x) = \partial_{\mathbf{n}} \psi(x) \quad \text{on } \{t = 0\} \times \partial G.$$

10.7.3   How to Use Wave Potentials in Representations of Solutions?

We introduced wave single- and double-layer potentials in Sect. 10.7.1.

Mixed Problems with Dirichlet Boundary Condition
In order to treat mixed problems with Dirichlet boundary condition we use wave double-layer potentials with unknown density. As in Sect. 8.4.2.1, jump relations for wave double-layer potentials after crossing ∂G are of interest. After some

calculations (see [70]) one can show for all $x_0 \in \partial G$ the relation

$$
\lim_{x \to x_0 \in \partial G} -\mathrm{div}_x \int_{\partial G} \frac{1}{4\pi r} \mu(t - r, y) \mathbf{n}_y \, d\sigma_y
$$

$$
= \lim_{x \to x_0 \in \partial G} \int_{\partial G} \frac{1}{4\pi} \left(\partial_{\mathbf{n}_y} \frac{1}{r} \right) \left(\mu(t - r, y) + r\mu_t(t - r, y) \right) d\sigma_y
$$

$$
= \pm \frac{1}{2} \mu(t, x_0) + \int_{\partial G} \frac{1}{4\pi} \left(\partial_{\mathbf{n}_y} \frac{1}{|x_0 - y|} \right)
$$

$$
\times \left(\mu(t - |x_0 - y|, y) + |x_0 - y| \mu_t(t - |x_0 - y|, y) \right) d\sigma_y.
$$

The negative sign appears in the case of the interior, the positive sign in the case of the exterior domain.

Mixed Problems with Neumann Boundary Condition

In order to treat mixed problems with Neumann boundary condition we use wave single-layer potentials with unknown density. As in Sect. 8.4.2.2, jump relations for wave single-layer potentials after crossing ∂G are of interest. After some calculations (see [70] or [135]) one can show for all $x_0 \in \partial G$ the relation

$$
\lim_{x \to x_0 \in \partial G} \partial_{\mathbf{n}_x} \int_{\partial G} \frac{1}{4\pi r} \mu(t - r, y) \, d\sigma_y
$$

$$
= \pm \frac{1}{2} \mu(t, x_0) + \int_{\partial G} \partial_{\mathbf{n}_x} \left(\frac{1}{4\pi r} \mu(t - r, y) \right) (t, x = x_0) \, d\sigma_y.
$$

The positive sign appears in the case of the interior, the negative sign in the case of the exterior domain.

10.7.4   Integral Equation for the Interior Dirichlet Problem for the Wave Equation

In Sect. 8.4.3 we showed how the Fredholm theory can be applied to treat the Fredholm integral equations of potential theory. In Sect. 9.4.3 we explained that successive approximation is applicable to treat Volterra integral equations of mixed problems for the heat equation (see the research project in Sect. 23.2).

Here we restrict ourselves to sketching how to treat the integral equation which appears in the study of the interior Dirichlet problem (see [70]).

Consider the interior Dirichlet problem for the wave equation. Setting the ansatz of solutions in the form of a wave double-layer potential in the Dirichlet condition and using the above jump relation we obtain for $(t, x) \in (0, \infty) \times \partial G$ the integral equation

$$
\mu(t, x) = \frac{1}{2\pi} \int_{\partial G} (1 + r\partial_t) \mu(t - r, y) \left(\partial_{\mathbf{n}_y} \frac{1}{r} \right) d\sigma_y - 2g(t, x).
$$

The general Fredholm theory (compare with Sect. 8.4.3) does not apply to this integral equation because we have μ on the left-hand side, but μ_t on the right-hand side. Starting successive approximations

$$\mu_0(t, x) = g(t, x),$$

$$\mu_{k+1}(t, x) = \frac{1}{2\pi} \int_{\partial G} (1 + r\partial_t)\mu_k(t - r, y)\left(\partial_{n_y} \frac{1}{r}\right) d\sigma_y - 2g(t, x),$$

for $k \geq 0$ requires that the data g is supposed to be infinitely times differentiable as a function of t. This we assume in the following. Then, we are able to express μ_k by a finite number of integrals, where in the integrands there appear the partial derivatives $\partial_t^l g$, $l = 0, 1, \cdots, k$. To get suitable estimates for μ_k it is reasonable to assume the following:

1. the boundary ∂G is sufficiently smooth
2. there is an a priori control of all derivatives $\partial_t^l g(t, x)$ for $(t, x) \in (-\infty, T] \times \partial G$, for example, g is continuous and $g(t, x) = 0$ for $t \leq 0$, all partial derivatives $\partial_t^l g$ are continuous, there exist positive constants C and $\delta \in (0, 1)$ such that $|\partial_t^l g(t, x)| \leq C^{l+1}(l + 1)^{(1+\delta)(l+1)}$ (see [70]).

Under these assumptions one can conclude the absolute and uniform convergence of the sequence $\{\mu_k = \mu_k(t, x)\}_{k \geq 0}$ to a function $\mu = \mu(t, x)$. This allows us to prove the following result.
The interior Dirichlet problem for the wave equation

$$u_{tt} - \Delta u = 0 \quad \text{in} \quad (0, \infty) \times G, \quad u(t, x) = g(t, x) \quad \text{on} \quad (0, \infty) \times \partial G,$$

$$u(0, x) = 0, \quad u_t(0, x) = 0 \quad \text{for} \quad x \in G,$$

$$\text{compatibility conditions:} \quad g(0, x) = 0, \quad g_t(0, x) = 0 \quad \text{on} \quad \{t = 0\} \times \partial G,$$

has a classical solution $u = u(t, x)$ in form of a wave double-layer potential with density $\mu = \mu(t, x)$. The density is the solution of the integral equation for the interior Dirichlet problem for the wave equation and can be found by successive approximation.

10.7.5 Final Comments to Mixed Problems

We define for a given interior domain $G \subset \mathbb{R}^n$ the so-called *local energy* (see also Chap. 11 for the discussion of the mathematical tool energy)

$$E_W(u, G)(t) := \frac{1}{2} \int_G \left(|u_t(t, x)|^2 + |\nabla u(t, x)|^2\right) dx.$$

In the following result we assume that G is a domain with smooth boundary such that the divergence theorem is applicable.

Theorem 10.7.1 (Conservation of the Local Energy) *Let $u \in C^2([0, \infty) \times \bar{G})$ be a solution to the wave equation*

$$u_{tt} - \Delta u = 0 \quad in \quad (0, \infty) \times G,$$

satisfying the initial conditions

$$u(0, x) = \varphi(x), \quad u_t(0, x) = \psi(x) \quad for \quad x \in G,$$

with data $\varphi \in H^1(\mathbb{R}^n)$ and $\psi \in L^2(\mathbb{R}^n)$ and one of the boundary conditions, either the homogeneous Dirichlet condition

$$u(t, x) = 0 \quad on \quad (0, \infty) \times \partial G,$$

or the homogeneous Neumann condition

$$\partial_\mathbf{n} u(t, x) = 0 \quad on \quad (0, \infty) \times \partial G.$$

Then the energy of u contained in G remains constant, i.e.,

$$E_W(u, G)(t) = E_W(u, G)(0) = \frac{1}{2}(\|\psi\|_{L^2}^2 + \|\nabla\varphi\|_{L^2}^2) \quad for\ all \quad t \geq 0.$$

Remark 10.7.1 By using a density argument (see the proof of Theorem 11.1.2) the conclusion of Theorem 10.7.1 is still true for energy solutions having the regularity $C([0, \infty), H^1(\mathbb{R}^n)) \cap C^1([0, \infty), L^2(\mathbb{R}^n))$.

Proof Let $Z_T = (0, T) \times G$, $T > 0$, be a cylinder and let S_T be the lateral surface of Z_T. The proof is based on the identity

$$0 = 2u_t \,\square u = -\nabla \cdot (2u_t \nabla u) + \partial_t(|\nabla u|^2 + u_t^2).$$

Here $\square$ denotes the d'Alembert operator $\partial_t^2 - \Delta$. It holds

$$0 = \int_{Z_T} \left(\nabla \cdot (2u_t \nabla u) - \partial_t(|\nabla u|^2 + u_t^2) \right) d(t, x).$$

The integrand is equal to the *divergence of the vector field*

$$\left(2u_t \nabla u, -(|\nabla u|^2 + u_t^2)\right).$$

After application of the *divergence theorem* it follows

$$0 = \int_{\partial Z_T} \left(2u_t \nabla u, -(|\nabla u|^2 + u_t^2)\right) \cdot \mathbf{n} \, d(t, \sigma_x),$$

where $\mathbf{n}$ is the outer unit normal vector to ∂Z_T. The surface ∂Z_T consists of three parts. We study how the above integral can be written on each of the three parts.

1. On the top of the cylinder $\{t = T\} \times G$, here $\mathbf{n} = (0, \cdots, 0, 1)$, the above integral is equal to

$$-\int_G \left(|\nabla u|^2 + u_t^2\right)\big|_{t=T} \, dx.$$

2. On the bottom of the cylinder $\{t = 0\} \times G$, here $\mathbf{n} = (0, 0, \cdots, 0, -1)$, the above integral is equal to

$$\int_G \left(|\nabla u|^2 + u_t^2\right)\big|_{t=0} \, dx.$$

3. On the lateral surface S_T, here $\mathbf{n} = (\mathbf{n}_x, 0)$, where $\mathbf{n}_x = (n_1, \cdots, n_n)$ is the outer unit normal vector to G, the above integral is equal to

$$\int_{S_T} \left(2u_t \nabla u, -(|\nabla u|^2 + u_t^2)\right) \cdot \mathbf{n} \, d(t, \sigma_x)$$

$$= \int_{S_T} 2u_t \nabla u \cdot \mathbf{n}_x \, d(t, \sigma_x) = \int_0^T \int_{\partial G} 2u_t \partial_{\mathbf{n}_x} u \, d\sigma_x \, dt.$$

If $\partial_{\mathbf{n}_x} u(t, x) = 0$ on $(0, \infty) \times \partial G$, then the last integral is obviously zero. If $u(t, x) = 0$ on $(0, \infty) \times \partial G$, then $\partial_t u = 0$ on ∂G for $t > 0$, too. Hence, the last integral is again zero. Summarizing, we have shown $E_W(u, G)(T) = E_W(u, G)(0)$ for all $T > 0$ and the statement is proved.

The *conservation of the local energy* is a basic tool for deriving uniqueness results for initial-boundary value problems for the wave equation.

Corollary 10.7.1 (Uniqueness) *Under the assumptions of Theorem 10.7.1 there is at most one classical solution of the initial-boundary value problem.*

Proof Let us assume the existence of two classical solutions u_1 and u_2. Then, the difference $u := u_1 - u_2$ is a solution with vanishing initial data. Since φ vanishes in G, then $\nabla \varphi$ must also vanish in G and Theorem 10.7.1 implies that $E_W(u, G)(t) = 0$ for all $t \geq 0$. Thus all first order partial derivatives of u must vanish and therefore u must be a constant. Since $u(0, x) = 0$ for $x \in G$, then $u \equiv 0$ on the region $[0, \infty) \times \bar{G}$ and the statement is proved.

We can employ, for example, the method of separation of variables and Fourier series to obtain a formal solution of the general initial-boundary value problem. This formal solution will involve series expansions of the initial data φ and ψ in terms of the eigenfunctions of an associated eigenvalue problem for the Laplace operator. For the sake of brevity we will not develop this approach, but the interested reader can find more on this topic in the textbooks [189] and [231].

A second possibility is to write mixed problems for the wave equation as Cauchy problems for the abstract wave equation $u_{tt} + A(t)u = f(t)$. This approach is developed in [105] and applied to linear (and even nonlinear) systems of generalized wave equations in interior domains of $\mathbb{R}^n$ with boundary conditions of Dirichlet, Neumann or Robin type. The main tools of the approach are Hilbert space theory, scales of Hilbert spaces, operators in Hilbert spaces and semigroup theory.

Exercises Relating to the Considerations of Chap. 10

Exercise 1 Explain the statement of Theorem 10.1.1 on the continuous dependence of the solution from the data by formulas.

Exercise 2 Let us consider the Cauchy problem

$$u_{tt} - u_{xx} = 0, \quad u(0, x) = \varphi(x), \quad u_t(0, x) = \psi(x)$$

with smooth data $\varphi = \psi = 0$ outside of the interval $[-l, l]$. Show that for each $x_0 \in \mathbb{R}^1$ there exist constants $T(x_0)$ and U with $u(t, x_0) = U$ for $t \geq T(x_0)$. Determine these constants.

Exercise 3 Derive the representation of solution

$$v(t, x) = \frac{1}{2} \int_0^t \left(\int_{x-(t-t')}^{x+(t-t')} F(t', x')dx' \right) dt'$$

for solutions to the Cauchy problem

$$u_{tt} - u_{xx} = F(t, x), \quad u(0, x) = 0, \quad u_t(0, x) = 0,$$

where the right-hand side F is supposed to belong to $L^1_{loc}([0, \infty) \times \mathbb{R}^1)$.

Exercise 4 Find the domain of dependence $\Omega(t_0, x_0)$ for solutions to the Cauchy problem of Exercise 3.

Exercise 5 Prove the statement of Lemma 10.3.1.

Exercise 6 Show that every classical radial symmetric solution $u = u(t, r)$ of $u_{tt} - \Delta u = 0$, $x \in \mathbb{R}^3$, has the following representation:

$$u(t, r) = \frac{u_1(r + t)}{r} + \frac{u_2(r - t)}{r}$$

with arbitrarily given twice differentiable functions u_1, u_2. Here $u_1 = u_1(r + t)$ is called *contracting wave* and $u_2 = u_2(r - t)$ is called *expanding wave*.

Exercise 7 Find the solution of the Cauchy problem

$$u_{tt} - \Delta u = 0, \quad u(0, x) = 1, \quad u_t(0, x) = \frac{1}{1 + |x|^2}, \quad x \in \mathbb{R}^3.$$

Try to find two different ways to derive the representation of the solution.

Exercise 8 (Duhamel's Principle)   Show that the solution u of

$$u_{tt} - \Delta u = F(t, x), \quad u(0, x) = u_t(0, x) = 0, \quad x \in \mathbb{R}^3,$$

is given by

$$u(t, x) = \int_0^t w(t, \tau, x) d\tau,$$

where $w = w(t, \tau, x)$ solves the following Cauchy problem:

$$w_{tt} - \Delta w = 0, \quad w(\tau, \tau, x) = 0, \quad w_t(\tau, \tau, x) = F(\tau, x).$$

Exercise 9 Use Duhamel's principle and Kirchhoff's representation of solution to derive a solution to the Cauchy problem

$$u_{tt} - \Delta u = F(t, x), \quad u(0, x) = u_t(0, x) = 0, \quad x \in \mathbb{R}^3.$$

We assume $F \in C^2([0, T], C^2(\mathbb{R}^3))$. Why?

Exercise 10 Study in the literature the method of descent.

Exercise 11 Model the propagation of sound in 3d in the half-space $\{(x, y, z) \in \mathbb{R}^3 : z > 0\}$. Hereby the sound is reflected on the plane $\{(x, y, z) \in \mathbb{R}^3 : z = 0\}$. Explain this reflection by solving the mixed problem.

Exercise 12 Determine the wave front set of $g(y)\delta_{x_0}$, where $g \in C_0^\infty(\mathbb{R}^m)$. Determine the wave front set of δ_{x_0} as a distribution on $\mathbb{R}_x^n \times \mathbb{R}_y^m$.

Chapter 11
The Notion of Energy of Solutions: One of the Most Important Quantities

The main issue of this chapter is the notion of energy of solutions, a very effective tool for the treatment of nonstationary or evolution models. We introduce energies for different models and explain conservation of energies, if possible. Sometimes, the choice of a suitable energy seems to be a miracle, sometimes one has different choices. Do we have energy conservation? Do we have blow up of the energy for $t \to \infty$? Do we have a decay of the energy for $t \to \infty$? These and related questions will be answered in this chapter. Moreover, it is shown how lower order terms may influence the choice and the long-time behavior of a suitable energy. Here we restrict ourselves to mass and different damping terms. Finally, we explain for several models the long-time behavior of local energies for solutions of mixed problems with Dirichlet condition in exterior domains.

11.1 Energies for Solutions to the Wave Equation

Let us come back to the wave equation $u_{tt} - \Delta u = 0$. We assume that

$$u \in C\big([0,\infty), H^1(\mathbb{R}^n)\big) \cap C^1\big([0,\infty), L^2(\mathbb{R}^n)\big)$$

is a given real-valued Sobolev solution. Then we denote by

$$E_W(u)(t) := \frac{1}{2} \int_{\mathbb{R}^n} \Big(|u_t(t,x)|^2 + |\nabla u(t,x)|^2 \Big) dx$$

$$= \frac{1}{2} \|u_t(t,\cdot)\|_{L^2}^2 + \frac{1}{2} \|\nabla u(t,\cdot)\|_{L^2}^2$$

the *energy, or total energy* of the solution u. The energy $E_W(u)(t)$ depends only on the time variable t. Here $\frac{1}{2}\|u_t(t,\cdot)\|_{L^2}^2$ models the *kinetic energy* (which appears

© Springer International Publishing AG 2018   147
M.R. Ebert, M. Reissig, *Methods for Partial Differential Equations*,
https://doi.org/10.1007/978-3-319-66456-9_11

due to the term u_{tt} in the wave model) and $\frac{1}{2}\|\nabla u(t, \cdot)\|_{L^2}^2$ models the *elastic energy* (which appears due to the term Δu in the wave model).

We define for a given set $K \subset \mathbb{R}^n$ (K is the closure of a domain) the so-called *local energy*

$$E_W(u, K)(t) := \frac{1}{2} \int_K \left(|u_t(t, x)|^2 + |\nabla u(t, x)|^2 \right) dx.$$

Let (t_0, x_0), $t_0 > 0$, be a fixed point in $\mathbb{R}^{n+1}$. Then, the set $\{(t, x) : |x - x_0| = |t - t_0|\}$ describes the lateral surface of a double cone with apex at (t_0, x_0). The *forward (backward) characteristic cone* is for $t \geq t_0$ ($t \leq t_0$), the upper (lower) cone with apex at (t_0, x_0). Let $T \in [0, t_0)$. The part of the plane $\{(t, x) : t = T\}$ lying inside the backward characteristic cone with apex at (t_0, x_0) will be denoted by $K(x_0, T)$. This part is a closed ball around the center $x = x_0$ with radius $t_0 - T$. The following remarkable statement holds for local energies $E_W(u, K(x_0, t))$:

Theorem 11.1.1 (Domain of Dependence Energy Inequality) *Let $(t_0, x_0) \in \mathbb{R}^{n+1}$ with $t_0 > 0$. We denote by Ω the conical domain bounded by the backward characteristic cone with apex at (t_0, x_0) and by the plane $t = 0$. Let $u \in C^2(\overline{\Omega})$ be a classical solution of the wave equation $u_{tt} - \Delta u = 0$. Then, the following domain of dependence inequality holds:*

$$E_W(u, K(x_0, t)) \leq E_W(u, K(x_0, 0)) \text{ for } t \in [0, t_0).$$

What consequences may we draw from this result?
The base of the conical domain Ω is exactly the domain of dependence with respect to the data of a solution u in the point (t_0, x_0). If we pose Cauchy conditions on $t = 0$, then the local energy $E_W(u, K(x_0, 0))$ can be calculated by using the data φ and ψ as follows:

$$E_W(u, K(x_0, 0)) = \frac{1}{2} \int_{K(x_0, 0)} \left(|\psi(x)|^2 + |\nabla \varphi(x)|^2 \right) dx.$$

Then, the local energy $E_W(u, K(x_0, t))$, $t \in (0, t_0)$, of the solution is estimated on the set $K(x_0, t)$ by $E_W(u, K(x_0, 0))$. The local energy may concentrate in $K(x_0, t)$, but its amount is never larger than $E_W(u, K(x_0, 0))$. The proof of this observation is similar to the proof of Theorem 10.7.1.

Proof Let Ω_T be the part of Ω below the plane $t = T$, $T \in (0, t_0)$, and let C_T be the lateral surface of Ω_T. Let u be a classical solution of the wave equation. The proof is based on the identity

$$0 = 2u_t \, \Box u = -\nabla \cdot \left(2u_t \nabla u \right) + \partial_t \left(|\nabla u|^2 + u_t^2 \right).$$

Here $\Box$ denotes the d'Alembert operator $\partial_t^2 - \Delta$. The *key idea of the multiplication method* is to multiply the left-hand side $\Box u$ by a function (here we multiply it by $2u_t$) such that we get on the right-hand side *the divergence of a vector field* (here we get the divergence of the vector field $(-2u_t \nabla u, |\nabla u|^2 + u_t^2)$). It holds that

$$0 = \int_{\Omega_T} \left(\nabla \cdot (2u_t \nabla u) - \partial_t(|\nabla u|^2 + u_t^2) \right) d(t, x).$$

Applying the *Divergence Theorem* we obtain

$$0 = \int_{\partial\Omega_T} \left(2u_t \, \nabla u, -(|\nabla u|^2 + u_t^2) \right) \cdot \mathbf{n} \, d\sigma,$$

where $\mathbf{n}$ is the outer unit normal vector to $\partial\Omega_T$. The surface $\partial\Omega_T$ consists of three parts. We study how the above integral can be written on each of the three parts.

(1) Top ball $K(x_0, T)$: $\mathbf{n} = (0, \cdots, 0, 1)$. The above integral is equal to $-\int_{K(x_0,T)} (|\nabla u|^2 + u_t^2) \, dx$.
(2) Bottom ball $K(x_0, 0)$: $\mathbf{n} = (0, 0, \cdots, 0, -1)$. The above integral is equal to $\int_{K(x_0,0)} (|\nabla u|^2 + u_t^2) \, dx$.
(3) Lateral surface C_T : $\mathbf{n} = \eta = (\eta_1, \cdots, \eta_{n+1})$ The above integral is equal to

$$\int_{C_T} (2u_t \, \nabla u, -(|\nabla u|^2 + u_t^2)) \cdot \eta \, d\sigma$$

$$= \sqrt{2} \int_{C_T} (2u_t u_{x_1} \eta_{n+1}\eta_1 + \cdots + 2u_t u_{x_n} \eta_{n+1}\eta_n$$

$$-(u_{x_1}^2 + \cdots + u_{x_n}^2 + u_t^2)\eta_{n+1}^2) \, d\sigma$$

$$= -\sqrt{2} \int_{C_T} ((u_{x_1}\eta_{n+1} - u_t\eta_1)^2 + \cdots + (u_{x_n}\eta_{n+1} - u_t\eta_n)^2) \, d\sigma \le 0.$$

Here we used $\eta_{n+1}^2 = \eta_1^2 + \cdots + \eta_n^2 = \frac{1}{2}$ and $\eta_{n+1} > 0$.

Summarizing, we have shown for all $T \in (0, t_0)$ the energy estimate

$$\int_{K(x_0,T)} \left(|\nabla u(x, T)|^2 + u_t(x, T)^2 \right) dx$$

$$\le \int_{K(x_0,0)} \left(|\nabla u(x, 0)|^2 + u_t(x, 0)^2 \right) dx.$$

Hence, the statement is proved.

The *domain of dependence energy inequality* is a basic tool for deriving uniqueness results for classical solutions to the Cauchy problems for the free wave equation from Theorems 10.5.1 and 10.5.2.

Corollary 11.1.1 *The Cauchy problem*

$$u_{tt} - \Delta u = 0, \quad u(0, x) = \varphi(x), \quad u_t(0, x) = \psi(x)$$

has a unique classical solution

$$u \in C^{k-n}\big([0, \infty) \times \mathbb{R}^{2n+1}\big), \quad u \in C^{k-n}\big([0, \infty) \times \mathbb{R}^{2n}\big),$$

respectively, for $k \geq n + 2$ and $n \geq 1$.

After learning the *domain of dependence energy inequality*, we ask for the behavior of the energy of solutions to Cauchy problems for the wave equation in the whole space $\mathbb{R}^n$. In $\mathbb{R}^n$ waves may propagate freely. We do not expect a loss of energy, consequently, we believe in energy conservation. This we shall explain in the following theorem.

Theorem 11.1.2 (Conservation of Energy) *Let*

$$u \in C\big([0, T], H^1(\mathbb{R}^n)\big) \cap C^1\big([0, T], L^2(\mathbb{R}^n)\big)$$

be a Sobolev solution of

$$u_{tt} - \Delta u = 0, \quad u(0, x) = \varphi(x), \quad u_t(0, x) = \psi(x)$$

with data $\varphi \in H^1(\mathbb{R}^n)$ and $\psi \in L^2(\mathbb{R}^n)$. It then holds that

$$E_W(u)(t) = E_W(u)(0) = \frac{1}{2}\big(\|\psi\|_{L^2}^2 + \|\nabla\varphi\|_{L^2}^2\big) \quad \text{for all} \ \ t \geq 0.$$

Proof Using the density of the function space $C_0^\infty(\mathbb{R}^n)$ in $H^1(\mathbb{R}^n) \subset L^2(\mathbb{R}^n)$ we are able to approximate the given data $\varphi \in H^1(\mathbb{R}^n)$ and $\psi \in L^2(\mathbb{R}^n)$ by sequences of data $\{\varphi_k\}_k$, $\{\psi_k\}_k$ with $\varphi_k, \psi_k \in C_0^\infty(\mathbb{R}^n)$. We consider the family of auxiliary Cauchy problems

$$u_{tt} - \Delta u = 0, \quad u(0, x) = \varphi_k(x), \quad u_t(0, x) = \psi_k(x).$$

From Theorems 10.5.1 and 10.5.2 we obtain for all $T > 0$ a unique solution $u_k \in C^\infty\big([0, T], C_0^\infty(\mathbb{R}^n)\big)$. Differentiating $E_W(u_k)(t)$ gives

$$E'_W(u_k)(t) = \int_{\mathbb{R}^n} \big(\partial_t u_k(t, x)\partial_t^2 u_k(t, x) + \nabla u_k(t, x) \cdot \nabla \partial_t u_k(t, x)\big)\, dx.$$

Taking into consideration the property of finite speed of propagation of perturbations (see Sects. 10.1.2 and 10.5), the function $u_k(t, \cdot)$ belongs to $C_0^\infty(\mathbb{R}^n)$ for each $t > 0$. After integration by parts (all boundary integrals vanish) we obtain immediately, after seeing that u_k is a classical solution to the wave equation, that

$$E'_W(u_k)(t) = \int_{\mathbb{R}^n} \big(\partial_t u_k(t, x)\Delta u_k(t, x) - \Delta u_k(t, x)\partial_t u_k(t, x)\big)\, dx = 0.$$

Hence,

$$E_W(u_k)(t) = E_W(u_k)(0) = \frac{1}{2}\left(\|\psi_k\|_{L^2}^2 + \|\nabla\varphi_k\|_{L^2}^2\right).$$

Together with the assumption for the data we have $\lim_{k\to\infty} E_W(u_k)(0) = E_W(u)(0)$. From the well-posedness of the Cauchy problem in Sobolev spaces (see Corollary 14.1.1) it follows $\lim_{k\to\infty} E_W(u_k)(t) = E_W(u)(t)$. This completes the proof.

Remark 11.1.1 We proved the energy conservation for the whole space $\mathbb{R}^n$. But the *energy conservation* remains true for classical solutions to the wave equation in *bounded domains* $G \subset \mathbb{R}^n$ if we prescribe some regime on the boundary. Such regimes are, for example, a homogeneous boundary condition of Dirichlet type (solution has to vanish on $(0, \infty) \times \partial G$) or Neumann type (the "flow" through $(0, \infty) \times \partial G$ has to vanish). The *energy conservation* holds also for unbounded domains G, for example, for exterior domains, if the initial data have a compact support and if classical solutions to the wave equation satisfy a homogeneous boundary condition of Dirichlet or Neumann type. In the proof we see that the initial data influence, due to the *finite speed of propagation of perturbations*, the solution only in the set $\{x \in G : |x| \le R + t, \ t \ge 0\}$. Here R denotes the radius of an arbitrarily chosen ball around the origin containing the support of the data.

11.2 Examples of Energies for Other Models

11.2.1 One Energy for Solutions to the Elastic Wave Equation

In the previous section we introduced $E_W(u)(t)$ as one energy for special Sobolev solutions to the wave equation. If we are interested in studying solutions to the elastic wave equation, then some modification of the energy $E_W(u)(t)$ would seem appropriate.

Let us consider the Cauchy problem for elastic waves

$$u_{tt} - a^2 \Delta u - (b^2 - a^2)\nabla \operatorname{div} u = 0, \quad u(0, x) = \varphi(x), \quad u_t(0, x) = \psi(x),$$

where the positive constants a^2 and b^2 are related to the Lamé constants and satisfy $b^2 > a^2$. We will explain one energy. The term u_{tt} suggests the kinetic energy $\frac{1}{2}\|u_t(t, \cdot)\|_{L^2}^2$, the term $a^2\Delta u$ suggests the elastic energy $\frac{1}{2}a^2\|\nabla u(t, \cdot)\|_{L^2}^2$ and, finally, the term $(b^2 - a^2)\nabla \operatorname{div} u$ gives the part $\frac{1}{2}(b^2 - a^2)\|\operatorname{div} u(t, \cdot)\|_{L^2}^2$. Summarizing, we obtain one energy $E_{EW}(u)(t)$ of solutions of the elastic wave equation as follows:

$$E_{EW}(u)(t) = \frac{1}{2}\|u_t(t, \cdot)\|_{L^2}^2 + \frac{1}{2}a^2\|\nabla u(t, \cdot)\|_{L^2}^2 + \frac{1}{2}(b^2 - a^2)\|\operatorname{div} u(t, \cdot)\|_{L^2}^2.$$

11.2.2   Energies for Solutions to the Heat Equation

Let us consider the Cauchy problem for the heat equation

$$u_t - \Delta u = 0, \quad u(0, x) = \varphi(x).$$

We will explain different energies, even those of higher order.
A first proposal for an energy could be $E_H(u)(t) := \frac{1}{2}\|u(t, \cdot)\|_{L^2}^2$. This energy arises from the term u_t in the heat equation. If we differentiate $E_H(u)(t)$ with respect to t, then integration by parts gives

$$E_H'(u)(t) \leq -\int_{\mathbb{R}^n} |\nabla u(t, x)|^2 \, dx.$$

Consequently, this energy is decreasing.
One can also define higher order energies $E_H^k(u)(t) := \frac{1}{2}\||D|^k u(t, \cdot)\|_{L^2}^2$ with $k \in \mathbb{N}$. Here, the action of the pseudodifferential operator $|D|^k$ on functions $u(t, \cdot)$ is defined by $|D|^k u(t, x) = F_{\xi \to x}^{-1}\big(|\xi|^k F_{y \to \xi}(u(t, y))\big)$ (see Sect. 24.1 for basics of the Fourier transformation).

11.2.3   Energies for Solutions to the Schrödinger Equation

Let us consider the Cauchy problem for the Schrödinger equation

$$\frac{1}{i}u_t \pm \Delta u = 0, \quad u(0, x) = \varphi(x).$$

Solutions are expected to be complex-valued. For this reason, the term u_t in the Schrödinger equation implies the energy

$$E_{Sch}(u)(t) := \frac{1}{2}\|u(t, \cdot)\|_{L^2}^2 = \frac{1}{2}(u, \bar{u})_{L^2}.$$

Differentiation with respect to t and integration by parts give immediately

$$E_{Sch}'(u)(t) := \Re(u_t, \bar{u}) = \Re(\mp i\Delta u, \bar{u}) = \Re(\pm i\nabla u, \overline{\nabla u}) = 0.$$

Thus, we have energy conservation. Higher order energies

$$E_{Sch}^k(u)(t) := \frac{1}{2}\||D|^k u(t, \cdot)\|_{L^2}^2 = \frac{1}{2}\big(|D|^k u, \overline{|D|^k u}\big)_{L^2}$$

satisfy $E_{Sch}^k(u)(t) = E_{Sch}^k(u)(0)$ because $|D|^k u$ solves the Schrödinger equation, too. Indeed, we have

$$\frac{1}{i}(|D|^k u)_t \pm \Delta |D|^k u = 0, \quad |D|^k u(0,x) = |D|^k \varphi(x)$$

if the data φ is sufficiently regular.

11.2.4 Energies for Solutions to the Plate Equation

Let us consider the Cauchy problem for the plate equation

$$u_{tt} + (-\Delta)^2 u = 0, \quad u(0,x) = \varphi(x), \quad u_t(0,x) = \psi(x).$$

We will explain one energy. From the term u_{tt} we get $\frac{1}{2}\|u_t(t,\cdot)\|_{L^2}^2$, the kinetic energy. The term $(-\Delta)^2 u$ implies $\frac{1}{2}\|\Delta u(t,\cdot)\|_{L^2}^2$, which is connected with the momentum. Summarizing, we get one energy

$$E_{PL}(u)(t) := \frac{1}{2}\|u_t(t,\cdot)\|_{L^2}^2 + \frac{1}{2}\|\Delta u(t,\cdot)\|_{L^2}^2.$$

This energy is reasonable if we take into account the fact that $u_t + i\Delta u$ is a distributional solution to the Schrödinger equation if u solves the plate equation in the distributional sense.

11.2.5 Energies for Solutions to Special Semilinear Wave Models

Let us consider the Cauchy problem for semilinear wave models with time-dependent speed of propagation

$$u_{tt} - a(t)^2 \Delta u = f(u), \quad u(0,x) = \varphi(x), \quad u_t(0,x) = \psi(x).$$

We will explain the definition of energies. The term u_{tt} yields the kinetic energy $\frac{1}{2}\|u_t(t,\cdot)\|_{L^2}^2$, the term $a(t)^2 \Delta u$ implies the elastic type energy (we should take into account the time-dependent coefficient) $\frac{1}{2}a(t)^2\|\nabla u(t,\cdot)\|_{L^2}^2$. But how does the nonlinear term $f(u)$ influence the definition of the energy? Let us define the primitive $F(u) = \int_0^u f(s)ds$. It could be an idea to include this term into the energy, so we could propose as a suitable energy

$$E_W(u)(t) = \frac{1}{2}\|u_t(t,\cdot)\|_{L^2}^2 + \frac{1}{2}a(t)^2\|\nabla u(t,\cdot)\|_{L^2}^2 + \int_{\mathbb{R}^n} -F(u)dx.$$

Is this a good idea? It depends heavily on $-F(u)$. The energy is supposed to be nonnegative, so we expect nonnegativity of $-F(u)$ for all u. But this is not always satisfied. For this reason we distinguish between (see Sects. 18.1 and 18.2 or 20.1 and 20.2)

1. *An absorbing nonlinearity* $f(u)$: $-F(u) \geq 0$ appears (is absorbed) in the definition of the energy. A typical example is $f(u) = -|u|^{p-1}u$, $p > 1$.
2. *A source nonlinearity* $f(u)$: $-F(u)$ is not nonnegative, thus, it does not appear in the definition of the energy. It should be treated as a source. Typical examples are $f(u) = |u|^{p-1}u$, $p > 1$, or $f(u) = |u|^p$, $p > 1$.

If we have an absorbing nonlinearity, then one can prove $E'_W(u)(t) = 0$, $E_W(u)(t) = E_W(u)(0)$, respectively, for all times t. Hence the energy is conserved. This allows us to control all three parts of the energy, among other things. We are able to control $\int_{\mathbb{R}^n} -F(u)dx \leq E_W(u)(0)$. Such a control is very helpful in the treatment of nonlinear wave models.

The main goals of such a treatment depend on the kind of power nonlinearity we find in the model. Let us explain this difference for global (in time) solutions for Cauchy problems to semilinear wave models.

1. If we have *an absorbing nonlinearity* $f(u) = -|u|^{p-1}u$, $p > 1$, for example, then the proof of global (in time) large data solutions is often a topic of interest. If the data in the initial condition $u(0, x) = \varphi(x)$ is large, then $f(\varphi)$ becomes large for large p. For this reason one is often able to prove such a global (in time) existence result only for some $p < p_0$, where p_0 depends on the concrete model (cf. with the considerations in Sects. 17.2, 18.2, 20.2, 21.3 and 21.4).
2. If we have *a source nonlinearity* $f(u) = |u|^{p-1}u$, $p > 1$, or $f(u) = |u|^p$, $p > 1$, for example, then the proof of global (in time) small data solutions is often a topic of interest. If the data in the initial condition $u(0, x) = \varphi(x)$ is small, then $f(\varphi)$ becomes small for large p. For this reason one is often able to prove such a global (in time) existence result only for some $p > p_0$, where p_0 depends on the concrete model (cf. with the considerations in Sects. 17.1, 18.1, 19.3 and 20.1).

11.2.6   How to Define Energies in General?

At the end of Sect. 11.2 we want to show by an example that the definition of suitable energies needs some experience. Sometimes it seems to be a miracle for beginners as the following example shows:

Example 11.2.1 Let us consider in $(0, \infty) \times G$, where $G \subset \mathbb{R}^n$ is a bounded domain, the nonlinear Kirchhoff type wave equation ($\beta > 0$ is a constant) with source nonlinearity

$$u_{tt} - \left(1 + \beta \int_G |\nabla u|^2 dx\right)\Delta u = u|u|^p \quad \text{in } (0, \infty) \times G,$$

with the initial conditions

$$u(0, x) = \varphi(x), \quad u_t(0, x) = \psi(x) \text{ in } G,$$

and the boundary condition

$$u(t, x) = 0 \text{ on } (0, \infty) \times \Gamma_0,$$

$$\partial_n u + \int_0^t k(t - s, x) u_t(s, x) ds + a(x) g(u_t) = 0 \text{ on } (0, \infty) \times \Gamma_1.$$

Here, the boundary $\Gamma = \partial G$ is supposed to be the union of $\Gamma_0 \cup \Gamma_1$, where the "interior" of Γ_0 and Γ_1 are disjoint. On Γ_0 we pose a homogeneous Dirichlet condition, on Γ_1 we pose a boundary condition with given linear memory term

$$\int_0^t k(t - s, x) u_t(s, x) ds$$

and given nonlinear dissipation term $a(x) g(u_t)$. Some mathematicians propose the following energy:

$$E(u)(t) := \frac{1}{2} \left(\int_G \left(|u_t(t, x)|^2 + |\nabla u(t, x)|^2 \right) dx \right) + \frac{\beta}{4} \left(\int_G |\nabla u(t, x)|^2 dx \right)^2$$

$$- \frac{1}{2} \left(1 + \beta \int_G |\nabla u(t, x)|^2 dx \right) \int_{\Gamma_1} \left(\int_0^t k_t(t - s, x) |u(t, x) - u(s, x)|^2 ds \right) d\sigma$$

$$+ \frac{1}{2} \left(1 + \beta \int_G |\nabla u(t, x)|^2 dx \right) \int_{\Gamma_1} k(t, x) |u(t, x) - u_0(x)|^2 d\sigma$$

$$- \frac{1}{p + 2} \int_\Omega |u(t, x)|^{p+2} dx.$$

But, here we pose the question as to whether this energy is nonnegative. Take into consideration that the nonlinearity has to be considered as a source nonlinearity.

11.3 Influences of Lower Order Terms to Qualitative Properties of Solutions

11.3.1 Wave Models with Terms of Lower Order

We are able to characterize terms of lower order in wave models. Klein (1927) and Gordon (1926) derived the following Klein-Gordon equation describing a charged particle in an electro-magnetic field:

$$u_{tt} - \Delta u + m^2 u = 0.$$

The term m^2u, m^2 is a positive constant and is called *mass or potential*. Another well-known equation is the *telegraph equation*

$$u_{tt} - u_{xx} + au_t + bu = 0,$$

where a and b are constants. This equation arises in the study of propagation of electric signals in a cable of transmission line, in the propagation of pressure waves in the study of pulsatile blood flow in arteries and in one-dimensional random motion of bugs along a hedge. Here bu is a *mass term* and au_t is a *damping* or *dissipation term*. A higher-dimensional generalization of the telegraph equation is

$$u_{tt} - \Delta u + au_t + bu = 0.$$

Finally, we mention wave equations with a *convection term* or a *transport term* $\sum_{k=1}^{n} a_k(t,x)\partial_{x_k} u$, that is,

$$u_{tt} - \Delta u + \sum_{k=1}^{n} a_k(t,x)\partial_{x_k} u = 0.$$

11.3.2   Classical Damped Wave Models

First of all, we have to say that the appearance of a damping term in the classical damped wave model has no influence on well-posedness results in Sobolev spaces. So, if we have a well-posedness result for the wave model, we expect the same for the classical damped wave model (see, for example, Theorem 14.2.1 and Remark 14.2.1). However, a damping term should influence the behavior of the wave type energy.

Let us turn to the Cauchy problem

$$u_{tt} - \Delta u + u_t = 0, \quad u(0,x) = \varphi(x), \quad u_t(0,x) = \psi(x).$$

As for solutions to the classical wave equation we introduce the total energy

$$E_W(u)(t) = \frac{1}{2} \int_{\mathbb{R}^n} \left(|u_t(t,x)|^2 + |\nabla u(t,x)|^2 \right) dx.$$

Then we are interested in estimates of $E_W(u)(t)$ following from differentiation of the energy $E_W(u)(t)$ with respect to t and integration by parts. We assume that all these steps can be carried out, that is, the data are supposed to be smooth enough (it is sufficient to assume for our purpose that the data (φ, ψ) belong to the so-called energy space $H^1(\mathbb{R}^n) \times L^2(\mathbb{R}^n))$. Then, we derive

$$E_W'(u)(t) = \frac{1}{2} \int_{\mathbb{R}^n} (2u_t u_{tt} + 2\nabla u \cdot \nabla u_t)\, dx$$

$$= \int_{\mathbb{R}^n} (u_t(\Delta u - u_t) + \nabla u \cdot \nabla u_t)\, dx = \int_{\mathbb{R}^n} -u_t(t,x)^2 dx \le 0.$$

Thus, the energy is decreasing for increasing t. We can not expect energy conservation. This seems to be no surprise because of the damping term. It raises the question for the behavior of the energy for $t \to \infty$. Of special interest is the question as to whether the energy $E_W(u)(t)$ tends to 0 for $t \to \infty$. Such behavior is called *decay*. By using phase space analysis we will study the decay behavior for $E_W(u)(t)$ in Sect. 14.2.2, Theorem 14.2.2.

11.3.3 Wave Models with Viscoelastic Damping

In the last 20 years mathematicians studied intensively *wave models with structural damping*

$$u_{tt} - \Delta u + (-\Delta)^\delta u_t = 0.$$

Here $\delta \in (0, 1]$. If $\delta = 1$, then we get the viscoelastic damped wave model. If $\delta = 0$, then we get the classical damped wave model.
Let us turn to the Cauchy problem for the wave model with viscoelastic damping

$$u_{tt} - \Delta u - \Delta u_t = 0, \quad u(0,x) = \varphi(x), \quad u_t(0,x) = \psi(x).$$

Firstly, let us understand the model itself.
What is the principal part of the viscoelastic damped wave operator? Using the usual definition it is $\Delta \partial_t$. What does it mean? It is better to interpret the term $-\Delta u_t$ as a damping term for the model. Thus, the wave operator is in some sense the "principal part". This interpretation proposes the use of the wave energy $E_W(u)(t)$. But how does it behave? If we follow the ideas from the previous section, then we get

$$E_W'(u)(t) = -\int_{\mathbb{R}^n} |\nabla u_t(t,x)|^2 dx \le 0$$

for data (φ, ψ) with suitable regularity. The energy is decreasing, but is it also decaying? What about well-posedness in Sobolev spaces? What regularity for the data do we propose? These and other questions will be studied in Sect. 14.3.

11.3.4   Klein-Gordon Equation

The Cauchy problem for the Klein-Gordon equation is

$$u_{tt} - \Delta u + m^2 u = 0, \quad u(0,x) = \varphi(x), \quad u_t(0,x) = \psi(x)$$

with a positive constant m^2.

How can we define the total energy of a solution?

The mass term, or potential, requires including, besides the *elastic and the kinetic energy*, a third component into the total energy. That is the so-called *potential energy* (which appears due to the term $m^2 u$). Thus, we define the total energy of a Sobolev solution belonging to

$$C\big([0,\infty), H^1(\mathbb{R}^n)\big) \cap C^1\big([0,\infty), L^2(\mathbb{R}^n)\big)$$

by

$$E_{KG}(u)(t) := \frac{1}{2} \int_{\mathbb{R}^n} \Big(|u_t(t,x)|^2 + |\nabla u(t,x)|^2 + m^2 |u(t,x)|^2 \Big)\, dx.$$

Repeating the proof of Theorem 11.1.2 one can show the following result.

Theorem 11.3.1 (Conservation of Energy) *Let*

$$u \in C\big([0,T], H^1(\mathbb{R}^n)\big) \cap C^1\big([0,T], L^2(\mathbb{R}^n)\big)$$

be a Sobolev solution of

$$u_{tt} - \Delta u + m^2 u = 0, \quad u(0,x) = \varphi(x), \quad u_t(0,x) = \psi(x)$$

with data $\varphi \in H^1(\mathbb{R}^n)$ and $\psi \in L^2(\mathbb{R}^n)$. Then, the conservation of energy holds. That is,

$$E_{KG}(u)(t) = E_{KG}(u)(0) = \frac{1}{2}\Big(\|\psi\|_{L^2}^2 + \|\nabla\varphi\|_{L^2}^2 + m^2 \|\varphi\|_{L^2}^2 \Big) \quad \text{for all } t \geq 0.$$

Do we have a corresponding statement to Theorem 11.1.1 if we are interested in *local energies*?

Yes, we have. Let $K \subset \mathbb{R}^n$ be the closure of a domain. We define the local energy as

$$E_{KG}(u,K)(t) := \frac{1}{2} \int_K \Big(|u_t(t,x)|^2 + |\nabla u(t,x)|^2 + m^2 |u(t,x)|^2 \Big)\, dx.$$

Using the same notations as in Sect. 11.1 the following remarkable result holds.

Theorem 11.3.2 (Domain of Dependence Energy Inequality) *Let* $(t_0, x_0) \in$ $\mathbb{R}^{n+1}$ *with* $t_0 > 0$. *We denote by* Ω *the conical domain bounded by the backward characteristic cone with apex at* (t_0, x_0) *and by the plane* $\{(t, x) : t = 0\}$. *Let* $u \in C^2(\overline{\Omega})$ *be a classical solution of the Klein-Gordon equation*

$$u_{tt} - \Delta u + m^2 u = 0.$$

Then, the following inequality holds:

$$E_{KG}(u, K(x_0, t)) \leq E_{KG}(u, K(x_0, 0)) \ for \ t \in [0, t_0].$$

The domain of dependence energy inequality implies the following uniqueness result.

Corollary 11.3.1 *The Cauchy problem*

$$u_{tt} - \Delta u + m^2 u = 0, \quad u(0, x) = \varphi(x), \quad u_t(0, x) = \psi(x)$$

has at most one classical solution $u \in C^2([0, \infty) \times \mathbb{R}^n)$ *if the data are supposed to be sufficiently smooth.*

Remark 11.3.1 The von Wahl's transformation allows for *transforming Cauchy problems for the Klein-Gordon equation to Cauchy problems for the wave equation.* With $x = (x_1, \cdots, x_n)$, let us consider the Cauchy problem

$$u_{tt} - \Delta u + m^2 u = 0, \quad u(0, x) = \varphi(x), \quad u_t(0, x) = \psi(x).$$

We introduce

$$v(t, x_1, \cdots, x_n, x_{n+1}) := e^{-i m x_{n+1}} u(t, x_1, \cdots, x_n).$$

Evidently, the function $v = v(t, x)$ is a solution to the Cauchy problem

$$v_{tt} - \Delta v = 0,$$

$$v(0, x) = v_0(x) := e^{-i m x_{n+1}} \varphi(x), \quad v_t(0, x) = v_1(x) := e^{-i m x_{n+1}} \psi(x)$$

but now with complex-valued data. If the data (φ, ψ) are chosen from the function space $C^k(\mathbb{R}^n) \times C^{k-1}(\mathbb{R}^n)$, then, after application of von Wahl's transformation, the complex-valued data (v_0, v_1) belong to the function space $C^k(\mathbb{R}^{n+1}) \times C^{k-1}(\mathbb{R}^{n+1})$. Hence, we are able to apply, for example, Theorems 10.5.1 and 10.5.2 to get complex-valued solutions $v = v(t, x)$. After the backward transformation one has to prove the optimality of the obtained results.

11.3.5   Plate Equations with Lower Order Terms

The Cauchy problem for plate models with mass and dissipation terms is

$$u_{tt} + (-\Delta)^2 u + m^2 u + u_t = 0, \quad u(0,x) = \varphi(x), \quad u_t(0,x) = \psi(x).$$

Model 1: $m^2 = 0$ Then, we introduce the total energy for solutions as follows:

$$E_{PL}(u)(t) = \frac{1}{2} \int_{\mathbb{R}^n} \left(|u_t(t,x)|^2 + |\Delta u(t,x)|^2 \right) dx.$$

Solutions having for all t the energy $E_{PL}(u)(t)$ are called energy solutions of the classical damped plate model. Some straight forward calculations imply (here we have to verify that for energy solutions it is allowed to carry out integration by parts)

$$E'_{PL}(u)(t) = -\int_{\mathbb{R}^n} u_t^2(t,x)\, dx \le 0.$$

Consequently, the energy decreases if t tends to infinity. But, does it decay?
Model 2: $m^2 > 0$ Then, we introduce the total energy for solutions as follows:

$$E_{PL,KG}(u)(t) = \frac{1}{2} \int_{\mathbb{R}^n} \left(|u_t(t,x)|^2 + |\Delta u(t,x)|^2 + m^2 |u(t,x)|^2 \right) dx.$$

Solutions having for all t the energy $E_{PL,KG}(u)(t)$ are called energy solutions of the classical damped plate model with mass term. Some straight forward calculations imply (here we have to verify that for energy solutions it is allowed to carry out integration by parts)

$$E'_{PL,KG}(u)(t) = -\int_{\mathbb{R}^n} u_t^2(t,x)\, dx \le 0.$$

Consequently, the energy decreases if t tends to infinity. But, does it decay?
   These and other issues will be studied in Chap. 15.

11.4   Behavior of Local Energies

In the last sections we have given several results on the behavior of total energies of solutions to different models. Now we come back to local energies (see, for example, Sect. 11.1). The authors' impression is that, in general, results on the behavior of local energies of solutions to different models do not belong to an expected basic knowledge on PDE's. For this reason, the main goal of this section is to list results on the behavior of local energies of solutions to mixed problems for quite different models. To have the possibility of comparing the results we choose the same boundary condition of Dirichlet type and exterior domains in all models. Sometimes we will only sketch the proofs of the results, sometimes we will even omit them. The interested reader will find for all models references which help in studying the topic of behavior of local energies in more detail.

First of all, let us introduce a mathematical model whose treatment uses local energy decay estimates. The following explanations are taken from [141].

Consider the following mixed problem in an exterior domain G:

$$u_{tt} - \Delta u + a(x)u_t = 0 \text{ in } (0,\infty) \times G, \quad u(t,x) = 0 \text{ on } (0,\infty) \times \partial G,$$

$$u(0,x) = \varphi(x), \quad u_t(0,x) = \psi(x) \text{ in } G,$$

where $G \subset \mathbb{R}^n$ is an exterior domain with compact obstacle $V := \mathbb{R}^n \setminus G$. The dissipation is supposed to only be effective near infinity. This assumption is described by the following condition:

The coefficient $a = a(x)$ is a bounded nonnegative function on G and satisfies

$$a(x) \geq C_0 \text{ for all } x : |x| \geq L$$

with some positive constant C_0 and some large positive constant L.
Here we may assume

$$V \subset B_L(0) := \{x \in \mathbb{R}^n : |x| \leq L\}.$$

So, the damping term behaves as in the case of a classical damped wave only outside of a large ball around the origin. But, we do not require any dissipative effect around the boundary ∂G (study in the literature the notions of trapping and non-trapping domains). The question of existence and uniqueness of solutions to the above mixed problem is of interest. The authors mention the existence and uniqueness of energy solutions belonging to the function space

$$C([0,\infty), W_{2,0}^1(G)) \cap C^1([0,\infty), L^2(G))$$

to given data $(\varphi, \psi) \in W_{2,0}^1(G) \times L^2(G)$ (see Definition 24.22). The data space for φ takes into consideration the homogeneous Dirichlet condition on $(0,\infty) \times \partial G$. Now, let us discuss energy estimates. We define the total wave type energy (for the given exterior domain G) by

$$E_W(u,G)(t) := \frac{1}{2} \int_G \left(|u_t(t,x)|^2 + |\nabla u(t,x)|^2 \right) dx.$$

Then, straight forward calculations imply

$$E_W'(u,G)(t) + \int_G a(x)|u_t(t,x)|^2 \, dx = 0.$$

It follows that $E_W(u, G)(t)$ is a decreasing function for $t \to \infty$. Therefore it seems to be of necessity to ask whether or not $E_W(u, G)(t)$ decays, that is, tends to 0 for $t \to \infty$. The main goal of [141] is to show that this really is true. The proof is based on the following steps:

1. Let $G_R := G \cap B_R(0)$ and let $E_W(u, G_R)(t)$ be the local energy for G_R. Then it holds $\lim_{t \to \infty} E_W(u, G_R)(t) = 0$.
2. Let $\phi = \phi(x)$ be a smooth function localizing to large values for x. So, we assume $\phi(x) = 1$ for $|x| \geq 2L$ and $\phi(x) = 0$ for $x \in G_L$. Then we get for $w := \phi u$ the wave equation with "effective damping term"

$$w_{tt} - \Delta w + a(x)w_t = -\nabla u \cdot \nabla \phi - u \, \Delta \phi.$$

Defining the energy $E_W(w)(t)$, there exists a sequence $\{t_k\}_k$ tending to infinity such that the sequence $\{E_W(w)(t_k)\}_k$ tends to 0.
3. Combining both results yields $\{E_W(u, G)(t_k)\}_k$ tends to 0 for $k \to \infty$. The decreasing behavior of $E_W(u, G)(t)$ implies the desired decay behavior $\lim_{t \to \infty} E_W(u, G)(t) = 0$.

In the following sections we list results on local energy decay for solutions to several models.

11.4.1   Behavior of Local Energies for Solutions to the Free Wave Equation

Consider the following mixed problem for the free wave equation:

$$u_{tt} - \Delta u = 0 \text{ in } (0, \infty) \times G, \quad u(t, x) = 0 \text{ on } (0, \infty) \times \partial G,$$

$$u(0, x) = \varphi(x), \quad u_t(0, x) = \psi(x) \text{ in } G,$$

where $G \subset \mathbb{R}^n$ is an exterior domain with C^∞ boundary and compact obstacle $V := \mathbb{R}^n \setminus G \subset B_{R_0}(0)$. Due to Remark 11.1.1, the total energy $E_W(u, G)(t)$ is conserved. So, it is suitable to ask for decay estimates of the local energies $E_W(u, G_R)(t)$, $G_R := G \cap B_R(0)$. From a physical point of view the energy propagates along the wave fronts (forward or backward). Hence, the motion stops after time passes unless the wave front is trapped in a bounded set. Consequently, geometrical conditions to the exterior domain G are expected. One reasonable condition to assume is that the domain is non-trapping. For example, if V is not star-shaped we can, in general, not expect any decay behavior of local energies. This was proved in [169]. It was shown there that if there is a closed ray solution, then there is no rate of decay. It is mentioned in [41] that the local energy decays exponentially fast if n is odd and polynomially fast if n is even for non-trapping exterior domains G. One of the first contributions is the paper [142]. The author proves the following result in dimensions $n = 2$ and $n = 3$.

Theorem 11.4.1 *Assume $n = 2$ or $n = 3$. The data (φ, ψ) are supposed to satisfy the condition $(\nabla\varphi, \psi) \in L^2(G) \times L^2(G)$. Moreover, the data have compact support in G_R with $R > R_0$. Finally, we assume the existence of a smooth convex function in G with positive normal derivative on ∂G and equal to the distance sufficiently far from the origin. Let u be an energy solution of the above mixed problem. Then, the following energy inequalities hold for large t:*

$$E_W(u, G_R)(t) \leq Ct^{-2}E_W(u, G_R)(0) \ \text{for} \ n = 2,$$
$$E_W(u, G_R)(t) \leq Ce^{-\delta t}E_W(u, G_R)(0) \ \text{for} \ n = 3,$$

where the positive constants C and δ depend on ∂G, G_R and the support of the data. One can find less precise decay estimates for local energies $E_W(u, G_R)(t)$ in higher dimensions $n \geq 4$, too. The following result is taken from [90].

Theorem 11.4.2 *Assume $n \geq 2$. The data (φ, ψ) are supposed to belong to the function space $W^1_{2,0}(G) \times L^2(G)$. Moreover, the data have compact support in G_R with $R > R_0$. Finally, we assume that V is star-shaped with respect to the origin. Let u be an energy solution of the above mixed problem. Then, the following energy inequality holds for large t:*

$$E_W(u, G_R)(t) \leq C(t - R)^{-1}E_W(u, G_R)(0) \ \text{for all} \ t > R,$$

where the positive constant C depends only on R and the data.

11.4.2   Behavior of Local Energies for Solutions to the Elastic Wave Equation

In Sect. 11.2.1 we introduced the total energy

$$E_{EW}(u)(t) = \frac{1}{2}\|u_t(t, \cdot)\|^2_{L^2} + \frac{1}{2}a^2\|\nabla u(t, \cdot)\|^2_{L^2} + \frac{1}{2}(b^2 - a^2)\|\text{div } u(t, \cdot)\|^2_{L^2}$$

for solutions to the elastic wave equation. The paper [102] is devoted to the following mixed problem for the elastic wave equation in an exterior domain $G \subset \mathbb{R}^3$:

$$u_{tt} - a^2\Delta u - (b^2 - a^2)\nabla \text{ div } u = 0 \ \text{in} \ (0, \infty) \times G,$$
$$u(t, x) = 0 \ \text{on} \ (0, \infty) \times \partial G, \ u(0, x) = \varphi(x), \ u_t(0, x) = \psi(x) \ \text{in} \ G,$$

where the positive constants a^2 and b^2 are related to the Lamé constants and satisfy $b^2 > a^2$. Here, $G \subset \mathbb{R}^3$ is an exterior domain with C^∞ boundary and compact obstacle $V := \mathbb{R}^3 \setminus G \subset B_{R_0}(0)$. The interior of V is supposed to be star-shaped with respect to the origin. We recall that if the interior of V is not star-shaped, then we can not expect any uniform decay rate of the energy (see [169]). In [102] the following modified energy to $E_{EW}(u)(t)$ is introduced:

$$E_{EW}^1(u, G)(t) = \frac{1}{2}\|u_t(t, \cdot)\|_{L^2(G)}^2 + \frac{1}{2}a^2\|\mathrm{rot}\, u(t, \cdot)\|_{L^2(G)}^2 + \frac{1}{2}b^2\|\mathrm{div}\, u(t, \cdot)\|_{L^2(G)}^2.$$

This energy is, in the case $G = \mathbb{R}^3$, equivalent to $E_{EW}(u)(t)$. Then, the energy conservation $E_{EW}^1(u, G)(t) = E_{EW}^1(u, G)(0)$ holds. So, it is appropriate to ask for decay properties of local energies. Using the local energy $E_{EW}^1(u, G_R)(t)$, $G_R := G \cap B_R(0)$, the following result was proved in [102]:

Theorem 11.4.3 *The data* $(\vec{\varphi}, \vec{\psi})$ *are supposed to belong to the function space* $W_{2,0}^1(G) \times L^2(G)$. *Moreover, the data have compact support in G_R with $R > R_0$. Let u be a weak energy (with respect to the energy $E_{EW}^1(u, G)(t)$) solution of the above mixed problem. Then, the following decay estimate of the local energy holds:*

$$E_{EW}^1(u, G_R)(t) \leq Ct^{-1} \ \textit{for large} \ t,$$

where the constant C depends on R.

11.4.3   Behavior of Local Energies for Solutions to the Klein-Gordon Equation

Consider the following mixed problem for the Klein-Gordon equation:

$$u_{tt} - \Delta u + m^2 u = 0 \ \text{ in } \ (0, \infty) \times G, \ \ u(t, x) = 0 \ \text{ on } \ (0, \infty) \times \partial G,$$

$$u(0, x) = \varphi(x), \ \ u_t(0, x) = \psi(x) \ \text{ in } \ G,$$

where m^2 is a positive constant, $G \subset \mathbb{R}^n$ is an exterior domain with C^∞ boundary and compact obstacle $V := \mathbb{R}^n \setminus G \subset B_{R_0}(0)$. If we define the total energy of Sobolev solutions to this mixed problem (cf. with Sect. 11.3.4) by

$$E_{KG}(u, G)(t) = \frac{1}{2}\big(\|\nabla u(t, \cdot)\|_{L^2(G)}^2 + \|u_t(t, \cdot)\|_{L^2(G)}^2 + m^2\|u(t, \cdot)\|_{L^2(G)}^2\big),$$

then, after using the homogeneous Dirichlet condition and integration by parts, we obtain the conservation of total energy, that is, $E_{KG}(u, G)(t) = E_{KG}(u, G)(0)$ for all $t > 0$. Consequently, the behavior of local energies is of interest.

To the authors' knowledge the literature does not provide estimates of the local energies $E_{KG}(u, G_R)(t)$, $G_R := G \cap B_R(0)$, in the case of large classes of exterior domains. But, we have such estimates for the domain $G = \mathbb{R}^n$ (see [155]), so we are interested in solutions to the Cauchy problem

$$u_{tt} - \Delta u + m^2 u = 0 \text{ in } (0, \infty) \times \mathbb{R}^n,$$

$$u(0, x) = \varphi(x), \quad u_t(0, x) = \psi(x) \text{ in } \mathbb{R}^n.$$

Theorem 11.4.4 *Assume $n \geq 1$. The data (φ, ψ) are supposed to belong to the function space $C_0^\infty(\mathbb{R}^n) \times C_0^\infty(\mathbb{R}^n)$ with compact support in a bounded domain $G_0 \subset \mathbb{R}^n$. Let u be a classical solution of the above Cauchy problem. Then, the following energy inequality holds for large $t \geq T_0 > d(G_0)$:*

$$E_{KG}(u, G_0)(t) \leq Ct^{-n} E_{KG}(u, G_0)(0),$$

where the positive constant C depends on G_0 and T_0. By $d(G_0)$ we denote the diameter of G_0.

The main tools used in proving this statement are explicit representations of classical solutions, the existence of forward wave fronts of solutions and the fact that for large times the integration is over the domain G_0 only and taking account of the compact support property of the data.

11.4.4 Behavior of Local Energies for Solutions to the Classical Damped Wave Equation

Consider the following mixed problem for the classical damped wave equation:

$$u_{tt} - \Delta u + u_t = 0 \text{ in } (0, \infty) \times G, \quad u(t, x) = 0 \text{ on } (0, \infty) \times \partial G,$$

$$u(0, x) = \varphi(x), \quad u_t(0, x) = \psi(x) \text{ in } G,$$

where $G \subset \mathbb{R}^n$ is an exterior domain with C^∞ boundary and compact obstacle $V := \mathbb{R}^n \setminus G \subset B_{R_0}(0)$. Now, the damping term is effective in the whole domain G. In [41] the authors proved the following result.

Theorem 11.4.5 *Assume $n \geq 2$. The data (φ, ψ) are supposed to belong to the function space $W_{2,0}^1(G) \times L^2(G)$. Moreover, the data have compact support in $G_R := G \cap B_R(0)$ with $R > R_0$. Let u be a weak solution of the above mixed problem. Then, the following energy inequality holds:*

$$E_{KG}(u, G_R)(t) \leq C(1 + t)^{-n} E_{KG}(u, G_R)(0)$$

with a constant $C = C(R)$.

Here the local energy of Klein-Gordon type is defined as in Sect. 11.3.4 for $m^2 = 1$ by

$$E_{KG}(u, K)(t) := \frac{1}{2} \int_K \left(|u_t(t, x)|^2 + |\nabla u(t, x)|^2 + |u(t, x)|^2 \right) dx$$

for all domains $K \subset \mathbb{R}^n$.
The proof relies on spectral (semi-group) methods.

Remark 11.4.1 Compare the statements of Theorems 11.4.5 and 14.2.2. As opposed to the estimates of Theorem 14.2.2, the dimension n appears in the local energy estimate of Theorem 11.4.5.

11.4.5   Behavior of Local Energies for Solutions to the Viscoelastic Damped Wave Equation

Consider the following mixed problem for the viscoelastic damped wave equation:

$$u_{tt} - \Delta u - \Delta u_t = 0 \text{ in } (0, \infty) \times G, \quad u(t, x) = 0 \text{ on } (0, \infty) \times \partial G,$$

$$u(0, x) = \varphi(x), \quad u_t(0, x) = \psi(x) \text{ in } G,$$

where $G \subset \mathbb{R}^n$ is an exterior domain with C^∞ boundary and compact obstacle $V := \mathbb{R}^n \setminus G \subset B_{R_0}(0)$ containing the origin. The total energy $E_W(u, G)(t)$ is decreasing taking into consideration the following result (see [89]).

Theorem 11.4.6 *Assume $n \geq 2$. The data (φ, ψ) are supposed to belong to the function space $(W_2^2(G) \cap W_{2,0}^1(G)) \times (W_2^2(G) \cap W_{2,0}^1(G))$. Then, there exists a unique solution of the above mixed problem. The solution belongs to $C^1([0, \infty), W_2^2(G) \cap W_{2,0}^1(G))$. Moreover, the total energy $E_W(u, G)(t)$ satisfies the relation*

$$E_W(u, G)(t) + \int_0^t \|\nabla u_t(s, \cdot)\|_{L^2(G)}^2 \, ds = E_W(u, G)(0).$$

But, does the total energy $E_W(u, G)(t)$ decay? If not, do we have at least some decay of local energies?
To the authors' knowledge there are no decay results for the behavior of local energies $E_W(u, G_R)(t)$, $G_R := G \cap B_R(0)$. But, there are contributions on the decay behavior of the total wave energy $E_W(u, G)(t)$ under special assumptions to the data (see [89]). To explain these special assumptions we need a function $d = d(|x|)$ which is defined as follows:

$$d(|x|) = |x| \text{ for } n \geq 3,$$

$$d(|x|) = |x| \log(B|x|) \text{ for } n = 2,$$

where B is a positive constant satisfying $B|x| \geq 2$ for all $x \in G$. Besides a certain regularity for the data φ, ψ (cf. with Sect. 14.3), we assume the compatibility-decay condition

$$\|d(\cdot)(\Delta\varphi - \psi)\|_{L^2(G)} < \infty.$$

The following statement was proved in [89]:

Theorem 11.4.7 *Assume $n \geq 2$. The data (φ, ψ) are supposed to belong to the function space $(W_2^2(G) \cap W_{2,0}^1(G)) \times (W_2^2(G) \cap W_{2,0}^1(G))$. Moreover, the data satisfy the above introduced compatibility-decay condition. Let u be an energy solution of the above mixed problem. Then, the following energy inequalities hold for all $t > 0$:*

$$(1 + t)E_W(u, G)(t) \leq C_1 + \frac{3}{2}C_2,$$

$$\int_0^t E_W(u, G)(s)\, ds \leq \frac{1}{2}C_2 + C_3, \quad \|u(t, \cdot)\|_{L^2(G)}^2 \leq 2C_2.$$

The constants C_1 to C_3 are defined as follows:

$$C_1 = \frac{3}{2}E_W(u, G)(0) - \frac{1}{2}(\varphi, \psi)_{L^2(G)},$$

$$C_2 = \frac{1}{2}\|\varphi\|_{L^2(G)}^2 + C_0\|d(\cdot)(\Delta\varphi - \psi)\|_{L^2(G)}^2,$$

$$C_3 = \frac{1}{2}E_W(u, G)(0) - \frac{1}{2}(\varphi, \psi)_{L^2(G)} + C_2,$$

where C_0 is a suitable positive constant.
The constant C_0 is related to the application of Hardy's inequality. It seems interesting to explain the sharpness of the compatibility-decay condition.

1. What about the behavior of the total energy if we use the assumption $(\Delta\varphi - \psi) \in L^2(G)$ only? Do there exist data (φ, ψ) such that the total energy does not decay?
2. What about the behavior of the total energy if we use the condition $d(\cdot)^a(\Delta\varphi - \psi) \in L^2(G)$ with $a \in (0, 1)$ only, assuming a weaker decay? Is there a relation between the decay rate in the compatibility-decay condition for the data and the decay rate of the total energy?

11.4.6 Behavior of Local Energies for Solutions to the Heat Equation

Consider the following mixed problem for the heat equation:

$$u_t - \Delta u = 0 \text{ in } (0, \infty) \times G, \quad u(t, x) = 0 \text{ on } (0, \infty) \times \partial G,$$

$$u(0, x) = \varphi(x) \text{ in } G,$$

where $G \subset \mathbb{R}^n$ is an exterior domain with C^∞ boundary and compact obstacle $V := \mathbb{R}^n \setminus G \subset B_{R_0}(0)$. Due to [20], for each $\varphi \in W^1_{2,0}(G)$ there exists a unique Sobolev solution

$$u \in C\big([0,\infty), W^1_{2,0}(G)\big) \cap C^1\big((0,\infty), L^2(G)\big) \cap C\big((0,\infty), W^2_2(G)\big)$$

to the above mixed problem.

If we define the total energy of Sobolev solutions to this mixed problem (cf. with Sect. 11.2.2) by

$$E_H(u, G)(t) = \frac{1}{2}\|u(t,\cdot)\|^2_{L^2(G)},$$

then, after using the homogeneous Dirichlet condition and integration by parts, we obtain the estimate

$$d_t E_H(u, G)(t) + \|\nabla u(t,\cdot)\|^2_{L^2(G)} = 0,$$

that is, the total energy $E_H(u, G)(t)$ is decreasing. But, does it decay? Here we refer to the paper [91]. To formulate the result we recall the function $d = d(x)$ of Sect. 11.4.5.

Theorem 11.4.8 *Assume $n \geq 2$. The data φ is supposed to belong to the function space $W^1_{2,0}(G)$ and satisfies $\|d(\cdot)\varphi\|_{L^2(G)} < \infty$. Then, the following decay estimates are valid for the total energy and for $t \geq 0$:*

$$E_H(u, G)(t) \leq \frac{C}{1+t}\|d(\cdot)\varphi\|_{L^2(G)}, \quad E_H(\nabla u, G)(t) \leq \frac{1}{1+t}\|\varphi\|_{W^1_2(G)},$$

where the constant C depends only on G.

It seems interesting to explain the necessity of the condition "$\varphi \in W^1_{2,0}(G)$ satisfies the additional condition $\|d(\cdot)\varphi\|_{L^2(G)} < \infty$."

1. What about the behavior of the total energy if we use the assumption $\varphi \in W^1_{2,0}(G)$ only?
2. What about the behavior of the total energy if we use the condition $\|d(\cdot)^a\varphi\|_{L^2(G)} < \infty$ with $a \in (0,1)$ only, assuming a weaker decay? Is there a relation between the decay rate for the data and the decay rate of the total energy?

To the authors' knowledge the literature does not provide any estimates for the local energies $E_H(u, G_R)(t)$, $G_R := G \cap B_R(0)$, in the case of exterior domains.

But, there might be a connection to local energy decay estimates for solutions to the nonstationary Stokes system in exterior domains (see [97]). Consider in an exterior domain $G \subset \mathbb{R}^n$ the mixed problem for the nonstationary Stokes system (see Sect. 2.8)

$$u_t - \Delta u + \nabla p = 0, \quad \operatorname{div} u = 0 \quad \text{in } (0,\infty) \times G,$$

$$u(t,x) = 0 \quad \text{on } (0,\infty) \times \partial G, \quad u(0,x) = \varphi(x) \quad \text{in } G,$$

where $G \subset \mathbb{R}^n$ is an exterior domain with C^∞ boundary and compact obstacle $V :=$ $\mathbb{R}^n \setminus G \subset B_{R_0}(0)$. Taking into consideration the compatibility between conservation of mass and initial condition the data φ is supposed to belong to the completion in $L^q(G)$ of the set $\{f \in (C_0^\infty(G))^n : \operatorname{div} f = 0\}$ (cf. with the Helmholtz decomposition of $L^q(G)$, [76]). The following result is taken from [97]:

Theorem 11.4.9 *Let $n \geq 3$ and $q \in (1, \infty)$. For any $R > R_0$ there exists a positive constant $C = C(q, R)$ such that the following local energy decay estimate is valid:*

$$\|u(t, \cdot)\|_{L^q(G_R)} \leq C(1 + t)^{-\frac{n}{2q}} \|\varphi\|_{L^q(G)}.$$

Remark 11.4.2 There are deeper relations between solutions to the heat equation and the Stokes system. If we restrict ourselves to $G = \mathbb{R}^n$, then some of the results of Chap. 12 can be generalized to solutions to the Stokes system (see [218]).

11.4.7 Behavior of Local Energies for Solutions to the Schrödinger Equation

Consider the following mixed problem for the free Schrödinger equation:

$$\frac{1}{i} u_t \pm \Delta u = 0 \text{ in } (0, \infty) \times G, \quad u(t, x) = 0 \text{ on } (0, \infty) \times \partial G,$$

$$u(0, x) = \varphi(x) \text{ in } G,$$

where $G \subset \mathbb{R}^n$ is an exterior domain with C^∞ boundary and compact obstacle $V := \mathbb{R}^n \setminus G \subset B_{R_0}(0)$. If we define the total energy of Sobolev solutions to this mixed problem (cf. with Sect. 11.2.3) by

$$E_{Sch}(u, G)(t) = \frac{1}{2} \|u(t, \cdot)\|_{L^2(G)}^2 = \frac{1}{2} (u, \bar{u})_{L^2(G)},$$

then after using the homogeneous Dirichlet condition and integration by parts we obtain the conservation of total energy, that is, we may conclude the relation $E_{Sch}(u, G)(t) = E_{Sch}(u, G)(0)$ for all $t > 0$. Consequently, the behavior of local energies is of interest. Here we refer to the paper [208].

The domain G is supposed to be "non-trapping". In this way we may exclude the existence of closed ray solutions. The "non-trapping" condition is described by Green's function of the wave equation in G. It is assumed that singularities of Green's function tend to infinity for t to infinity. This condition is satisfied in the case where the interior of V is convex.

The following result is proved in [208].

Theorem 11.4.10 *Assume $n \geq 3$. The data φ is supposed to belong to the function space $L^2(G)$. Moreover, φ has compact support in G_R with $R > R_0$ and $G_R :=$*

$G \cap B_R(0)$. *Finally, we assume the above conditions for the domain G. Let u be an energy solution of the above mixed problem. Then, the following energy inequality holds for large t:*

$$E_{Sch}(u, G_{2R})(t) \leq Ct^{-n} E_{Sch}(u, G_R)(0),$$

where the positive constant C depends on ∂G, G, R and n.

The proof relies on an explicit representation of the evolution operator associated with solutions of the mixed problem and resolvent estimates for the operator ($is \pm \Delta)^{-1}$.

Exercises Relating to the Considerations of Chap. 11

Exercise 1 Show that the energies of higher order $E_H^k(u)(t)$ of solutions to the Cauchy problem

$$u_t - \Delta u = 0, \quad u(0, x) = \varphi(x)$$

decay faster and faster, that is, an increasing k gives a faster decay.

Exercise 2 Prove the statement of Theorem 11.3.1.

Exercise 3 Prove the statement of Theorem 11.3.2.

Exercise 4 Prove the equivalence of the energies $E_{EW}(u, \mathbb{R}^3)(t)$ and $\overset{1}{E}_{EW}(u, \mathbb{R}^3)(t)$ introduced in Sect. 11.4.2.

Part III

Chapter 12
Phase Space Analysis for the Heat Equation

This chapter explains in an elementary way via the Cauchy problem for the heat equation without and with mass term how phase space analysis and interpolation techniques can be used to prove $L^p - L^q$ estimates on and away from the conjugate line $\frac{1}{p} + \frac{1}{q} = 1, p \in [1, \infty]$. Here we distinguish between $L^p - L^q$ estimates for low regular and for large regular data.

12.1 The Classical Heat Equation

Let us consider the Cauchy problem for the heat equation

$$u_t - \Delta u = 0, \quad u(0, x) = \varphi(x).$$

Applying the partial Fourier transformation yields for $v(t, \xi) = F_{x \to \xi}(u)(t, \xi)$ the Cauchy problem

$$v_t + |\xi|^2 v = 0, \quad v(0, \xi) = F(\varphi)(\xi).$$

Its solution is given by

$$v(t, \xi) = e^{-|\xi|^2 t} F(\varphi)(\xi).$$

After application of inverse Fourier transformation we obtain

$$u(t, x) = F^{-1}_{\xi \to x}\left(e^{-|\xi|^2 t} F(\varphi)(\xi)\right).$$

M.R. Ebert, M. Reissig, *Methods for Partial Differential Equations*,
https://doi.org/10.1007/978-3-319-66456-9_12

12.1.1 $L^2 - L^2$ Estimates

Theorem 12.1.1 *We study the Cauchy problem*

$$u_t - \Delta u = 0, \quad u(0, x) = \varphi(x).$$

Then, we have the following estimates for the derivatives $\partial_t^k \partial_x^\alpha u(t, \cdot)$ of the solution u $(k + |\alpha| \geq 0)$:

$$\|\partial_t^k \partial_x^\alpha u(t, \cdot)\|_{L^2} \leq C_{k,\alpha} t^{-k - \frac{|\alpha|}{2}} \|\varphi\|_{L^2},$$

$$\|\partial_t^k \partial_x^\alpha u(t, \cdot)\|_{L^2} \leq C_{k,\alpha} (1 + t)^{-k - \frac{|\alpha|}{2}} \|\varphi\|_{H^{2k+|\alpha|}}.$$

Proof Using the properties of the Fourier transformation it holds

$$\partial_t^k \partial_x^\alpha u(t, x) = F_{\xi \to x}^{-1} \left((-1)^k i^{|\alpha|} |\xi|^{2k} \xi^\alpha e^{-|\xi|^2 t} F(\varphi)(\xi) \right).$$

By Parseval-Plancharel formula (Remark 24.1.2) we get for $t > 0$

$$\|\partial_t^k \partial_x^\alpha u(t, \cdot)\|_{L^2}^2 = \left\| |\xi|^{2k} \xi^\alpha e^{-|\xi|^2 t} F(\varphi)(\xi) \right\|_{L^2}^2 \leq \left\| |\xi|^{2k+|\alpha|} e^{-|\xi|^2 t} F(\varphi)(\xi) \right\|_{L^2}^2$$

$$= \left\| \frac{|\xi|^{2k+|\alpha|} t^{k + \frac{|\alpha|}{2}}}{t^{k + \frac{|\alpha|}{2}}} e^{-|\xi|^2 t} F(\varphi)(\xi) \right\|_{L^2}^2.$$

The term

$$|\xi|^{2k+|\alpha|} t^{k + \frac{|\alpha|}{2}} e^{-|\xi|^2 t} = \left(|\xi|^2 t \right)^{k + \frac{|\alpha|}{2}} e^{-|\xi|^2 t}$$

is uniformly bounded for a given k and α for all positive t and $|\xi|$ by a constant $C_{k,\alpha}$. Consequently,

$$\|\partial_t^k \partial_x^\alpha u(t, \cdot)\|_{L^2}^2 \leq C_{k,\alpha} t^{-2k-|\alpha|} \|F(\varphi)\|_{L^2}^2 = C_{k,\alpha} t^{-2k-|\alpha|} \|\varphi\|_{L^2}^2.$$

This implies the first desired estimate.

Now we use the above estimate for $t \geq 1$ only. We change our argument for small times $t \in (0, 1]$ as follows. We assume further regularity of φ and conclude the estimate

$$\|\partial_t^k \partial_x^\alpha u(t, \cdot)\|_{L^2}^2 = \left\| |\xi|^{2k} \xi^\alpha e^{-|\xi|^2 t} F(\varphi)(\xi) \right\|_{L^2}^2 \leq \left\| |\xi|^{2k+|\alpha|} e^{-|\xi|^2 t} F(\varphi)(\xi) \right\|_{L^2}^2$$

$$\leq \left\| |\xi|^{2k+|\alpha|} F(\varphi)(\xi) \right\|_{L^2}^2 \leq \|\varphi\|_{H^{2k+|\alpha|}}^2.$$

This implies the second desired estimate.

Remark 12.1.1 Let us discuss the difference between both estimates. The first estimate requires only L^2 regularity for φ. We get for large t the decay $t^{-k-\frac{|\alpha|}{2}}$. But, this term becomes unbounded for $t \to +0$. To avoid this singular behavior we assume additional regularity $H^{2k+|\alpha|}$. Then, the decay is described by $(1+t)^{-k-\frac{|\alpha|}{2}}$, a function that remains bounded for $t \to +0$.

12.1.2 $L^p - L^q$ Estimates on the Conjugate Line

To derive $L^1 - L^\infty$ estimates we use the representation of solution

$$u(t,x) = F_{\xi \to x}^{-1}\big(e^{-|\xi|^2 t} F(\varphi)(\xi)\big) = F_{\xi \to x}^{-1}\big(e^{-|\xi|^2 t}\big) * \varphi.$$

Here we applied the convolution rule and supposed the Fourier inversion formula for φ. Using Young's inequality of Proposition 24.5.2 and the statements of Theorems 24.2.1, 24.2.2 and 24.2.3 yield

$$\|u(t,\cdot)\|_{L^\infty} \le \big\|F_{\xi \to x}^{-1}\big(e^{-|\xi|^2 t}\big)\big\|_{L^\infty} \|\varphi\|_{L^1} \le C t^{-\frac{n}{2}} \|\varphi\|_{L^1}$$

for data $\varphi \in L^1(\mathbb{R}^n)$. In the same way we conclude together with the statements of Corollaries 24.2.1, 24.2.2 and 24.2.3 the following estimates for the derivatives:

$$\big\|\partial_t^k \partial_x^\alpha u(t,\cdot)\big\|_{L^\infty} \le \big\|F_{\xi \to x}^{-1}\big(|\xi|^{2k} \xi^\alpha e^{-|\xi|^2 t}\big)\big\|_{L^\infty} \|\varphi\|_{L^1} \le C_{k,\alpha} t^{-\frac{n+2k+|\alpha|}{2}} \|\varphi\|_{L^1}.$$

This estimate and the $L^2 - L^2$ estimate of Theorem 12.1.1 allow us to apply Proposition 24.5.1 to conclude a $L^p - L^q$ decay estimate on the conjugate line.

Theorem 12.1.2 *We study the Cauchy problem*

$$u_t - \Delta u = 0, \quad u(0,x) = \varphi(x).$$

Then, we have the following estimates for the derivatives $\partial_t^k \partial_x^\alpha u(t,\cdot)$ of the solution u ($k + |\alpha| \ge 0$):

$$\|\partial_t^k \partial_x^\alpha u(t,\cdot)\|_{L^q} \le C_{k,\alpha} t^{-k-\frac{|\alpha|}{2}-\frac{n}{2}\left(\frac{1}{p}-\frac{1}{q}\right)} \|\varphi\|_{L^p}$$

for $p \in [1,2]$ and $\frac{1}{p} + \frac{1}{q} = 1$.

12.1.3 $L^p - L^q$ Estimates Away of the Conjugate Line

In the previous section we restricted our considerations to $L^p - L^q$ decay estimates on the conjugate line. But, the range of admissible p and q is determined due to Young's inequality taking account of the results of Sect. 24.2.2. This issue is discussed as follows. By the approach of the previous section we arrive immediately at the following result.

Theorem 12.1.3 *We study the Cauchy problem*

$$u_t - \Delta u = 0, \quad u(0, x) = \varphi(x).$$

Then, we have the following estimates for the derivatives $\partial_t^k \partial_x^\alpha u(t, \cdot)$ of the solution u ($k + |\alpha| \geq 0$):

$$\|\partial_t^k \partial_x^\alpha u(t, \cdot)\|_{L^q} \leq C_{k,\alpha} t^{-k - \frac{|\alpha|}{2} - \frac{n}{2}(1 - \frac{1}{r})} \|\varphi\|_{L^p}$$

for all $1 \leq p \leq q \leq \infty$ and $1 + \frac{1}{q} = \frac{1}{r} + \frac{1}{p}$.
This result contains the interesting special case $p = q$.

Corollary 12.1.1 *Under the assumptions of Theorem 12.1.3 we conclude the estimates*

$$\|\partial_t^k \partial_x^\alpha u(t, \cdot)\|_{L^q} \leq C_{k,\alpha} t^{-k - \frac{|\alpha|}{2}} \|\varphi\|_{L^q}$$

for all $1 \leq q \leq \infty$.
This corollary is used to derive, for example, $L^p - L^q$ decay estimates with decay function $1 + t$ instead of t (cf. Theorem 12.1.1). On the one hand, $\partial_t^k \partial_x^\alpha u$ solves the Cauchy problem for the heat equation if u does, supposing that the data $\Delta^k \partial_x^\alpha \varphi$ have the required regularity. While on the other hand we use the embedding of $W_1^n(\mathbb{R}^n)$ into $L^\infty(\mathbb{R}^n)$. Both together give, with the previous corollary,

$$\|\partial_t^k \partial_x^\alpha u(t, \cdot)\|_{L^\infty} \leq C \|\Delta^k \partial_x^\alpha \varphi\|_{L^\infty} \leq C \|\Delta^k \partial_x^\alpha \varphi\|_{W_1^n} \leq C \|\varphi\|_{W_1^{n+2k+|\alpha|}}$$

for all $t \in (0, 1]$. For $t \geq 1$ we apply Theorem 12.1.3 for $q = \infty$ and $p = 1$. Summarizing implies the following estimate:

$$\|\partial_t^k \partial_x^\alpha u(t, \cdot)\|_{L^\infty} \leq C_{k,\alpha} (1 + t)^{-k - \frac{|\alpha|}{2} - \frac{n}{2}} \|\varphi\|_{W_1^{n+2k+|\alpha|}}.$$

Due to Proposition 24.5.3, interpolation between this estimate and the $L^2 - L^2$ estimate of Theorem 12.1.1 leads to the following $L^p - L^q$ decay estimate on the conjugate line.

Corollary 12.1.2 *Under the assumptions of Theorem 12.1.3 we conclude the estimates*

$$\|\partial_t^k \partial_x^\alpha u(t, \cdot)\|_{L^q} \leq C_{k,\alpha} (1 + t)^{-k - \frac{|\alpha|}{2} - \frac{n}{2}(\frac{1}{p} - \frac{1}{q})} \|\varphi\|_{W_p^{N_p + 2k + |\alpha|}}$$

for all $2 \leq q \leq \infty$, $\frac{1}{p} + \frac{1}{q} = 1$, $N_p > n(\frac{1}{p} - \frac{1}{q})$ for $q \in (2, \infty)$, $N_2 = 0$ and $N_1 = n$.

We are able to derive corresponding $L^p - L^q$ decay estimates away from the conjugate line. On the one hand we have the estimate of Theorem 12.1.3. Interpolating the estimate of Corollary 12.1.1 for $|\alpha| = k = 0$ and $q = \infty$ with the embedding

$$W_p^{N_p}(\mathbb{R}^n) \hookrightarrow L^\infty(\mathbb{R}^n) \text{ for } N_p > \frac{n}{p}, p \in (1,\infty), \ N_p = \frac{n}{p}, p \in \{1;\infty\},$$

gives (see Definition 24.8 for the Sobolev-Slobodeckij space $W_p^{N_p}$)

$$\|u(t,\cdot)\|_{L^\infty} \leq C\|\varphi\|_{W_p^{N_p}}, \quad \|\partial_t^k \partial_x^\alpha u(t,\cdot)\|_{L^\infty} \leq C\|\varphi\|_{W_p^{N_p+2k+|\alpha|}},$$

respectively, for all $p \in [1,\infty]$ with $N_p > \frac{n}{p}$ for $p \in (1,\infty)$ and $N_p = \frac{n}{p}$ for $p \in \{1;\infty\}$.

For $q \in [1,\infty)$ we may conclude directly by using embedding theorems only. Thanks to Remark 24.3.3 and the identity modulo norms (see Remark 24.3.1)

$$B_{p,p}^s(\mathbb{R}^n) = W_p^s(\mathbb{R}^n) \text{ for } p \in [1,\infty) \text{ and } s \notin \mathbb{Z}$$

we have the embedding

$$W_p^{n(\frac{1}{p}-\frac{1}{q})}(\mathbb{R}^n) \hookrightarrow L^q(\mathbb{R}^n) \text{ for } 1 \leq p \leq q < \infty.$$

Using the $L^q - L^q$ decay estimate from Theorem 12.1.3 gives

$$\|u(t,\cdot)\|_{L^q} \leq C\|\varphi\|_{W_p^{n(\frac{1}{p}-\frac{1}{q})}}, \quad \|\partial_t^k \partial_x^\alpha u(t,\cdot)\|_{L^q} \leq C\|\varphi\|_{W_p^{n(\frac{1}{p}-\frac{1}{q})+2k+|\alpha|}},$$

respectively, for $1 \leq p \leq q < \infty$. Summarizing all these estimates implies the following statement.

Theorem 12.1.4 *We study the Cauchy problem*

$$u_t - \Delta u = 0, \quad u(0,x) = \varphi(x).$$

Then, we have the following estimates for the derivatives $\partial_t^k \partial_x^\alpha u(t,\cdot)$ of the solution u ($k + |\alpha| \geq 0$):

$$\|\partial_t^k \partial_x^\alpha u(t,\cdot)\|_{L^q} \leq C_{k,\alpha}(1+t)^{-k-\frac{|\alpha|}{2}-\frac{n}{2}(1-\frac{1}{r})}\|\varphi\|_{W_p^{n(1-\frac{1}{r})+2k+|\alpha|}}$$

for all $1 \leq p \leq q < \infty$ and $1 + \frac{1}{q} = \frac{1}{r} + \frac{1}{p}$.

We complete this section with a corresponding result to Theorem 12.1.4 for Bessel potential spaces $H_p^s(\mathbb{R}^n)$ (see Definition 24.9). If $p \in (1,\infty)$ and $s_2 > s_1 > 0$, then we can choose the following chain of embeddings (see [179] and the Definitions 24.8, 24.9, 24.11):

$$H_p^{s_2}(\mathbb{R}^n) = F_{p,2}^{s_2}(\mathbb{R}^n) \hookrightarrow F_{p,\infty}^{s_2}(\mathbb{R}^n) \hookrightarrow F_{p,p}^{s_1}(\mathbb{R}^n) = W_p^{s_1}(\mathbb{R}^n).$$

The embedding of $H_p^{s_2}(\mathbb{R}^n)$ into $W_p^{s_1}(\mathbb{R}^n)$ remains valid for nonnegative integers s_1. For this reason the following result is a consequence of Theorem 12.1.4.

Corollary 12.1.3 *We study the Cauchy problem*

$$u_t - \Delta u = 0, \quad u(0, x) = \varphi(x).$$

Then, we have the following estimates for the derivatives $\partial_t^k \partial_x^\alpha u(t, \cdot)$ of the solution u $(k + |\alpha| \geq 0)$:

$$\|\partial_t^k \partial_x^\alpha u(t, \cdot)\|_{L^q} \leq C_{k,\alpha}(1 + t)^{-k - \frac{|\alpha|}{2} - \frac{n}{2}(1 - \frac{1}{r})}\|\varphi\|_{H_p^{N_r + 2k + |\alpha|}}$$

for all $1 < p \leq q < \infty$, $1 + \frac{1}{q} = \frac{1}{r} + \frac{1}{p}$ and $N_r > n(1 - \frac{1}{r})$.

12.2   The Classical Heat Equation with Mass

Now we study the Cauchy problem for the heat equation with a mass term. We are interested in understanding the influence of the *mass term* on $L^p - L^q$ estimates. Let us turn to the Cauchy problem for the heat equation with mass term

$$u_t - \Delta u + m^2 u = 0, \quad u(0, x) = \varphi(x),$$

where m^2 is a positive constant. Applying the partial Fourier transformation yields for $v(t, \xi) = F_{x \to \xi}(u)(t, \xi)$ the Cauchy problem

$$v_t + (|\xi|^2 + m^2)v = 0, \quad v(0, \xi) = F(\varphi)(\xi).$$

Introducing the notation $\langle \xi \rangle_m := \sqrt{|\xi|^2 + m^2}$, its solution is

$$v(t, \xi) = e^{-\langle \xi \rangle_m^2 t} F(\varphi)(\xi).$$

After application of inverse Fourier transformation we obtain

$$u(t, x) = F_{\xi \to x}^{-1}\left(e^{-\langle \xi \rangle_m^2 t} F(\varphi)(\xi)\right).$$

Theorem 12.2.1 *We study the Cauchy problem*

$$u_t - \Delta u + m^2 u = 0, \quad u(0, x) = \varphi(x).$$

Then, we have the following estimates for the derivatives $\partial_t^k \partial_x^\alpha u(t, \cdot)$ *of the solution* u $(k + |\alpha| \geq 0)$:

$$\|\partial_t^k \partial_x^\alpha u(t, \cdot)\|_{L^2} \leq C_{k,\alpha} t^{-k-\frac{|\alpha|}{2}} e^{-\frac{m^2}{2}t} \|\varphi\|_{L^2},$$

$$\|\partial_t^k \partial_x^\alpha u(t, \cdot)\|_{L^2} \leq C_{k,\alpha} e^{-m^2 t} \|\varphi\|_{H^{2k+|\alpha|}}.$$

Proof We follow the proof to Theorem 12.1.1 to estimate

$$\|\partial_t^k \partial_x^\alpha u(t, \cdot)\|_{L^2} = \left\| \langle \xi \rangle_m^{2k} \xi^\alpha e^{-\langle \xi \rangle_m^2 t} F(\varphi)(\xi) \right\|_{L^2}$$

$$\leq e^{-\frac{1}{2}m^2 t} \left\| \langle \xi \rangle_m^{2k} \xi^\alpha e^{-\frac{1}{2}\langle \xi \rangle_m^2 t} F(\varphi)(\xi) \right\|_{L^2}.$$

Remark 12.2.1 Comparing the estimates of Theorems 12.1.1 and 12.2.1 we see that the mass term has an improving influence on $L^2 - L^2$ estimates. We have an exponential decay for $t \to \infty$ in comparison with the potential decay of Theorem 12.1.1.

We expect this exponential decay in more general $L^p - L^q$ decay estimates, too (see the research project in Sect. 23.3).

Exercises Relating to the Considerations of Chap. 12

Exercise 1 Verify that the solution $u = u(t, x)$ to the Cauchy problem for the heat equation with mass

$$u_t - \Delta u + m^2 u = 0, \quad u(0, x) = \varphi(x)$$

can be given by $u(t, x) = e^{-m^2 t} w(t, x)$, where $w = w(t, x)$ solves the Cauchy problem

$$w_t - \Delta w = 0, \quad w(0, x) = \varphi(x).$$

Then derive $L^p - L^q$ decay estimates not necessarily on the conjugate line for the heat equation with mass.

Exercise 2 Let

$$K(t, \cdot) = F_{\xi \to x}^{-1} \left(|\xi|^a e^{-t|\xi|^2} \right).$$

Verify that for all $r \in [1, \infty]$ and $a \geq 0$ it holds

$$\|K(t, \cdot)\|_{L^r} = t^{-\frac{n}{2}\left(1-\frac{1}{r}\right)-\frac{a}{2}} \|K(1, \cdot)\|_{L^r}.$$

Hence, applying Young's inequality to conclude the statement of Theorem 12.1.3, it is sufficient to prove that $\|K(1, \cdot)\|_{L^r}$ is finite.

Chapter 13
Phase Space Analysis and Smoothing for Schrödinger Equations

Consider this chapter as a brief introduction to some properties of solutions to the classical Schrödinger equation with or without mass term. We continue the discussion of $L^p - L^q$ estimates for this special example of a dispersive equation. In particular, we explain differences between expected results for the Schrödinger equation and for those of the heat equation. In addition to these applications of phase space analysis we discuss the topic "Smoothing effect" for solutions, local and global smoothing as well.

Many thanks to Mitsuru Sugimoto (Nagoya) for useful discussions on the content of this chapter.

13.1 $L^p - L^q$ Estimates

13.1.1 The Classical Schrödinger Equation

Let us consider the Cauchy problem for the Schrödinger equation

$$D_t u - \Delta u = 0, \quad u(0, x) = \varphi(x).$$

We recall that $D_t := -i\partial_t$ (see Sect. 3.1). Then one can show (cf. with the representation of solution for the heat equation of Sect. 9.3) that

$$u(t, x) = \frac{1}{(2\sqrt{\pi it})^n} \int_{\mathbb{R}^n} e^{i\frac{|x-y|^2}{4t}} \varphi(y) \, dy$$

is a solution of this Cauchy problem for a suitable class of data, let us say, for data φ from $L^1(\mathbb{R}^n)$.

© Springer International Publishing AG 2018
M.R. Ebert, M. Reissig, *Methods for Partial Differential Equations*,
https://doi.org/10.1007/978-3-319-66456-9_13

One can derive a second representation of solutions by using Fourier multipliers. Applying the partial Fourier transformation yields for $v(t, \xi) = F_{x\to\xi}(u)(t, \xi)$ the Cauchy problem

$$v_t + i|\xi|^2 v = 0, \quad v(0, \xi) = F(\varphi)(\xi).$$

Its solution is

$$v(t, \xi) = e^{-i|\xi|^2 t} F(\varphi)(\xi).$$

After application of inverse Fourier transformation (we assume the validity of Fourier's inversion formula) we obtain

$$u(t, x) = F^{-1}_{\xi\to x}\left(e^{-i|\xi|^2 t} F(\varphi)(\xi)\right).$$

13.1.1.1 $L^2 - L^2$ Estimates

Theorem 13.1.1 *We study the Cauchy problem*

$$D_t u - \Delta u = 0, \quad u(0, x) = \varphi(x).$$

Then, we have the following estimates for the derivatives $\partial_t^k \partial_x^\alpha u(t, \cdot)$ of the solution u ($k + |\alpha| \geq 0$):

$$\|\partial_t^k \partial_x^\alpha u(t, \cdot)\|_{L^2} = \||D|^{2k} \partial_x^\alpha \varphi\|_{L^2} \leq \|\varphi\|_{H^{2k+|\alpha|}}.$$

Proof Here we use the energy conservation $E_{Sch}(u)(t) = E_{Sch}(u)(0)$ of Sect. 11.2.3 and the property that $\partial_t^k \partial_x^\alpha u$ solves the Cauchy problem

$$D_t\left(\partial_t^k \partial_x^\alpha u\right) - \Delta\left(\partial_t^k \partial_x^\alpha u\right) = 0, \quad \left(\partial_t^k \partial_x^\alpha u\right)(0, x) = i^k \Delta^k \partial_x^\alpha \varphi(x)$$

under the assumption that $\Delta^k \partial_x^\alpha \varphi$ belongs to $L^2(\mathbb{R}^n)$. The proof is complete.

Remark 13.1.1 Let us discuss the difference between the statements of Theorems 12.1.1 and 13.1.1. In Theorem 12.1.1 we explain the so-called *parabolic effect*, that is, higher order partial derivatives decay faster and faster for $t \to \infty$, where the order of derivatives appears in the decay rate. We can not expect such an effect for solutions to the classical Schrödinger equation. One of the reasons is that the conservation of energy $E_{Sch}(u)(t)$ of Sect. 11.2.3 is in opposition to the energy $E_H(u)(t)$ of Sect. 11.2.2, which is decreasing. Thus solutions to Schrödinger equations behave similar to solutions to wave equations where higher order derivatives of solutions do not decay faster (see Theorem 16.1.1 in Sect. 16.1).

13.1.1.2 $L^p - L^q$ Estimates on the Conjugate Line

First we derive $L^1 - L^\infty$ estimates.

Theorem 13.1.2 *We study the Cauchy problem*

$$D_t u - \Delta u = 0, \quad u(0, x) = \varphi(x).$$

Then, we have the following $L^1 - L^\infty$ estimates for the derivatives $\partial_t^k \partial_x^\alpha u(t, \cdot)$ of the solution u $(k + |\alpha| \geq 0)$:

$$\|\partial_t^k \partial_x^\alpha u(t, \cdot)\|_{L^\infty} \leq Ct^{-\frac{n}{2}} \||D|^{2k} \partial_x^\alpha \varphi\|_{L^1}.$$

Proof Again we use $\partial_t^k \partial_x^\alpha u$ which solves the Cauchy problem

$$D_t \left(\partial_t^k \partial_x^\alpha u\right) - \Delta \left(\partial_t^k \partial_x^\alpha u\right) = 0, \quad \left(\partial_t^k \partial_x^\alpha u\right)(0, x) = i^k \Delta^k \partial_x^\alpha \varphi(x)$$

under the assumption that $\Delta^k \partial_x^\alpha \varphi$ has the required regularity. So it remains to derive the desired estimates only for u itself. Using the representation of solution

$$u(t, x) = \frac{1}{(2\sqrt{\pi i t})^n} \int_{\mathbb{R}^n} e^{i\frac{|x-y|^2}{4t}} \varphi(y) \, dy$$

we immediately obtain $\|u(t, \cdot)\|_{L^\infty} \leq Ct^{-\frac{n}{2}} \|\varphi\|_{L^1}$. The proof is complete.

Using the interpolation argument in Proposition 24.5.1, the statements of Theorems 13.1.1 and 13.1.2 imply the following result on $L^p - L^q$ estimates on the conjugate line for solutions to the Cauchy problem for the classical Schrödinger equation.

Theorem 13.1.3 *We study the Cauchy problem*

$$D_t u - \Delta u = 0, \quad u(0, x) = \varphi(x).$$

Then, we have the following $L^p - L^q$ estimates on the conjugate line for the derivatives $\partial_t^k \partial_x^\alpha u(t, \cdot)$ of the solution u $(k + |\alpha| \geq 0)$ for $q \in [2, \infty]$:

$$\|\partial_t^k \partial_x^\alpha u(t, \cdot)\|_{L^q} \leq Ct^{-\frac{n}{2}(\frac{1}{p} - \frac{1}{q})} \||D|^{2k} \partial_x^\alpha \varphi\|_{L^p},$$

where $\frac{1}{p} + \frac{1}{q} = 1$.

Remark 13.1.2 By using a Littman type lemma (see Theorem 16.8.1) we may, at least for $p > 1$, avoid in the last theorem the singular estimate at $t = 0$ by asking for higher regularity for the data φ, (see Theorem 16.8.4).

Remark 13.1.3 Contrary to the heat equation, one may not expect $L^p - L^p$ estimates for $p \neq 2$ for solutions to the Schrödinger equation (see page 63 of [122]).

The last remark is one reason to introduce other classes of estimates, so-called *Strichartz estimates*.

Remark 13.1.4 The solution to the Cauchy problem

$$D_t u - \Delta u = 0, \quad u(0, x) = \varphi(x)$$

satisfies the Strichartz estimates

$$\|u\|_{L^r(\mathbb{R}^1, L^q(\mathbb{R}^n))} \leq C\|\varphi\|_{L^2(\mathbb{R}^n)},$$

provided that

$$\frac{2}{r} + \frac{n}{q} \leq \frac{n}{2}, \quad 2 \leq r \leq \infty, \quad (n, r) \neq (2, 2).$$

Every such pair (r, q) is called an admissible pair, sharp admissible pair if the equality holds (see, for example, [109], the authors proved Strichartz estimates for solutions to wave equations, too). Such estimates describe a certain regularity for the solutions in terms of summability, but they do not contain any gain of regularity (see next section). If we choose $r = \infty$ and $q = 2$, then we get the same estimate as in Theorem 13.1.1 for $|\alpha| = k = 0$.

Remark 13.1.5 One has also Strichartz estimates for solutions to the Cauchy problem

$$D_t u - \Delta u = F(t, x), \quad u(0, x) = 0$$

(see [109]). For any admissible pairs (r, q) and $(\tilde{r}, \tilde{q})$ we have the estimates

$$\|u\|_{L^r(\mathbb{R}^1, L^q(\mathbb{R}^n))} \leq C\|F\|_{L^{\tilde{r}'}(\mathbb{R}^1, L^{\tilde{q}'}(\mathbb{R}^n))},$$

where $\frac{1}{r} + \frac{1}{\tilde{r}'} = 1$ and $\frac{1}{q} + \frac{1}{\tilde{q}'} = 1$.

13.1.2 The Classical Schrödinger Equation with Mass

Let us consider the Cauchy problem for the classical Schrödinger equation with mass term

$$D_t u - \Delta u + m^2 u = 0, \quad u(0, x) = \varphi(x), \quad m^2 > 0.$$

Defining the energies $E^k_{Sch}(u)(t)$ we obtain, as in Sect. 11.2.3, the energy conservation $E^k_{Sch}(u)(t) = E^k_{Sch}(u)(0)$. If we introduce the change of variables $v := e^{im^2 t}u$, then the above Cauchy problem is transformed into

$$D_t v - \Delta v = 0, \quad v(0, x) = \varphi(x).$$

Consequently, we arrive at the following representations of solutions:

$$u(t, x) = e^{-im^2 t} \frac{1}{(2\sqrt{\pi i t})^n} \int_{\mathbb{R}^n} e^{i\frac{|x-y|^2}{4t}} \varphi(y)\, dy,$$

$$u(t, x) = F^{-1}_{\xi \to x}\left(e^{-i\langle\xi\rangle^2_m t} F(\varphi)(\xi)\right).$$

These representations allow us to apply the statements of Theorems 13.1.1 to 13.1.3. In this way we may conclude the following result.

Theorem 13.1.4 *We study the Cauchy problem*

$$D_t u - \Delta u + m^2 u = 0, \quad u(0, x) = \varphi(x), \quad m^2 > 0.$$

Then, we have the following $L^p - L^q$ estimates on the conjugate line for the derivatives $\partial_t^k \partial_x^\alpha u(t, \cdot)$ of the solution u $(k + |\alpha| \geq 0)$ for $q \in (2, \infty)$:

$$\|\partial_t^k \partial_x^\alpha u(t, \cdot)\|_{L^q} \leq C t^{-\frac{n}{2}\left(\frac{1}{p}-\frac{1}{q}\right)} \|\langle D\rangle^{2k}_m \partial_x^\alpha \varphi\|_{L^p},$$

where $\frac{1}{p} + \frac{1}{q} = 1$.

13.2 Smoothing Effect for Solutions

In Sect. 9.3.3 we explained the smoothing effect for solutions to the Cauchy problem for the heat equation

$$u_t - \Delta u = 0, \quad u(0, x) = \varphi(x).$$

There exist several ideas describing the smoothing effect. Using the representation of solution

$$u(t, x) = \frac{1}{(4\pi t)^{\frac{n}{2}}} \int_{\mathbb{R}^n} e^{-\frac{|x-y|^2}{4t}} \varphi(y)\, dy$$

is one way to describe the differentiability behavior of $u = u(t, x)$ in a neighborhood of a point (t_0, x_0) with $t_0 > 0$ and $x_0 \in \mathbb{R}^n$.

In the treatment of nonlinear models one is interested in a *global smoothing behavior*, that is, we expect some behavior of $u(t, \cdot)$ in some function spaces of smoother functions than the given data. Such a global smoothing behavior was introduced already in Lemma 9.3.1. We will discuss these issues for solutions to the Cauchy problem for the classical Schrödinger equation in the next two sections.

13.2.1   Local Smoothing Properties of Solutions

The paper [187] is one of the first studying this issue. One can find the smoothing behavior $\langle D \rangle^{\frac{1}{2}} u \in L^2_{loc}(\mathbb{R}^{n+1})$ (cf. with Definition 24.20) for solutions to the Cauchy problem

$$D_t u - \Delta u = 0, \quad u(0,x) = \varphi(x)$$

if the data φ is supposed to belong to $L^2(\mathbb{R}^n)$. This result was, among other things, independently obtained in [25] and can be described in the following form:

$$\| \chi \langle D \rangle^{\frac{1}{2}} u \|_{L^2} \le C_\chi \| \varphi \|_{L^2}$$

for all test functions $\chi = \chi(t,x) \in C_0^\infty(\mathbb{R}^{n+1})$.
If we assume more regularity for the data φ, then one can find, for example in [68], the estimate (cf. with Definition 24.12)

$$\sup_{R \in (0,\infty)} \frac{1}{R} \int_{-\infty}^{\infty} \int_{B_R} |\nabla u(t,x)|^2 \, dx \, dt \le C \| \varphi \|_{\dot{H}^{\frac{1}{2}}}^2 .$$

Here B_R denotes the ball around the origin of radius R. All these results hint to local smoothing with an extra gain of regularity $\frac{1}{2}$ with respect to spatial variables.

13.2.2   Global Smoothing Properties of Solutions

On the one hand we have the energy conservation $E_{Sch}(u)(t) = E_{Sch}(u)(0)$ in Sect. 11.2.3 for all $t > 0$. On the other hand, if we restrict ourselves to the one-dimensional case $n = 1$, then after integrating the solution in t we conclude

$$\sup_{x \in \mathbb{R}} \big\| |D|^{\frac{1}{2}} u(\cdot, x) \big\|_{L^2(\mathbb{R}^1)} \le C \| \varphi \|_{L^2(\mathbb{R}^1)}$$

for any fixed $x \in \mathbb{R}^1$. One can find this result in [111], see also [199]. The reader can find an overview of results for global smoothing in dimension $n \ge 2$ in [181]. Among other things, the following estimates are given:

1. $\big\| \langle x \rangle^{-s} |D|^{\frac{1}{2}} u \big\|_{L^2(\mathbb{R}^{n+1})} \le C \| \varphi \|_{L^2(\mathbb{R}^n)}$ for $s > \frac{1}{2}$.
   This type of estimate was given in [6] for $n \ge 3$ and in [23] for $n \ge 2$.

2. $\left\| \langle x \rangle^{-s} \langle D \rangle^{\frac{1}{2}} u \right\|_{L^2(\mathbb{R}^{n+1})} \le C \|\varphi\|_{L^2(\mathbb{R}^n)}$ for $s \ge 1$ and $n \ge 3$ or for $s > 1$ and $n = 2$. This type of estimate was given in [106] for $n \ge 3$ and in [213] for $n \ge 2$. Here one can also find the necessity of the conditions for s. That is, the estimate is not true for $s < 1$ in the case $n \ge 3$ and for $s \le 1$ in the case $n = 2$.

3. $\left\| |x|^{\alpha-1} |D|^{\alpha} u \right\|_{L^2(\mathbb{R}^{n+1})} \le C \|\varphi\|_{L^2(\mathbb{R}^n)}$ for $\alpha \in (1 - \frac{n}{2}, \frac{1}{2})$. This type of estimate was given in [106] for $n \ge 3$, $\alpha \in [0, \frac{1}{2})$ or $n = 2$, $\alpha \in (0, \frac{1}{2})$ and in [199] for $n \ge 2$ for all admissible values of α. In [216] it is shown that the estimate is not true for $\alpha = \frac{1}{2}$.

Remark 13.1 Let us explain some tools for proving the above estimates. One possible tool is the so-called restriction property for Fourier transforms.

Suppose that M is a given smooth submanifold of $\mathbb{R}^n$ and that $d\sigma$ is its induced Lebesgue measure. We say that the L^p *restriction property* for the Fourier transform is valid for M if there exists a parameter $q = q(p)$ so that the inequality

$$\|F(f)\|_{L^q(M_0)} \le A_{p,q}(M_0) \|f\|_{L^p(\mathbb{R}^n)}$$

holds for all Schwartz functions f whenever M_0 is an open subset of M with compact closure in M.

Authors are interested in proving restriction results as

$$\|F(A^* f)\|_{L^2(\rho S^{n-1})} \le C \sqrt{\rho} \|f\|_{L^2(\mathbb{R}^n)}$$

for $\rho > 0$, where

$$\rho S^{n-1} = \{\xi \in \mathbb{R}^n : |\xi| = \rho\}.$$

This estimate implies the dual estimate to

$$\sup_{x \in \mathbb{R}^1} \|Au(\cdot, x)\|_{L^2(\mathbb{R}^1)} \le C \|\varphi\|_{L^2(\mathbb{R}^1)},$$

where A^* is the adjoint operator to the operator A appearing on the left-hand sides of the above explained estimates for global smoothing.

Another important tool is *resolvent estimates* of the type

$$\sup_{\Im \zeta > 0} \left| (R(\zeta) A^* f, A^* f) \right| \le C \|f\|_{L^2(\mathbb{R}^n)}^2,$$

where $R(\zeta) = (-\Delta - \zeta)^{-1}$ is the Laplace transform of the solution operator $e^{i\Delta}$ of $D_t u - \Delta u = 0$, that is,

$$R(\zeta) = \frac{1}{i} \int_0^{\infty} e^{i\zeta t} e^{i\Delta} \, dt \quad \text{for } \Im \zeta > 0.$$

Other tools can be found, for example, in [181].

For applications to nonlinear models it is useful to have smoothing results for solutions to the Cauchy problem for the inhomogeneous Schrödinger equation

$$\partial_t u - i\Delta u = f(t,x), \quad u(0,x) = 0.$$

We explain only some results of [199]. Using Duhamel's principle we get the representation of solution

$$u(t,x) = \int_0^t e^{i(t-\tau)\Delta} f(\tau,x) \, d\tau.$$

This representation implies immediately the estimate

$$\|u(t,\cdot)\|_{L^2(\mathbb{R}^n)} \leq \int_0^t \|f(s,\cdot)\|_{L^2(\mathbb{R}^n)} \, ds \text{ for any fixed } t.$$

Integrating the solution in t leads in the one-dimensional case $n = 1$ to the following estimate (see [112]):

$$\sup_{x \in \mathbb{R}^1} \|D_x u(\cdot,x)\|_{L^2(\mathbb{R}^1)} \leq \int_{\mathbb{R}^1} \|f(\cdot,x)\|_{L^2(\mathbb{R}^1)} \, dx.$$

If $n \geq 2$, then we have the following global smoothing result (see [199]):

$$\left\| |x|^{\alpha-1} |D|^{s+\alpha} u \right\|_{L^2(\mathbb{R}^{n+1})} \leq C \left\| |x|^{1-s} f \right\|_{L^2(\mathbb{R}^n)}$$

for $\alpha, s \in (1 - \frac{n}{2}, \frac{1}{2})$.

Remark 13.2.1 The last estimate explains the effect "decay implies smoothing". The solution may gain regularity of positive order s if we assume a suitable decay order of the source term f. Moreover, we may gain an additional regularity of positive order α if we measure the regularity of solution in a suitable weighted L^2 space.

The considerations of smoothing properties of solutions are not only restricted to the model

$$D_t u - \Delta u = D_t u + \sum_{k=1}^n D_{x_k}^2 u = f,$$

but are also done for more general models of Schrödinger type

$$D_t u + a(D_x) u = f,$$

where the operator $a(D_x)$ is supposed to be elliptic. For instance, the interested reader can find smoothing results in [212] for $a(D_x) = |D_x|^p$ with $p > 0$. In [199] one can find results for $a(D_x) = p(D_x)^2$, where $p(\xi) > 0$ is positively homogeneous of order 1 and belongs to the class $C^\infty(\mathbb{R}^n \setminus \{0\})$. Moreover, a geometrical condition for the level set $\Sigma_p = \{\xi \in \mathbb{R}^n : p(\xi) = 1\}$ (cf. with the stationary phase method of Chap. 16) is supposed to be satisfied.

Finally, we mention further studies in this direction. We have already remarked that estimates listed in 1. to 3. above are not true for critical indices, but sometimes are true under a structural condition for smoothing factors. For example, on the one hand the estimate

$$\left\| |x|^{-\frac{1}{2}} |D|^{\frac{1}{2}} u \right\|_{L^2(\mathbb{R}^{n+1})} \leq C \|\varphi\|_{L^2(\mathbb{R}^n)}$$

is not true, but on the other hand the estimate

$$\left\| |x|^{-\frac{3}{2}} (x \wedge D) |D|^{-\frac{1}{2}} u \right\|_{L^2(\mathbb{R}^{n+1})} \leq C \|\varphi\|_{L^2(\mathbb{R}^n)}$$

is true for solutions to the Cauchy problem

$$D_t u - \Delta u = 0, \quad u(0, x) = \varphi(x).$$

Here $x \wedge D = (x_j D_k - x_k D_j)_{j<k}$ denotes the rotating vector fields. This observation was first found in [83] and a generalization for the Cauchy problem

$$D_t u - a(D_x) u = 0, \quad u(0, x) = \varphi(x)$$

is also discussed in [181].

Exercises Relating to the Considerations of Chap. 13

Exercise 1 Determine a suitable class of data such that

$$u(t, x) = \frac{1}{(2\sqrt{\pi it})^n} \int_{\mathbb{R}^n} e^{i \frac{|x-y|^2}{4t}} \varphi(y) \, dy$$

is a solution of the Cauchy problem

$$D_t u - \Delta u = 0, \quad u(0, x) = \varphi(x)$$

(cf. with Exercise 4 of Chap. 9).

Exercise 2 Prove the statement of Theorem 13.1.4.

Chapter 14
Phase Space Analysis for Wave Models

This chapter treats in detail phase space analysis for wave models. We begin with the elementary proof of well-posedness results for solutions to the Cauchy problem for the free wave equation. The classical damped wave model serves as an example of how to prove sharp decay estimates for the wave energy of solutions by using phase space analysis.

Moreover, the so-called diffusion phenomenon is explained. This phenomenon allows for a relation between classical damped wave models and heat models from the point of view of decay estimates. From this point of view classical damped waves are parabolic-like models. For the same model we show how additional regularity of the data implies a better decay, an idea that is often used in the literature.

Then, the same issues are discussed for Klein-Gordon models and viscoelastic damped wave models. Viscoelastic damping is introduced as a special structural damping.

In the concluding remarks we describe some recent results of the authors, too. These results are devoted to the question or the influence of time-dependent coefficients in wave models on energy estimates.

14.1 The Classical Wave Model

We are interested in the Cauchy problem

$$u_{tt} - \Delta u = 0, \quad u(0,x) = \varphi(x), \quad u_t(0,x) = \psi(x), \quad x \in \mathbb{R}^n, \quad n \geq 1.$$

After application of *partial Fourier transformation* $\big(v(t,\xi) := F_{x \to \xi}(u(t,x))\big)$, we get the auxiliary Cauchy problem

$$v_{tt} + |\xi|^2 v = 0, \quad v(0,\xi) = F(\varphi)(\xi), \quad v_t(0,\xi) = F(\psi)(\xi)$$

© Springer International Publishing AG 2018
M.R. Ebert, M. Reissig, *Methods for Partial Differential Equations*,
https://doi.org/10.1007/978-3-319-66456-9_14

for an *ordinary differential equation depending on the parameter* $\xi \in \mathbb{R}^n$. For $\xi \neq 0$ we have the general solution

$$v(t, \xi) = c_1(\xi)e^{-i|\xi|t} + c_2(\xi)e^{i|\xi|t}.$$

The Cauchy conditions imply

$$c_1(\xi) + c_2(\xi) = F(\varphi)(\xi), \quad -i|\xi|c_1(\xi) + i|\xi|c_2(\xi) = F(\psi)(\xi).$$

It follows then that

$$c_1(\xi) = \frac{1}{2} F(\varphi)(\xi) - \frac{1}{2i|\xi|} F(\psi)(\xi), \quad c_2(\xi) = \frac{1}{2} F(\varphi)(\xi) + \frac{1}{2i|\xi|} F(\psi)(\xi).$$

Setting these coefficients into the general solution gives

$$v(t, \xi) = \cos(|\xi|t)F(\varphi)(\xi) + \frac{\sin(|\xi|t)}{|\xi|} F(\psi)(\xi).$$

Supposing for a moment the validity of the *Fourier inversion formula (see Sect. 24.1.4)* $u(t, x) = F^{-1}_{\xi \to x}(F_{x \to \xi}(u(t, x))$ (the validity must be checked at the end of our considerations) we arrive at the following representation for u:

$$u(t, x) = F^{-1}_{\xi \to x}\left(\cos(|\xi|t)F(\varphi)(\xi) \right) + F^{-1}_{\xi \to x}\left(\frac{\sin(|\xi|t)}{|\xi|} F(\psi)(\xi) \right).$$

Sometimes we shall use instead the equivalent representation

$$u(t, x) = F^{-1}_{\xi \to x}\left(e^{-i|\xi|t} \frac{1}{2} F(\varphi)(\xi) \right) - F^{-1}_{\xi \to x}\left(e^{-i|\xi|t} \frac{1}{2i|\xi|} F(\psi)(\xi) \right)$$

$$+ F^{-1}_{\xi \to x}\left(e^{i|\xi|t} \frac{1}{2} F(\varphi)(\xi) \right) + F^{-1}_{\xi \to x}\left(e^{i|\xi|t} \frac{1}{2i|\xi|} F(\psi)(\xi) \right).$$

This representation consists of so-called *Fourier multipliers*

$$F^{-1}_{\xi \to x}\left(e^{i\phi(t,\xi)} a(t, \xi) F(u_0)(\xi) \right).$$

Here $\phi = \phi(t, \xi)$ is part of the so-called *phase function* and $a = a(t, \xi)$ is the so-called *amplitude function*.

For the given data we assume $\varphi \in H^s(\mathbb{R}^n)$ and $\psi \in H^{s-1}(\mathbb{R}^n)$ with $s \geq 1$ (see Definition 24.9). Taking into consideration Theorem 11.1.2 we know that any solution (if it exists) possesses an energy $E_W(u)(t)$ for all $t \geq 0$.

Theorem 14.1.1 *Let* $\varphi \in H^s(\mathbb{R}^n)$ *and* $\psi \in H^{s-1}(\mathbb{R}^n)$, $s \geq 1$, $n \geq 1$ *in the Cauchy problem*

$$u_{tt} - \Delta u = 0, \quad u(0, x) = \varphi(x), \quad u_t(0, x) = \psi(x).$$

Then, there exists a uniquely determined energy solution

$$u \in C\big([0, T], H^s(\mathbb{R}^n)\big) \cap C^1\big([0, T], H^{s-1}(\mathbb{R}^n)\big) \text{ for all } T > 0.$$

Proof The uniqueness follows by Theorem 11.1.2. A solution is given in the form

$$u(t, x) = F_{\xi \to x}^{-1}\big(\cos(|\xi|t)F(\varphi)(\xi)\big) + F_{\xi \to x}^{-1}\Big(\frac{\sin(|\xi|t)}{|\xi|}F(\psi)(\xi)\Big).$$

For this reason it remains to verify that this solution satisfies the desired regularity. First, we explain how the assumptions for the data are transmitted to the Fourier transforms of the data. We have

$$F(\varphi) = F(\varphi)(\xi) \in L^{2,s}(\mathbb{R}^n), \quad \text{that is,} \quad \langle \xi \rangle^s F(\varphi) \in L^2(\mathbb{R}^n),$$

$$F(\psi) = F(\psi)(\xi) \in L^{2,s-1}(\mathbb{R}^n), \quad \text{that is,} \quad \langle \xi \rangle^{s-1} F(\psi) \in L^2(\mathbb{R}^n).$$

In the further considerations we use the following estimates:

- $|\cos(|\xi|t)| \leq 1$,
- $|\sin(|\xi|t)| \leq |\xi|t \leq |\xi|T$ for $|\xi| \leq \varepsilon$ and $t \in [0, T]$,
- $|\sin(|\xi|t)| \leq 1$ for $|\xi| \geq \varepsilon$ and $t \in [0, T]$.

Thus, we may conclude

$$|v(t, \xi)| \leq |F(\varphi)(\xi)| + C(\varepsilon, T)\frac{|F(\psi)(\xi)|}{\langle \xi \rangle},$$

$$\langle \xi \rangle^s |v(t, \xi)| \leq \langle \xi \rangle^s |F(\varphi)(\xi)| + C(\varepsilon, T)\langle \xi \rangle^{s-1}|F(\psi)(\xi)|,$$

respectively. These estimates lead to $v \in L^\infty\big((0, T), L^{2,s}(\mathbb{R}^n)\big)$. In the same way we are able to derive the property $\partial_t v \in L^\infty\big((0, T), L^{2,s-1}(\mathbb{R}^n)\big)$. It remains to prove

$$v \in C\big([0, T], L^{2,s}(\mathbb{R}^n)\big) \cap C^1\big([0, T], L^{2,s-1}(\mathbb{R}^n)\big).$$

The property $v \in C\big([0, T], L^{2,s}(\mathbb{R}^n)\big)$ follows from the relation

$$\lim_{t_1 \to t_2} \|v(t_1, \cdot) - v(t_2, \cdot)\|_{L^{2,s}} = 0 \text{ for } t_1, t_2 \in [0, T].$$

Using the explicit representation of solution we conclude as follows:

$$\int_{\mathbb{R}^n} |v(t_1, \xi) - v(t_2, \xi)|^2 \langle\xi\rangle^{2s} \, d\xi$$

$$\leq \int_{\mathbb{R}^n} \left| \sin\left(\frac{|\xi|(t_1 + t_2)}{2}\right) \sin\left(\frac{|\xi|(t_1 - t_2)}{2}\right) \right|^2 |F(\varphi)(\xi)|^2 \langle\xi\rangle^{2s} \, d\xi$$

$$+ \int_{\mathbb{R}^n} \left| \cos\left(\frac{|\xi|(t_1 + t_2)}{2}\right) \sin\left(\frac{|\xi|(t_1 - t_2)}{2}\right) \right|^2 \frac{1}{|\xi|^2} |F(\psi)(\xi)|^2 \langle\xi\rangle^{2s} \, d\xi.$$

Let $K_R(0) \subset \mathbb{R}^n$ be a sufficiently large ball around the origin with radius R. We divide the integral $\int_{\mathbb{R}^n}$ in or into the two integrals $\int_{K_R(0)} + \int_{\mathbb{R}^n \setminus K_R(0)}$. Using the above estimates it holds

$$\int_{\mathbb{R}^n} \left| \sin\left(\frac{|\xi|(t_1 + t_2)}{2}\right) \sin\left(\frac{|\xi|(t_1 - t_2)}{2}\right) \right|^2 |F(\varphi)(\xi)|^2 \langle\xi\rangle^{2s} \, d\xi$$

$$= \int_{K_R(0)} \left| \sin\left(\frac{|\xi|(t_1 + t_2)}{2}\right) \sin\left(\frac{|\xi|(t_1 - t_2)}{2}\right) \right|^2 |F(\varphi)(\xi)|^2 \langle\xi\rangle^{2s} \, d\xi$$

$$+ \int_{\mathbb{R}^n \setminus K_R(0)} \left| \sin\left(\frac{|\xi|(t_1 + t_2)}{2}\right) \sin\left(\frac{|\xi|(t_1 - t_2)}{2}\right) \right|^2 |F(\varphi)(\xi)|^2 \langle\xi\rangle^{2s} \, d\xi$$

$$\leq \int_{K_R(0)} \frac{|\xi|^2 (t_1 - t_2)^2}{4} |F(\varphi)(\xi)|^2 \langle\xi\rangle^{2s} d\xi + \int_{\mathbb{R}^n \setminus K_R(0)} |F(\varphi)(\xi)|^2 \langle\xi\rangle^{2s} \, d\xi$$

for $|t_1 - t_2| < \varepsilon(R)$. The first integral at the right-hand side is estimated by $C_R(t_1 - t_2)^2 \|F(\varphi)\|_{L^{2,s}}^2$. Using the continuity of the Lebesgue measure, the second integral is estimated by $\widetilde{\varepsilon}(R)$, where $\widetilde{\varepsilon}(R)$ tends to 0 for R to infinity. Summarizing we obtain

$$\lim_{t_1 \to t_2} \int_{\mathbb{R}^n} \left| \sin\left(\frac{|\xi|(t_1 + t_2)}{2}\right) \sin\left(\frac{|\xi|(t_1 - t_2)}{2}\right) \right|^2 |F(\varphi)(\xi)|^2 \langle\xi\rangle^{2s} \, d\xi$$

$$\leq \lim_{t_1 \to t_2} C_R(t_1 - t_2)^2 \|F(\varphi)\|_{L^{2,s}}^2 + \widetilde{\varepsilon}(R) = \widetilde{\varepsilon}(R).$$

Taking account of $\widetilde{\varepsilon}(R) \to 0$ for $R \to \infty$ we conclude

$$\lim_{t_1 \to t_2} \int_{\mathbb{R}^n} \left| \sin\left(\frac{|\xi|(t_1 + t_2)}{2}\right) \sin\left(\frac{|\xi|(t_1 - t_2)}{2}\right) \right|^2 |F(\varphi)(\xi)|^2 \langle\xi\rangle^{2s} \, d\xi = 0.$$

Repeating this approach yields

$$\lim_{t_1 \to t_2} \int_{\mathbb{R}^n} \left| \cos\left(\frac{|\xi|(t_1 + t_2)}{2}\right) \sin\left(\frac{|\xi|(t_1 - t_2)}{2}\right) \right|^2 \frac{|F(\psi)(\xi)|^2}{|\xi|^2} \langle\xi\rangle^{2s} \, d\xi = 0.$$

Consequently, we have shown $v \in C([0,T], L^{2,s}(\mathbb{R}^n))$. The validity of the Fourier inversion formula for functions from $H^s(\mathbb{R}^n)$ (see Sect. 24.1.4) implies $u \in C([0,T], H^s(\mathbb{R}^n))$. The same reasoning implies

$$v \in C^1([0,T], L^{2,s-1}(\mathbb{R}^n)), \quad u \in C^1([0,T], H^{s-1}(\mathbb{R}^n)),$$

respectively. Here we again use the Fourier inversion formula.

The above considerations imply that the Cauchy problem for the free wave equation is H^s well-posed, even for all $s \in \mathbb{R}^1$.

Corollary 14.1.1 *The Cauchy problem*

$$u_{tt} - \Delta u = 0, \quad u(0,x) = \varphi(x), \quad u_t(0,x) = \psi(x), \quad x \in \mathbb{R}^n, \quad n \geq 1$$

is H^s well-posed, $s \in \mathbb{R}^1$, that is, for given data $\varphi \in H^s(\mathbb{R}^n)$, $\psi \in H^{s-1}(\mathbb{R}^n)$ there exists a uniquely determined (in general) distributional solution

$$u \in C([0,T], H^s(\mathbb{R}^n)) \cap C^1([0,T], H^{s-1}(\mathbb{R}^n)) \quad \text{for all } T > 0.$$

The solution depends continuously on the data, that is, to each $\varepsilon > 0$ there exists a $\delta(\varepsilon)$ such that

$$\|\varphi_1 - \varphi_2\|_{H^s} + \|\psi_1 - \psi_2\|_{H^{s-1}} < \delta \text{ implies}$$

$$\|u_1 - u_2\|_{C([0,T],H^s)\cap C^1([0,T],H^{s-1})} < \varepsilon.$$

Up to now we have become aquainted with two different ways to represent solutions of wave equations. On the one hand we know the representations from Theorems 10.1.1, 10.5.1 and 10.5.2. On the other hand we got to know representations consisting of Fourier multipliers. Is it possible to transfer one representation to another?

In the $1d$ case we have the representation of solution

$$u(t,x) = F_{\xi \to x}^{-1}\left(\left(e^{i\xi t} + e^{-i\xi t}\right)\frac{1}{2}F(\varphi)(\xi)\right)$$

$$+ F_{\xi \to x}^{-1}\left(\left(e^{i\xi t} - e^{-i\xi t}\right)\frac{1}{2i\xi}F(\psi)(\xi)\right).$$

How can we derive from this representation formula the d'Alembert's representation formula from Sect. 10.1? To answer this question the reader should solve Exercise 2. What about generalizations to higher dimensions? From the representation

$$v(t,\xi) = \cos(|\xi|t)F(\varphi)(\xi) + \frac{\sin(|\xi|t)}{|\xi|}F(\psi)(\xi)$$

$$= \partial_t\left(\frac{\sin(|\xi|t)}{|\xi|}F(\varphi)(\xi)\right) + \frac{\sin(|\xi|t)}{|\xi|}F(\psi)(\xi)$$

it follows

$$u(t,x) = \partial_t F_{\xi \to x}^{-1}\Big(\frac{\sin(|\xi|t)}{|\xi|}F(\varphi)(\xi)\Big) + F_{\xi \to x}^{-1}\Big(\frac{\sin(|\xi|t)}{|\xi|}F(\psi)(\xi)\Big).$$

Therefore, we only have to study

$$F_{\xi \to x}^{-1}\Big(\frac{\sin(|\xi|t)}{|\xi|}F(\psi)(\xi)\Big).$$

What is the main difficulty in the discussion of the last Fourier multiplier? Which methods do we find in the literature to overcome these difficulties?

What are the advantages and disadvantages of the application of the method of Fourier transformation to study wave equations?

Choosing data from Sobolev spaces we have *no loss of regularity* (see Theorem 14.1.1) with respect to the spatial variables. The approach is independent of the spatial dimension n. But, special qualitative properties of solutions of the wave equation as existence of forward or backward wave front, or finite speed of propagation of perturbations or domain of dependence are much more difficult to get by using Fourier multipliers in the representation of solution.

14.2　The Classical Damped Wave Model

First of all, we mention that solutions to classical damped wave equations have qualitative properties as existence of a forward wave front, finite propagation speed of perturbations or existence of a domain of dependence.

14.2.1　Representation of Solutions by Using Fourier Multipliers

Let us turn to the Cauchy problem

$$u_{tt} - \Delta u + u_t = 0, \quad u(0,x) = \varphi(x), \quad u_t(0,x) = \psi(x).$$

Step 1　Transformation of the dissipation term into a mass term

We introduce a new function $w = w(t,x)$ by $w(t,x) := e^{\frac{1}{2}t}u(t,x)$. Then, w satisfies the Cauchy problem

$$w_{tt} - \Delta w - \frac{1}{4}w = 0, \quad w(0,x) = \varphi(x), \quad w_t(0,x) = \frac{1}{2}\varphi(x) + \psi(x).$$

Opposite to the Klein-Gordon equation now appears a *negative mass term*. This negative mass needs some special considerations.

Step 2   Application of partial Fourier transformation

The application of partial Fourier transformation gives the following ordinary differential equation for $v = v(t, \xi) := F_{x \to \xi}(w(t, x))(t, \xi)$:

$$v_{tt} + \left(|\xi|^2 - \frac{1}{4} \right) v = 0, \quad v(0, \xi) = v_0(\xi) := F(\varphi)(\xi),$$

$$v_t(0, \xi) = v_1(\xi) := \frac{1}{2} F(\varphi)(\xi) + F(\psi)(\xi).$$

We make a distinction of cases for $\{\xi \in \mathbb{R}^n : |\xi| > \frac{1}{2}\}$ (the coefficient $|\xi|^2 - \frac{1}{4}$ is positive) and for $\{\xi \in \mathbb{R}^n : |\xi| < \frac{1}{2}\}$ (the coefficient $|\xi|^2 - \frac{1}{4}$ is negative).

Case 1 $\{\xi : |\xi| > \frac{1}{2}\}$

Using $|\xi|^2 > \frac{1}{4}$ we define a new positive variable $|\eta|$ by $|\eta|^2 := |\xi|^2 - \frac{1}{4} > 0$. So, we get the ordinary differential equation $v_{tt} + |\eta|^2 v = 0$. Taking account of the results from Sect. 14.1 we obtain immediately the following representation of solution $v(t, \xi)$:

$$v(t, \xi) = \cos \left(\sqrt{|\xi|^2 - \frac{1}{4}} \, t \right) v_0(\xi) + \frac{\sin \left(\sqrt{|\xi|^2 - \frac{1}{4}} \, t \right)}{\sqrt{|\xi|^2 - \frac{1}{4}}} v_1(\xi).$$

Case 2 $\{\xi : |\xi| < \frac{1}{2}\}$

The solution to the transformed differential equation is

$$v(t, \xi) = \left(\frac{v_0(\xi)}{2} - \frac{v_1(\xi)}{\sqrt{1 - 4|\xi|^2}} \right) e^{-\frac{1}{2}\sqrt{1 - 4|\xi|^2} \, t}$$

$$+ \left(\frac{v_0(\xi)}{2} + \frac{v_1(\xi)}{\sqrt{1 - 4|\xi|^2}} \right) e^{\frac{1}{2}\sqrt{1 - 4|\xi|^2} \, t}$$

$$= v_0(\xi) \cosh \left(\frac{1}{2} \sqrt{1 - 4|\xi|^2} \, t \right) + \frac{2 v_1(\xi)}{\sqrt{1 - 4|\xi|^2}} \sinh \left(\frac{1}{2} \sqrt{1 - 4|\xi|^2} \, t \right).$$

If we consider the Cauchy problem

$$u_{tt} - \Delta u + u_t = 0, \quad u(0, x) = \varphi(x), \quad u_t(0, x) = \psi(x)$$

with data $\varphi \in H^s(\mathbb{R}^n)$ and $\psi \in H^{s-1}(\mathbb{R}^n)$, then we conclude from the above representations of solutions the next result after taking into consideration that only the *behavior for large frequencies is important for the regularity of solutions*. The continuity with respect to t is proved as in the proof to Theorem 14.1.1.

Theorem 14.2.1 *Let the data $\varphi \in H^s(\mathbb{R}^n)$ and $\psi \in H^{s-1}(\mathbb{R}^n)$, $s \in \mathbb{R}^1$, $n \geq 1$ be given in the Cauchy problem*

$$u_{tt} - \Delta u + u_t = 0, \quad u(0,x) = \varphi(x), \quad u_t(0,x) = \psi(x).$$

Then, there exists for all $T > 0$ a uniquely determined (in general) distributional solution

$$u \in C\big([0,T], H^s(\mathbb{R}^n)\big) \cap C^1\big([0,T], H^{s-1}(\mathbb{R}^n)\big).$$

We have the a priori estimate

$$\|u(t,\cdot)\|_{H^s} + \|u_t(t,\cdot)\|_{H^{s-1}} \leq C(T)\big(\|\varphi\|_{H^s} + \|\psi\|_{H^{s-1}}\big).$$

Finally, the solution depends continuously on the data.

Remark 14.2.1 The statements of Corollary 14.1.1 and Theorem 14.2.1 coincide. The dissipation term has no influence on the regularity of solutions. Dissipation terms have an essential influence on energy estimates as they produce a *decay of the energy*. This will be explained in the next section.

14.2.2   Decay Behavior and Decay Rate of the Wave Energy

From the considerations in Sect. 11.3.2 we know that the wave energy $E_W(u)(t)$ of Sobolev solutions to the Cauchy problem

$$u_{tt} - \Delta u + u_t = 0, \quad u(0,x) = \varphi(x), \quad u_t(0,x) = \psi(x)$$

is a decreasing function if $E_W(u)(0)$ is finite.
Applying phase space analysis allows us verify that the energy $E_W(u)(t)$ is even decaying for $t \to \infty$. We are able to derive for $E_W(u)(t)$ an *optimal decay behavior* with an optimal *decay rate*.

Theorem 14.2.2 *The solution to the Cauchy problem*

$$u_{tt} - \Delta u + u_t = 0, \quad u(0,x) = \varphi(x), \quad u_t(0,x) = \psi(x)$$

with data $\varphi \in H^1(\mathbb{R}^n)$ and $\psi \in L^2(\mathbb{R}^n)$ satisfies the following estimates for $t \geq 0$:

$$\|u(t, \cdot)\|_{L^2} \leq C(\|\varphi\|_{L^2} + \|\psi\|_{H^{-1}}),$$

$$\|\nabla u(t, \cdot)\|_{L^2} \leq C(1 + t)^{-\frac{1}{2}}(\|\varphi\|_{H^1} + \|\psi\|_{L^2}),$$

$$\|u_t(t, \cdot)\|_{L^2} \leq C(1 + t)^{-1}(\|\varphi\|_{H^1} + \|\psi\|_{L^2}).$$

Consequently, the wave energy satisfies the estimate

$$E_W(u)(t) \leq C(1 + t)^{-1}(\|\varphi\|_{H^1}^2 + \|\psi\|_{L^2}^2).$$

Remark 14.2.2   We see that the kinetic energy decays faster than the elastic energy. To get these estimates we suppose for the data (φ, ψ) the regularity $H^1(\mathbb{R}^n) \times L^2(\mathbb{R}^n)$ which is stronger than the regularity $\dot{H}^1(\mathbb{R}^n) \times L^2(\mathbb{R}^n)$. The last regularity guarantees a Sobolev solution becoming an energy solution. Try to understand what kind of estimates we would have in the case of data belonging to $\dot{H}^1(\mathbb{R}^n) \times L^2(\mathbb{R}^n)$. In which step of the following proof do we use the assumption $\varphi \in L^2(\mathbb{R}^n)$?

Proof   Step 1   Transformation of energy in the phase space

Let $\hat{u}$ be the Fourier transform of u, that is, $\hat{u}(t, \xi) = F_{x \to \xi}(u(t, x))(t, \xi)$. We transfer the energy in the phase space as follows:

$$E_W(u)(t) = \frac{1}{2}\left(\|\nabla u(t, \cdot)\|_{L^2}^2 + \|u_t(t, \cdot)\|_{L^2}^2\right)$$

$$= \frac{1}{2}\left(\||\xi|\hat{u}(t, \cdot)\|_{L^2}^2 + \|\hat{u}_t(t, \cdot)\|_{L^2}^2\right).$$

Here we applied the formula of Parseval-Plancherel from Remark 24.1.2. After introducing $u(t, x) = e^{-\frac{1}{2}t}w(t, x)$ and $v(t, \xi) = F_{x \to \xi}(w)(t, \xi)$, it follows $\hat{u}(t, \xi) = e^{-\frac{1}{2}t}v(t, \xi)$. For the elastic energy we will use

$$|\xi|\hat{u}(t, \xi) = e^{-\frac{1}{2}t}|\xi|v(t, \xi),$$

for the kinetic energy we will use

$$\hat{u}_t(t, \xi) = e^{-\frac{1}{2}t}\left(v_t(t, \xi) - \frac{1}{2}v(t, \xi)\right).$$

Step 2   Estimate of the solution itself

We divide the phase space $\mathbb{R}_\xi^n$ into several regions.

Case 1 $\{\xi : |\xi| \geq 1\}$

The representation of solution yields

$$|\hat{u}(t, \xi)| \leq Ce^{-\frac{t}{2}}\left(|v_0(\xi)| + \frac{|v_1(\xi)|}{|\xi|}\right).$$

Case 2 $\{\xi : |\xi| \in (\frac{1}{4}, 1)\}$

The representation of solution yields

$$|\hat{u}(t, \xi)| \leq Ce^{-\delta t}\left(|v_0(\xi)| + |v_1(\xi)|\right)$$

with a suitable positive constant δ.

Case 3 $\{\xi : |\xi| \leq \frac{1}{4}\}$

The representation of solution yields

$$|\hat{u}(t, \xi)| \leq C\left(|v_0(\xi)| + |v_1(\xi)|\right).$$

Summarizing, we conclude the first estimate.

Step 3   Estimate of the elastic energy

We divide the phase space $\mathbb{R}^n_\xi$ into several regions already motivated in Sect. 14.2.1 (cf. with Cases 1 and 2 there). We shall use the notations from the previous section.

Case 1 $\{\xi : |\xi| > \frac{1}{2}\}$

First we notice

$$|\xi|\hat{u}(t, \xi) = e^{-\frac{1}{2}t}\left(\cos\left(\sqrt{|\xi|^2 - \frac{1}{4}}\, t\right) |\xi|v_0(\xi) + t\, \frac{\sin\left(\sqrt{|\xi|^2 - \frac{1}{4}}\, t\right)}{\sqrt{|\xi|^2 - \frac{1}{4}}\, t} |\xi|v_1(\xi)\right).$$

Using the formula of Parseval-Plancherel from Remark 24.1.2 helps us to estimate the elastic energy $\|\nabla u(t, \cdot)\|_{L^2}^2$. We have

$$\||\xi|\hat{u}(t, \xi)\|_{L^2\{|\xi|>\frac{1}{2}\}}^2 = \int_{|\xi|>\frac{1}{2}} |\xi|^2|\hat{u}(t, \xi)|^2\, d\xi \leq 2\left(\int_{|\xi|>\frac{1}{2}} e^{-t}|\xi|^2|v_0(\xi)|^2\, d\xi\right.$$

$$+ \int_{\frac{1}{2}<|\xi|\leq 1} \underbrace{\frac{\sin^2\left(\sqrt{|\xi|^2 - \frac{1}{4}}\, t\right)}{\left(\sqrt{|\xi|^2 - \frac{1}{4}}\, t\right)^2}}_{\frac{\sin^2\alpha}{\alpha^2}\, \leq C}\, t^2 e^{-t}|\xi|^2|v_1(\xi)|^2\, d\xi$$

$$+ \int_{|\xi|>1} \underbrace{\frac{1}{|\xi|^2 - \frac{1}{4}}}_{\leq C} |\xi|^2 e^{-t} |v_1(\xi)|^2 \, d\xi \Big)$$

$$\leq 2e^{-t} \int_{\mathbb{R}^n} |\xi|^2 |v_0(\xi)|^2 \, d\xi \;+\; Ct^2 e^{-t} \int_{\mathbb{R}^n} |v_1(\xi)|^2 \, d\xi \;+\; Ce^{-t} \int_{\mathbb{R}^n} |v_1(\xi)|^2 \, d\xi.$$

Summarizing, *we obtain an exponential decay for large frequencies*. It holds

$$\int_{|\xi|>\frac{1}{2}} |\xi|^2 |\hat{u}(t,\xi)|^2 \, d\xi \leq Ct^2 e^{-t} \int_{\mathbb{R}^n} \left(|\xi|^2 |v_0(\xi)|^2 + |v_1(\xi)|^2 \right) d\xi.$$

We need the regularity $\dot{H}^1(\mathbb{R}^n) \times L^2(\mathbb{R}^n)$ for the data (φ, ψ).

Case 2 $\{\xi : |\xi| < \frac{1}{2}\}$

To estimate the elastic energy we use

$$|\xi| \hat{u}(t,\xi) = |\xi| e^{-\frac{1}{2}t} \left(\left(\frac{v_0(\xi)}{2} - \frac{v_1(\xi)}{\sqrt{1 - 4|\xi|^2}} \right) e^{-\frac{1}{2}\sqrt{1-4|\xi|^2}\, t} \right.$$

$$\left. + \left(\frac{v_0(\xi)}{2} + \frac{v_1(\xi)}{\sqrt{1 - 4|\xi|^2}} \right) e^{\frac{1}{2}\sqrt{1-4|\xi|^2}\, t} \right)$$

$$= v_0(\xi) |\xi| \cosh \left(\frac{1}{2}\sqrt{1 - 4|\xi|^2}\, t \right) e^{-\frac{1}{2}t}$$

$$+ \frac{2 v_1(\xi) |\xi|}{\sqrt{1 - 4|\xi|^2}} \sinh \left(\frac{1}{2}\sqrt{1 - 4|\xi|^2}\, t \right) e^{-\frac{1}{2}t}.$$

We divide the interval $(0, \frac{1}{2})$ for $|\xi|$ into two subintervals.

Case 2a $\{\xi : |\xi| \in [\frac{1}{4}, \frac{1}{2})\}$:

Here we estimate as follows:

$$|\xi| |\hat{u}(t,\xi)| = \Bigg| v_0(\xi) |\xi| \underbrace{\cosh \left(\frac{1}{2}\sqrt{1 - 4|\xi|^2}\, t \right)}_{\leq \cosh(\frac{\sqrt{3}}{4} t)} e^{-\frac{1}{2}t}$$

$$+ \underbrace{\frac{\sinh \left(\frac{1}{2}\sqrt{1 - 4|\xi|^2}\, t \right)}{\frac{1}{2}\sqrt{1 - 4|\xi|^2}\, t}}_{\leq Ct \cosh(\frac{\sqrt{3}}{4} t)} t \, v_1(\xi) |\xi| e^{-\frac{1}{2}t} \Bigg|$$

$$\leq \Bigg| v_0(\xi) |\xi| \underbrace{\cosh \left(\frac{\sqrt{3}}{4} t \right) e^{-\frac{1}{2}t}}_{\leq e^{-\delta t},\, \delta > 0} + C \underbrace{v_1(\xi) |\xi|}_{\leq |v_1(\xi)|} \underbrace{\cosh \left(\frac{\sqrt{3}}{4} t \right) t e^{-\frac{1}{2}t}}_{\leq e^{-\delta t},\, \delta > 0} \Bigg|,$$

and obtain with a suitable positive constant δ the estimate

$$\int_{\frac{1}{4}\leq|\xi|<\frac{1}{2}} |\xi|^2 |\hat{u}(t,\xi)|^2 \, d\xi \leq Ce^{-\delta t} \int_{\mathbb{R}^n} \left(|\xi|^2 |v_0(\xi)|^2 + |v_1(\xi)|^2 \right) d\xi.$$

Here we get an exponential decay and use again the regularity $\dot{H}^1(\mathbb{R}^n) \times L^2(\mathbb{R}^n)$ of the data (φ, ψ).

Case 2b $\{\xi : |\xi| \in (0, \frac{1}{4})\}$:

By using the property

$$\sqrt{x+y} \leq \sqrt{x} + \frac{y}{2\sqrt{x}} \quad \text{for any } x > 0 \ \text{ and } \ y \geq -x,$$

it follows the inequality

$$-4|\xi|^2 \leq -1 + \sqrt{1 - 4|\xi|^2} \leq -2|\xi|^2 \quad \text{for } |\xi| < \frac{1}{2}.$$

With this inequality we proceed as follows:

$$\int_{|\xi|<\frac{1}{4}} |\xi|^2 |\hat{u}(t,\xi)|^2 \, d\xi$$

$$\leq \int_{|\xi|<\frac{1}{4}} \left(|v_0(\xi)|^2 |\xi|^2 + |v_1(\xi)|^2 |\xi|^2 \right) \Big(\underbrace{e^{-t-\sqrt{1-4|\xi|^2}\,t}}_{\leq e^{-t}} + \underbrace{e^{-t+\sqrt{1-4|\xi|^2}\,t}}_{\leq e^{-2|\xi|^2 t}} \Big) d\xi$$

$$\leq Ce^{-t} \int_{|\xi|<\frac{1}{4}} \left(|v_0(\xi)|^2 |\xi|^2 + |v_1(\xi)|^2 |\xi|^2 \right) d\xi$$

$$+ C \int_{|\xi|<\frac{1}{4}} \left(|v_0(\xi)|^2 + |v_1(\xi)|^2 \right) |\xi|^2 e^{-2|\xi|^2 t} \, d\xi.$$

For $t \geq 1$, we may estimate the second term on the right-hand side of the last inequality by

$$C \int_{|\xi|<\frac{1}{4}} \left(|v_0(\xi)|^2 + |v_1(\xi)|^2 \right) |\xi|^2 e^{-2|\xi|^2 t} \, d\xi$$

$$\leq C \sup_{|\xi|<\frac{1}{4},\, t\geq 1} \frac{t|\xi|^2}{t} e^{-2|\xi|^2 t} \int_{\mathbb{R}^n} \left(|v_0(\xi)|^2 + |v_1(\xi)|^2 \right) d\xi$$

$$\leq C \frac{1}{t} \underbrace{\sup_{|\xi|<\frac{1}{4},\, t\geq 1} t|\xi|^2 e^{-2|\xi|^2 t}}_{\leq C} \int_{\mathbb{R}^n} \left(|v_0(\xi)|^2 + |v_1(\xi)|^2 \right) d\xi.$$

For $t \in [0, 1]$, we use $|\xi|^2 e^{-2|\xi|^2 t} \leq C$. Summarizing, we have shown for small frequencies

$$\int_{|\xi|<\frac{1}{4}} |\xi|^2 |\hat{u}(t, \xi)|^2 \, d\xi \leq C(1 + t)^{-1} \int_{\mathbb{R}^n} \left(|v_0(\xi)|^2 + |v_1(\xi)|^2 \right) d\xi.$$

In this case we need the regularity $L^2(\mathbb{R}^n) \times L^2(\mathbb{R}^n)$ for the data (φ, ψ). Summarizing all estimates of the Cases 1 to 2b we may conclude the desired decay estimate for the elastic energy.

Step 4   Estimate of the kinetic energy

We use the identity $\|u_t(t, \cdot)\|^2_{L^2(\mathbb{R}^n_x)} = \|\hat{u}_t(t, \cdot)\|^2_{L^2(\mathbb{R}^n_\xi)}$ with

$$\hat{u}_t(t, \xi) = e^{-\frac{1}{2}t} \left(v_t(t, \xi) - \frac{1}{2} v(t, \xi) \right).$$

Case 1 $\{\xi : |\xi| > \frac{1}{2}\}$

Using

$$v_t(t, \xi) = -\sqrt{|\xi|^2 - \frac{1}{4}} \sin\left(\sqrt{|\xi|^2 - \frac{1}{4}}\, t\right) v_0(\xi) + \cos\left(\sqrt{|\xi|^2 - \frac{1}{4}}\, t\right) v_1(\xi)$$

we obtain

$$\hat{u}_t(t, \xi) = e^{-\frac{1}{2}t} \left(v_1(\xi) \left(\cos\left(\sqrt{|\xi|^2 - \frac{1}{4}}\, t\right) - \frac{1}{2} \frac{\sin\left(\sqrt{|\xi|^2 - \frac{1}{4}}\, t\right)}{\sqrt{|\xi|^2 - \frac{1}{4}}} \right) \right.$$

$$\left. - v_0(\xi) \left(\frac{1}{2} \cos\left(\sqrt{|\xi|^2 - \frac{1}{4}}\, t\right) + \sqrt{|\xi|^2 - \frac{1}{4}} \sin\left(\sqrt{|\xi|^2 - \frac{1}{4}}\, t\right) \right) \right).$$

Repeating the approach for estimating the elastic energy gives

$$\|\hat{u}_t(t, \cdot)\|^2_{L^2\{|\xi|>\frac{1}{2}\}}$$

$$\leq C \underbrace{\int_{|\xi|>\frac{1}{2}} e^{-t} |v_1(\xi)|^2 \left(\cos\left(\sqrt{|\xi|^2 - \frac{1}{4}}\, t\right) - \frac{1}{2} \frac{\sin\left(\sqrt{|\xi|^2 - \frac{1}{4}}\, t\right)}{\sqrt{|\xi|^2 - \frac{1}{4}}} \right)^2 d\xi}_{\leq C(1+t)^2}$$

$$+ C \int_{|\xi| > \frac{1}{2}} e^{-t} |v_0(\xi)|^2 \left(\sqrt{|\xi|^2 - \frac{1}{4}} \sin \left(\sqrt{|\xi|^2 - \frac{1}{4}} t \right) \right.$$

$$\left. + \frac{1}{2} \cos \left(\sqrt{|\xi|^2 - \frac{1}{4}} t \right) \right)^2 d\xi.$$

$$\underbrace{\phantom{+ \frac{1}{2} \cos \left(\sqrt{|\xi|^2 - \frac{1}{4}} t \right)}}_{\leq C}$$

The inequality $(|\xi|^2 - \frac{1}{4}) \sin^2 \left(\sqrt{|\xi|^2 - \frac{1}{4}} t \right) \leq |\xi|^2$ implies for $\{\xi : |\xi| > \frac{1}{2}\}$

$$\int_{|\xi| > \frac{1}{2}} |\hat{u}_t(t, \xi)|^2 d\xi \leq C(1 + t)^2 e^{-t} \int_{\mathbb{R}^n} \left(|\xi|^2 |v_0(\xi)|^2 + |v_1(\xi)|^2 \right) d\xi.$$

We need the regularity $\dot{H}^1(\mathbb{R}^n) \times L^2(\mathbb{R}^n)$ for the data (φ, ψ) and obtain an exponential decay in time.

Case 2 $\{\xi : |\xi| < \frac{1}{2}\}$

We get immediately

$$\hat{u}_t(t, \xi)$$

$$= \frac{1}{2} e^{-\frac{1}{2}t} \left(\sqrt{1 - 4|\xi|^2} \sinh \left(\frac{1}{2} \sqrt{1 - 4|\xi|^2} t \right) - \cosh \left(\frac{1}{2} \sqrt{1 - 4|\xi|^2} t \right) \right) v_0(\xi)$$

$$+ e^{-\frac{1}{2}t} \left(\cosh \left(\frac{1}{2} \sqrt{1 - 4|\xi|^2} t \right) - \frac{1}{\sqrt{1 - 4|\xi|^2}} \sinh \left(\frac{1}{2} \sqrt{1 - 4|\xi|^2} t \right) \right) v_1(\xi).$$

Again we divide the interval $[0, \frac{1}{2})$ into two subintervals.

Case 2a $\{\xi : |\xi| \in [\frac{1}{4}, \frac{1}{2})\}$:

Here we show the exponential decay of the kinetic energy. On the one hand we use

$$\cosh \left(\frac{1}{2} \sqrt{1 - 4|\xi|^2} t \right) + \sinh \left(\frac{1}{2} \sqrt{1 - 4|\xi|^2} t \right) \leq 2 \cosh \left(\frac{\sqrt{3}}{4} t \right),$$

on the other hand we use

$$\left| \frac{1}{\sqrt{1 - 4|\xi|^2}} \sinh \left(\frac{1}{2} \sqrt{1 - 4|\xi|^2} t \right) \right| \leq C_\varepsilon t \quad \text{for} \quad \frac{1}{2} \sqrt{1 - 4|\xi|^2} t \leq \varepsilon.$$

Both estimates lead to

$$\|\hat{u}_t(t, \cdot)\|^2_{L^2\{|\xi| \in [\frac{1}{4}, \frac{1}{2})\}} \leq C e^{-\delta t} \int_{\mathbb{R}^n} \left(|\xi|^2 |v_0(\xi)|^2 + |v_1(\xi)|^2 \right) d\xi$$

with a suitable positive δ. Here we need the regularity $\dot{H}^1(\mathbb{R}^n) \times L^2(\mathbb{R}^n)$ for the data (φ, ψ). Moreover, we derived an exponential decay in time.

Case 2b $\{\xi : |\xi| < \frac{1}{4}\}$:

In this case we obtain

$$\hat{u}_t(t, \xi) = \Big(\frac{v_0(\xi)}{4} + \frac{v_1(\xi)}{2\sqrt{1 - 4|\xi|^2}}\Big)\big(\sqrt{1 - 4|\xi|^2} - 1\big)e^{-\frac{1}{2}t + \frac{1}{2}\sqrt{1 - 4|\xi|^2}\,t}$$

$$- \Big(\frac{v_0(\xi)}{4} - \frac{v_1(\xi)}{2\sqrt{1 - 4|\xi|^2}}\Big)\big(\sqrt{1 - 4|\xi|^2} + 1\big)e^{-\frac{1}{2}t - \frac{1}{2}\sqrt{1 - 4|\xi|^2}\,t}.$$

Hence, we can estimate as follows:

$$|\hat{u}_t(t, \xi)| \le \Big|\Big(\frac{v_1(\xi)}{2\sqrt{1 - 4|\xi|^2}} + \frac{v_0(\xi)}{4}\Big)\underbrace{\Big(\sqrt{1 - 4|\xi|^2} - 1\Big)}_{\le -2|\xi|^2}\underbrace{e^{-\frac{1}{2}t + \frac{1}{2}\sqrt{1 - 4|\xi|^2}\,t}}_{\le e^{-|\xi|^2 t},\ |\xi| < \frac{1}{4}}\Big|.$$

Recalling the estimates for the elastic energy, a similar approach leads to

$$\|\hat{u}_t(t, \cdot)\|^2_{L^2\{|\xi| < \frac{1}{4}\}} \le C\int_{|\xi| < \frac{1}{4}} \big(|v_0(\xi)|^2 + |v_1(\xi)|^2\big)|\xi|^4\big(e^{-t} + e^{-2|\xi|^2 t}\big)\,d\xi$$

$$\le Ce^{-t}\int_{|\xi| < \frac{1}{4}} \big(|v_0(\xi)|^2 + |v_1(\xi)|^2\big)\,d\xi$$

$$+ C\int_{|\xi| < \frac{1}{4}} \big(|v_0(\xi)|^2 + |v_1(\xi)|^2\big)|\xi|^4 e^{-2|\xi|^2 t}\,d\xi$$

$$\le Ce^{-t}\int_{|\xi| < \frac{1}{4}} \big(|v_0(\xi)|^2 + |v_1(\xi)|^2\big)\,d\xi$$

$$+ C\frac{1}{t^2}\underbrace{\sup_{|\xi| < \frac{1}{4},\, t \ge 1} t^2|\xi|^4 e^{-2|\xi|^2 t}}_{\le c}\int_{|\xi| < \frac{1}{4}} \big(|v_0(\xi)|^2 + |v_1(\xi)|^2\big)\,d\xi$$

$$\le Ce^{-t}\int_{|\xi| < \frac{1}{4}} \big(|v_0(\xi)|^2 + |v_1(\xi)|^2\big)\,d\xi$$

$$+ \frac{C}{(1 + t)^2}\int_{|\xi| < \frac{1}{4}} \big(|v_0(\xi)|^2 + |v_1(\xi)|^2\big)\,d\xi$$

$$\le \frac{C}{(1 + t)^2}\int_{\mathbb{R}^n} \big(|v_0(\xi)|^2 + |v_1(\xi)|^2\big)\,d\xi.$$

Here we need the regularity $L^2(\mathbb{R}^n) \times L^2(\mathbb{R}^n)$ for the data (φ, ψ). Summarizing all the estimates from the Cases 1 to 2b we have proved the third inequality for the kinetic energy.

Thus, all statements from the theorem are proved.

Which part of the phase space does the decay behavior of the energy influence?

The decay behavior is influenced by the small frequencies. But, which properties of solutions do the large frequencies influence? The large frequencies influence the necessary regularity of the data.

The reader can find a detailed discussion on the classical damped wave model in [131].

14.2.3 The Diffusion Phenomenon for Damped Wave Models

We discussed in Chaps. 9 and 10 properties of solutions to heat and wave equations, respectively. At first glance the properties of solutions to heat or wave models are completely different. One does not expect any relation between heat and wave models. In general this is true, but, already Exercise 5 at the end of this chapter hints at something.

So, to what does this exercise hint?

From time to time mathematicians use instead of the heat equation $w_t - w_{xx} = 0$, which solutions possess an *infinite speed of propagation*, the damped wave equation $\varepsilon^2 u_{tt} - u_{xx} + u_t = 0$, $\varepsilon^2 > 0$ is small. Now, solutions have a *finite speed of propagation*. The speed depends on ε. One can prove the relation $\lim_{\varepsilon \to 0} u(t, x, \varepsilon) = w(t, x)$.

The main result of this section is a relation between solutions of the heat and of the classical damped wave equations. Recalling estimates from Chaps. 9 and 10 motivates this desired relation. On the one hand we have from Theorem 12.1.1 the a priori estimate

$$\|w(t, \cdot)\|_{L^2} \le C\|\varphi\|_{L^2} \text{ for solutions to } w_t - \Delta w = 0, \ w(0, x) = \varphi(x).$$

On the other hand we have from Theorem 14.2.2 the a priori estimate

$$\|u(t, \cdot)\|_{L^2} \le C\big(\|\varphi\|_{L^2} + \|\psi\|_{H^{-1}}\big) \text{ for solutions to}$$

$$u_{tt} - \Delta u + u_t = 0, \ u(0, x) = \varphi(x), \ u_t(0, x) = \psi(x).$$

Problem Let (φ, ψ) be given data in the Cauchy problem for the classical damped wave equation. Can we find a data $\tilde{\varphi}$ in the Cauchy problem for the heat equation such that the difference of the corresponding solutions $u(t, \cdot) - w(t, \cdot)$ decays for $t \to \infty$ in the L^2-norm? Take into consideration that the estimates for $\|u(t, \cdot)\|_{L^2}$ and $\|w(t, \cdot)\|_{L^2}$ are optimal, thus we can not expect any decay for $t \to \infty$. If we show that the difference decays in the L^2-norm, then it is said that the *asymptotic behavior* of both solutions coincide for $t \to \infty$.

In the following we give a positive answer to this last question. This effect is called *diffusion phenomenon*, which was originally observed in [88] and was, for example, studied among other things in the papers [150, 228], and for an abstract model in [93].

To apply the above a priori estimates for solutions to the Cauchy problem for the heat and for the wave equation as well we assume $\varphi, \psi \in L^2(\mathbb{R}^n)$. Then let us turn to the Cauchy problems

$$
\begin{aligned}
u_{tt} - \Delta u + u_t = 0 \\
u(0, x) = \varphi(x), \quad u_t(0, x) = \psi(x)
\end{aligned}
\quad \text{and} \quad
\begin{aligned}
w_t - \Delta w = 0 \\
w(0, x) = \varphi(x) + \psi(x).
\end{aligned}
$$

We introduce a cut-off function $\chi \in C_0^\infty(\mathbb{R}^n)$ with $\chi(s) = 1$ for $|s| \le \frac{\varepsilon}{2} \ll 1$ and $\chi(s) = 0$ for $|s| \ge \varepsilon$ which localizes to small frequencies. Then, we have the following remarkable result.

Theorem 14.2.3 *The difference of solutions to the above Cauchy problems satisfies the following estimate:*

$$
\left\| F_{\xi \to x}^{-1} \Big(\chi(\xi) F_{x \to \xi} \big(u(t, x) - w(t, x) \big) \Big) \right\|_{L^2} \le C(1 + t)^{-1} \| (\varphi, \psi) \|_{L^2}.
$$

Proof We use for small frequencies $|\xi| < \frac{1}{2}$ the following representation for the solution $u = u(t, x)$ from Sect. 14.2.1:

$$
\begin{aligned}
F_{x \to \xi}(u)(t, \xi) \\
= e^{-\frac{1}{2}t} \Bigg(\Big(\frac{1}{2} F(\varphi)(\xi) - \frac{\frac{1}{2} F(\varphi)(\xi) + F(\psi)(\xi)}{\sqrt{1 - 4|\xi|^2}} \Big) e^{-\frac{1}{2}\sqrt{1 - 4|\xi|^2}\, t} \\
+ \Big(\frac{1}{2} F(\varphi)(\xi) + \frac{\frac{1}{2} F(\varphi)(\xi) + F(\psi)(\xi)}{\sqrt{1 - 4|\xi|^2}} \Big) e^{\frac{1}{2}\sqrt{1 - 4|\xi|^2}\, t} \Bigg).
\end{aligned}
$$

We have for $w = w(t, x)$ the representation of solution

$$
F_{x \to \xi}(w)(t, \xi) = e^{-|\xi|^2 t} \big(F(\varphi)(\xi) + F(\psi)(\xi) \big).
$$

Taking into consideration the relations

$$
\sqrt{1 + s} = 1 + \frac{s}{2} - \frac{s^2}{8} + O(s^3)
$$

$$
\text{and} \quad \frac{1}{\sqrt{1 + s}} = 1 - \frac{s}{2} + O(s^2) \quad \text{for } s \to +0
$$

we get

$$\sqrt{1 - 4|\xi|^2} = 1 - 2|\xi|^2 - 2|\xi|^4 + O(|\xi|^6)$$

$$\text{and} \quad \frac{1}{\sqrt{1 - 4|\xi|^2}} = 1 + 2|\xi|^2 + O(|\xi|^4) \quad \text{for} \ |\xi| \to +0.$$

These relations allow us to conclude

$$\left\| F_{\xi \to x}^{-1}\left(\chi(\xi) F_{x \to \xi}\big(u(t,x) - w(t,x)\big) \right) \right\|_{L^2} = \left\| \chi(\xi) F_{x \to \xi}\big(u(t,x) - w(t,x)\big) \right\|_{L^2}$$

$$= \left\| \chi(\xi) \left(\left(\frac{1}{2} F(\varphi)(\xi) - \left(\frac{1}{2} F(\varphi)(\xi) + F(\psi)(\xi) \right) \right. \right.$$

$$+ \left(\frac{1}{2} F(\varphi)(\xi) + F(\psi)(\xi) \right) O(|\xi|^2) \right) e^{-\frac{1}{2}t + O(|\xi|^2)t} e^{-\frac{1}{2}t}$$

$$+ \left(\frac{1}{2} F(\varphi)(\xi) + \left(\frac{1}{2} F(\varphi)(\xi) + F(\psi)(\xi) \right) + \left(F(\varphi)(\xi) + 2F(\psi)(\xi) \right) |\xi|^2 \right.$$

$$+ \left(\frac{1}{2} F(\varphi)(\xi) + F(\psi)(\xi) \right) O(|\xi|^4) \right) e^{\frac{1}{2}t - |\xi|^2 t - |\xi|^4 t + O(|\xi|^6)t} e^{-\frac{1}{2}t}$$

$$\left. - e^{-|\xi|^2 t}\big(F(\varphi)(\xi) + F(\psi)(\xi) \big) \right) \right\|_{L^2}.$$

On the one hand we have

$$\left\| \chi(\xi) \left(-F(\psi)(\xi) + \left(\frac{1}{2} F(\varphi)(\xi) + F(\psi)(\xi) \right) O(|\xi|^2) \right) e^{(-1 + O(|\xi|^2))t} \right\|_{L^2}$$

$$\leq C e^{-ct} \| (\varphi, \psi) \|_{L^2}$$

with a positive constant $c < 1$ depending on the support of χ. On the other hand we have

$$\left\| \chi(\xi) \left(\big(F(\varphi)(\xi) + F(\psi)(\xi) + (F(\varphi)(\xi) + 2F(\psi)(\xi)) |\xi|^2 \right. \right.$$

$$+ \left(\frac{1}{2} F(\varphi)(\xi) + F(\psi)(\xi) \right) O(|\xi|^4) \right)$$

$$\times e^{-|\xi|^2 t - |\xi|^4 t + O(|\xi|^6)t} - e^{-|\xi|^2 t}\big(F(\varphi)(\xi) + F(\psi)(\xi) \big) \right) \bigg\|_{L^2}$$

$$\leq \left\| \chi(\xi)\big(F(\varphi)(\xi) + F(\psi)(\xi) \big) \left(e^{-|\xi|^2 t - |\xi|^4 t + O(|\xi|^6)t} - e^{-|\xi|^2 t} \right) \right\|_{L^2}$$

$$+ \left\| \chi(\xi) \left((F(\varphi)(\xi) + 2F(\psi)(\xi)) |\xi|^2 \right. \right.$$

$$+ \left(\frac{1}{2} F(\varphi)(\xi) + F(\psi)(\xi) \right) O(|\xi|^4) \right) e^{-|\xi|^2 t + O(|\xi|^4)t} \bigg\|_{L^2}.$$

We denote the two norms on the right-hand side of the last inequality by J_1 and by J_2. Let us assume $t \geq 1$. So, we obtain the estimates

$$J_1 = \left\| \chi(\xi)\big(F(\varphi)(\xi) + F(\psi)(\xi)\big)(-|\xi|^4 t + O(|\xi|^6) t)e^{-|\xi|^2 t} \right.$$

$$\times \underbrace{\int_0^1 e^{(-|\xi|^4 t + O(|\xi|^6) t)s} ds}_{\leq 1} \Bigg\|_{L^2}$$

$$\leq C \left\| \chi(\xi)\big(F(\varphi)(\xi) + F(\psi)(\xi)\big)\frac{|\xi|^4 t^2}{t} e^{-|\xi|^2 t} \right\|_{L^2} \leq C t^{-1} \|(\varphi, \psi)\|_{L^2}$$

and

$$J_2 \leq C \left\| \chi(\xi)\big(|F(\varphi)(\xi)| + |F(\psi)(\xi)|\big)\frac{|\xi|^2 t}{t} e^{-c|\xi|^2 t} \right\|_{L^2} \leq C t^{-1} \|(\varphi, \psi)\|_{L^2}.$$

For $t \in (0, 1]$ and for $k = 1, 2$ we have

$$J_k \leq C \|(\varphi, \psi)\|_{L^2}.$$

Summarizing all derived estimates gives

$$J_k \leq C(1 + t)^{-1} \|(\varphi, \psi)\|_{L^2} \text{ for } k = 1, 2.$$

The proof is complete.

The diffusion phenomenon contains the information that the solution to the Cauchy problem

$$u_{tt} - \Delta u + u_t = 0, \quad u(0, x) = \varphi(x), \quad u_t(0, x) = \psi(x)$$

has asymptotically a *parabolic structure* (compare with the behavior of solutions to the Cauchy problem for the heat equation) from the point of view of L^2-estimates for the solution itself.

Why do we consider the diffusion phenomenon only for small frequencies?

Due to the considerations from Sects. 12.1.1 and 14.2.2 we conclude for large frequencies and for large times t the estimates

$$\left\| F_{\xi \to x}^{-1}\big((1 - \chi(\xi))\hat{w}(t, \cdot)\big) \right\|_{L^2} \leq C_0 e^{-C_2 t} \|(\varphi, \psi)\|_{L^2}$$

and

$$\left\| F_{\xi \to x}^{-1}\big((1 - \chi(\xi))\hat{u}(t, \cdot)\big) \right\|_{L^2} \leq C_0 e^{-C_1 t} \|(\varphi, \psi)\|_{L^2}$$

with some positive constants C_1 and C_2.

Thus we already have an exponential decay. This is optimal. There is not any reason to study in detail the difference of Fourier transforms localized to large frequencies for a better decay than the exponential one.

14.2.4 Decay Behavior Under Additional Regularity of Data

We learned in Theorem 14.2.2 that the energy of solutions to classical damped wave models decays. This decay becomes faster under additional regularity of the data (φ, ψ). Let us turn again to the Cauchy problem

$$u_{tt} - \Delta u + u_t = 0, \quad u(0, x) = \varphi(x), \quad u_t(0, x) = \psi(x)$$

under the additional regularity assumption $(\varphi, \psi) \in L^m(\mathbb{R}^n) \times L^m(\mathbb{R}^n)$, $m \in [1, 2)$. In the following we restrict ourselves to explaining modifications in the treatment, in particular, how to use this additional regularity. For large frequencies we do not change our approach because under regularity $\dot{H}^1(\mathbb{R}^n) \times L^2(\mathbb{R}^n)$ of the data we have an exponential decay. But, for small frequencies the additional regularity $L^m(\mathbb{R}^n) \times L^m(\mathbb{R}^n)$ leads to better decay estimates.
Setting

$$\frac{1}{2} = \frac{1}{r} + \frac{1}{m'}$$

and after using Hölder's inequality we get

$$\||\xi| \hat{u}(t, \xi)\|^2_{L^2\{|\xi| < \frac{1}{4}\}} \le C \int_{|\xi| < \frac{1}{4}} |\xi|^2 e^{-|\xi|^2 t} \left(|v_0(\xi)|^2 + |v_1(\xi)|^2 \right) d\xi$$

$$\le C \left(\|v_0\|^2_{L^{m'}} + \|v_1\|^2_{L^{m'}} \right) \left(\int_{|\xi| < \frac{1}{4}} \left(|\xi|^2 e^{-|\xi|^2 t} \right)^{\frac{r}{2}} d\xi \right)^{\frac{2}{r}}$$

$$\le C \left(\|\varphi\|^2_{L^m} + \|\psi\|^2_{L^m} \right) \left(\int_{|\xi| < \frac{1}{4}} \left(|\xi|^2 e^{-|\xi|^2 t} \right)^{\frac{m}{2-m}} d\xi \right)^{\frac{2-m}{m}}.$$

Here, $m' \in (2, \infty]$ is the conjugate exponent to $m \in [1, 2)$ (see also Sect. 24.1.2.1). Let us only estimate the integral on the right-hand side. By using polar coordinates we obtain for large t the estimate

$$\int_{|\xi| < \frac{1}{4}} |\xi|^{\frac{2m}{2-m}} e^{-|\xi|^2 \frac{tm}{2-m}} d\xi = C \int_0^{\frac{1}{4}} r^{\frac{2m}{2-m}} e^{-r^2 \frac{tm}{2-m}} r^{n-1} dr$$

$$\le C \left(\frac{2-m}{tm} \right)^{\frac{n}{2} + \frac{m}{2-m}} \int_0^{\infty} s^{n-1 + \frac{2m}{2-m}} e^{-s^2} ds \le C \left(\frac{1+tm}{2-m} \right)^{-\frac{n}{2} - \frac{m}{2-m}}.$$

Summarizing implies

$$\||\xi|\hat{u}(t,\xi)\|^2_{L^2\{|\xi|<\frac{1}{4}\}} \le C\Big(\frac{1+tm}{2-m}\Big)^{-\frac{n(2-m)}{2m}-1}\big(\|\varphi\|^2_{L^m}+\|\psi\|^2_{L^m}\big)$$

$$\le C_m(1+t)^{-\frac{n(2-m)}{2m}-1}\big(\|\varphi\|^2_{L^m}+\|\psi\|^2_{L^m}\big).$$

Mapping properties of the Fourier transformation (see Remarks 24.1.2 and 24.1.3) explains why we suppose additional regularity $L^m(\mathbb{R}^n)$ for $m \in [1, 2)$, only. Similar estimates can be derived for $\|\partial_t^j \hat{u}(t,\xi)\|^2_{L^2\{|\xi|<\frac{1}{4}\}}$ with $j = 0, 1$. All these estimates together imply the following result.

Theorem 14.2.4 *The solution to the Cauchy problem*

$$u_{tt} - \Delta u + u_t = 0, \quad u(0,x) = \varphi(x), \quad u_t(0,x) = \psi(x)$$

satisfies the following estimates for $t \ge 0$:

$$\|u(t,\cdot)\|_{L^2} \le C_m(1+t)^{-\frac{n(2-m)}{4m}}\big(\|\varphi\|_{H^1 \cap L^m}+\|\psi\|_{L^2 \cap L^m}\big),$$

$$\|\nabla u(t,\cdot)\|_{L^2} \le C_m(1+t)^{-\frac{1}{2}-\frac{n(2-m)}{4m}}\big(\|\varphi\|_{H^1 \cap L^m}+\|\psi\|_{L^2 \cap L^m}\big),$$

$$\|u_t(t,\cdot)\|_{L^2} \le C_m(1+t)^{-1-\frac{n(2-m)}{4m}}\big(\|\varphi\|_{H^1 \cap L^m}+\|\psi\|_{L^2 \cap L^m}\big).$$

Consequently, the energy satisfies the estimate

$$E_W(u)(t) \le C_m(1+t)^{-1-\frac{n(2-m)}{2m}}\big(\|\varphi\|^2_{H^1 \cap L^m}+\|\psi\|^2_{L^2 \cap L^m}\big).$$

Remark 14.2.3 The statement of the last theorem coincides for $m = 2$ (we suppose no additional regularity for the data) with the statement of Theorem 14.2.2.

14.3 Viscoelastic Damped Wave Model

Here we are interested in the Cauchy problem for the viscoelastic damped wave model

$$u_{tt} - \Delta u - \Delta u_t = 0, \quad u(0,x) = \varphi(x), \quad u_t(0,x) = \psi(x).$$

In the following we interpret the term $-\Delta u_t$ as a damping term for the wave operator. Does the viscoelastic damped wave equation fit into the classification of partial differential equations of Sect. 3.1?

Of course not. This becomes clear after writing the given equation as

$$D_t^2 u - |D|^2 u - i|D|^2 D_t u = 0, \quad |D|^2 := \sum_{k=1}^{n} D_{x_k}^2.$$

The principal part of the operator, $-i|D|^2 D_t$, is treated as a damping producing term.

The main goal of this section is to compare the influence of the classical damping term u_t with that of the viscoelastic damping term $-\Delta u_t$ on qualitative properties of solutions to wave models. The reader can find a detailed discussion on the above viscoelastic damped wave model in [184].

14.3.1 Representation of Solutions by Using Fourier Multipliers

A formal application of the partial Fourier transformation gives the following ordinary differential equation for $v = v(t, \xi) = F_{x \to \xi}(u(t, x))$ depending on the parameter $|\xi|$:

$$v_{tt} + |\xi|^2 v_t + |\xi|^2 v = 0, \quad v(0, \xi) = v_0(\xi), \quad v_t(0, \xi) = v_1(\xi),$$

where $v_0(\xi) := F(\varphi)(\xi)$ and $v_1(\xi) := F(\psi)(\xi)$. The solution is given on the set $\{\xi \in \mathbb{R}^n : |\xi| > 2\}$ by

$$
\begin{aligned}
v(t, \xi) &= e^{-\frac{|\xi|^2 t}{2}} \left(\frac{v_0(\xi)\left(|\xi|\sqrt{|\xi|^2 - 4} + |\xi|^2\right) + 2v_1(\xi)}{2|\xi|\sqrt{|\xi|^2 - 4}} \, e^{\frac{|\xi|\sqrt{|\xi|^2 - 4} \, t}{2}} \right. \\
&\quad + \left. \frac{v_0(\xi)\left(|\xi|\sqrt{|\xi|^2 - 4} - |\xi|^2\right) - 2v_1(\xi)}{2|\xi|\sqrt{|\xi|^2 - 4}} \, e^{\frac{-|\xi|\sqrt{|\xi|^2 - 4} \, t}{2}} \right) \\
&= e^{-\frac{|\xi|^2 t}{2}} \left(\frac{|\xi|^2 v_0(\xi) + 2v_1(\xi)}{|\xi|\sqrt{|\xi|^2 - 4}} \sinh\left(\frac{|\xi|\sqrt{|\xi|^2 - 4}}{2} t \right) \right. \\
&\quad + \left. v_0(\xi) \cosh\left(\frac{|\xi|\sqrt{|\xi|^2 - 4}}{2} t \right) \right)
\end{aligned}
$$

and on the set $\{\xi \in \mathbb{R}^n : |\xi| < 2\}$ by

$$v(t, \xi) = e^{-\frac{|\xi|^2 t}{2}} \left(\frac{|\xi|^2 v_0(\xi) + 2v_1(\xi)}{|\xi| \sqrt{4 - |\xi|^2}} \sin \left(\frac{|\xi| \sqrt{4 - |\xi|^2}}{2} t \right) \right.$$

$$\left. + v_0(\xi) \cos \left(\frac{|\xi| \sqrt{4 - |\xi|^2}}{2} t \right) \right).$$

In these representation formulas there appear the two characteristic roots

$$\lambda_{1,2}(\xi) := -\frac{|\xi|^2}{2} \pm \frac{|\xi| \sqrt{|\xi|^2 - 4}}{2}.$$

For our further considerations the asymptotic behavior of the roots for small and large frequencies is of interest.

Lemma 14.3.1 *The characteristic roots $\lambda_{1,2}$ behave as follows:*

1. $\Re \lambda_{1,2} = -\frac{|\xi|^2}{2}$ *for small frequencies,*
2. $\lambda_1 \sim -1$, $\lambda_2 \sim -|\xi|^2$ *for large frequencies.*

To prove H^s well-posedness we consider the representation for large frequencies only. Repeating the proofs of Corollary 14.1.1 and Theorem 14.2.1 implies immediately the following result.

Theorem 14.3.1 *Let the data $\varphi \in H^s(\mathbb{R}^n)$ and $\psi \in H^{s-2}(\mathbb{R}^n)$, $s \in \mathbb{R}^1$, $n \geq 1$ be given in the Cauchy problem*

$$u_{tt} - \Delta u - \Delta u_t = 0, \quad u(0, x) = \varphi(x), \quad u_t(0, x) = \psi(x).$$

Then, there exists for all $T > 0$ a uniquely determined (in general) distributional solution

$$u \in C([0, T], H^s(\mathbb{R}^n)) \cap C^1([0, T], H^{s-2}(\mathbb{R}^n)).$$

We have the a priori estimate

$$\|u(t, \cdot)\|_{H^s} + \|u_t(t, \cdot)\|_{H^{s-2}} \leq C(T) (\|\varphi\|_{H^s} + \|\psi\|_{H^{s-2}}).$$

Finally, the solution depends continuously on the data.

Remark 14.3.1 There is a difference between the statements of Theorems 14.3.1 and 14.2.1 or Corollary 14.1.1. In the classical wave or classical damped wave model the difference of order of Sobolev regularity of data is 1. In the viscoelastic damped wave model the difference is 2.

14.3.2   Decay Behavior and Decay Rate of the Wave Energy

To study the decay behavior of the wave energy $E_W(u)(t)$ (see Sect. 11.1) we can follow the proof to Theorem 14.2.2 by taking account of the representation of solutions for small and large frequencies as well. From Lemma 14.3.1 we conclude for large frequencies an exponential type decay for $t \to \infty$ of the elastic energy if we suppose data $(\varphi, \psi) \in H^1(\mathbb{R}^n) \times H^{-1}(\mathbb{R}^n)$. If we study $v_t(t, \xi)$ for large frequencies, then we conclude an exponential type decay for $t \to \infty$ of the kinetic energy, but now we assume for the data $(\varphi, \psi) \in L^2 \times L^2$. Here we use the relation

$$\lambda_1(\xi)\lambda_2(\xi) \sim \lambda_1(\xi) - \lambda_2(\xi) \text{ for large frequencies.}$$

More interesting is the study of the behavior of $|\xi|v(t, \xi)$ or $v_t(t, \xi)$ for small frequencies. This behavior will determine the decay behavior (cf. with the end of Sect. 14.2.2). As in the case of classical damped waves, the behavior of characteristic roots for small frequencies of Lemma 14.3.1 coupled with suitable powers of $|\xi|$ gives the desired decay. Some straight forward calculations lead to the following estimates for small frequencies:

$$|\xi||v(t, \xi)| \le Ce^{-\frac{|\xi|^2 t}{2}}\big(|\xi||v_0(\xi)| + |v_1(\xi)|\big),$$

$$|v_t(t, \xi)| \le Ce^{-\frac{|\xi|^2 t}{2}}\big(|\xi||v_0(\xi)| + |v_1(\xi)|\big).$$

Combining all these considerations from different regions of the phase space we immediately get the desired result.

Theorem 14.3.2 *Every energy solution (with respect to the wave energy $E_W(u)(t)$) to the Cauchy problem*

$$u_{tt} - \Delta u - \Delta u_t = 0, \quad u(0, x) = \varphi(x), \quad u_t(0, x) = \psi(x)$$

satisfies the following estimates for $t \ge 0$:

$$\|\nabla u(t, \cdot)\|_{L^2} \le C\big((1 + t)^{-\frac{1}{2}}\|\varphi\|_{H^1} + \|\psi\|_{L^2}\big),$$

$$\|u_t(t, \cdot)\|_{L^2} \le C\big((1 + t)^{-\frac{1}{2}}\|\varphi\|_{L^2} + \|\psi\|_{L^2}\big).$$

Consequently, the wave energy satisfies the estimate

$$E_W(u)(t) \le C\big((1 + t)^{-1}\|\varphi\|_{H^1}^2 + \|\psi\|_{L^2}^2\big).$$

Remark 14.3.2 Opposite to the decay behavior for the wave energy of solutions to the Cauchy problem for classical damped waves we can, in general, not expect a decay behavior of the energy of solutions to the Cauchy problem for viscoelastic damped waves. If $\psi \equiv 0$, then we have a decay behavior for both energies as well, even with the same decay behavior. But, in the above estimates we require less regularity of φ in the estimate of the kinetic energy than in the estimate of the elastic energy.

In the same way we obtain the following result for the elastic and kinetic energies of higher order.

Corollary 14.3.1 *The solution to the Cauchy problem*

$$u_{tt} - \Delta u - \Delta u_t = 0, \quad u(0, x) = \varphi(x), \quad u_t(0, x) = \psi(x)$$

satisfies the following estimates for the elastic energies and kinetic energies of higher order for $t \geq 0$:

$$\||D|^k u(t, \cdot)\|_{L^2}^2 \leq C\big((1 + t)^{-k}\|\varphi\|_{\dot{H}^k}^2 + (1 + t)^{-(k-1)}\|\psi\|_{\dot{H}^{k-1}}^2\big) \ for \ k \geq 2,$$

$$\||D|^k u_t(t, \cdot)\|_{L^2}^2 \leq C\big((1 + t)^{-(k+1)}\|\varphi\|_{\dot{H}^k}^2 + (1 + t)^{-k}\|\psi\|_{\dot{H}^k}^2\big) \ for \ k \geq 1.$$

Remark 14.3.3 From the last theorem we conclude the parabolic effect for solutions to the viscoelastic damped wave model. This means the wave energy of higher order of solutions decays faster and faster with increasing order (cf. with Theorem 12.1.1 for the heat equation). The reader can also observe this parabolic effect for solutions to the classical damped wave model.

14.3.3 Decay Behavior Under Additional Regularity of Data

We learned in the previous section that the wave energy, in general, does not decay. There exist several ideas of generating a decay under additional assumptions. What we have in mind is to suppose additional regularity of the data (φ, ψ) (cf. with Sect. 14.2.4). Let us turn again to the Cauchy problem

$$u_{tt} - \Delta u - \Delta u_t = 0, \quad u(0, x) = \varphi(x), \quad u_t(0, x) = \psi(x)$$

under the additional assumption $(\varphi, \psi) \in L^m(\mathbb{R}^n) \times L^m(\mathbb{R}^n)$, $m \in [1, 2)$. In the following we restrict ourselves to explaining modifications in the treatment, in particular, how to use this additional regularity. For large frequencies we do not change our approach because under the assumption that the data (φ, ψ) belong to $\dot{H}^1(\mathbb{R}^n) \times L^2(\mathbb{R}^n)$ we arrive at an exponential decay. But, for small frequencies the additional regularity leads to decay estimates.

By using Hölder's inequality we verify that

$$\||\xi| v(t,\xi)\|^2_{L^2\{|\xi|<1\}} \le C \int_{|\xi|<1} e^{-|\xi|^2 t}\big(|\xi|^2 |v_0(\xi)|^2 + |v_1(\xi)|^2\big)\, d\xi$$

$$\le C \|v_0\|^2_{L^{m'}} \Big(\int_{|\xi|<1} \big(|\xi|^2 e^{-|\xi|^2 t}\big)^{\frac{m}{2-m}}\, d\xi \Big)^{\frac{2-m}{m}}$$

$$+ \|v_1\|^2_{L^{m'}} \Big(\int_{|\xi|<1} \big(e^{-|\xi|^2 t}\big)^{\frac{m}{2-m}}\, d\xi \Big)^{\frac{2-m}{m}}$$

$$\le C \|\varphi\|^2_{L^m} \Big(\int_{|\xi|<1} \big(|\xi|^2 e^{-|\xi|^2 t}\big)^{\frac{m}{2-m}}\, d\xi \Big)^{\frac{2-m}{m}}$$

$$+ \|\psi\|^2_{L^m} \Big(\int_{|\xi|<1} e^{-|\xi|^2 \frac{tm}{2-m}}\, d\xi \Big)^{\frac{2-m}{m}}.$$

Here, $m' \in (2,\infty]$ is the conjugate exponent to $m \in [1,2)$. Let us only estimate the last integral on the right-hand side. By using polar coordinates we have for large t

$$\int_{|\xi|<1} e^{-|\xi|^2 \frac{tm}{2-m}}\, d\xi = C \int_0^1 e^{-r^2 \frac{tm}{2-m}} r^{n-1}\, dr$$

$$\le C \Big(\frac{2-m}{tm}\Big)^{\frac{n}{2}} \int_0^\infty s^{n-1} e^{-s^2}\, ds \le C \Big(\frac{1+tm}{2-m}\Big)^{-\frac{n}{2}}.$$

In the same way we estimate the first integral on the right-hand side. Summarizing implies

$$\||\xi| v(t,\xi)\|^2_{L^2\{|\xi|<1\}}$$

$$\le C \Big(\frac{1+tm}{2-m}\Big)^{-\frac{n(2-m)}{2m}-1} \|\varphi\|^2_{L^m} + \Big(\frac{1+tm}{2-m}\Big)^{-\frac{n(2-m)}{2m}} \|\psi\|^2_{L^m}$$

$$\le C_m (1+t)^{-\frac{n(2-m)}{2m}-1} \|\varphi\|^2_{L^m} + C_m (1+t)^{-\frac{n(2-m)}{2m}} \|\psi\|^2_{L^m}.$$

Mapping properties of the Fourier transformation (see Remarks 24.1.2 and 24.1.3) explain why we suppose additional regularity L^m, $m \in [1,2)$, only. All these estimates together imply the following result.

Theorem 14.3.3 *Let us suppose for the data (φ, ψ) to belong to the function space $(H^1 \cap L^m) \times (L^2 \cap L^m)$, $m \in [1,2)$. Then, every energy solution (with respect to the wave energy $E_W(u)(t)$) to the Cauchy problem*

$$u_{tt} - \Delta u - \Delta u_t = 0, \quad u(0,x) = \varphi(x), \quad u_t(0,x) = \psi(x)$$

satisfies the following estimates for $t \geq 0$:

$$\|\nabla u(t,\cdot)\|_{L^2} \leq C_m\big((1+t)^{-\frac{1}{2}-\frac{n(2-m)}{4m}}\|\varphi\|_{H^1\cap L^m} + (1+t)^{-\frac{n(2-m)}{4m}}\|\psi\|_{L^2\cap L^m}\big),$$

$$\|u_t(t,\cdot)\|_{L^2} \leq C_m\big((1+t)^{-\frac{1}{2}-\frac{n(2-m)}{4m}}\|\varphi\|_{L^2\cap L^m} + (1+t)^{-\frac{n(2-m)}{4m}}\|\psi\|_{L^2\cap L^m}\big).$$

Consequently, the wave energy satisfies the decay estimate

$$E_W(u)(t) \leq C_m\big((1+t)^{-1-\frac{n(2-m)}{2m}}\|\varphi\|^2_{H^1\cap L^m} + (1+t)^{-\frac{n(2-m)}{2m}}\|\psi\|^2_{L^2\cap L^m}\big).$$

14.4   Klein-Gordon Model

The Cauchy problem for the Klein-Gordon equation is

$$u_{tt} - \Delta u + m^2 u = 0, \quad u(0,x) = \varphi(x), \quad u_t(0,x) = \psi(x)$$

with a constant $m^2 > 0$. The mass term or potential forces us to include into the total energy, apart from the *elastic and the kinetic energy*, a third component, namely the *potential energy*. Thus, we define the total energy for a solution to the above Cauchy problem as

$$E_{KG}(u)(t) := \frac{1}{2}\int_{\mathbb{R}^n} \big(|\nabla u(t,x)|^2 + |u_t(t,x)|^2 + m^2|u(t,x)|^2\big)\, dx.$$

This we already learned in Sect. 11.3.4. The total energy is conserved (Theorem 11.3.1). Moreover, a domain of dependence energy inequality holds (Theorem 11.3.2). Finally, solutions have the qualitative properties such as existence of a forward wave front, finite propagation speed of perturbations or existence of a domain of dependence.

Now let us apply phase space analysis to study other quantitative and qualitative properties of solutions to the Cauchy problem for Klein-Gordon equations.

14.4.1   Representation of Solutions by Using Fourier Multipliers

Applying the *partial Fourier transformation* $(v(t,\xi) = F_{x\to\xi}(u(t,x)))$ we obtain the auxiliary Cauchy problem

$$v_{tt} + \langle \xi \rangle^2_m v = 0, \quad v(0,\xi) = F(\varphi)(\xi), \quad v_t(0,\xi) = F(\psi)(\xi).$$

Following the approach of Sect. 14.1 implies

$$v(t, \xi) = \cos(\langle \xi \rangle_m t) F(\varphi)(\xi) + \frac{\sin(\langle \xi \rangle_m t)}{\langle \xi \rangle_m} F(\psi)(\xi).$$

Supposing for the moment the validity of *Fourier's inversion formula*
$u(t, x) = F_{\xi \to x}^{-1}(F_{x \to \xi}(u(t, x)))$ (this we check at the end of our considerations, see
Theorem 24.1.7) brings

$$u(t, x) = F_{\xi \to x}^{-1}\left(\cos(\langle \xi \rangle_m t) F(\varphi)(\xi) \right) + F_{\xi \to x}^{-1}\left(\frac{\sin(\langle \xi \rangle_m t)}{\langle \xi \rangle_m} F(\psi)(\xi) \right).$$

This is the desired *representation of solutions* by Fourier multipliers.

Theorem 14.4.1 *Let $\varphi \in H^s(\mathbb{R}^n)$ and $\psi \in H^{s-1}(\mathbb{R}^n)$, $s \geq 1$, $n \geq 1$ in the Cauchy
problem*

$$u_{tt} - \Delta u + m^2 u = 0, \quad u(0, x) = \varphi(x), \quad u_t(0, x) = \psi(x).$$

Then, there exists a uniquely determined energy solution

$$u \in C([0, T], H^s(\mathbb{R}^n)) \cap C^1([0, T], H^{s-1}(\mathbb{R}^n)) \text{ for all } T > 0.$$

Proof The uniqueness follows from Theorem 11.3.1. A solution is defined by

$$u(t, x) = F_{\xi \to x}^{-1}\left(\cos(\langle \xi \rangle_m t) F(\varphi)(\xi) \right) + F_{\xi \to x}^{-1}\left(\frac{\sin(\langle \xi \rangle_m t)}{\langle \xi \rangle_m} F(\psi)(\xi) \right).$$

If this solution satisfies the desired regularity, then we can follow the proof to
Theorem 14.1.1 step by step. The proof is simpler because we do not have to take
into special consideration the behavior of Fourier multipliers for $\xi \to 0$. Here the
relation $\langle \xi \rangle_m \geq m > 0$ helps. In this way we complete the proof.
We obtain a corresponding statement to Corollary 14.1.1.

Corollary 14.4.1 *The Cauchy problem*

$$u_{tt} - \Delta u + m^2 u = 0, \quad u(0, x) = \varphi(x), \quad u_t(0, x) = \psi(x), \quad x \in \mathbb{R}^n, \quad n \geq 1,$$

*is H^s well-posed, $s \in \mathbb{R}^1$, that is, to given data $\varphi \in H^s(\mathbb{R}^n)$, $\psi \in H^{s-1}(\mathbb{R}^n)$ there
exists a uniquely determined (in general) distributional solution*

$$u \in C([0, T], H^s(\mathbb{R}^n)) \cap C^1([0, T], H^{s-1}(\mathbb{R}^n)) \text{ for all } T > 0.$$

The solution depends continuously on the data, that is, to each ε > 0 there exists a
δ(ε) such that

$$\|\varphi_1 - \varphi_2\|_{H^s} + \|\dot{\psi}_1 - \psi_2\|_{H^{s-1}} < \delta \ implies$$

$$\|u_1 - u_2\|_{C([0,T],H^s) \cap C^1([0,T],H^{s-1})} < \varepsilon.$$

Remark 14.4.1   The statements of Theorems 14.1.1 and 14.4.1, Corollaries 14.1.1
and 14.4.1, respectively, coincide. So, the mass term or potential has no important
influence on the regularity of solutions. But, mass terms have an influence on energy
estimates, as one can see in Theorems 11.1.2 and 11.3.1 or 11.1.1 and 11.3.2. We are
able to control the elastic and kinetic energy in classical wave models. Additionally,
we are able to control the potential energy in Klein-Gordon models.

14.5   Klein-Gordon Model with External Dissipation

In this section we study quantitative properties of solutions for the Cauchy problem
to a classical damped Klein-Gordon model

$$u_{tt} - \Delta u + m^2 u + u_t = 0, \quad u(0,x) = \varphi(x), \quad u_t(0,x) = \psi(x).$$

What kind of results do we expect?
The total energy of solutions is the Klein-Gordon energy $E_{KG}(u)(t)$ of Sect. 14.4.
Taking account of the statements of Theorems 14.2.1 and 14.4.1 we expect the
following result.

Theorem 14.5.1   *The Cauchy problem*

$$u_{tt} - \Delta u + m^2 u + u_t = 0, \quad u(0,x) = \varphi(x), \quad u_t(0,x) = \psi(x), \quad x \in \mathbb{R}^n, \quad n \geq 1,$$

is H^s well-posed, $s \in \mathbb{R}^1$, that is, to given data $\varphi \in H^s(\mathbb{R}^n)$, $\psi \in H^{s-1}(\mathbb{R}^n)$ there
exists a uniquely determined (in general) distributional solution

$$u \in C([0,T], H^s(\mathbb{R}^n)) \cap C^1([0,T], H^{s-1}(\mathbb{R}^n)) \ for \ all \ T > 0.$$

The solution depends continuously on the data, that is, to each ε > 0 there exists a
δ(ε) such that

$$\|\varphi_1 - \varphi_2\|_{H^s} + \|\psi_1 - \psi_2\|_{H^{s-1}} < \delta \ implies$$

$$\|u_1 - u_2\|_{C([0,T],H^s) \cap C^1([0,T],H^{s-1})} < \varepsilon.$$

The proof of this result is part of the research project in Sect. 23.5. What about the
decay behavior of the Klein-Gordon energy? We expect exponential decay of the

energy $E_{KG}(u)(t)$. Let us explain why. For the partial Fourier transform $\hat{u}$ of u we obtain the Cauchy problem

$$\hat{u}_{tt} + \langle \xi \rangle_m^2 \hat{u} + \hat{u}_t = 0, \quad \hat{u}(0, \xi) = F(\varphi)(\xi), \quad \hat{u}_t(0, \xi) = F(\psi)(\xi).$$

The characteristic roots are $\lambda_{1,2} = -\frac{1}{2} \pm \sqrt{\frac{1}{4} - \langle \xi \rangle_m^2}$. We have, in general, the following representation of solution $u = u(t, x)$:

$$u(t, x) = F_{\xi \to x}^{-1}\big(c_1(\xi)e^{\lambda_1(\xi)t} + c_2(\xi)e^{\lambda_2(\xi)t}\big),$$

where $c_k(\xi)$, $k = 1, 2$, are determined by the characteristic roots, $F(\varphi)$ and $F(\psi)$. Consequently,

1. If $m^2 \geq \frac{1}{4}$, then for all frequencies $|\xi| > 0$ the characteristic roots are complex conjugate with $\Re \lambda_{1,2} = -\frac{1}{2}$, this hints to an exponential type decay.
2. If $m^2 < \frac{1}{4}$, then for large frequencies $|\xi|$ we have $\Re \lambda_{1,2} = -\frac{1}{2}$, for small frequencies we use similar as in Case 2b in the proof to Theorem 14.2.2 the equivalence $-\frac{1}{2} \pm \sqrt{\frac{1}{4} - \langle \xi \rangle_m^2} \sim -\langle \xi \rangle_m^2$. Hence, there exists from the point of view of decay estimates for solutions a relation between the classical damped Klein-Gordon model and the heat model with mass term of Sect. 12.2.

Theorem 12.2.1 contains an exponential decay of L^2-norms of all partial derivatives of solutions. Hence, the above sketched relation hints to a corresponding result for solutions to the Klein-Gordon model with classical dissipation (see Sect. 23.5).

14.6   Klein-Gordon Model with Viscoelastic Dissipation

In this section we will study quantitative properties of solutions for the Cauchy problem to a viscoelastic damped Klein-Gordon model

$$u_{tt} - \Delta u + m^2 u - \Delta u_t = 0, \quad u(0, x) = \varphi(x), \quad u_t(0, x) = \psi(x).$$

What kind of results do we expect?
The total energy of solutions is the Klein-Gordon energy $E_{KG}(u)(t)$ of Sect. 14.4. Accounting for the statements of Theorem 14.3.1 we expect the following result.

Theorem 14.6.1   *The Cauchy problem*

$$u_{tt} - \Delta u + m^2 u - \Delta u_t = 0, \quad u(0, x) = \varphi(x), \quad u_t(0, x) = \psi(x), \quad x \in \mathbb{R}^n, \quad n \geq 1,$$

is H^s well-posed, $s \in \mathbb{R}^1$, that is, to given data $\varphi \in H^s(\mathbb{R}^n)$, $\psi \in H^{s-2}(\mathbb{R}^n)$ there exists a uniquely determined (in general) distributional solution

$$u \in C\big([0, T], H^s(\mathbb{R}^n)\big) \cap C^1\big([0, T], H^{s-2}(\mathbb{R}^n)\big) \text{ for all } T > 0.$$

The solution depends continuously on the data, that is, to each $\varepsilon > 0$ there exists a $\delta(\varepsilon)$ *such that*

$$\|\varphi_1 - \varphi_2\|_{H^s} + \|\psi_1 - \psi_2\|_{H^{s-2}} < \delta \ \ implies$$

$$\|u_1 - u_2\|_{C([0,T],H^s) \cap C^1([0,T],H^{s-2})} < \varepsilon.$$

The proof of this result is similar to the proof of Theorem 14.3.1.
What about the decay behavior of $E_{KG}(u)(t)$? For the partial Fourier transform $\hat{u}$ of u we obtain the Cauchy problem

$$\hat{u}_{tt} + \langle \xi \rangle_m^2 \hat{u} + |\xi|^2 \hat{u}_t = 0, \ \ \hat{u}(0, \xi) = F(\varphi)(\xi), \ \ \hat{u}_t(0, \xi) = F(\psi)(\xi).$$

The characteristic roots are

$$\lambda_{1,2} = -\frac{|\xi|^2}{2} \pm \frac{|\xi|^2}{2} \sqrt{1 - 4\frac{\langle \xi \rangle_m^2}{|\xi|^4}}.$$

We have, in general, the following representation of solution $u = u(t, x)$:

$$u(t, x) = F_{\xi \to x}^{-1} \left(c_1(\xi) e^{\lambda_1(\xi)t} + c_2(\xi) e^{\lambda_2(\xi)t} \right),$$

where $c_k(\xi)$, $k = 1, 2$, are determined by the characteristic roots, by $F(\varphi)$ and $F(\psi)$. Thus we may observe that:

1. For small frequencies the characteristic roots are complex conjugate with $\Re\lambda_{1,2} = -\frac{|\xi|^2}{2}$.
2. For large frequencies the characteristic root λ_1 behaves as $-|\xi|^2 + 1$, the characteristic root λ_2 behaves as -1, both hinting towards an exponential type decay.

Consequently, the decay will be determined by the behavior of the characteristic roots $\Re\lambda_{1,2} = -\frac{|\xi|^2}{2}$ for the small frequencies as we observed for viscoelastic damped waves in Sect. 14.3.2. All these explanations lead to a corresponding result to Theorem 14.3.2, at least for the kinetic and elastic energy. The reader may solve Exercise 7 below.

14.7 Concluding Remarks

During the last decade a lot of progress has been made in the treatment of wave models with time-dependent mass or dissipation. This development has been caused by relations among this theory and other fields of mathematics, like harmonic analysis, for instance, and by a strong connection with problems of mathematical physics.

The study of the long time asymptotical behavior of solutions and, moreover, the study of the long time behavior of suitable energies of solutions has been a topic of interest in the recent years. In this section we focus on models with time-dependent coefficients, shedding light on new phenomena which appear in dealing with time-dependent coefficients. These phenomena may also help getting a better understanding of classical results for constant coefficient models, as well, as giving hints or possible generalizations.

In order to obtain energy estimates for solutions to the Cauchy problem for linear wave models, particular attention is devoted to phase space analysis, and it is convenient to distinguish between low and high frequencies, in general. To derive estimates on the L^2 basis for linear wave models with time-dependent coefficients one may use Fourier transformation with respect to spatial variables and the formula of Parseval-Plancherel (see Remark 24.1.2) reducing our problems in estimating Fourier multipliers in function spaces related to Sobolev spaces. It is convenient to choose a partition of the extended phase space $\{(t, \xi) \in [0, \infty) \times \mathbb{R}^n\}$ into zones. One may use different approaches in different zones taking advantage of properties of the zones themselves. Under suitable assumptions to the smoothness of the time-dependent coefficients, this partition is given by smooth curves in $\{(t, |\xi|) \in [0, \infty) \times \mathbb{R}^1_+\}$, which shapes are related to the coefficients. Therefore, we have to distinguish between low and high frequencies with respect to suitable time-dependent functions.

Let us consider the following Cauchy problem for a wave equation with time-dependent propagation speed $a = a(t)$:

$$u_{tt} - a(t)^2 \Delta u = 0, \quad u(0, x) = \varphi(x), \quad u_t(0, x) = \psi(x).$$

The wave type energy of weak solutions $u = u(t, x)$ is given by

$$E_{W,a}(u)(t) := \frac{1}{2}\left(a(t)^2 \|\nabla u(t, \cdot)\|_{L^2}^2 + \|u_t(t, \cdot)\|_{L^2}^2\right).$$

One can observe many different effects for the behavior of $E_{W,a}(u)(t)$ as $t \to \infty$ according to properties of the speed of propagation $a(t)$. If $0 < a_0 \le a(t) \le a_1 < \infty$ for any $t \ge 0$, then the energy $E_{W,a}(u)(t)$ is equivalent to

$$E_{W,1}(u)(t) = E_W(u)(t) = \frac{1}{2}\left(\|\nabla u(t, \cdot)\|_{L^2}^2 + \|u_t(t, \cdot)\|_{L^2}^2\right).$$

Although $E_W(u)(t)$ is a conserved quantity for the classical wave equation (see Theorem 11.1.2), oscillations of the time-dependent coefficient $a(t)$ may have a deteriorating influence on the energy behavior of solutions (see [24] and [176]). Interested readers can try to understand by themselves (see [176]) the influence of oscillations in the time-dependent coefficient $a = a(t)$ on a possible blow up behavior of $E_{W,a}(u)(t)$ for $t \to \infty$ for solutions to the Cauchy problem

$$u_{tt} - (2 + \sin t)^2 \Delta u = 0, \quad u(0, x) = \varphi(x), \quad u_t(0, x) = \psi(x).$$

For this reason, it is necessary to control the oscillating behavior of the coefficients. Such a classification was proposed in [176] and can be used to control suitable energies.

If, for example, $a \in C^2[0, \infty)$ and

$$\left|a^{(k)}(t)\right| \leq C_k(1 + t)^{-k} \text{ for } k = 1, 2$$

(due to the proposed classification only very slow oscillations are allowed), then the so-called *generalized energy conservation property* holds (see [174] or [77]). This means that there exist positive constants C_0 and C_1 such that the inequalities

$$C_0 E_{W,a}(u)(0) \leq E_{W,a}(u)(t) \leq C_1 E_{W,a}(u)(0)$$

are valid for all $t \in (0, \infty)$ and all energy solutions where the positive constants C_0 and C_1 are independent of the data. This generalized energy conservation property excludes decay and blow up of $E_{W,a}(u)(t)$ for $t \to \infty$.

If $a(t) \geq a_0 > 0$ is an increasing function satisfying a suitable control on the oscillations, then one can prove the estimate (cf. with [17])

$$E_{W,a}(u)(t) \leq C_1 a(t) \left(E_{W,a}(u)(0) + \|\varphi\|_{L^2}^2\right).$$

We remark that in the case of an increasing $a = a(t)$ in the derived energy estimate for $E_{W,a}(u)(t)$ the H^1-norm of φ appears, not only the L^2-norm of its gradient as in the case of bounded $a = a(t)$.

One can allow for faster oscillations of the coefficients if further structural properties of the coefficients are supposed. One possibility is a so-called *stabilization condition* [77] and, possibly, more regularity of the coefficients [78]. The situation becomes more complicated if one considers an unbounded propagation speed $a(t)$, but it does remain possible to prove the boundedness of suitable energies [80].

The control of oscillations of a bounded or unbounded propagation speed $a = a(t)$ is also relevant if one is interested in $L^p - L^q$ estimates (on the conjugate line) where $q = \frac{p}{p-1}$ is the Sobolev conjugate of $p \in [1, 2)$ [132, 176, 177]. By this notation we mean that we estimate the L^q-norm of the solution and its derivatives by the L^p-norm of the data (φ, ψ) with some loss of regularity as also happens for the classical wave equation (see, for example, [180]).

On the other hand, if we add a time-dependent damping term in the classical wave equation,

$$u_{tt} - \Delta u + b(t)u_t = 0, \quad u(0, x) = \varphi(x), \quad u_t(0, x) = \psi(x),$$

then the asymptotic behavior of solutions and their wave energy $E_W(u)(t)$ change accordingly to the positive coefficient $b = b(t)$ in the damping term. The dissipative term is said to be *effective* when the solution has the same decay behavior as the solution of the corresponding parabolic Cauchy problem

$$-\Delta v + b(t)v_t = 0, \quad v(0, x) = \chi(x),$$

for a suitable choice of initial data χ (see [224]). In this case we say that a *diffusion phenomenon* appears (cf. with Sect. 14.2.3). This latter effect has been extensively studied for the case $b \equiv 1$ in [128, 145, 151]. Moreover, in the case $b \equiv 1$, the energy for the solution to the damped wave equation decays as $t \to \infty$ and higher order energies decay with faster speed (the so-called *parabolic effect*) as do the solutions to the corresponding parabolic Cauchy problem. In some *effective* cases, with suitable control on oscillations, one may expect the same effect for effective damping terms $b(t)u_t$, in particular, if $b(t) = \mu(1 + t)^{-\alpha}$ with $\mu > 0$ and $\alpha \in (-1, 1)$.

However, in the limit case $b(t) = \mu(1 + t)^{-1}$, which generates a *scale-invariant damping* $b(t)u_t$, the *parabolic effect* only appears for large values of μ [221], that is, the dissipation term becomes less *effective*. If $b(t)$ decays faster than $(1 + t)^{-1}$ as $t \to \infty$, the dissipation becomes *non-effective* [223]. In particular, if $b \in L^1(\mathbb{R}^1_+)$, then the damping term has no effect on the energy behavior anymore in the sense that scattering to free waves can be proved. In the case of *non-effective* dissipation, a *stabilization condition* may be used to control stronger oscillations [79].

As in the case $b \equiv 1$, if the damping term is *effective*, then the decay behavior becomes faster under additional regularity of the data. These estimates have been derived in [38] and applied to prove results on the global existence (in time) of small data solutions to semilinear Cauchy problems (cf. with Chaps. 18 and 19).

In [17] the authors proposed a classification of the damping term $b(t)u_t$ in terms of an increasing speed of propagation $a = a(t)$ for the model

$$u_{tt} - a(t)^2 \Delta u + b(t)u_t = 0, \quad u(0, x) = \varphi(x), \quad u_t(0, x) = \psi(x).$$

Besides others things, they proved energy estimates in the cases of *non-effective* and *effective* dissipation, respectively. In the case of *non-effective* dissipation, energy estimates were also proved in [31] under the weaker condition $\frac{a'(t)}{a(t)} + b(t) \geq 0$ with $a = a(t) \notin L^1(\mathbb{R}^1_+)$.

In the paper [46] the authors treated a completely new case. They assume the propagation speed $a \in L^1(\mathbb{R}^1_+)$, then introduce a classification of damping terms $b(t)u_t$ in terms of the propagation speed. They derive estimates for a suitable wave type energy and show optimality of these decay estimates. Moreover, a new phenomenon for these models is observed, the so-called loss of regularity appearing in the derived energy estimates.

If we add a time-dependent potential in the classical wave equation,

$$u_{tt} - \Delta u + m(t)^2 u = 0, \quad u(0, x) = \varphi(x), \quad u_t(0, x) = \psi(x),$$

then the asymptotic behavior of the solution and a related energy change accordingly to the positive coefficient $m(t)^2$ in the mass term. For the classical Klein-Gordon equation, i.e. $m(t) \equiv 1$, $L^p - L^q$ estimates were proved in [129]. In [10] the authors proved $L^p - L^q$ estimates for the scale-invariant case $m(t) = \mu(1 + t)^{-1}$. In [11] and [172] the authors proved energy and $L^p - L^q$ estimates for solutions to a Klein-

Gordon model with a so-called *effective* mass term, i.e., the solutions have the same asymptotical behavior as those of the classical Klein-Gordon model. Recently, in [47] the authors proved some scattering result to free waves and explain qualitative properties of the energy for Klein-Gordon models with non-effective masses, i.e., a class that is "below" the scale-invariant mass term but does not allow for proving scattering to free waves. Finally, in [156] $L^p - L^q$ estimates are derived for non-effective time-dependent potentials. Also, as was done for the wave equation with a dissipative term [79], the assumptions to the oscillations of the mass term $m(t)^2 u$ may be weakened if one assumes a *stabilization condition* and, possibly, more regularity for the coefficients.

Exercises Relating to the Considerations of Chap. 14

Exercise 1 Complete the step

$$\lim_{t_1 \to t_2} \int_{\mathbb{R}^n} \left| \cos\left(\frac{|\xi|(t_1 + t_2)}{2} \right) \sin\left(\frac{|\xi|(t_1 - t_2)}{2} \right) \right|^2 \frac{|F(\psi)(\xi)|^2}{|\xi|^2} \langle \xi \rangle^{2s} \, d\xi = 0$$

in the proof to Theorem 14.1.1.

Exercise 2 How can we derive d'Alembert's representation formula (see Sect. 10.1) from the following representation of solutions by Fourier multipliers:

$$u(t,x) = F_{\xi \to x}^{-1}\left(\left(e^{i\xi t} + e^{-i\xi t} \right) \frac{1}{2} F(\varphi)(\xi) \right)$$

$$+ F_{\xi \to x}^{-1}\left(\left(e^{i\xi t} - e^{-i\xi t} \right) \frac{1}{2i\xi} F(\psi)(\xi) \right).$$

Exercise 3 Let us consider the Cauchy problem for a very large damped membrane

$$u_{tt} - c^2 \Delta u + k u_t = 0, \quad u(0,x) = \varphi(x), \quad u_t(0,x) = \psi(x), \quad x \in \mathbb{R}^2.$$

Solve this problem with the aid of the following changes of variables:

$$u(t,x) = \exp\left(-\frac{kt}{2} \right) w(t,x), \quad v(t,x_1,x_2,x_3) = w(t,x_1,x_2) \exp\left(\frac{kx_3}{2c} \right).$$

Exercise 4 We are interested in the Cauchy problem

$$u_{tt} - u_{xx} + \varepsilon u_t = 0, \quad u(0,x,\varepsilon) = \varphi(x), \quad u_t(0,x,\varepsilon) = \psi(x), \quad x \in \mathbb{R}^1$$

with sufficiently smooth data φ and ψ. Let $u = u(t,x,\varepsilon)$ be the unique solution of this Cauchy problem. Show that we have for every fixed (t,x) the relation $\lim_{\varepsilon \to 0} u(t,x,\varepsilon) = w(t,x)$, where $w = w(t,x)$ solves the Cauchy problem

$$w_{tt} - w_{xx} = 0, \quad w(0,x) = \varphi(x), \quad w_t(0,x) = \psi(x), \quad x \in \mathbb{R}^1.$$

Exercise 5 Let us consider the mixed problem

$$\varepsilon^2 u_{tt} - u_{xx} + u_t = 0, \quad u(0, x, \varepsilon) = \varphi(x), \quad u_t(0, x, \varepsilon) = \psi(x), \quad x \in (0, L),$$

$$u(t, 0, \varepsilon) = u(t, L, \varepsilon) = 0 \text{ for } t > 0,$$

with sufficiently smooth data φ and ψ. We assume that the compatibility conditions are satisfied. Let $u = u(t, x, \varepsilon)$ be the unique (distributional) solution of this mixed problem (without explaining the precise regularity). Prove that for every fixed (t, x) the following relation holds: $\lim_{\varepsilon \to 0} u(t, x, \varepsilon) = w(t, x)$, where $w = w(t, x)$ solves the mixed problem

$$w_t - w_{xx} = 0, \quad w(0, x) = \varphi(x), \quad x \in (0, L),$$

$$w(t, 0) = w(t, L) = 0 \text{ for } t > 0.$$

Exercise 6 Prove the statements of Corollary 14.3.1.

Exercise 7 Let us consider the Cauchy problem to the viscoelastic damped Klein-Gordon model

$$u_{tt} - \Delta u + m^2 u - \Delta u_t = 0, \quad u(0, x) = \varphi(x), \quad u_t(0, x) = \psi(x).$$

Derive an estimate for the Klein-Gordon energy $E_{KG}(u)(t)$.

Exercise 8 Consider the Cauchy problems

$$\begin{array}{ccc} u_{tt} - \Delta u + u_t = 0 & & w_t - \Delta w = 0 \\ u(0, x) = \varphi(x), \quad u_t(0, x) = \psi(x) & \text{and} & w(0, x) = \varphi(x) + \psi(x), \end{array}$$

with $\varphi, \psi \in L^1(\mathbb{R}^n)$. By following the proof of Theorem 14.2.3, prove that the difference of solutions to the above Cauchy problems satisfies

$$\left\| F_{\xi \to x}^{-1} \left(\chi(\xi) F_{x \to \xi} \left(u(t, \cdot) - w(t, \cdot) \right) \right) \right\|_{L^2} \leq C(1 + t)^{-\frac{n}{4} - 1} \|(\varphi, \psi)\|_{L^1},$$

where $\chi \in C_0^\infty(\mathbb{R}^n)$ is a cut-off function which localizes to small frequencies. Moreover, if $|D|\varphi, \psi \in L^2(\mathbb{R}^n)$, with $\widehat{|D|\varphi}(\xi) = |\xi|\widehat{\varphi}(\xi)$, conclude that

$$\left\| u(t, \cdot) - w(t, \cdot) \right\|_{L^2} \leq C(1 + t)^{-\frac{n}{4} - 1} \left(\|\varphi\|_{H^1 \cap L^1} + \|\psi\|_{L^2 \cap L^1} \right).$$

Chapter 15
Phase Space Analysis for Plate Models

This chapter is devoted to the application of phase space analysis to plate models. Our main focus is on the discussion of different damped plate models. On the one hand we study classical damped plate models with and without additional regularity of the data and derive estimates, even for the energies of higher order. On the other hand, we study a viscoelastic damped plate model and show a different influence of this special case of structural damping in comparison with the influence of a classical damping term on qualitative properties of solutions. Finally, we discuss the interaction of several terms in plate models such as mass and classical damping or mass and viscoelastic damping.

15.1   The Classical Plate Model

In this section we are interested in the Cauchy problem

$$u_{tt} + (-\Delta)^2 u = 0, \quad u(0,x) = \varphi(x), \quad u_t(0,x) = \psi(x), \quad x \in \mathbb{R}^n, \quad n \geq 1.$$

After application of *partial Fourier transformation* $\big(v(t,\xi) := F_{x \to \xi}(u(t,x))\big)$ we get the auxiliary Cauchy problem

$$v_{tt} + |\xi|^4 v = 0, \quad v(0,\xi) = F(\varphi)(\xi), \quad v_t(0,\xi) = F(\psi)(\xi)$$

for an *ordinary differential equation depending on the parameter* $\xi \in \mathbb{R}^n$. Repeating the approach of Sect. 14.1 we arrive at the following representation of solutions by using Fourier multipliers:

$$u(t,x) = F_{\xi \to x}^{-1}\big(\cos(|\xi|^2 t)F(\varphi)(\xi)\big) + F_{\xi \to x}^{-1}\Big(\frac{\sin(|\xi|^2 t)}{|\xi|^2} F(\psi)(\xi)\Big).$$

© Springer International Publishing AG 2018
M.R. Ebert, M. Reissig, *Methods for Partial Differential Equations*,
https://doi.org/10.1007/978-3-319-66456-9_15

Instead though, we shall use the equivalent representation

$$u(t,x) = F_{\xi \to x}^{-1}\left(e^{-i|\xi|^2 t}\frac{1}{2} F(\varphi)(\xi)\right) - F_{\xi \to x}^{-1}\left(e^{-i|\xi|^2 t}\frac{1}{2i|\xi|^2} F(\psi)(\xi)\right)$$

$$+ F_{\xi \to x}^{-1}\left(e^{i|\xi|^2 t}\frac{1}{2} F(\varphi)(\xi)\right) + F_{\xi \to x}^{-1}\left(e^{i|\xi|^2 t}\frac{1}{2i|\xi|^2} F(\psi)(\xi)\right).$$

Remark 15.1.1 The last relations hint towards a connection of plate operators to Schrödinger operators. The functions $\pm i|\xi|^2 t$ are part of the phase function in the representation of solutions to the Schrödinger equation (see Chap. 13)

$$\frac{1}{i}\partial_t u \pm \Delta u = 0.$$

By the approach of Sect. 14.1 we are able to prove the following result:

Theorem 15.1.1 *The Cauchy problem*

$$u_{tt} + (-\Delta)^2 u = 0, \quad u(0,x) = \varphi(x), \quad u_t(0,x) = \psi(x), \quad x \in \mathbb{R}^n, \quad n \geq 1$$

is H^s well-posed, $s \in \mathbb{R}^1$, that is, to given data $\varphi \in H^s(\mathbb{R}^n)$ and $\psi \in H^{s-2}(\mathbb{R}^n)$ there exists a uniquely determined (in general) distributional solution

$$u \in C\big([0,T], H^s(\mathbb{R}^n)\big) \cap C^1\big([0,T], H^{s-2}(\mathbb{R}^n)\big) \text{ for all } T > 0.$$

The solution depends continuously on the data, that is, to each $\varepsilon > 0$ there exists a positive $\delta(\varepsilon)$ such that

$$\|\varphi_1 - \varphi_2\|_{H^s} + \|\psi_1 - \psi_2\|_{H^{s-2}} < \delta \text{ implies}$$

$$\|u_1 - u_2\|_{C([0,T],H^s) \cap C^1([0,T],H^{s-2})} < \varepsilon.$$

The solution is a Sobolev solution for $s \geq 0$, an energy solution for $s \geq 2$ (see the definition of the energy $E_{PL}(u)(t)$ of Sect. 11.2.4) and a classical solution for $s > \frac{n}{2} + 4$.

15.2   The Classical Damped Plate Model

15.2.1   *Representation of Solutions by Using Fourier Multipliers*

Let us turn to the Cauchy problem

$$u_{tt} + (-\Delta)^2 u + u_t = 0, \quad u(0,x) = \varphi(x), \quad u_t(0,x) = \psi(x).$$

We follow the approach of Sect. 14.2.1. The dissipative transformation $w(t,x) := e^{\frac{1}{2}t}u(t,x)$ transfers the above Cauchy problem to

$$w_{tt} + (-\Delta)^2 w - \frac{1}{4}w = 0, \quad w(0,x) = \varphi(x), \quad w_t(0,x) = \frac{1}{2}\varphi(x) + \psi(x).$$

The application of partial Fourier transformation gives the following ordinary differential equation for $v = v(t,\xi) = F_{x\to\xi}(w(t,x))(t,\xi)$:

$$v_{tt} + \left(|\xi|^4 - \frac{1}{4}\right)v = 0, \quad v(0,\xi) = v_0(\xi) := F(\varphi)(\xi),$$

$$v_t(0,\xi) = v_1(\xi) := \frac{1}{2}F(\varphi)(\xi) + F(\psi)(\xi).$$

We use a similar distinction of cases if the coefficient $|\xi|^4 - \frac{1}{4}$ in the mass term is positive or negative, respectively.

Case 1 $\{\xi : |\xi|^4 > \frac{1}{4}\}$

We obtain the following representation of solution $v(t,\xi)$:

$$v(t,\xi) = \cos\left(\sqrt{|\xi|^4 - \frac{1}{4}}\, t\right) v_0(\xi) + \frac{\sin\left(\sqrt{|\xi|^4 - \frac{1}{4}}\, t\right)}{\sqrt{|\xi|^4 - \frac{1}{4}}} v_1(\xi).$$

Case 2 $\{\xi : |\xi|^4 < \frac{1}{4}\}$

The solution to the transformed differential equation is

$$v(t,\xi) = \left(\frac{v_0(\xi)}{2} - \frac{v_1(\xi)}{\sqrt{1-4|\xi|^4}}\right)e^{-\frac{1}{2}\sqrt{1-4|\xi|^4}\,t}$$

$$+\left(\frac{v_0(\xi)}{2} + \frac{v_1(\xi)}{\sqrt{1-4|\xi|^4}}\right)e^{\frac{1}{2}\sqrt{1-4|\xi|^4}\,t}$$

$$= v_0(\xi)\cosh\left(\frac{1}{2}\sqrt{1-4|\xi|^4}\,t\right) + \frac{2v_1(\xi)}{\sqrt{1-4|\xi|^4}}\sinh\left(\frac{1}{2}\sqrt{1-4|\xi|^4}\,t\right).$$

To prove H^s well-posedness we use the representation for large frequencies only. Analogous to the considerations in Sect. 14.2.1, we can prove that for the Cauchy problem for classical damped plates the statements of Theorem 15.1.1 remain true. This we expected. The damping term has only an influence on an energy decay.

15.2.2 Decay Behavior and Decay Rate of a Suitable Energy

As in Sect. 11.3.2, we are able to prove for the energy $E_{PL}(u)(t)$ of Sect. 11.2.4 the estimate

$$E'_{PL}(u)(t) = \int_{\mathbb{R}^n} -u_t(t,x)^2 \, dx \le 0$$

if we assume a suitable regularity of the data. Hence, the energy $E_{PL}(u)(t)$ is decreasing in time. Therefore, it is reasonable to ask whether the energy is decaying. To give an answer we may follow the approach of Sect. 14.2.2. We get the following representations:

Case 1 $\{\xi : |\xi|^4 > \frac{1}{4}\}$

We obtain for the two parts of the energy in the phase space

$$|\xi|^2 \hat{u}(t, \xi)$$

$$= e^{-\frac{1}{2}t} \left(\cos\left(\sqrt{|\xi|^4 - \frac{1}{4}} \, t \right) |\xi|^2 v_0(\xi) + t \, \frac{\sin\left(\sqrt{|\xi|^4 - \frac{1}{4}} \, t \right)}{\sqrt{|\xi|^4 - \frac{1}{4}} \, t} \, |\xi|^2 v_1(\xi) \right),$$

$$\hat{u}_t(t, \xi) = e^{-\frac{1}{2}t} \left(v_1(\xi) \left(\cos\left(\sqrt{|\xi|^4 - \frac{1}{4}} \, t \right) - \frac{1}{2} \frac{\sin\left(\sqrt{|\xi|^4 - \frac{1}{4}} \, t \right)}{\sqrt{|\xi|^4 - \frac{1}{4}}} \right) \right.$$

$$\left. - v_0(\xi) \left(\frac{1}{2} \cos\left(\sqrt{|\xi|^4 - \frac{1}{4}} \, t \right) + \sqrt{|\xi|^4 - \frac{1}{4}} \sin\left(\sqrt{|\xi|^4 - \frac{1}{4}} \, t \right) \right) \right).$$

Case 2 $\{\xi : |\xi|^4 < \frac{1}{4}\}$

We obtain for the two parts of the energy in the phase space

$$|\xi|^2 \hat{u}(t, \xi) = |\xi|^2 e^{-\frac{1}{2}t} \left(\left(\frac{v_0(\xi)}{2} - \frac{v_1(\xi)}{\sqrt{1 - 4|\xi|^4}} \right) e^{-\frac{1}{2}\sqrt{1 - 4|\xi|^4} \, t} \right.$$

$$\left. + \left(\frac{v_0(\xi)}{2} + \frac{v_1(\xi)}{\sqrt{1 - 4|\xi|^4}} \right) e^{\frac{1}{2}\sqrt{1 - 4|\xi|^4} \, t} \right)$$

$$= v_0(\xi) |\xi|^2 \cosh\left(\frac{1}{2} \sqrt{1 - 4|\xi|^4} \, t \right) e^{-\frac{1}{2}t}$$

$$+ \frac{2 v_1(\xi) |\xi|^2}{\sqrt{1 - 4|\xi|^4}} \sinh\left(\frac{1}{2} \sqrt{1 - 4|\xi|^4} \, t \right) e^{-\frac{1}{2}t},$$

$\hat{u}_t(t, \xi)$

$$= \frac{1}{2}e^{-\frac{1}{2}t}\left(\sqrt{1 - 4|\xi|^4}\sinh\left(\frac{1}{2}\sqrt{1 - 4|\xi|^4}\,t\right) - \cosh\left(\frac{1}{2}\sqrt{1 - 4|\xi|^4}\,t\right)\right)v_0(\xi)$$

$$+ e^{-\frac{1}{2}t}\left(\cosh\left(\frac{1}{2}\sqrt{1 - 4|\xi|^4}\,t\right) - \frac{1}{\sqrt{1 - 4|\xi|^4}}\sinh\left(\frac{1}{2}\sqrt{1 - 4|\xi|^4}\,t\right)\right)v_1(\xi).$$

We can follow all the steps of the proof to Theorem 14.2.2 and conclude immediately the following result:

Theorem 15.2.1 *The energy solution*

$$u \in C\big([0, \infty), H^2(\mathbb{R}^n)\big) \cap C^1\big([0, \infty), L^2(\mathbb{R}^n)\big)$$

to the Cauchy problem

$$u_{tt} + (-\Delta)^2 u + u_t = 0, \quad u(0, x) = \varphi(x), \quad u_t(0, x) = \psi(x)$$

with data $\varphi \in H^2(\mathbb{R}^n)$ and $\psi \in L^2(\mathbb{R}^n)$ satisfies the following decay estimates for $t \geq 0$:

$$\|\Delta u(t, \cdot)\|_{L^2} \leq C(1 + t)^{-\frac{1}{2}}\big(\|\varphi\|_{H^2} + \|\psi\|_{L^2}\big),$$

$$\|u_t(t, \cdot)\|_{L^2} \leq C(1 + t)^{-1}\big(\|\varphi\|_{H^2} + \|\psi\|_{L^2}\big).$$

Consequently, the energy $E_{PL}(u)(t)$ satisfies the decay estimate

$$E_{PL}(u)(t) \leq C(1 + t)^{-1}\big(\|\varphi\|_{H^2}^2 + \|\psi\|_{L^2}^2\big).$$

15.2.3   Energies of Higher Order

In Sect. 11.2.2 we introduced energies of higher order $E_H^k(u)(t)$ for solutions to the heat equation. We encouraged readers to prove in Exercise 1 of Chap. 11 the parabolic effect, that is, higher order energies decay with increasing order faster and faster. In Sect. 11.2.3 we introduced the energies of higher order $E_{Sch}^k(u)(t)$ for solutions to Schrödinger equations. We have no parabolic effect any more and have even conservation of all these energies of higher order.

The situation is similar for waves or plates with or without classical damping. For classical waves or classical plates one can prove conservation of energies of higher order, too. But, an additional classical damping term allows for one to prove a parabolic effect. The diffusion phenomenon of Sect. 14.2.3 hints at this property for classical damped waves. In the following we restrict ourselves to classical damped plates, only (see also Sect. 23.6). There exist different possibilities of introducing energies of higher order for plates or classical damped plates, for example,

$$E_{PL,1}^k(u)(t) = \frac{1}{2}\left\|\Delta^k u_t(t,\cdot)\right\|_{L^2}^2 + \frac{1}{2}\left\|\Delta^{k+1} u(t,\cdot)\right\|_{L^2}^2 \quad \text{or}$$

$$E_{PL,2}^k(u)(t) = \frac{1}{2}\left\||D|^k u_t(t,\cdot)\right\|_{L^2}^2 + \frac{1}{2}\left\||D|^k \Delta u(t,\cdot)\right\|_{L^2}^2 \quad \text{for } k \geq 0.$$

We shall consider only $E_{PL,2}^k(u)(t)$. After writing the representation of both parts of the energy in the phase space and following the proof to Theorem 14.2.2 we obtain the next statement.

Theorem 15.2.2 *The energy solution to the Cauchy problem*

$$u_{tt} + (-\Delta)^2 u + u_t = 0, \quad u(0,x) = \varphi(x), \quad u_t(0,x) = \psi(x)$$

with data $\varphi \in H^{k+2}(\mathbb{R}^n)$ *and* $\psi \in H^k(\mathbb{R}^n)$, $k \geq 0$, *satisfies the following decay estimates for* $t \geq 0$:

$$\left\||D|^k \Delta u(t,\cdot)\right\|_{L^2} \leq C(1+t)^{-\frac{k+2}{4}}\left(\|\varphi\|_{H^{k+2}} + \|\psi\|_{H^k}\right),$$

$$\left\||D|^k u_t(t,\cdot)\right\|_{L^2} \leq C(1+t)^{-\frac{k+4}{4}}\left(\|\varphi\|_{H^{k+2}} + \|\psi\|_{H^k}\right).$$

Consequently, the energies of higher order satisfy the decay estimates

$$E_{PL,2}^k(u)(t) \leq C(1+t)^{-\frac{k+2}{2}}\left(\|\varphi\|_{H^{k+2}}^2 + \|\psi\|_{H^k}^2\right).$$

Remark 15.2.1 We observe the parabolic effect, that is, higher order energies $E_{PL,2}^k(u)(t)$ decay faster and faster with increasing order $k \geq 0$.

15.3   The Viscoelastic Damped Plate Model

In this section we turn to the Cauchy problem

$$u_{tt} + (-\Delta)^2 u + (-\Delta)^2 u_t = 0, \quad u(0,x) = \varphi(x), \quad u_t(0,x) = \psi(x)$$

for plates with viscoelastic damping. We do not expect essential new difficulties in the treatment in comparison to the viscoelastic damped wave models of Sect. 14.3. For this reason we formulate only the main results and sketch briefly their proofs. It seems interesting to compare the result for viscoelastic damped waves with those of viscoelastic damped plates.

15.3.1 Representation of Solutions by Using Fourier Multipliers

Formal application of partial Fourier transformation gives the following ordinary differential equation for $v = v(t, \xi) = F_{x \to \xi}(u(t, x))$ depending on the parameter $\xi \in \mathbb{R}^n$:

$$v_{tt} + |\xi|^4 v_t + |\xi|^4 v = 0, \quad v(0, \xi) = v_0(\xi), \quad v_t(0, \xi) = v_1(\xi),$$

where $v_0(\xi) := F(\varphi)(\xi)$ and $v_1(\xi) := F(\psi)(\xi)$.
The solution is given on the set $\{\xi : |\xi| > \sqrt{2}\}$ by

$$v(t, \xi) = e^{-\frac{|\xi|^4 t}{2}} \left(\frac{v_0(\xi)\left(|\xi|^2 \sqrt{|\xi|^4 - 4} + |\xi|^4\right) + 2v_1(\xi)}{2|\xi|^2 \sqrt{|\xi|^4 - 4}} e^{\frac{|\xi|^2 \sqrt{|\xi|^4 - 4} t}{2}} \right.$$

$$\left. + \frac{v_0(\xi)\left(|\xi|^2 \sqrt{|\xi|^4 - 4} - |\xi|^4\right) - 2v_1(\xi)}{2|\xi|^2 \sqrt{|\xi|^4 - 4}} e^{\frac{-|\xi|^2 \sqrt{|\xi|^4 - 4} t}{2}} \right)$$

$$= e^{-\frac{|\xi|^4 t}{2}} \left(\frac{|\xi|^4 v_0(\xi) + 2v_1(\xi)}{|\xi|^2 \sqrt{|\xi|^4 - 4}} \sinh\left(\frac{|\xi|^2 \sqrt{|\xi|^4 - 4}}{2} t\right) \right.$$

$$\left. + v_0(\xi) \cosh\left(\frac{|\xi|^2 \sqrt{|\xi|^4 - 4}}{2} t\right)\right),$$

and on the set $\{\xi : |\xi| < \sqrt{2}\}$ by

$$v(t, \xi) = e^{-\frac{|\xi|^4 t}{2}} \left(\frac{|\xi|^4 v_0(\xi) + 2v_1(\xi)}{|\xi|^2 \sqrt{4 - |\xi|^4}} \sin\left(\frac{|\xi|^2 \sqrt{4 - |\xi|^4}}{2} t\right) \right.$$

$$\left. + v_0(\xi) \cos\left(\frac{|\xi|^2 \sqrt{4 - |\xi|^4}}{2} t\right)\right).$$

There appear the two characteristic roots

$$\lambda_{1,2}(\xi) := -\frac{|\xi|^4}{2} \pm \frac{|\xi|^2 \sqrt{|\xi|^4 - 4}}{2}$$

in these representation formulas. The asymptotic behavior of the roots for small and large frequencies is of interest for our further considerations.

Lemma 15.3.1 *The characteristic roots $\lambda_{1,2}$ behave as follows:*

1. $\Re \lambda_{1,2} = -\frac{|\xi|^4}{2}$ *for small frequencies,*
2. $\lambda_1 \sim -1$, $\lambda_2 \sim -|\xi|^4$ *for large frequencies.*

15.3.2   Conclusions from the Representation Formulas

Repeating the considerations of Sects. 14.3.1 to 14.3.3 we arrive at the following results:

Theorem 15.3.1 (H^s **Well-Posedness**) *Let the data* $\varphi \in H^s(\mathbb{R}^n)$ *and* $\psi \in H^{s-4}(\mathbb{R}^n)$, $s \in \mathbb{R}^1$, $n \geq 1$ *be given in the Cauchy problem*

$$u_{tt} + (-\Delta)^2 u + (-\Delta)^2 u_t = 0, \quad u(0, x) = \varphi(x), \quad u_t(0, x) = \psi(x).$$

Then, there exists for all $T > 0$ *a uniquely determined (in general) distributional solution*

$$u \in C([0, T], H^s(\mathbb{R}^n)) \cap C^1([0, T], H^{s-4}(\mathbb{R}^n)).$$

We have the a priori estimate

$$\|u(t, \cdot)\|_{H^s} \leq C(T)(\|\varphi\|_{H^s} + \|\psi\|_{H^{s-4}}).$$

Finally, the solution depends continuously on the data.

Theorem 15.3.2 (Decay of the Energy $E_{PL}(u)(t)$**)** *The energy solution to the Cauchy problem*

$$u_{tt} + (-\Delta)^2 u + (-\Delta)^2 u_t = 0, \quad u(0, x) = \varphi(x), \quad u_t(0, x) = \psi(x)$$

with data $\varphi \in H^2(\mathbb{R}^n)$ *and* $\psi \in L^2(\mathbb{R}^n)$ *satisfies the following estimates for* $t \geq 0$:

$$\|\Delta u(t, \cdot)\|_{L^2} \leq C((1 + t)^{-\frac{1}{2}}\|\varphi\|_{H^2} + \|\psi\|_{L^2}),$$

$$\|u_t(t, \cdot)\|_{L^2} \leq C((1 + t)^{-\frac{1}{2}}\|\varphi\|_{L^2} + \|\psi\|_{L^2}).$$

Consequently, the energy $E_{PL}(u)(t)$ *satisfies the estimate*

$$E_{PL}(u)(t) \leq C((1 + t)^{-1}\|\varphi\|_{H^2}^2 + \|\psi\|_{L^2}^2).$$

The energy $E_{PL}(u)(t)$ decays only under the assumption $\psi \equiv 0$. But, an additional regularity of the data may imply a decay behavior of the energy $E_{PL}(u)(t)$.

Theorem 15.3.3 (Additional Regularity of the Data) *The energy solution to the Cauchy problem*

$$u_{tt} + (-\Delta)^2 u + (-\Delta)^2 u_t = 0, \quad u(0, x) = \varphi(x), \quad u_t(0, x) = \psi(x)$$

with data $\varphi \in H^2(\mathbb{R}^n) \cap L^m(\mathbb{R}^n)$ and $\psi \in L^2(\mathbb{R}^n) \cap L^m(\mathbb{R}^n)$, $m \in [1, 2)$, satisfies the following decay estimates for $t \geq 0$:

$$\|\Delta u(t, \cdot)\|_{L^2} \leq C_m \Big((1 + t)^{-\frac{1}{2} - \frac{n(2-m)}{8m}} \|\varphi\|_{H^2 \cap L^m} + (1 + t)^{-\frac{n(2-m)}{8m}} \|\psi\|_{L^2 \cap L^m} \Big),$$

$$\|u_t(t, \cdot)\|_{L^2} \leq C_m \Big((1 + t)^{-\frac{1}{2} - \frac{n(2-m)}{8m}} \|\varphi\|_{L^2 \cap L^m} + (1 + t)^{-\frac{n(2-m)}{8m}} \|\psi\|_{L^2 \cap L^m} \Big).$$

Consequently, the energy $E_{PL}(u)(t)$ satisfies the decay estimate

$$E_{PL}(u)(t) \leq C_m \Big((1 + t)^{-1 - \frac{n(2-m)}{4m}} \|\varphi\|_{H^2 \cap L^m}^2 + (1 + t)^{-\frac{n(2-m)}{4m}} \|\psi\|_{L^2 \cap L^m}^2 \Big).$$

15.4   The Classical Plate Model with Mass

In Sect. 11.3.5 we introduced the energy

$$E_{PL,KG}(u)(t) := \frac{1}{2} \int_{\mathbb{R}^n} \Big(|u_t(t, x)|^2 + |\Delta u(t, x)|^2 + m^2 |u(t, x)|^2 \Big) dx.$$

This is a suitable energy for solutions to the plate model with mass

$$u_{tt} + (-\Delta)^2 u + m^2 u = 0, \quad u(0, x) = \varphi(x), \quad u_t(0, x) = \psi(x).$$

Following the proof to Theorem 11.1.2 leads to the conservation of energy

$$E_{PL,KG}(u)(t) = E_{PL,KG}(u)(0)$$

(cf. with Theorem 11.3.1). If we define for $k \geq 0$ the higher order energies

$$E_{PL,KG}^k(u)(t)$$

$$:= \frac{1}{2} \int_{\mathbb{R}^n} \Big(|D|^{2k} |u_t(t, x)|^2 + |D|^{2k} |\Delta u(t, x)|^2 + m^2 |D|^{2k} |u(t, x)|^2 \Big) dx,$$

then energy conservation remains true after taking into consideration that $|D|^k u$ is a solution of the above plate equation with mass term with data

$$(|D|^k u)(0, x) = |D|^k \varphi(x) \quad \text{and} \quad (|D|^k u)_t(0, x) = |D|^k \psi(x).$$

The benefit of the mass term is that the potential energy can be controlled for all times.

Now let us say something about the application of phase space analysis to the above model. Here we follow the considerations of Sect. 14.4.1. We may conclude the representation of solution

$$u(t,x) = F_{\xi \to x}^{-1}\Big(\cos\Big(\sqrt{|\xi|^4 + m^2}\, t \Big) F(\varphi)(\xi) \Big)$$

$$+ F_{\xi \to x}^{-1}\Big(\frac{\sin\Big(\sqrt{|\xi|^4 + m^2}\, t \Big)}{\sqrt{|\xi|^4 + m^2}} \, F(\psi)(\xi) \Big).$$

The same arguments as in the proof of Theorem 14.4.1 allow us to prove the following statement.

Theorem 15.4.1 *The Cauchy problem*

$$u_{tt} + (-\Delta)^2 u + m^2 u = 0, \quad u(0,x) = \varphi(x), \quad u_t(0,x) = \psi(x), \quad x \in \mathbb{R}^n, \quad n \geq 1,$$

is H^s well-posed, $s \in \mathbb{R}^1$, that is, to given data $\varphi \in H^s(\mathbb{R}^n)$ and $\psi \in H^{s-2}(\mathbb{R}^n)$ there exists a uniquely determined (in general) distributional solution

$$u \in C\big([0,T], H^s(\mathbb{R}^n)\big) \cap C^1\big([0,T], H^{s-2}(\mathbb{R}^n)\big) \text{ for all } T > 0.$$

The solution depends continuously on the data, that is, to each $\varepsilon > 0$ there exists a positive constant $\delta(\varepsilon)$ such that

$$\|\varphi_1 - \varphi_2\|_{H^s} + \|\psi_1 - \psi_2\|_{H^{s-2}} < \delta \text{ implies}$$

$$\|u_1 - u_2\|_{C([0,T],H^s) \cap C^1([0,T],H^{s-2})} < \varepsilon.$$

15.5   The Classical Plate Model with Mass and Dissipation

The Cauchy problem for the general damped plate model with mass reads as follows:

$$u_{tt} + (-\Delta)^2 u + m^2 u + (-\Delta)^\delta u_t = 0, \quad u(0,x) = \varphi(x), \quad u_t(0,x) = \psi(x),$$

where $\delta \in [0,2]$. If $\delta = 0$, then we have the classical damped plate model with mass. If $\delta \in (0,2]$, then we get structurally damped plate models with mass. Among these structurally damped plate models with mass we shall discuss only the case of $\delta = 2$, the viscoelastic damped plate model with mass.

15.5.1   The Classical Damped Plate Model with Mass

The mass term $m^2 u$ and the dissipation term u_t have no influence on the H^s well-posedness. So, we get the same result as in Theorem 15.4.1.

Moreover, we can expect an exponential decay of the energy $E_{PL,KG}(u)(t)$. Let us explain why. For the partial Fourier transform $v := F_{x \to \xi}(u)$ of u we obtain the Cauchy problem

$$v_{tt} + (|\xi|^4 + m^2)v + v_t = 0, \quad v(0, \xi) = F(\varphi)(\xi), \quad v_t(0, \xi) = F(\psi)(\xi).$$

The characteristic roots are $\lambda_{1,2} = -\frac{1}{2} \pm \sqrt{\frac{1}{4} - (|\xi|^4 + m^2)}$.

1. If $m^2 \geq \frac{1}{4}$, then for all frequencies $|\xi| > 0$ the characteristic roots are complex conjugate with $\Re\lambda_{1,2} = -\frac{1}{2}$, this hints at an exponential type decay.
2. If $m^2 < \frac{1}{4}$, then for large frequencies $|\xi|$ we have $\Re\lambda_{1,2} = -\frac{1}{2}$, for small frequencies we use similarly to the Case 2b in the proof to Theorem 14.2.2 the equivalence $-\frac{1}{2} \pm \sqrt{\frac{1}{4} - (|\xi|^4 + m^2)} \sim -(|\xi|^4 + m^2)$. This hints for small frequencies at an exponential type decay, too.

15.5.2 The Classical Plate Model with Mass and Viscoelastic Dissipation

The mass term m^2u and the dissipation term $(-\Delta)^2u_t$ have no influence on the H^s well-posedness. So, we expect the same result as in Theorem 15.3.1.

Theorem 15.5.1 *The Cauchy problem*

$$u_{tt} + (-\Delta)^2u + m^2u + (-\Delta)^2u_t = 0,$$

$$u(0, x) = \varphi(x), \quad u_t(0, x) = \psi(x), \quad x \in \mathbb{R}^n, \quad n \geq 1,$$

is H^s well-posed, $s \in \mathbb{R}^1$, that is, to given data $\varphi \in H^s(\mathbb{R}^n)$ and $\psi \in H^{s-4}(\mathbb{R}^n)$ there exists a uniquely determined (in general) distributional solution

$$u \in C([0, T], H^s(\mathbb{R}^n)) \cap C^1([0, T], H^{s-4}(\mathbb{R}^n)) \quad \text{for all } T > 0.$$

The solution depends continuously on the data, that is, to each $\varepsilon > 0$ there exists a positive constant $\delta(\varepsilon)$ such that

$$\|\varphi_1 - \varphi_2\|_{H^s} + \|\psi_1 - \psi_2\|_{H^{s-4}} < \delta \quad \text{implies}$$

$$\|u_1 - u_2\|_{C([0,T],H^s) \cap C^1([0,T],H^{s-4})} < \varepsilon.$$

The proof of this result is similar to the proof of Theorem 15.3.1.
What about a decay behavior for a suitable energy? For the partial Fourier transform $v := F_{x \to \xi}(u)$ of u we obtain the Cauchy problem

$$v_{tt} + (|\xi|^4 + m^2)v + |\xi|^4v_t = 0, \quad v(0, \xi) = F(\varphi)(\xi), \quad v_t(0, \xi) = F(\psi)(\xi).$$

The characteristic roots are

$$\lambda_{1,2} = -\frac{|\xi|^4}{2} \pm \frac{|\xi|^4}{2}\sqrt{1 - 4\frac{|\xi|^4 + m^2}{|\xi|^8}}.$$

We have, in general, the following representation of solution $u = u(t,x)$:

$$u(t,x) = F_{\xi \to x}^{-1}\left(c_1(\xi)e^{\lambda_1(\xi)t} + c_2(\xi)e^{\lambda_2(\xi)t}\right),$$

where $c_k(\xi)$, $k = 1, 2$, are determined by the characteristic roots, by $F(\varphi)$ and $F(\psi)$. Thus, we may observe the following.

1. For small frequencies the characteristic roots are complex conjugate with $\Re\lambda_{1,2} = -\frac{|\xi|^4}{2}$.
2. For large frequencies the characteristic root λ_1 behaves as $-|\xi|^4 + 1$, the characteristic root λ_2 behaves as -1, each one hints at an exponential type decay.

Consequently, the decay will be determined by the behavior of the real parts $\Re\lambda_{1,2} = -\frac{|\xi|^4}{2}$ of the characteristic roots $\lambda_{1,2}$ for the small frequencies as we observed for viscoelastic damped plates in Sect. 15.3. All these explanations lead to a corresponding result to Theorem 15.3.2, at least for the kinetic energy and the energy related to momentum. Solve Exercise 9.

Exercises Relating to the Considerations of Chap. 15

Exercise 1 Prove the statement of Theorem 15.1.1.

Exercise 2 Let us consider the Cauchy problem for the classical damped plate model of Sect. 15.2. Which Sobolev regularity of the data φ and ψ do we need to prove the energy estimate $E'_{PL}(u)(t) \leq 0$? Here $E_{PL}(u)(t)$ is defined as in Sect. 11.2.4.

Exercise 3 Prove the statement of Theorem 15.2.2.

Exercise 4 Prove the statement of Theorem 15.3.1.

Exercise 5 Prove the statement of Theorem 15.3.2.

Exercise 6 Prove the statement of Theorem 15.3.3.

Exercise 7 In Sect. 14.5 we explained from the point of view of decay estimates a relation between the classical damped Klein-Gordon model and the heat model with mass term of Sect. 12.2. What about a reference model to the classical damped plate model? Explain the relation from the point of view of decay estimates.

Exercise 8 Verify that if u solves

$$u_{tt} + (-\Delta)^2 u = 0, \quad u(0,x) = \varphi(x), \quad u_t(0,x) = \psi(x), \quad x \in \mathbb{R}^n, \quad n \geq 1,$$

then $w := u_t + i\Delta u$ solves

$$w_t - i\Delta w = 0, \quad w(0, x) = \psi(x) + i\Delta\varphi(x),$$

i.e., w is a solution of the linear Schrödinger equation (see Sect. 13.1). By using Theorem 13.1.3 conclude that

$$\|(\Delta u, u_t)(t, \cdot)\|_{L^q} \leq Ct^{-\frac{n}{2}\left(\frac{1}{p} - \frac{1}{q}\right)}\|(\Delta\varphi, \psi)\|_{L^p}$$

for all $2 \leq q \leq \infty$ and $\frac{1}{p} + \frac{1}{q} = 1$.

Exercise 9 Let us consider the Cauchy problem to a viscoelastic damped plate model with mass

$$u_{tt} + (-\Delta)^2 u + m^2 u + (-\Delta)^2 u_t = 0, \quad u(0, x) = \varphi(x), \quad u_t(0, x) = \psi(x).$$

Derive an estimate for the energy $E_{PL,KG}(u)(t)$.

Chapter 16
The Method of Stationary Phase and Applications

In Sects. 12.1.2 and 12.1.3 we derived $L^p - L^q$ decay estimates on and away from the conjugate line for solutions to the Cauchy problem for the heat equation. The basic tools of the approach are tools from the theory of Fourier multipliers, Young's inequality and embedding theorems. This approach can not be applied to the free wave equation. The goal to derive $L^p - L^q$ decay estimates for solutions to the Cauchy problem for the wave equation requires a deeper understanding of oscillating integrals with localized amplitudes in different parts of the extended phase space. In particular, $L^\infty - L^\infty$ estimates of such integrals are of interest. One basic tool to get such estimates is the method of stationary phase. We will apply this method to prove $L^p - L^q$ decay estimates for solutions to the Cauchy problems for the free wave equation, for the Schrödinger equation and for the plate equation. The key lemmas are Littman-type lemmas in the form of Theorems 16.3.1 and 16.8.1. All these tools and interpolation arguments together yield $L^p - L^q$ estimates on the conjugate line.

Many thanks to Karen Yagdjian (Edinburg, Texas) for useful discussions on the stationary phase method and the content of this chapter.

16.1 $L^2 - L^2$ Estimates

First we explain $L^2 - L^2$ estimates.

Theorem 16.1.1 *We study the Cauchy problem*

$$u_{tt} - \Delta u = 0, \quad u(0, x) = \varphi(x), \quad u_t(0, x) = \psi(x).$$

We assume that the derivatives $\nabla \Delta^l \partial_x^\alpha \varphi$, $\Delta^{l+1} \partial_x^\alpha \varphi$, $\Delta^l \partial_x^\alpha \psi$, $\nabla \Delta^l \partial_x^\alpha \psi$ *of the data* φ, ψ *belong to* $L^2(\mathbb{R}^n)$. *Then, we have the following energy identities for the*

© Springer International Publishing AG 2018

M.R. Ebert, M. Reissig, *Methods for Partial Differential Equations*,
https://doi.org/10.1007/978-3-319-66456-9_16

derivatives $\left(\partial_t^k \partial_x^\alpha u\right)(t, \cdot)$ of the solution u $(k + |\alpha| \geq 0)$:

$$Ew(\partial_t^{2l} \partial_x^\alpha u)(t) = \frac{1}{2}\left(\|\nabla \Delta^l \partial_x^\alpha \varphi\|_{L^2}^2 + \|\Delta^l \partial_x^\alpha \psi\|_{L^2}^2\right),$$

$$Ew(\partial_t^{2l+1} \partial_x^\alpha u)(t) = \frac{1}{2}\left(\|\Delta^{l+1} \partial_x^\alpha \varphi\|_{L^2}^2 + \|\nabla \Delta^l \partial_x^\alpha \psi\|_{L^2}^2\right)$$

for $|\alpha|, l \geq 0$.

Proof Here we use the energy conservation

$$Ew(u)(t) = Ew(u)(0)$$

of Sect. 11.1 and the property that $\partial_t^{2l} \partial_x^\alpha u$ and $\partial_t^{2l+1} \partial_x^\alpha u$ solve the wave equation, that is,

$$\left(\partial_t^{2l} \partial_x^\alpha u\right)_{tt} - \Delta\left(\partial_t^{2l} \partial_x^\alpha u\right) = 0, \quad \left(\partial_t^{2l+1} \partial_x^\alpha u\right)_{tt} - \Delta\left(\partial_t^{2l+1} \partial_x^\alpha u\right) = 0,$$

with data

$$\left(\partial_t^{2l} \partial_x^\alpha u\right)(0, x) = \Delta^l \partial_x^\alpha \varphi(x), \quad \left(\partial_t^{2l} \partial_x^\alpha u\right)_t(0, x) = \Delta^l \partial_x^\alpha \psi(x),$$

$$\left(\partial_t^{2l+1} \partial_x^\alpha u\right)(0, x) = \Delta^l \partial_x^\alpha \psi(x), \quad \left(\partial_t^{2l+1} \partial_x^\alpha u\right)_t(0, x) = \Delta^{l+1} \partial_x^\alpha \varphi(x).$$

The proof is complete.

16.2   Philosophy of Our Approach to Derive $L^p - L^q$ Estimates on the Conjugate Line for Solutions to the Wave Equation

We use the following representation of solutions for the Cauchy problem for the free wave equation of Sect. 14.1:

$$u(t, x) = F_{\xi \to x}^{-1}\left(\left(e^{i|\xi|t} + e^{-i|\xi|t}\right)\frac{1}{2}F(\varphi)(\xi)\right)$$

$$+ F_{\xi \to x}^{-1}\left(\left(e^{i|\xi|t} - e^{-i|\xi|t}\right)\frac{1}{2i|\xi|}F(\psi)(\xi)\right).$$

Let us turn to the model Fourier multiplier

$$F_{\xi \to x}^{-1}\left(e^{-i|\xi|t}F(\varphi)(\xi)\right).$$

What kind of tools do we apply to derive $L^p - L^q$ estimates on the conjugate line? Let us apply the convolution theorem of Sect. 24.1.3. Then we get

$$F_{\xi \to x}^{-1}\left(e^{-i|\xi|t}\right) * \varphi.$$

Even if we choose φ to be very smooth, there is no hope of applying Young's inequality as in Sect. 12.1.2 because of the first term. This term belongs for all positive t to $S'(\mathbb{R}^n)$ only. We should modify our approach to derive some decay estimates. The main ideas are as follows:

1. We add an amplitude function in our model Fourier multiplier. Our model Fourier multiplier instead takes the form

$$F_{\xi \to x}^{-1}\left(e^{-i|\xi|t}\frac{1}{|\xi|^{2r}}F(\varphi)(\xi)\right),$$

   where the parameter r is determined later.
2. We decompose the extended phase space $(0, \infty) \times \mathbb{R}_{\xi}^n$ into two zones. For this reason we introduce a function $\chi \in C^{\infty}(\mathbb{R}_{\xi}^n)$ satisfying $\chi(\xi) \equiv 0$ for $|\xi| \leq \frac{1}{2}$, $\chi(\xi) \equiv 1$ for $|\xi| \geq \frac{3}{4}$, and $\chi(\xi) \in [0, 1]$. Then we define the *pseudo-differential zone*

$$Z_{pd} = \{(t, \xi) \in (0, \infty) \times \mathbb{R}_{\xi}^n : t|\xi| \leq 1\},$$

   and the *hyperbolic zone*

$$Z_{hyp} = \{(t, \xi) \in (0, \infty) \times \mathbb{R}_{\xi}^n : t|\xi| \geq 1\}.$$

3. In the next steps we shall study model Fourier multipliers. In both multipliers the phase function

$$\phi = \phi(t, x, \xi) := x \cdot \xi - |\xi|t$$

   appears. *The stationary points of the phase function* are of importance. These points are determined by the equation

$$\nabla_{\xi}\phi(t, x, \xi) = x - t\frac{\xi}{|\xi|} = 0.$$

So, we have to distinguish both the following cases:

a. If $\frac{x}{t} =: y$ is lying on the unit sphere in $\mathbb{R}^n$, then there exist non-isolated stationary points, namely, the ray $\{\xi \in \mathbb{R}^n : \xi = \lambda y \text{ for all } \lambda > 0\}$. But in the set of all stationary points, that is, on the ray, there is only one which is lying on the unit sphere.

b. If $\frac{x}{t} =: y$ is not lying on the unit sphere in $\mathbb{R}^n$, then there do not exist any stationary points.

4. In Sect. 16.4 we shall derive $L^p - L^q$ estimates for the model Fourier multiplier

$$F_{\xi \to x}^{-1}\left(e^{-i|\xi|t}\frac{1 - \chi(t|\xi|)}{|\xi|^{2r}}F(\varphi)(\xi)\right).$$

The amplitude function

$$\frac{1 - \chi(t|\xi|)}{|\xi|^{2r}}$$

vanishes in the hyperbolic zone.

5. In Sect. 16.5 we derive $L^p - L^q$ estimates on the conjugate line for the model Fourier multiplier

$$F_{\xi \to x}^{-1}\left(e^{-i|\xi|t}\frac{\chi(t|\xi|)}{|\xi|^{2r}}F(\varphi)(\xi)\right).$$

The amplitude function

$$\frac{\chi(t|\xi|)}{|\xi|^{2r}}$$

vanishes in a large part of the pseudo-differential zone.

6. After all these preparations we are able to derive in Sect. 16.6 the desired $L^p - L^q$ estimates on the conjugate line for solutions to the Cauchy problem for the wave equation.

Remark 16.2.1 Decompositions of the phase space $\mathbb{R}_\xi^n$ or extended phase space $(0, \infty) \times \mathbb{R}_\xi^n$, respectively, are used very often in modern analysis. Here we use a very rough decomposition into two zones only. We take into consideration the behavior of the term $e^{-i|\xi|t}$ in the corresponding Fourier multipliers. This term allows for symbol-like estimates in the pseudo-differential zone. So, $-|\xi|t$ has no contribution to the phase function of the Fourier multiplier, but contributes to the amplitude function instead. In the hyperbolic zone $-|\xi|t$ is part of the phase function of the Fourier multiplier.

16.3   A Littman Type Lemma

Littmann type lemmas are a key tool in getting time decay in the expected $L^p - L^q$ estimates on the conjugate line for solutions to the Cauchy problem for the free wave equation. The classical Littman lemma can be found in [126]. The proof of this lemma is a bit complicated for most readers, so we prefer to present a Littman type lemma which is related to our model Fourier multiplier and give a simpler proof in this special case. The main tools in the proof are *stationary phase method* and *Morse lemma*.

Theorem 16.3.1 (A Littman Type Lemma)   *Let us consider for $\tau \geq \tau_0$, τ_0 is a large positive number, the oscillating integral*

$$F_{\eta \to x}^{-1}\left(e^{-i\tau|\eta|}v(\eta)\right).$$

The amplitude function $v = v(\eta)$ is supposed to belong to $C_0^\infty(\mathbb{R}^n)$ with support in $\{\eta \in \mathbb{R}^n : |\eta| \in [\frac{1}{2}, 2]\}$. Then, the following $L^\infty - L^\infty$ estimate holds:

$$\left\|F_{\eta \to x}^{-1}\left(e^{-i\tau|\eta|}v(\eta)\right)\right\|_{L^\infty(\mathbb{R}_x^n)} \leq C(1+\tau)^{-\frac{n-1}{2}} \sum_{|\alpha| \leq s} \|D_\eta^\alpha v(\eta)\|_{L^\infty(\mathbb{R}_\eta^n)},$$

where $s > \frac{n+3}{2}$.

Proof The method of stationary phase is explained, in [190] and [193] for example. We follow the proof of Theorem 4.1 from [177]. The proof is divided into several steps.

Step 1:   Change of variables and a property of the kernel

After the change of variables $y := \frac{x}{\tau}$ we have to estimate

$$\sup_{y \in \mathbb{R}^n} \left| \int_{\mathbb{R}^n} e^{i\tau(y \cdot \eta - |\eta|)} v(\eta)\, d\eta \right|.$$

We choose the operator $L = L(\tau, y, \eta, \partial_\eta)$ in such a way that it reproduces the kernel, that is,

$$L e^{i\tau(y \cdot \eta - |\eta|)} = e^{i\tau(y \cdot \eta - |\eta|)}.$$

Introducing the vector-valued function

$$\Phi = \Phi(\tau, y, \eta) = \tau\left(y - \frac{\eta}{|\eta|}\right)$$

the operator L is defined as follows:

$$L := \frac{1}{|\Phi(\tau, y, \eta)|^2} \sum_{j=1}^n \Phi_j(\tau, y, \eta) \frac{1}{i} \partial_{\eta_j}.$$

Here Φ_j denotes the j-th component of Φ.

Step 2:   Application of integration by parts

If y satisfies $|y| \geq 1 + \delta$ or $|y| \leq 1 - \delta$, where δ is a small positive constant, then $|y - \frac{\eta}{|\eta|}| \geq \delta$ for all $\eta \in \mathbb{R}^n \setminus \{0\}$. So, we may apply an arbitrary number N of integration by parts. Here we use the smoothness and support assumption for v to conclude

$$\sup_{|y| \in [0,\infty) \setminus [1-\delta,1+\delta]} \left| \int_{\mathbb{R}^n} e^{i\tau(y \cdot \eta - |\eta|)} v(\eta) \, d\eta \right|$$

$$= \sup_{|y| \in [0,\infty) \setminus [1-\delta,1+\delta]} \left| \int_{\mathbb{R}^n} e^{i\tau(y \cdot \eta - |\eta|)} (L^T)^N v(\eta) \, d\eta \right|.$$

Every application of the transposed operator L^T to v generates the decay τ^{-1}. This follows by the estimate $|\Phi(\tau, y, \eta)| \geq \tau\delta$. Summarizing, we get the estimate

$$\sup_{|y| \in [0,\infty) \setminus [1-\delta,1+\delta]} \left| \int_{\mathbb{R}^n} e^{i\tau(y \cdot \eta - |\eta|)} v(\eta) \, d\eta \right|$$

$$\leq C\tau^{-N} \sum_{|\alpha| \leq N} \|D_\eta^\alpha v(\eta)\|_{L^\infty(\mathbb{R}^n_\eta)}.$$

The constant C depends only on δ, which later will be fixed.

Step 3:   A useful property of the phase function

Let us discuss the above integral for $y \in \{y \in \mathbb{R}^n : |y| \in [1 - \delta, 1 + \delta]\}$. It is sufficient to estimate it for vectors $y = (s, 0, \cdots, 0)$ with $s \in [1 - \delta, 1 + \delta]$. If this is done, then a rotation of the spherical shell $\{\eta \in \mathbb{R}^n : |\eta| \in [\frac{1}{2}, 2]\}$ does not essentially change the above oscillating integral. The special structure of the phase function allows us to transfer this rotation to a rotation of y without any changes in the phase function. In this way we derive estimates of the integral for all y.

Step 4:   Application of Morse lemma

We study for $s \in [1 - \delta, 1 + \delta]$ the integral

$$\int_{|\eta| \in [\frac{1}{2},2]} e^{i\tau(s\eta_1 - |\eta|)} v(\eta) \, d\eta.$$

Here we distinguish two cases:

Case 1   The vector η does not belong to a conical neighborhood K_δ of the vector $y = (s, 0, \cdots, 0)$. Here δ is a small positive constant which can be chosen as small as necessary. Then we can apply the approach of step 2. Integration by parts and the fact that $|y - \frac{\eta}{|\eta|}| \geq c > 0$, where the constant c is independent of all admissible η outside of K_δ, implies the estimate

$$\left| \int_{|\eta| \in \{[\frac{1}{2},2] \setminus K_\delta\}} e^{i\tau(s\eta_1 - |\eta|)} v(\eta) \, d\eta \right| \le C\tau^{-N} \sum_{|\alpha| \le N} \|D_\eta^\alpha v(\eta)\|_{L^\infty(\mathbb{R}_\eta^n)}.$$

Case 2   The vector η belongs to K_δ. Then η_1 is positive and we can apply Taylor's formula to obtain for all $\eta =: (\eta_1, \eta') \in K_\delta$ the relation

$$|\eta| = |(\eta_1, \eta_2, \cdots, \eta_n)| = |(\eta_1, 0, \cdots, 0)| + (\nabla_{\eta'} |\eta|)(\eta_1, 0, \cdots, 0) \cdot \eta'$$

$$+ \frac{1}{2} \eta' H_{\eta'}(|\eta|)(\eta_1, 0, \cdots, 0)\eta'^T + R(\eta).$$

Here $H_{\eta'}$ denotes the Hessian with respect to η' and $R(\eta)$ is the remainder in Taylor's formula. Taking account of

$$(\nabla_{\eta'} |\eta|)(\eta_1, 0, \cdots, 0) = 0, \quad \eta' H_{\eta'}(\eta_1, 0, \cdots, 0)\eta'^T = \frac{1}{\eta_1} |\eta'|^2$$

we get

$$\int_{|\eta| \in K_\delta} e^{i\tau(s\eta_1 - |\eta|)} v(\eta) \, d\eta = \int_{|\eta| \in K_\delta} e^{i\tau \left(s\eta_1 - |\eta_1| - \frac{1}{2\eta_1} |\eta'|^2 + R(\eta) \right)} v(\eta) \, d\eta$$

$$= \int_{\frac{1}{2}}^{2} \left(\int_{M_\delta(r)} e^{i\tau \left(s\eta_1 - |\eta_1| - \frac{1}{2r} |\eta'|^2 + R(\eta) \right)} v(\eta) \, d\sigma_r \right) dr.$$

By $M_\delta(r)$ we denote the intersection of K_δ with the sphere $\{\eta \in \mathbb{R}^n : |\eta| = r\}$, $d\sigma_r$ is the surface measure of $M_\delta(r)$. Now, let us turn for a fixed $r \in [\frac{1}{2}, 2]$ to the integral

$$\int_{M_\delta(r)} e^{i\tau \left(-\frac{1}{2r} |\eta'|^2 + R(\eta) \right)} v(\eta) \, d\sigma_r.$$

There exists a coordinate system with variables $(\rho, \omega) := (\rho, \omega_1, \cdots, \omega_{n-1})$ centered at $(r, 0, \cdots, 0)$ such that the tangent plane to $M_\delta(r)$ is given by $\rho = 0$ and $M_\delta(r)$ can be represented in the following way:

$$M_\delta(r) := \left\{ (\rho, \omega_1, \cdots, \omega_{n-1}) \in \mathbb{R}^n : \rho = \sum_{k=1}^{n-1} a_k \omega_k^2 + \tilde{R}(\omega), \ |\omega| \le \varepsilon(\delta) \right\},$$

where $\varepsilon(\delta)$ is small for small δ and the small remainder $\tilde{R}(\omega)$ behaves as $O(|\omega|^3)$. The quadratic form $\sum_{k=1}^{n-1} a_k \omega_k^2$ is positive definite (it is related to the curvature of

$M_\delta(r)$). Using this change of variables we obtain

$$\int_{M_\delta(r)} e^{i\tau\left(-\frac{1}{2r}|\eta'|^2+R(\eta)\right)} v(\eta)\, d\sigma_r$$

$$= \int_{|\omega|\le\varepsilon} e^{i\tau\left(-\frac{1}{2r}|\eta'(\rho,\omega)|^2+R(r,\rho,\omega)\right)} v(\eta(\rho,\omega))J_r(\rho,\omega)\, d\omega.$$

The surface element is transformed as $d\sigma_r = J_r(\rho,\omega)d\omega$. Now, the Morse lemma (see [143]) comes into play, explaining a suitable change of variables for the phase function

$$-\frac{1}{2r}|\eta'(\rho,\omega)|^2 + R(r,\rho,\omega).$$

Due to Taylor's formula the remainder $R(r,\rho,\omega)$ is only a small perturbation of $-\frac{1}{2r}|\eta'(\rho,\omega)|^2$. Using the above representation formula for $M_\delta(r)$ by coordinates ρ,ω there exist a small neighborhood $V = \{|\omega| \le \varepsilon\}$ of the origin in $\mathbb{R}^{n-1}_\omega$ (if necessary we can choose a smaller ε) and a small neighborhood $U = \{|y| \le \kappa\}$ of the origin in $\mathbb{R}^{n-1}_y$ (the smallness of ε guarantees the smallness of κ, too) and a diffeomorphism

$$H = H_{r,\rho}(\omega) : \omega \in V \to y = H_{r,\rho}(\omega) \in U$$

such that the phase function is transformed to $\frac{1}{2}|y|^2$, $y = (y_1,\cdots,y_{n-1})$. All the terms appearing in this change of variables $y = H_{r,\rho}(\omega)$ can be estimated uniformly with respect to the variables ρ, r from the admissible sets introduced above. This change of variables transforms the above integral into

$$\int_{|y|\le\kappa} e^{i\tau\frac{|y|^2}{2}} u(r,\rho,y)\, dy = \int_{\mathbb{R}^{n-1}_y} e^{i\tau\frac{|y|^2}{2}} u(r,\rho,y)\, dy$$

after using the compact support property of u with respect to y.

Step 5:   Conclusion

The last integral is equal to $\left(e^{i\tau\frac{|D|^2}{2}} u(r,\rho,y)\right)(x=0)$ because we have

$$\left(e^{i\tau\frac{|D|^2}{2}} u(r,\rho,y)\right)(x) = c_{n-1}(\tau)\int_{\mathbb{R}^{n-1}} e^{i\tau\frac{|y|^2}{2}} u(r,\rho,x+y)\, dy,$$

where

$$c_{n-1}(\tau) = (2\pi)^{\frac{n-1}{2}} e^{i\frac{\pi(n-1)}{4}} \tau^{-\frac{n-1}{2}}.$$

Then, we approximate $e^{i\tau \frac{|D|^2}{2}}$ by a differential operator of order 0. We use

$$\left| \left(e^{i\tau \frac{|D|^2}{2}} u(r, \rho, y) \right)(x = 0) - u(r, \rho, 0) \right|$$

$$\leq |c_{n-1}(\tau)| \tau^{-1} \sum_{|\alpha| \leq s} \left\| \frac{1}{2} \Delta D_y^\alpha u(y) \right\|_{L^\infty(\mathbb{R}_y^{n-1})},$$

where $s > \frac{n-1}{2}$. In the second step we need only $N \geq \frac{n-1}{2}$ derivatives to have the desired decay. Thus, totally, we need $s > \frac{n+3}{2}$ derivatives of u with respect to y and of v with respect to η, respectively. This completes the proof.

16.4 $L^p - L^q$ Estimates for Fourier Multipliers with Amplitudes Localized in the Pseudo-Differential Zone

Now we estimate the Fourier multiplier

$$F_{\xi \to x}^{-1} \left(e^{-i|\xi| t} \frac{1 - \chi(t|\xi|)}{|\xi|^{2r}} F(\varphi)(\xi) \right)$$

with an amplitude function localized in the pseudo-differential zone. The main tool is a Hardy-Littlewood type inequality.

Theorem 16.4.1 *Let us consider for $\varphi \in C_0^\infty(\mathbb{R}^n)$ the Fourier multiplier*

$$F_{\xi \to x}^{-1} \left(e^{-i|\xi| t} \frac{1 - \chi(t|\xi|)}{|\xi|^{2r}} F(\varphi)(\xi) \right)$$

from Sect. 16.2. We assume $1 < p \leq 2 \leq q < \infty$ and $0 \leq 2r \leq n(\frac{1}{p} - \frac{1}{q})$. Then, we have the $L^p - L^q$ estimates

$$\left\| F_{\xi \to x}^{-1} \left(e^{-i|\xi| t} \frac{1 - \chi(t|\xi|)}{|\xi|^{2r}} F(\varphi)(\xi) \right) \right\|_{L^q(\mathbb{R}^n)} \leq C t^{2r - n(\frac{1}{p} - \frac{1}{q})} \|\varphi\|_{L^p(\mathbb{R}^n)}$$

for all admissible p, q. The constant C depends on p and q.

Proof In some steps we follow the approach of [12] and [177]. We introduce the notation

$$I_0 := \left\| F_{\xi \to x}^{-1} \left(e^{-i|\xi| t} \frac{1 - \chi(t|\xi|)}{|\xi|^{2r}} F(\varphi)(\xi) \right) \right\|_{L^q}^q.$$

The change of variables $\eta := t\xi$ and $tz := x$ implies

$$
I_0 = t^{2rq-nq+n} \left\| F_{\eta \to z}^{-1} \left(e^{-i|\eta|} \frac{1 - \chi(|\eta|)}{|\eta|^{2r}} F(\varphi) \left(\frac{\eta}{t} \right) \right) \right\|_{L^q}^q
$$

$$
= t^{2rq-nq+n} \left\| F_{\eta \to z}^{-1} \left(e^{-i|\eta|} \frac{1 - \chi(|\eta|)}{|\eta|^{2r}} \right) * F_{\eta \to z}^{-1} \left(F(\varphi) \left(\frac{\eta}{t} \right) \right) \right\|_{L^q}^q.
$$

For the oscillating integral

$$
T := F_{\eta \to z}^{-1} \left(e^{-i|\eta|} \frac{1 - \chi(|\eta|)}{|\eta|^{2r}} \right)
$$

we have with an arbitrarily chosen positive real l the estimate

$$
\text{meas} \left\{ \eta \in \mathbb{R}^n : |F_{z \to \eta}(T)| \geq l \right\} \leq \text{meas} \left\{ \eta \in \mathbb{R}^n : |\eta| \leq l^{-\frac{1}{2r}} \right\} \leq C l^{-\frac{n}{2r}}.
$$

Due to Theorem 24.1.6 we may conclude $F(T) \in M_p^q$ (cf. with Definition 24.3) for all $1 < p \leq 2 \leq q < \infty$ satisfying $2r \leq n \left(\frac{1}{p} - \frac{1}{q} \right)$. Hence, $T \in L_p^q$ (cf. with Definition 24.2) and

$$
\left\| T * F_{\eta \to z}^{-1} \left(F(\varphi) \left(\frac{\eta}{t} \right) \right) \right\|_{L^q} \leq C_p t^n \| \varphi(yt) \|_{L^p} \leq C_p t^{n - \frac{n}{p}} \| \varphi \|_{L^p}.
$$

Summarizing, we have shown

$$
I_0 \leq C_p t^{2rq - n(\frac{q}{p} - 1)} \| \varphi \|_{L^p}^q.
$$

This leads immediately to the desired Hardy-Littlewood type inequality

$$
\left\| F_{\xi \to x}^{-1} \left(e^{-i|\xi|t} \frac{1 - \chi(t|\xi|)}{|\xi|^{2r}} F(\varphi)(\xi) \right) \right\|_{L^q} \leq C_{p,q} t^{2r - n(\frac{1}{p} - \frac{1}{q})} \| \varphi \|_{L^p}
$$

for all admissible $1 < p \leq 2 \leq q < \infty$. The proof is complete.

16.5 $L^p - L^q$ Estimates on the Conjugate Line for Fourier Multipliers with Amplitudes Localized in the Hyperbolic Zone

Now we estimate the Fourier multiplier

$$
F_{\xi \to x}^{-1} \left(e^{-i|\xi|t} \frac{\chi(t|\xi|)}{|\xi|^{2r}} F(\varphi)(\xi) \right)
$$

with an amplitude function which is more or less localized in the hyperbolic zone. What is important is only that the amplitude vanishes for $t|\xi| \leq \frac{1}{2}$. The main tools are a refined decomposition of the extended phase space, the Littman type lemma in the form of Theorem 16.3.1 and an interpolation argument.

Theorem 16.5.1 *Let us consider for* $\varphi \in C_0^\infty(\mathbb{R}^n)$ *the Fourier multiplier*

$$F_{\xi \to x}^{-1}\left(e^{-i|\xi|t} \frac{\chi(t|\xi|)}{|\xi|^{2r}} F(\varphi)(\xi)\right)$$

from Sect. 16.2. We assume $\frac{n+1}{2}\left(\frac{1}{p} - \frac{1}{q}\right) \leq 2r$. *Then, we have the following* $L^p - L^q$ *estimates on the conjugate line:*

$$\left\| F_{\xi \to x}^{-1}\left(e^{-i|\xi|t} \frac{\chi(t|\xi|)}{|\xi|^{2r}} F(\varphi)(\xi)\right) \right\|_{L^q(\mathbb{R}^n)} \leq C t^{2r-n\left(\frac{1}{p}-\frac{1}{q}\right)} \|\varphi\|_{L^p(\mathbb{R}^n)}$$

for all admissible $1 < p \leq 2$, $\frac{1}{p} + \frac{1}{q} = 1$.

Proof In some steps we follow the approach of [12] and [177]. To estimate

$$\left\| F_{\xi \to x}^{-1}\left(e^{-i|\xi|t} \frac{\chi(t|\xi|)}{|\xi|^{2r}} F(\varphi)(\xi)\right) \right\|_{L^q}$$

we choose a nonnegative infinitely differentiable function $\phi = \phi(\xi)$ having compact support in $\{\xi \in \mathbb{R}^n : |\xi| \in [\frac{1}{2}, 2]\}$. We set $\phi_k(\xi) := \phi(2^{-k}\xi)$ for $k \geq 1$ and $\phi_0(\xi) := 1 - \sum_{k=1}^\infty \phi_k(\xi)$.

$L^1 - L^\infty$ *estimates:*

Due to Theorem 24.1.3 we have for all $k \leq k_0$ the estimates

$$\left\| F_{\xi \to x}^{-1}\left(e^{-i|\xi|t} \frac{\chi(t|\xi|)\phi_k(t|\xi|)}{|\xi|^{2r}} F(\varphi)(\xi)\right) \right\|_{L^\infty} \leq C t^{2r-n} \|\varphi\|_{L^1}.$$

Here k_0 can be chosen arbitrarily large. To estimate for all $k \geq k_0$

$$\left\| F_{\xi \to x}^{-1}\left(e^{-i|\xi|t} \frac{\chi(t|\xi|)\phi_k(t|\xi|)}{|\xi|^{2r}} F(\varphi)(\xi)\right) \right\|_{L^\infty}$$

we introduce the change of variables $t\xi =: 2^k \eta$. Then, we obtain

$$\left\| F_{\xi \to x}^{-1}\left(e^{-i|\xi|t} \frac{\chi(t|\xi|)\phi_k(t|\xi|)}{|\xi|^{2r}}\right) \right\|_{L^\infty}$$

$$= 2^{k(n-2r)} t^{2r-n} \left\| F_{\eta \to x}^{-1}\left(e^{-i2^k|\eta|} \frac{\phi_k(2^k|\eta|)}{|\eta|^{2r}}\right) \right\|_{L^\infty}$$

$$= 2^{k(n-2r)} t^{2r-n} \left\| F_{\eta \to x}^{-1}\left(e^{-i2^k|\eta|} \frac{\phi(|\eta|)}{|\eta|^{2r}}\right) \right\|_{L^\infty}.$$

The last oscillating integral is estimated by the Littman type lemma in the form of Theorem 16.3.1 with $\tau := 2^k$. Taking into account the properties of ϕ (smoothness and compact support) we may conclude

$$\left\| F_{\eta \to x}^{-1} \left(e^{-i2^k |\eta|} \frac{\phi(|\eta|)}{|\eta|^{2r}} \right) \right\|_{L^\infty} \leq C(1 + 2^k)^{-\frac{n-1}{2}}.$$

Summarizing all estimates gives, together with Young's inequality of Proposition 24.5.2, the desired $L^1 - L^\infty$ estimate

$$\left\| F_{\xi \to x}^{-1} \left(e^{-i|\xi|t} \frac{\chi(t|\xi|)\phi_k(t|\xi|)}{|\xi|^{2r}} F(\varphi)(\xi) \right) \right\|_{L^\infty} \leq C 2^{k(\frac{1}{2}(n+1)-2r)} t^{2r-n} \|\varphi\|_{L^1}.$$

$L^2 - L^2$ *estimates:*
After application of the formula of Parseval-Plancherel from Remark 24.1.2 we immediately get the desired $L^2 - L^2$ estimate

$$\left\| F_{\xi \to x}^{-1} \left(e^{-i|\xi|t} \frac{\chi(t|\xi|)\phi_k(t|\xi|)}{|\xi|^{2r}} F(\varphi)(\xi) \right) \right\|_{L^2} \leq C 2^{-k \cdot 2r} t^{2r} \|\varphi\|_{L^2}.$$

$L^p - L^q$ *estimates on the conjugate line:*
Using the $L^1 - L^\infty$ and $L^2 - L^2$ estimates, the application of the Riesz-Thorin interpolation theorem Proposition 24.5.1 yields the desired $L^p - L^q$ estimates on the conjugate line

$$\left\| F_{\xi \to x}^{-1} \left(e^{-i|\xi|t} \frac{\chi(t|\xi|)\phi_k(t|\xi|)}{|\xi|^{2r}} F(\varphi)(\xi) \right) \right\|_{L^q} \leq C 2^{k(\frac{1}{2}(n+1)(\frac{1}{p}-\frac{1}{q})-2r)} t^{2r-n(\frac{1}{p}-\frac{1}{q})} \|\varphi\|_{L^p},$$

where $p \in (1, 2]$.
After choosing $\frac{n+1}{2}(\frac{1}{p} - \frac{1}{q}) \leq 2r$, summarizing all these estimates, applying Theorem 24.1.4 and Remark 24.1.4, the desired $L^p - L^q$ estimates follow. This completes the proof.

16.6 $L^p - L^q$ Estimates on the Conjugate Line for Solutions to the Wave Equation

Now let us return to the representation of solutions to the Cauchy problem for the free wave equation

$$u(t, x) = F_{\xi \to x}^{-1} \left(\left(e^{i|\xi|t} + e^{-i|\xi|t} \right) \frac{1}{2} F(\varphi)(\xi) \right)$$

$$+ F_{\xi \to x}^{-1} \left(\left(e^{i|\xi|t} - e^{-i|\xi|t} \right) \frac{1}{2i|\xi|} F(\psi)(\xi) \right).$$

We explain how to proceed with the Fourier multiplier

$$F_{\xi \to x}^{-1}\big(e^{-i|\xi|t}F(\varphi)(\xi)\big).$$

Case 1 $t \in (0, 1]$

We need regularity of the data φ. We add in the amplitude function the term $\dfrac{|\xi|^{2r_1}}{|\xi|^{2r_1}}$ and study instead

$$F_{\xi \to x}^{-1}\big(e^{-i|\xi|t}F(\varphi)(\xi)\big) = F_{\xi \to x}^{-1}\Big(e^{-i|\xi|t}\frac{|\xi|^{2r_1}}{|\xi|^{2r_1}}F(\varphi)(\xi)\Big)$$

$$= F_{\xi \to x}^{-1}\Big(e^{-i|\xi|t}\frac{1}{|\xi|^{2r_1}}F(|D|^{2r_1}\varphi)(\xi)\Big).$$

Here we used (cf. with Theorem 24.1.1) the relation

$$|\xi|^{2r_1}F(\varphi)(\xi) = F(|D|^{2r_1}\varphi).$$

To avoid a singular behavior of $t^{2r_1 - n(\frac{1}{p} - \frac{1}{q})}$ at $t = 0$ we choose the parameter $2r_1 = n(\frac{1}{p} - \frac{1}{q})$ in Theorems 16.4.1 and 16.5.1. Both statements imply immediately the $L^p - L^q$ estimate

$$\big\|F_{\xi \to x}^{-1}\big(e^{-i|\xi|t}F(\varphi)(\xi)\big)\big\|_{L^q} \leq C\big\||D|^{n(\frac{1}{p} - \frac{1}{q})}\varphi\big\|_{L^p}$$

for all admissible $1 < p \leq 2$, $\frac{1}{p} + \frac{1}{q} = 1$.

Case 2 $t \in [1, \infty)$

We need regularity of the data φ. We add in the amplitude function the term $\dfrac{|\xi|^{2r_2}}{|\xi|^{2r_2}}$ and study instead

$$F_{\xi \to x}^{-1}\big(e^{-i|\xi|t}F(\varphi)(\xi)\big) = F_{\xi \to x}^{-1}\Big(e^{-i|\xi|t}\frac{|\xi|^{2r_2}}{|\xi|^{2r_2}}F(\varphi)(\xi)\Big)$$

$$= F_{\xi \to x}^{-1}\Big(e^{-i|\xi|t}\frac{1}{|\xi|^{2r_2}}F(|D|^{2r_2}\varphi)(\xi)\Big).$$

Here we used (cf. with Theorem 24.1.1) the relation

$$|\xi|^{2r_2}F(\varphi)(\xi) = F(|D|^{2r_2}\varphi).$$

To obtain the fastest decay behavior we choose the parameter $2r_2 = \frac{n+1}{2}(\frac{1}{p} - \frac{1}{q})$ in Theorems 16.4.1 and 16.5.1. Both statements imply immediately the $L^p - L^q$

estimate

$$\left\| F_{\xi \to x}^{-1} \left(e^{-i|\xi|t} F(\varphi)(\xi) \right) \right\|_{L^q} \leq C(1+t)^{-\frac{n-1}{2}(\frac{1}{p}-\frac{1}{q})} \||D|^{\frac{n+1}{2}(\frac{1}{p}-\frac{1}{q})} \varphi \|_{L^p}$$

for all admissible $1 < p \leq 2$, $\frac{1}{p} + \frac{1}{q} = 1$.

Summarizing the last estimates and taking account of Remark 24.3.1 we arrive at the following statement.

Theorem 16.6.1 *Let φ belong to $C_0^\infty(\mathbb{R}^n)$. Then, the Fourier multiplier*

$$F_{\xi \to x}^{-1} \left(e^{-i|\xi|t} F(\varphi)(\xi) \right)$$

satisfies the following $L^p - L^q$ decay estimates on the conjugate line:

$$\left\| F_{\xi \to x}^{-1} \left(e^{-i|\xi|t} F(\varphi)(\xi) \right) \right\|_{L^q(\mathbb{R}^n)} \leq C(1+t)^{-\frac{n-1}{2}(\frac{1}{p}-\frac{1}{q})} \|\varphi\|_{H_p^{M_p}(\mathbb{R}^n)},$$

where the real number $M_p \geq n(\frac{1}{p} - \frac{1}{q})$, $1 < p \leq 2$, $\frac{1}{p} + \frac{1}{q} = 1$.

In the same way we can treat the other Fourier multipliers appearing in the representation of solutions to the Cauchy problem for the free wave equation. For example, by using the same approach we get the following result.

Theorem 16.6.2 *Let ψ belong to $C_0^\infty(\mathbb{R}^n)$. Then, the Fourier multiplier*

$$F_{\xi \to x}^{-1} \left(e^{-i|\xi|t} \frac{F(\psi)(\xi)}{|\xi|} \right)$$

satisfies the following $L^p - L^q$ decay estimates on the conjugate line:

$$\left\| F_{\xi \to x}^{-1} \left(e^{-i|\xi|t} \frac{F(\psi)(\xi)}{|\xi|} \right) \right\|_{L^q(\mathbb{R}^n)} \leq C(1+t)^{-\frac{n-1}{2}(\frac{1}{p}-\frac{1}{q})} \|\psi\|_{H_p^{M_p}(\mathbb{R}^n)},$$

where $n(\frac{1}{p} - \frac{1}{q}) \geq 1$, $1 < p \leq 2$, $\frac{1}{p} + \frac{1}{q} = 1$, and the real number $M_p \geq n(\frac{1}{p} - \frac{1}{q}) - 1$.

Remark 16.6.1 In Theorem 16.6.2 appears the restriction $n(\frac{1}{p} - \frac{1}{q}) \geq 1$ due to the fact that we are applying the Riesz potential operator

$$I_1 \psi = F_{\xi \to x}^{-1} \left(\frac{F(\psi)(\xi)}{|\xi|} \right).$$

However, one can remove this condition by using some cancelation property for the multiplier

$$\frac{e^{i|\xi|t} - e^{-i|\xi|t}}{2i|\xi|}$$

and allowing some loss of decay in time in the expected estimates.

Theorem 16.6.3 *Let ψ belong to $C_0^\infty(\mathbb{R}^n)$. Then, we have the following $L^p - L^q$ decay estimates on the conjugate line:*

$$\left\| F_{\xi\to x}^{-1}\left(\frac{e^{i|\xi|t} - e^{-i|\xi|t}}{2i|\xi|} F(\psi)(\xi)\right)\right\|_{L^q(\mathbb{R}^n)} \le Ct(1+t)^{-\frac{n-1}{2}(\frac{1}{p}-\frac{1}{q})}\|\psi\|_{H_p^{M_p}(\mathbb{R}^n)},$$

where $M_p \ge n(\frac{1}{p} - \frac{1}{q})$, $1 < p \le 2$ and $\frac{1}{p} + \frac{1}{q} = 1$.

Proof Due to the fact that for $|\xi|t \le 1$ we have

$$\left|\frac{e^{i|\xi|t} - e^{-i|\xi|t}}{2i|\xi|}\right| \le Ct.$$

After following the proof to Theorem 16.4.1 we get in the pseudodifferential zone the $L^p - L^q$ estimates

$$\left\| F_{\xi\to x}^{-1}\left(\frac{(e^{i|\xi|t} - e^{-i|\xi|t})}{2i|\xi|} \frac{(1 - \chi(t|\xi|))}{|\xi|^{2r}} |\xi|^{2r} F(\psi)(\xi)\right)\right\|_{L^q}$$

$$\le Ct^{1+2r-n(\frac{1}{p}-\frac{1}{q})}\||D|^{2r}\psi\|_{L^p}$$

for all $1 < p \le 2 \le q < \infty$ and $0 \le 2r \le n(\frac{1}{p} - \frac{1}{q})$.
In the hyperbolic zone, applying Theorem 16.5.1, we get

$$\left\| F_{\xi\to x}^{-1}\left(\frac{(e^{i|\xi|t} - e^{-i|\xi|t})}{2i|\xi|} \frac{\chi(t|\xi|)}{|\xi|^{2r}} |\xi|^{2r} F(\psi)(\xi)\right)\right\|_{L^q}$$

$$\le Ct^{1+2r-n(\frac{1}{p}-\frac{1}{q})}\||D|^{2r}\psi\|_{L^p}$$

for all $1 \le p \le 2 \le q \le \infty$, $\frac{1}{p} + \frac{1}{q} = 1$ and $2r + 1 \ge \frac{n+1}{2}(\frac{1}{p} - \frac{1}{q})$.
For $t \in (0, 1]$ we apply the derived estimates in both zones with $2r = n(\frac{1}{p} - \frac{1}{q})$ to verify

$$\left\| F_{\xi\to x}^{-1}\left(\frac{(e^{i|\xi|t} - e^{-i|\xi|t})}{2i|\xi|} F(\psi)(\xi)\right)\right\|_{L^q} \le Ct\||D|^{n(\frac{1}{p}-\frac{1}{q})}\psi\|_{L^p}$$

for all $1 < p \le 2$ and $\frac{1}{p} + \frac{1}{q} = 1$.

Now, for $t \in [1, \infty)$ we apply the derived estimates in both zones with $2r = \frac{n+1}{2}(\frac{1}{p} - \frac{1}{q})$ to verify

$$\left\| F_{\xi \to x}^{-1} \left(\frac{(e^{i|\xi|t} - e^{-i|\xi|t})}{2i|\xi|} F(\psi)(\xi) \right) \right\|_{L^q}$$

$$\leq C(1 + t)^{1 - \frac{n-1}{2}(\frac{1}{p} - \frac{1}{q})} \left\| |D|^{\frac{n+1}{2}(\frac{1}{p} - \frac{1}{q})} \psi \right\|_{L^p}$$

for all $1 < p \leq 2$ and $\frac{1}{p} + \frac{1}{q} = 1$. This completes the proof.

Summarizing the estimates of Theorems 16.6.1 and 16.6.2 we are able to conclude the following result.

Theorem 16.6.4 *Let φ and ψ belong to $C_0^\infty(\mathbb{R}^n)$. Then, the solution u to the Cauchy problem for the free wave equation satisfies the following $L^p - L^q$ decay estimates on the conjugate line:*

$$\|u(t, \cdot)\|_{L^q} \leq C(1 + t)^{-\frac{n-1}{2}(\frac{1}{p} - \frac{1}{q})} \left(\|\varphi\|_{H_p^{M_p}} + \|\psi\|_{H_p^{M_p - 1}} \right),$$

where $n(\frac{1}{p} - \frac{1}{q}) \geq 1$, $1 < p \leq 2$, $\frac{1}{p} + \frac{1}{q} = 1$, and the real number M_p satisfies $M_p \geq n(\frac{1}{p} - \frac{1}{q})$.

Applying the derived estimates in Theorems 16.6.1 and 16.6.3 we are able to conclude the following result.

Theorem 16.6.5 *Let φ and ψ belong to $C_0^\infty(\mathbb{R}^n)$. Then, the solution u to the Cauchy problem for the free wave equation satisfies the following $L^p - L^q$ decay estimates on the conjugate line:*

$$\|u(t, \cdot)\|_{L^q} \leq C(1 + t)^{-\frac{n-1}{2}(\frac{1}{p} - \frac{1}{q})} \left(\|\varphi\|_{H_p^{M_p}} + t\|\psi\|_{H_p^{M_p}} \right),$$

where $1 < p \leq 2$, $\frac{1}{p} + \frac{1}{q} = 1$, and the real number M_p satisfies $M_p \geq n(\frac{1}{p} - \frac{1}{q})$.

Analogously, we are able to conclude $L^p - L^q$ decay estimates for $\partial_t u$ and ∇u which are defined by

$$\partial_t u(t, x) = F_{\xi \to x}^{-1} \left(-\left(e^{i|\xi|t} - e^{-i|\xi|t} \right) \frac{|\xi|}{2i} F(\varphi)(\xi) \right)$$

$$+ F_{\xi \to x}^{-1} \left(\left(e^{i|\xi|t} + e^{-i|\xi|t} \right) \frac{1}{2} F(\psi)(\xi) \right),$$

$$\partial_{x_k} u(t, x) = F_{\xi \to x}^{-1} \left(i\xi_k \left(e^{i|\xi|t} + e^{-i|\xi|t} \right) \frac{1}{2} F(\varphi)(\xi) \right)$$

$$+ F_{\xi \to x}^{-1} \left(i\xi_k \left(e^{i|\xi|t} - e^{-i|\xi|t} \right) \frac{1}{2i|\xi|} F(\psi)(\xi) \right).$$

Theorem 16.6.6 *Let φ and ψ belong to $C_0^\infty(\mathbb{R}^n)$. Then the partial derivatives $\partial_t u$ and ∇u of the solution to the Cauchy problem for the free wave equation satisfy the following $L^p - L^q$ decay estimates on the conjugate line:*

$$\|u_t(t, \cdot)\|_{L^q} \le C(1+t)^{-\frac{n-1}{2}(\frac{1}{p} - \frac{1}{q})} \left(\|\varphi\|_{H_p^{M_p}} + \|\psi\|_{H_p^{M_p-1}} \right),$$

$$\|\nabla u(t, \cdot)\|_{L^q} \le C(1+t)^{-\frac{n-1}{2}(\frac{1}{p} - \frac{1}{q})} \left(\|\varphi\|_{H_p^{M_p}} + \|\psi\|_{H_p^{M_p-1}} \right),$$

where the real number M_p fulfils $M_p \ge n(\frac{1}{p} - \frac{1}{q}) + 1$, $1 < p \le 2$, $\frac{1}{p} + \frac{1}{q} = 1$.

16.7 $L^p - L^q$ Estimates Away from the Conjugate Line

Let us consider the Cauchy problem for the free wave equation

$$u_{tt} - \Delta u = 0, \quad u(0, x) = \varphi(x), \quad u_t(0, x) = \psi(x).$$

From Sects. 16.4 and 16.5, by taking $\varphi \equiv 0$ and without asking for additional regularity of the data, one still may expect some singular $L^p - L^q$ estimates. Indeed, by putting $r = \frac{1}{2}$ into the statements of Theorems 16.4.1 and 16.5.1 we conclude the following $L^p - L^q$ estimates on the conjugate line:

$$\|u(t, \cdot)\|_{L^q} \le C\, t^{1 - \frac{n}{p} + \frac{n}{q}} \|\psi\|_{L^p}$$

uniformly for any $t > 0$ and for

$$\frac{n+1}{2} \left(\frac{1}{p} - \frac{1}{q} \right) \le 1 \le n \left(\frac{1}{p} - \frac{1}{q} \right).$$

In particular, it is true for $(\frac{1}{p}, \frac{1}{q}) = P_1 := \left(\frac{1}{2} + \frac{1}{n+1}, \frac{1}{2} - \frac{1}{n+1} \right)$. Since these estimates are true for $(\frac{1}{p}, \frac{1}{q}) = P_0 := \left(\frac{1}{2}, \frac{1}{2} \right)$, by interpolation we conclude it on the line with end points P_0 and P_1.

More in general, the estimates in [161] and [198] imply that the solution to the above Cauchy problem satisfies $L^p - L^q$ estimates if and only if the point $(\frac{1}{p}, \frac{1}{q})$ belongs to the closed triangle with vertices

$$P_1 = \left(\tfrac{1}{2} + \tfrac{1}{n+1}, \tfrac{1}{2} - \tfrac{1}{n+1} \right), \quad P_2 = \left(\tfrac{1}{2} - \tfrac{1}{n-1}, \tfrac{1}{2} - \tfrac{1}{n-1} \right),$$

$$\text{and } P_3 = \left(\tfrac{1}{2} + \tfrac{1}{n-1}, \tfrac{1}{2} + \tfrac{1}{n-1} \right).$$

In the case $n = 1$ or $n = 2$ we define $P_2 = (0, 0)$ and $P_3 = (1, 1)$. Moreover, the asymptotic behavior in t follows by homogeneity, namely, that there exists a positive constant C such that the $L^p - L^q$ estimates

$$\|u(t,\cdot)\|_{L^q} \leq C \, t^{1-\frac{n}{p}+\frac{n}{q}} \|\psi\|_{L^p}$$

hold uniformly for any $t > 0$.

If we ask for additional regularity of the data, besides avoiding singular estimates at $t = 0$ as was done in Theorem 16.6.4, we can also enlarge the admissible range for p, q in the $L^p - L^q$ estimates. For instance, combining results from [188] and [137], the estimates

$$\|u(t,\cdot)\|_{L^p} \leq C(1+t)^{(n-1)|\frac{1}{p}-\frac{1}{2}|} \|\varphi\|_{H_p^s} + t(1+t)^{\max\{(n-1)|\frac{1}{p}-\frac{1}{2}|-1,0\}} \|\psi\|_{H_p^r}$$

hold for $p \in (1, \infty)$ if and only if

$$(n-1)\left|\frac{1}{p} - \frac{1}{2}\right| \leq s \quad \text{and} \quad (n-1)\left|\frac{1}{p} - \frac{1}{2}\right| \leq r+1, \, r \geq 0.$$

Therefore, apart from the case $p = 2$, in general, one can not expect $L^p - L^p$ estimates for the solutions of the free wave equation.

It is interesting to compare $L^p - L^q$ estimates for the solution to the Cauchy problem for the free wave equation with the ones for the Klein-Gordon equation

$$v_{tt} - \Delta v + v = 0, \quad v(0,x) = \varphi(x), \quad v_t(0,x) = \psi(x).$$

In [129] the authors proved that for every $t > 0$ the operator $T_t : (\varphi, \psi) \to v(t,\cdot)$ ($\varphi \equiv 0$) is bounded from $L^p(\mathbb{R}^n)$ to $L^q(\mathbb{R}^n)$ if, and only if, the point $(\frac{1}{p}, \frac{1}{q})$ belongs to the same closed triangle $P_1 P_2 P_3$.

16.8 $L^p - L^q$ Estimates on the Conjugate Line for Solutions to the Schrödinger Equation

Theorem 13.1.3 contains $L^p - L^q$ estimates on the conjugate line for solutions to the Cauchy problem for the Schrödinger equation under the assumption of low regularity of the data. In the following we are interested in deriving such estimates for high regular data. We use the representation

$$u(t,x) = F_{\xi \to x}^{-1}\left(e^{-i|\xi|^2 t} F(\varphi)(\xi)\right)$$

of solutions to the Cauchy problem

$$D_t u - \Delta u = 0, \quad u(0,x) = \varphi(x)$$

under a suitable regularity assumption for the data φ.

16.8.1 Philosophy of Our Approach to Derive $L^p - L^q$ Estimates on the Conjugate Line for Solutions to the Schrödinger Equation

We can follow, with some modifications, the approach explained in Sect. 16.2.

1. We add an amplitude function in our Fourier multiplier. Instead, our Fourier multiplier takes the form

$$F_{\xi \to x}^{-1}\left(e^{-i|\xi|^2 t}\frac{1}{|\xi|^{2r}}F(\varphi)(\xi)\right),$$

   where the parameter r is determined later.
2. We decompose the extended phase space $(0, \infty) \times \mathbb{R}_\xi^n$ into two zones. We define the *pseudodifferential zone*

$$Z_{pd} = \{(t, \xi) \in (0, \infty) \times \mathbb{R}_\xi^n : t|\xi|^2 \leq 1\},$$

   and the *evolution zone*

$$Z_{ev} = \{(t, \xi) \in (0, \infty) \times \mathbb{R}_\xi^n : t|\xi|^2 \geq 1\}.$$

3. In the Fourier multiplier there appears the phase function

$$\phi = \phi(t, x, \xi) = x \cdot \xi - |\xi|^2 t.$$

   The stationary points of the phase function are determined by

$$\nabla_\xi \phi(t, x, \xi) = x - 2t\xi = 0.$$

   So, we have only one case.
   If we introduce the notation $\frac{x}{2t} =: y$, then we have the isolated stationary point $\xi = y$.
4. In Sect. 16.8.3 we shall derive $L^p - L^q$ estimates for the Fourier multiplier

$$F_{\xi \to x}^{-1}\left(e^{-i|\xi|^2 t}\frac{1 - \chi(t|\xi|^2)}{|\xi|^{2r}}F(\varphi)(\xi)\right).$$

   The amplitude function

$$\frac{1 - \chi(t|\xi|^2)}{|\xi|^{2r}}$$

   vanishes in the evolution zone.

5. In Sect. 16.8.4 we shall derive $L^p - L^q$ estimates on the conjugate line for the model Fourier multiplier

$$F_{\xi\to x}^{-1}\left(e^{-i|\xi|^2 t}\frac{\chi(t|\xi|^2)}{|\xi|^{2r}}F(\varphi)(\xi)\right).$$

The amplitude function

$$\frac{\chi(t|\xi|^2)}{|\xi|^{2r}}$$

vanishes in a large part of the pseudodifferential zone.
6. After all these preparations we are able to derive in Sect. 16.8.5 the desired $L^p - L^q$ estimates on the conjugate line for solutions to the Cauchy problem for the Schrödinger equation.

16.8.2   A Littman Type Lemma

In further considerations we shall use the following Littman type lemma related to the phase function ϕ of the previous section.

Theorem 16.8.1 (A Littman Type Lemma)   *Let us consider for $\tau \geq \tau_0$, τ_0 is a large positive number, the oscillating integral*

$$F_{\eta\to x}^{-1}\left(e^{-i\tau|\eta|^2}v(\eta)\right).$$

The amplitude function $v = v(\eta)$ is supposed to belong to $C_0^\infty(\mathbb{R}^n)$ with support in $\{\eta \in \mathbb{R}^n : |\eta| \in [\frac{1}{2}, 2]\}$. Then, the following $L^\infty - L^\infty$ estimate holds:

$$\left\|F_{\eta\to x}^{-1}\left(e^{-i\tau|\eta|^2}v(\eta)\right)\right\|_{L^\infty(\mathbb{R}_x^n)} \leq C(1+\tau)^{-\frac{n}{2}}\sum_{|\alpha|\leq s}\|D_\eta^\alpha v(\eta)\|_{L^\infty(\mathbb{R}_\eta^n)},$$

where $s > \frac{n+4}{2}$.

Proof The method of stationary phase is explained, for example, in [190] and [193]. We can follow the steps of the proof to Theorem 16.3.1. For this reason we only sketch modifications in the steps of the proof.

Step 1:   After the change of variables $y := \frac{x}{\tau}$ we have to estimate

$$\sup_{y\in\mathbb{R}^n}\left|\int_{\mathbb{R}^n}e^{i\tau(y\cdot\eta-|\eta|^2)}v(\eta)\,d\eta\right|.$$

Introducing the vector-valued function

$$\Phi = \Phi(\tau, y, \eta) = \tau(y - 2\eta)$$

the operator L is defined as follows:

$$L := \frac{1}{|\Phi(\tau, y, \eta)|^2} \sum_{j=1}^{n} \Phi_j(\tau, y, \eta) \frac{1}{i} \partial_{\eta_j}.$$

Here Φ_j denotes the j-th component of Φ.

Step 2: Let us fix $y \in \mathbb{R}^n$. If $\eta \notin B_\varepsilon(\frac{y}{2})$, where $B_\varepsilon(\frac{y}{2})$ denotes the ball centered at $\eta = \frac{y}{2}$ with radius ε, then $|\Phi| \geq 2\tau\varepsilon$ for all $\eta \in \mathbb{R}^n \setminus B_\varepsilon(\frac{y}{2})$. We may apply an arbitrary number N of integration by parts. Here we use the smoothness and support assumption for v to conclude

$$\sup_{y \in \mathbb{R}^n} \left| \int_{\mathbb{R}^n \setminus B_\varepsilon(\frac{y}{2})} e^{i\tau(y \cdot \eta - |\eta|^2)} v(\eta) \, d\eta \right| \leq C\tau^{-N} \sum_{|\alpha| \leq N} \|D_\eta^\alpha v(\eta)\|_{L^\infty(\mathbb{R}_\eta^n)}.$$

The constant C is independent of $y \in \mathbb{R}^n$ and depends only on ε, which is later fixed. We do not need Step 3.

Step 4: It remains to estimate

$$\sup_{y \in \mathbb{R}^n} \left| \int_{B_\varepsilon(\frac{y}{2})} e^{i\tau(y \cdot \eta - |\eta|^2)} v(\eta) \, d\eta \right|.$$

For a given y the phase function has the only stationary point $\eta = \frac{y}{2}$. Using

$$y \cdot \eta - |\eta|^2 = \frac{|y|^2}{4} - \left| \eta - \frac{y}{2} \right|^2$$

(in fact this follows from application of Taylor's formula around the stationary point) it remains to estimate the integral

$$\int_{B_\varepsilon(\frac{y}{2})} e^{-i\tau|\eta - \frac{y}{2}|^2} v(\eta) \, d\eta = \int_{B_\varepsilon(0)} e^{-i\tau|\eta|^2} v\left(\eta + \frac{y}{2} \right) d\eta$$

$$= \int_{\mathbb{R}^n} e^{-i\tau|\eta|^2} v\left(\eta + \frac{y}{2} \right) d\eta$$

for all $y \in \mathbb{R}^n$.

Step 5:   The last integral is equal to $\left(e^{-i\tau \frac{|D|^2}{2}} u(r, \rho, y)\right)(x = 0)$ taking into
consideration

$$\left(e^{-i\tau \frac{|D|^2}{2}} u(r, \rho, y)\right)(x) = c_n(\tau) \int_{\mathbb{R}^n} e^{-i\tau \frac{|y|^2}{2}} u(r, \rho, x + y)\, dy,$$

where

$$c_n(\tau) = (2\pi)^{\frac{n}{2}} e^{i\frac{\pi n}{4}} \tau^{-\frac{n}{2}}.$$

Then, we approximate $e^{-i\tau \frac{|D|^2}{2}}$ by a differential operator of order 0. We use

$$\left| \left(e^{-i\tau \frac{|D|^2}{2}} u(r, \rho, y)\right)(x = 0) - u(r, \rho, 0) \right|$$

$$\leq |c_n(\tau)| \tau^{-1} \sum_{|\alpha| \leq s} \left\| \frac{1}{2} \Delta D_y^\alpha u(y) \right\|_{L^\infty(\mathbb{R}_y^n)},$$

where $s > \frac{n}{2}$. In the second step we need only $N \geq \frac{n}{2}$ derivatives to have the desired
decay. Thus, totally, we need $s > \frac{n+4}{2}$ derivatives of u with respect to y and of v
with respect to η, respectively. This completes the proof.

16.8.3 $L^p - L^q$ Estimates for Fourier Multipliers with Amplitudes Localized in the Pseudodifferential Zone

We can follow step by step the proof of Theorem 16.4.1 to derive $L^p - L^q$ estimates
for the Fourier multiplier

$$F_{\xi \to x}^{-1}\left(e^{-i|\xi|^2 t} \frac{1 - \chi(t|\xi|^2)}{|\xi|^{2r}} F(\varphi)(\xi)\right)$$

with localized amplitude in the pseudodifferential zone. In this way we may
conclude the following result.

Theorem 16.8.2   *Let us consider for $\varphi \in C_0^\infty(\mathbb{R}^n)$ the Fourier multiplier*

$$F_{\xi \to x}^{-1}\left(e^{-i|\xi|^2 t} \frac{1 - \chi(t|\xi|^2)}{|\xi|^{2r}} F(\varphi)(\xi)\right)$$

from Sect. 16.8.1. We assume $1 < p \leq 2 \leq q < \infty$ *and* $0 \leq 2r \leq n(\frac{1}{p} - \frac{1}{q})$. *Then, we have the* $L^p - L^q$ *estimates*

$$\left\| F_{\xi \to x}^{-1} \left(e^{-i|\xi|^2 t} \frac{1 - \chi(t|\xi|^2)}{|\xi|^{2r}} F(\varphi)(\xi) \right) \right\|_{L^q(\mathbb{R}^n)} \leq Ct^{-\frac{r}{2}(\frac{1}{p} - \frac{1}{q})} \|\varphi\|_{L^p(\mathbb{R}^n)}$$

for all admissible p, q. *The constant* C *depends on* p *and* q.

16.8.4 $L^p - L^q$ Estimates on the Conjugate Line for Fourier Multipliers with Amplitudes Localized in the Evolution Zone

Now we estimate the Fourier multiplier

$$F_{\xi \to x}^{-1} \left(e^{-i|\xi|^2 t} \frac{\chi(t|\xi|^2)}{|\xi|^{2r}} F(\varphi)(\xi) \right)$$

with an amplitude function localized more or less in the evolution zone. It is only important that the amplitude vanishes if $t|\xi|^2 \leq \frac{1}{2}$. The main tools are a refined decomposition of the extended phase space, the Littman type lemma in the form of Theorem 16.8.1 and an interpolation argument.

Theorem 16.8.3 *Let us consider for* $\varphi \in C_0^\infty(\mathbb{R}^n)$ *the Fourier multiplier*

$$F_{\xi \to x}^{-1} \left(e^{-i|\xi|^2 t} \frac{\chi(t|\xi|^2)}{|\xi|^{2r}} F(\varphi)(\xi) \right)$$

from Sect. 16.8.1. We assume $0 \leq r$. *Then, we have the* $L^p - L^q$ *estimates on the conjugate line*

$$\left\| F_{\xi \to x}^{-1} \left(e^{-i|\xi|^2 t} \frac{\chi(t|\xi|^2)}{|\xi|^{2r}} F(\varphi)(\xi) \right) \right\|_{L^q(\mathbb{R}^n)} \leq Ct^{-\frac{n}{2}(\frac{1}{p} - \frac{1}{q})} \|\varphi\|_{L^p(\mathbb{R}^n)}$$

for all admissible $1 < p \leq 2$, $\frac{1}{p} + \frac{1}{q} = 1$.

Proof We can follow, with some modifications, the proof to Theorem 16.5.1. So, we sketch the modifications only.
$L^1 - L^\infty$ *estimates:*
We introduce the change of variables $t^{\frac{1}{2}} \xi =: 2^{\frac{k}{2}} \eta$ to estimate the L^∞-norm of the oscillating integral

$$F_{\xi \to x}^{-1} \left(e^{-i|\xi|^2 t} \frac{\chi(t|\xi|^2) \phi_k(t|\xi|^2)}{|\xi|^{2r}} \right).$$

Then we obtain

$$\left\| F_{\xi \to x}^{-1} \left(e^{-i|\xi|^2 t} \frac{\chi(t|\xi|^2)\phi_k(t|\xi|^2)}{|\xi|^{2r}} \right) \right\|_{L^\infty}$$

$$= 2^{k(\frac{n}{2}-r)} t^{r-\frac{n}{2}} \left\| F_{\eta \to x}^{-1} \left(e^{-i2^k|\eta|^2} \frac{\phi(|\eta|^2)}{|\eta|^{2r}} \right) \right\|_{L^\infty}.$$

The last oscillating integral is estimated by the Littman type lemma in the form of Theorem 16.8.1 with $\tau := 2^k$. Taking into account the properties of ϕ (smoothness and compact support) we may conclude

$$\left\| F_{\eta \to x}^{-1} \left(e^{-i2^k|\eta|^2} \frac{\phi(|\eta|^2)}{|\eta|^{2r}} \right) \right\|_{L^\infty} \leq C(1+2^k)^{-\frac{n}{2}}.$$

Summarizing all estimates gives, together with Young's inequality of Proposition 24.5.2, the desired $L^1 - L^\infty$ estimate

$$\left\| F_{\xi \to x}^{-1} \left(e^{-i|\xi|^2 t} \frac{\chi(t|\xi|^2)\phi_k(t|\xi|^2)}{|\xi|^{2r}} F(\varphi)(\xi) \right) \right\|_{L^\infty} \leq C 2^{-kr} t^{r-\frac{n}{2}} \|\varphi\|_{L^1}.$$

$L^2 - L^2$ *estimates:*
After application of the formula of Parseval-Plancherel from Remark 24.1.2 we immediately get the desired $L^2 - L^2$ estimate

$$\left\| F_{\xi \to x}^{-1} \left(e^{-i|\xi|^2 t} \frac{\chi(t|\xi|^2)\phi_k(t|\xi|^2)}{|\xi|^{2r}} F(\varphi)(\xi) \right) \right\|_{L^2} \leq C 2^{-kr} t^r \|\varphi\|_{L^2}.$$

$L^p - L^q$ *estimates on the conjugate line:*
Using the $L^1 - L^\infty$ and $L^2 - L^2$ estimates the application of the Riesz-Thorin interpolation theorem Proposition 24.5.1 yields the desired $L^p - L^q$ estimates on the conjugate line

$$\left\| F_{\xi \to x}^{-1} \left(e^{-i|\xi|^2 t} \frac{\chi(t|\xi|^2)\phi_k(t|\xi|^2)}{|\xi|^{2r}} F(\varphi)(\xi) \right) \right\|_{L^q} \leq C 2^{-kr} t^{r-\frac{n}{2}(\frac{1}{p}-\frac{1}{q})} \|\varphi\|_{L^p},$$

where $p \in (1, 2]$.
Summarizing all these estimates, applying Theorem 24.1.4 and Remark 24.1.4, the desired $L^p - L^q$ estimates follow. This completes the proof.

16.8.5 $L^p - L^q$ Estimates on the Conjugate Line for Solutions to the Schrödinger Equation

Now, let us come back to the representation of solutions to the Cauchy problem for the Schrödinger equation

$$u(t,x) = F_{\xi \to x}^{-1}\left(e^{-i|\xi|^2 t} F(\varphi)(\xi)\right).$$

We explain how to proceed with the Fourier multiplier

$$F_{\xi \to x}^{-1}\left(e^{-i|\xi|^2 t} F(\varphi)(\xi)\right).$$

Case 1: $t \in (0, 1]$

We follow the approach of Sect. 16.6. We need regularity of the data φ. To avoid a singular behavior of $t^{r_1 - \frac{n}{2}(\frac{1}{p} - \frac{1}{q})}$ at $t = 0$ we choose $2r_1 = n(\frac{1}{p} - \frac{1}{q})$ in Theorems 16.8.2 and 16.8.3. Both statements imply immediately the $L^p - L^q$ estimate

$$\left\|F_{\xi \to x}^{-1}\left(e^{-i|\xi|^2 t} F(\varphi)(\xi)\right)\right\|_{L^q} \leq C \||D|^{n(\frac{1}{p} - \frac{1}{q})} \varphi\|_{L^p}$$

for all admissible $1 < p \leq 2$, $\frac{1}{p} + \frac{1}{q} = 1$.

Case 2: $t \in [1, \infty)$

We do not need higher regularity of the data φ. To obtain the fastest decay behavior we choose $r = 0$ in Theorems 16.8.2 and 16.8.3. Both statements imply immediately the $L^p - L^q$ estimate

$$\left\|F_{\xi \to x}^{-1}\left(e^{-i|\xi|^2 t} F(\varphi)(\xi)\right)\right\|_{L^q} \leq C(1 + t)^{-\frac{n}{2}(\frac{1}{p} - \frac{1}{q})} \|\varphi\|_{L^p}$$

for all admissible $1 < p \leq 2$, $\frac{1}{p} + \frac{1}{q} = 1$.

Summarizing the last estimates and taking account of Remark 24.3.1 we arrive at the following statement.

Theorem 16.8.4 *Let φ belong to $C_0^\infty(\mathbb{R}^n)$. Then, the solution u to the Cauchy problem for the Schrödinger equation satisfies the following $L^p - L^q$ decay estimates on the conjugate line:*

$$\|u(t, \cdot)\|_{L^q} \leq C(1 + t)^{-\frac{n}{2}(\frac{1}{p} - \frac{1}{q})} \|\varphi\|_{H_p^{M_p}},$$

where the real number $M_p \geq n(\frac{1}{p} - \frac{1}{q})$, $1 < p \leq 2$ and $\frac{1}{p} + \frac{1}{q} = 1$.

Remark 16.8.1 Let us compare the estimates of Theorems 13.1.3 and 16.8.4. If we choose $\alpha = 0$ and $k = 0$ in Theorem 13.1.3, then we get on the conjugate line the

$L^p - L^q$ estimate ($p \in (1, 2]$)

$$\|u(t, \cdot)\|_{L^q} \leq C t^{-\frac{n}{2}(\frac{1}{p} - \frac{1}{q})} \|\varphi\|_{L^p}.$$

This inequality explains a decay behavior of $\|u(t, \cdot)\|_{L^q}$ for $t \to \infty$ under L^p regularity for the data φ. But, for $t \to +0$ the L^q-norm of the solution may blow up. To avoid such possible blow up behavior we measure the data in fractional Sobolev spaces of higher regularity. This implies the $L^p - L^q$ decay estimate

$$\|u(t, \cdot)\|_{L^q} \leq C(1 + t)^{-\frac{n}{2}(\frac{1}{p} - \frac{1}{q})} \|\varphi\|_{H_p^{M_p}}.$$

In both estimates the decay functions are equivalent for $t \to \infty$.

16.9 $L^p - L^q$ Estimates on the Conjugate Line for Solutions to the Plate Equation

The considerations of the last sections allow us to conclude immediately $L^p - L^q$ decay estimates on the conjugate line for solutions to the Cauchy problem for the plate equation

$$u_{tt} + (-\Delta)^2 u = 0, \quad u(0, x) = \varphi(x), \quad u_t(0, x) = \psi(x), \quad x \in \mathbb{R}^n, \quad n \geq 1.$$

We recall the representation of solutions

$$u(t, x) = F_{\xi \to x}^{-1}\left(e^{-i|\xi|^2 t} \frac{1}{2} F(\varphi)(\xi)\right) - F_{\xi \to x}^{-1}\left(e^{-i|\xi|^2 t} \frac{1}{2i|\xi|^2} F(\psi)(\xi)\right)$$

$$+ F_{\xi \to x}^{-1}\left(e^{i|\xi|^2 t} \frac{1}{2} F(\varphi)(\xi)\right) + F_{\xi \to x}^{-1}\left(e^{i|\xi|^2 t} \frac{1}{2i|\xi|^2} F(\psi)(\xi)\right).$$

The statements of Theorems 16.8.2 and 16.8.3 yield the following result.

Theorem 16.9.1 *Let φ and ψ belong to $C_0^\infty(\mathbb{R}^n)$. Then, the solution u to the Cauchy problem for the plate equation satisfies the following $L^p - L^q$ decay estimates on the conjugate line:*

$$\|u(t, \cdot)\|_{L^q} \leq C(1 + t)^{-\frac{n}{2}(\frac{1}{p} - \frac{1}{q})}\left(\|\varphi\|_{H_p^{M_p}} + \|\psi\|_{H_p^{M_p - 2}}\right),$$

where $n(\frac{1}{p} - \frac{1}{q}) \geq 2$, $1 < p \leq 2$, $\frac{1}{p} + \frac{1}{q} = 1$, and the real number M_p satisfies $M_p \geq n(\frac{1}{p} - \frac{1}{q})$.
Following the goal of Theorem 16.6.3 and the steps of its proof one may conclude the following result.

Theorem 16.9.2 *Let* ψ *belong to* $C_0^\infty(\mathbb{R}^n)$. *Then, we have the following* $L^p - L^q$ *decay estimates on the conjugate line:*

$$\left\| F_{\xi \to x}^{-1} \left(\frac{e^{i|\xi|^2 t} - e^{-i|\xi|^2 t}}{2i|\xi|^2} F(\psi)(\xi) \right) \right\|_{L^q} \le Ct(1+t)^{-\frac{n}{2}(\frac{1}{p} - \frac{1}{q})} \|\psi\|_{H_p^{M_p}},$$

where $M_p \ge n(\frac{1}{p} - \frac{1}{q})$, $1 < p \le 2$ *and* $\frac{1}{p} + \frac{1}{q} = 1$.
As a consequence of Theorems 16.8.2, 16.8.3 and 16.9.2 we have the following statement.

Theorem 16.9.3 *Let* φ *and* ψ *belong to* $C_0^\infty(\mathbb{R}^n)$. *Then, the solution* u *to the Cauchy problem for the plate equation satisfies the following* $L^p - L^q$ *decay estimates on the conjugate line:*

$$\|u(t, \cdot)\|_{L^q} \le C(1+t)^{-\frac{n}{2}(\frac{1}{p} - \frac{1}{q})} \left(\|\varphi\|_{H_p^{M_p}} + t \|\psi\|_{H_p^{M_p}} \right),$$

where $1 < p \le 2$, $\frac{1}{p} + \frac{1}{q} = 1$, *and the real number* M_p *satisfies* $M_p \ge n(\frac{1}{p} - \frac{1}{q})$.
Analogously we are able to conclude $L^p - L^q$ decay estimates for $\partial_t u$ and Δu (terms appearing in the energy $E_{PL}(u)(t)$, see Sect. 11.2.4) which are defined by

$$\partial_t u(t, x) = F_{\xi \to x}^{-1} \left(-e^{-i|\xi|^2 t} \frac{i|\xi|^2}{2} F(\varphi)(\xi) \right) + F_{\xi \to x}^{-1} \left(e^{-i|\xi|^2 t} \frac{1}{2} F(\psi)(\xi) \right)$$

$$+ F_{\xi \to x}^{-1} \left(e^{i|\xi|^2 t} \frac{i|\xi|^2}{2} F(\varphi)(\xi) \right) + F_{\xi \to x}^{-1} \left(e^{i|\xi|^2 t} \frac{1}{2} F(\psi)(\xi) \right),$$

$$\Delta u(t, x) = F_{\xi \to x}^{-1} \left(e^{-i|\xi|^2 t} \frac{|\xi|^2}{2} F(\varphi)(\xi) \right) - F_{\xi \to x}^{-1} \left(e^{-i|\xi|^2 t} \frac{1}{2i} F(\psi)(\xi) \right)$$

$$+ F_{\xi \to x}^{-1} \left(e^{i|\xi|^2 t} \frac{|\xi|^2}{2} F(\varphi)(\xi) \right) + F_{\xi \to x}^{-1} \left(e^{i|\xi|^2 t} \frac{1}{2i} F(\psi)(\xi) \right).$$

Theorem 16.9.4 *Let* φ *and* ψ *belong to* $C_0^\infty(\mathbb{R}^n)$. *Then, the derivatives* $\partial_t u$ *and* Δu *of the solution to the Cauchy problem for the plate equation satisfy the following* $L^p - L^q$ *decay estimates on the conjugate line:*

$$\|u_t(t, \cdot)\|_{L^q} \le C(1+t)^{-\frac{n}{2}(\frac{1}{p} - \frac{1}{q})} \left(\|\varphi\|_{H_p^{M_p}} + \|\psi\|_{H_p^{M_p - 2}} \right),$$

$$\|\Delta u(t, \cdot)\|_{L^q} \le C(1+t)^{-\frac{n}{2}(\frac{1}{p} - \frac{1}{q})} \left(\|\varphi\|_{H_p^{M_p}} + \|\psi\|_{H_p^{M_p - 2}} \right),$$

where the real number M_p *fulfils* $M_p \ge n(\frac{1}{p} - \frac{1}{q}) + 2$, $1 < p \le 2$ *and* $\frac{1}{p} + \frac{1}{q} = 1$.

16.10   Concluding Remarks

16.10.1   Littman's Lemma

For the method of stationary phase bases on the so-called Littman's lemma please see [126] and Theorem 16.10.1 below. In this paper W. Littman studied the linear transformation

$$T : f \in C(\mathbb{R}^{n+1}) \rightarrow Tf := \int_M f(y - x)\mu(x)d\sigma_x \in C(\mathbb{R}^{n+1}),$$

where M is a given n-dimensional surface (possibly with boundary) embedded in $\mathbb{R}^{n+1}$ and μ is a mass density on the surface, vanishing near the boundary. He discusses the question under which assumptions for f, μ and M the following $L^p - L^p$ estimate is true:

$$\|\partial_y^\alpha Tf\|_{L^p} \le C_\alpha \|f\|_{L^p}.$$

The answer to this question is related to the behavior at ∞ of the Fourier transform of the measure μ. For this reason the following result is proved in [126].

Theorem 16.10.1 *Let M be a sufficiently smooth compact n-dimensional surface (possibly with boundary) embedded in $\mathbb{R}^{n+1}$, μ a sufficiently smooth mass distribution on M vanishing near the boundary of M. Suppose that at each point of M, k of the n principal curvatures are different from zero. Then,*

$$\int_M e^{y \cdot x} \mu(x) d\sigma_x = O(|y|^{-\frac{k}{2}}) \ \text{for} \ y \rightarrow \infty.$$

This result was very often used in the study of nonlinear partial differential equations. For example, in [160] one can find the following result which can be concluded from Theorem 16.10.1.

Theorem 16.10.2 *Let us consider for $\tau \ge \tau_0$, τ_0 is a large positive number, the oscillating integral*

$$F_{\eta \rightarrow x}^{-1}\left(e^{-i\tau p(\eta)} v(\eta)\right).$$

The amplitude function $v = v(\eta)$ is supposed to belong to $C_0^\infty(\mathbb{R}^n)$ with support in $\{\eta \in \mathbb{R}^n : |\eta| \in [\frac{1}{2}, 2]\}$. The function $p = p(\eta)$ is C^∞ in a neighborhood of the support of v. Moreover, the rank of the Hessian $H_p(\eta)$ is supposed to satisfy the assumption rank $H_p(\eta) \ge k$ on the support of v. Then, the following $L^\infty - L^\infty$ estimate holds:

$$\left\|F_{\eta \rightarrow x}^{-1}\left(e^{-i\tau p(\eta)} v(\eta)\right)\right\|_{L^\infty(\mathbb{R}_x^n)} \le C(1 + \tau)^{-\frac{k}{2}} \sum_{|\alpha| \le L} \|D_\eta^\alpha v(\eta)\|_{L^\infty(\mathbb{R}_\eta^n)},$$

where L is a suitable entire number.

In Theorems 16.3.1 and 16.8.1 we determine the number L in the special cases $p(\eta) = |\eta|^l$, $l = 1, 2$, and hope to give for the beginners a proof in detail for the desired $L^\infty - L^\infty$ estimates.

Exercises Relating to the Considerations of Chap. 16

Exercise 1 Study the approach from [168] to derive $L^p - L^q$ estimates of solutions to the Cauchy problems for the Schrödinger and for the plate equation.

Exercise 2 Let $(\varphi, \psi) \in H^s(\mathbb{R}^n) \times H^{s-1}(\mathbb{R}^n)$ and $f \in L^1((0, T), H^{s-1}(\mathbb{R}^n))$ with $s \geq 1$. Here we denote the function space $H_2^s(\mathbb{R}^n)$ by $H^s(\mathbb{R}^n)$. Prove that there exists a uniquely determined energy solution

$$u \in C([0, T], H^s(\mathbb{R}^n)) \cap C^1([0, T], H^{s-1}(\mathbb{R}^n))$$

to the Cauchy problem for the non-homogeneous free wave equation

$$u_{tt} - \Delta u = f(t, x), \quad u(0, x) = \varphi(x), \quad u_t(0, x) = \psi(x)$$

which satisfies for each $t \in [0, T]$ the estimates

$$\|u(t, \cdot)\|_{H^s} + \|u_t(t, \cdot)\|_{H^{s-1}}$$

$$\leq C(1 + t)\left(\|\varphi\|_{H^s} + \|\psi\|_{H^{s-1}} + \int_0^t \|f(s, \cdot)\|_{H^{s-1}}\, ds\right).$$

Moreover, for each $t \in [0, T]$ and $s \geq 1$ the solution satisfies

$$\|u(t, \cdot)\|_{\dot{H}^s} + \|u_t(t, \cdot)\|_{\dot{H}^{s-1}} \leq C\left(\|\varphi\|_{\dot{H}^s} + \|\psi\|_{\dot{H}^{s-1}} + \int_0^t \|f(s, \cdot)\|_{\dot{H}^{s-1}}\, ds\right).$$

Suggestion: Putting $K(t, 0, x) = F_{\xi \to x}^{-1}\left(\frac{\sin(|\xi|t)}{|\xi|}\right)$, one may write

$$u(t, \cdot) = K_t(t, 0, \cdot) * \varphi + K(t, 0, \cdot) * \psi + \int_0^t K(t - s, 0, \cdot) * f(s, \cdot)\, ds.$$

Part IV

Chapter 17
Semilinear Heat Models

In this chapter we consider the semilinear heat model with power nonlinearity

$$u_t - \Delta u = \pm |u|^{p-1}u, \quad u(0,x) = \varphi(x).$$

Here $\pm |u|^{p-1}u$ is an example of a source nonlinearity (positive sign) and of an absorbing nonlinearity (negative sign) (see, for example, [153]).

First of all we show how the Fujita exponent $p_{Fuj}(n) = 1 + \frac{2}{n}$ appears. In the case of a source power nonlinearity this exponent is the threshold between global existence of small data solutions for exponents larger and blow up behavior of solutions for exponents smaller or equal to the Fujita exponent. In both results the sign of a functional of the data plays a fundamental role. One main goal is to discuss the issue of global existence (in time) of solutions. Besides global existence, we explain blow up of solutions, too. The situation is quite different in the case of an absorbing power nonlinearity. In this case, we have well-posedness results even for large data because the right-hand side can be included in the definition of a suitable energy. The Fujita exponent influences the profile of a solution. For exponents above the Fujita exponent, the profile of solutions is close to the Gauss kernel. For exponents below the Fujita exponent the profile of solutions is close to that of a self-similar solution. Decay estimates for solutions complete this chapter.

17.1 Semilinear Heat Models with Source Nonlinearity

In this section we consider the Cauchy problem

$$u_t - \Delta u = |u|^{p-1}u, \quad u(0,x) = \varphi(x).$$

M.R. Ebert, M. Reissig, *Methods for Partial Differential Equations*,
https://doi.org/10.1007/978-3-319-66456-9_17

In Sect. 11.2.5 we learned that the nonlinear term $|u|^{p-1}u$ is a source nonlinearity. For this reason the global existence (in time) of small data solutions is of interest. This means, that we try to prove that the steady-state solution $u \equiv 0$ of the Cauchy problem with homogeneous data is stable. Small perturbations of data in suitable Banach spaces preserve the property of the Cauchy problem to have global (in time) solutions. It turns out that there exists a critical exponent p_{crit}, a threshold between global and non-global existence of small data solutions. For the above semilinear heat model this critical exponent is the *Fujita exponent* $p_{Fuj}(n) = 1 + \frac{2}{n}$.

17.1.1 Fujita Discovered the Critical Exponent

In his pioneering paper (see [56]) Fujita proved the following two results.

Theorem 17.1.1 *Let us consider the Cauchy problem*

$$u_t - \Delta u = u^p, \quad u(0,x) = \varphi(x).$$

The data φ does not vanish identically and is supposed to be nonnegative and to belong to the function space $B^2(\mathbb{R}^n)$. Let $p \in (1, 1 + \frac{2}{n})$. Then, there is no global (in time) classical solution satisfying for any $T > 0$ the estimate

$$|u(t,x)| \le M_T \exp(|x|^\beta) \text{ for all } (t,x) \in [0,T] \times \mathbb{R}^n, \ \beta \in (0,2).$$

Here $B^2(\mathbb{R}^n)$ denotes the space of functions with continuous and bounded derivatives up to order 2.

Theorem 17.1.2 *Let us consider the Cauchy problem*

$$u_t - \Delta u = u^p, \quad u(0,x) = \varphi(x),$$

where the data φ is supposed to be nonnegative and to belong to the function space $B^2(\mathbb{R}^n)$. Let $p \in (1 + \frac{2}{n}, \infty)$. Take any positive number γ. Then, there exists a positive number δ with the following property.

If $\varphi(x) \le \delta G_n(\gamma, x)$, then there is a global (in time) classical solution satisfying the estimate

$$0 \le u(t,x) \le M G_n(t + \gamma, x) \text{ for all } (t,x) \in [0,T] \times \mathbb{R}^n.$$

Here

$$G_n(t,x) = \frac{1}{(4\pi t)^{\frac{n}{2}}} \exp\left(-\frac{|x|^2}{4t}\right)$$

is the Gauss kernel.

Both Theorems 17.1.1 and 17.1.2 imply that $p_{Fuj}(n)$ is really the critical exponent. The case $p = p_{Fuj}(n)$ remained open. A blow up result for $p = p_{Fuj}(n)$ has since been proved in [75] or in [116].

17.1.2   Self-Similar Solutions

The following explanations for the semilinear heat model with source nonlinearity are taken from [21, 74] and [107]. Let us come back to the Cauchy problem

$$u_t - \Delta u = a|u|^{p-1}u, \quad u(0, x) = \varphi(x),$$

where $p > 1$ and $a = 1$ or $a = -1$. If $u = u(t, x)$ is some classical solution, then so is $u_\lambda = u_\lambda(t, x) := \lambda^{\frac{2}{p-1}}u(\lambda^2 t, \lambda x)$ for any fixed positive λ. In [74] the authors looked for such solutions invariant under this transformation, that is, $u = u_\lambda$ for all positive λ. Such solutions are called self-similar solutions. Letting $\lambda = t^{-\frac{1}{2}}$, then a necessary and sufficient condition for $u = u_\lambda$ is

$$u(t, x) = t^{-\frac{1}{p-1}}u\Big(1, \frac{x}{\sqrt{t}}\Big).$$

For this reason, the authors have chosen the ansatz

$$u(t, x) = t^{-\frac{1}{p-1}}f\Big(\frac{x}{\sqrt{t}}\Big),$$

where $f = f(y)$, $y := \frac{x}{\sqrt{t}}$, satisfies the semilinear elliptic equation

$$-\Delta f - \frac{1}{2}y \cdot \nabla f = a|f|^{p-1}f + \frac{1}{p-1}f, \quad f > 0,$$

on $\mathbb{R}_y^n$. Moreover, if f has spherical symmetry, that is, if $f(x) = g(|x|)$, then g satisfies

$$-g'' - \Big(\frac{r}{2} + \frac{n-1}{r}\Big)g' - a|g|^{p-1}g = \frac{1}{p-1}g, \quad g'(0) = 0.$$

In order for $f(x) = g(|x|)$ to be well behaved at $x = 0$ it is intuitively clear that we need $g'(0) = 0$. If $p > p_{Fuj}(n)$, Then, there exist due to [74] positive solutions of the form

$$u(t, x) = t^{-\frac{1}{p-1}}f\Big(\frac{x}{\sqrt{t}}\Big).$$

Remark 17.1.1 Here we used the notion of a self-similar solution. These are solutions being invariant under the dilation scaling $u(t, x) = \lambda^\alpha u(\lambda t, \lambda^\beta x)$ with suitable α and β. Such special solutions are useful in the treatment of linear or even nonlinear partial differential equations (cf., for example, with Theorem 17.2.8). They depend on variables which interlink the given variables in space-time. Therefore, some symmetry or homogeneous asymmetry in the problem is used. Often self-similar solutions solve "easier" partial differential equations or ordinary differential equations (see the above considerations).

If $u = u(t, x)$ is such a self-similar solution, then $u = u(t + \tau, x)$ is for any positive τ a self-similar solution, too. So, we have an infinite number of self-similar solutions. Later it was observed in [217] and in [50] that the semilinear elliptic equation has infinitely many solutions which are dominated by the Gaussian. But among these there is a positive one if and only if $p > p_{Fuj}(n)$.

17.1.3 A Useful Change of Variables

We introduce the operator

$$Lf := -\Delta f - \frac{1}{2}y \cdot \nabla f = -\frac{1}{K(y)}\nabla \cdot \big(K(y)\nabla f\big),$$

where

$$K(y) := \exp\Big(\frac{|y|^2}{4}\Big).$$

The domain $D(L)$ of L is contained in the weighted L^2 space

$$L^2(K)(\mathbb{R}^n) := \{f \in L^2(\mathbb{R}^n) : \|f\sqrt{K}\|_{L^2} < \infty\}.$$

Then, there is an interesting relation between the solutions of the Cauchy problems

$$v_s + Lv = |v|^{p-1}v + \lambda v, \quad v(0, y) = v_0(y), \quad (s, y) \in (0, \infty) \times \mathbb{R}^n,$$

$$u_t - \Delta u = |u|^{p-1}u, \quad u(0, x) = \varphi(x), \quad (t, x) \in (0, \infty) \times \mathbb{R}^n.$$

If $u = u(t, x)$ is a classical solution, then

$$v(s, y) = \exp\Big(\frac{s}{p-1}\Big)u\Big(\exp s - 1, \exp\Big(\frac{s}{2}\Big)y\Big)$$

is a classical solution with $\lambda = \frac{1}{p-1}$ and $v_0 = \varphi$. Conversely, if $v = v(s, y)$ is a classical solution with $\lambda = \frac{1}{p-1}$, then

$$u(t, x) = (1 + t)^{-\frac{1}{p-1}}v\Big(\log(1 + t), \frac{x}{\sqrt{1+t}}\Big)$$

is a classical solution with $\varphi = v_0$.

Remark 17.1.2 The interest in this change of variables lies in the fact that the operator L has in suitable function spaces a compact inverse, is self-adjoint and its least eigenvalue is $\frac{n}{2}$. Therefore, the problem for v can be studied in the same manner as the problem for u, but in a bounded domain $G \subset \mathbb{R}^n$.

17.1.4 Blow Up Via Global Existence

We turn now to the interplay blow up via global existence. First we prove a blow up result (see [107]).

Theorem 17.1.3 *Let us assume $p \in (1, p_{Fuj}(n))$. Then, any positive classical solution of*

$$v_s + Lv = |v|^{p-1}v + \frac{1}{p-1}v, \quad v(0, y) = v_0(y), \quad (s, y) \in (0, \infty) \times \mathbb{R}^n,$$

blows up in finite time.

Proof The function

$$\phi_1 = \phi_1(y) := \exp\left(-\frac{|y|^2}{4}\right)$$

is an eigenfunction of the operator L to the eigenvalue $\frac{n}{2}$, that is, $L\phi_1 = \frac{n}{2}\phi_1$. Then we define $\psi = \psi(y) = c(\varepsilon)\phi_1^{1+\varepsilon}$. We choose ε in such a way that

$$\int_{\mathbb{R}^n} \psi(y)K(y)\,dy = 1.$$

Then,

$$L\psi \le (1+\varepsilon)\frac{n}{2}\psi.$$

After multiplying the differential equation by $\psi(y)K(y)$, integrating over $\mathbb{R}^n$ and integrating by parts we get

$$d_s \int_{\mathbb{R}^n} v(s, y)\psi(y)K(y)\,dy = \int_{\mathbb{R}^n} v(s, y)^p \psi(y)K(y)\,dy$$

$$+ \int_{\mathbb{R}^n} v(s, y)\left(\frac{1}{p-1}\psi(y) - L\psi(y)\right)K(y)\,dy.$$

By Jensen's inequality and the fact that

$$\frac{1}{p-1}\psi(y) - L\psi(y) \ge \left(\frac{1}{p-1} - (1+\varepsilon)\frac{n}{2}\right)\psi(y)$$

we may conclude

$$d_s \int_{\mathbb{R}^n} v(s,y)\psi(y)K(y)\, dy \ge \left(\int_{\mathbb{R}^n} v(s,y)\psi(y)K(y)\, dy\right)^p$$
$$+ \left(\frac{1}{p-1} - (1+\varepsilon)\frac{n}{2}\right)\int_{\mathbb{R}^n} v(s,y)\psi(y)K(y)\, dy.$$

Taking account of $p \in (1, p_{Fuj}(n))$ a sufficiently small ε implies

$$\frac{1}{p-1} - (1+\varepsilon)\frac{n}{2} \ge 0.$$

Introducing the functional

$$F(s) := \int_{\mathbb{R}^n} v(s,y)\psi(y)K(y)\, dy$$

we derived the differential inequality

$$F'(s) \ge F(s)^p, \quad F(0) > 0.$$

Let us assume that the positive solution $v = v(s,y)$ exists on every time interval $[0, L]$. Then, with $p > 1$ and $F(s) > 0$ for all s we obtain from the differential inequality

$$\frac{1}{p-1}F(0)^{1-p} \ge \int_0^L \frac{F'(s)}{F(s)^p}\, ds \ge L.$$

This is a contradiction. So, we may conclude a blow up behavior. The proof is complete.

Using the relation between both Cauchy problems of Sect. 17.1.3 we get the same blow up result for positive solutions to the Cauchy problem

$$u_t - \Delta u = |u|^{p-1}u, \quad u(0,x) = \varphi(x), \quad (t,x) \in (0, \infty) \times \mathbb{R}^n.$$

Now we are interested in the large time behavior or the asymptotic behavior of solutions u to the last Cauchy problem. Again we consider instead the asymptotic behavior of solutions to the evolution equation for v. We use weighted Sobolev spaces

$$H^2(K)(\mathbb{R}^n) = \{f \in H^2(\mathbb{R}^n) : \partial_y^\alpha f \in L^2(K) \text{ for } |\alpha| \le 2\}$$

with $K(y) = \exp\left(\frac{|y|^2}{4}\right)$ and $L^2(K)(\mathbb{R}^n)$ as introduced in Sect. 17.1.3. The domain of the operator L is $H^2(K)(\mathbb{R}^n)$. For

$$v \in L^{p+1}(K)(\mathbb{R}^n) \cap H^1(K)(\mathbb{R}^n)$$

and fixed real λ we introduce the functional

$$E_\lambda(v) := \frac{1}{2} \int_{\mathbb{R}^n} |\nabla v(y)|^2 K(y)\, dy - \frac{\lambda}{2} \int_{\mathbb{R}^n} |v(y)|^2 K(y)\, dy$$

$$- \frac{1}{p+1} \int_{\mathbb{R}^n} |v(y)|^{p+1} K(y)\, dy.$$

This functional is also called energy, although now the energy might become negative. To formulate the next results, let $\lambda_1 := \frac{n}{2}$ be the least eigenvalue of L on $H^2(K)(\mathbb{R}^n)$. The life span T^* of a solution v belonging to

$$C\big([0, T), L^{p+1}(K)(\mathbb{R}^n) \cap H^1(K)(\mathbb{R}^n)\big) \cap C^1\big([0, T), L^2(K)(\mathbb{R}^n)\big)$$

is defined by

$$T^* := \sup_{T>0} \big\{v \in C\big([0, T), L^{p+1}(K)(\mathbb{R}^n) \cap H^1(K)(\mathbb{R}^n)\big) \cap C^1\big([0, T), L^2(K)(\mathbb{R}^n)\big)\big\}.$$

Finally, for $p \in (1, \frac{n+2}{n-2}]$ and fixed λ we define

$$a_\lambda = a_\lambda(p) := \inf_{f \in H^1(K)} \Big\{ \int_{\mathbb{R}^n} |\nabla f(y)|^2 K(y)\, dy - \lambda \int_{\mathbb{R}^n} |f(y)|^2 K(y)\, dy$$

$$: \int_{\mathbb{R}^n} |f(y)|^{p+1} K(y)\, dy = 1 \Big\}.$$

Theorem 17.1.4 *Assume that* $\lambda < \lambda_1$. *Let us choose a non-vanishing data* $v_0 \in L^{p+1}(K)(\mathbb{R}^n) \cap H^1(K)(\mathbb{R}^n)$ *in the Cauchy problem*

$$v_s + Lv = |v|^{p-1}v + \lambda v, \quad v(0, y) = v_0(y), \quad (s, y) \in (0, \infty) \times \mathbb{R}^n.$$

Then, the following statements hold:

1. *If $E_\lambda(v_0) \leq 0$, then T^* is finite, so we have blow up.*
2. *If $E_\lambda(v_0) \in \left(0, \frac{p-1}{2(p+1)}a_\lambda^{\frac{p+1}{p-1}}\right)$ and*

$$\int_{\mathbb{R}^n} |\nabla v_0(y)|^2 K(y)\, dy - \lambda \int_{\mathbb{R}^n} |v_0(y)|^2 K(y)\, dy < a_\lambda^{\frac{p+1}{p-1}},$$

then $T^* = \infty$, so we have global (in time) existence.

Theorem 17.1.5 *Assume that $\lambda > \lambda_1$ and the non-vanishing data v_0 is nonnegative in the Cauchy problem*

$$v_s + Lv = |v|^{p-1}v + \lambda v, \quad v(0, y) = v_0(y), \quad (s, y) \in (0, \infty) \times \mathbb{R}^n.$$

Then every solution

$$v \in C\big([0, T), L^{p+1}(K)(\mathbb{R}^n) \cap H^1(K)(\mathbb{R}^n)\big) \cap C^1\big([0, T), L^2(K)(\mathbb{R}^n)\big)$$

blows up in finite time T^*.

Both of the last theorems imply immediately the following two corollaries due to the change of variables of Sect. 17.1.3.

Corollary 17.1.1 *Assume that $p > p_{Fuj}(n)$. Let us choose a non-vanishing data $\varphi \in L^{p+1}(K)(\mathbb{R}^n) \cap H^1(K)(\mathbb{R}^n)$ in the Cauchy problem*

$$u_t - \Delta u = |u|^{p-1}u, \quad u(0, x) = \varphi(x), \quad (t, x) \in (0, \infty) \times \mathbb{R}^n.$$

Then, the following statements hold with $\lambda = \frac{1}{p-1}$:

1. *If $E_\lambda(\varphi) \leq 0$, then T^* is finite, so we have blow up.*
2. *If $E_\lambda(\varphi) \in \left(0, \frac{p-1}{2(p+1)}a_\lambda^{\frac{p+1}{p-1}}\right)$ and*

$$\int_{\mathbb{R}^n} |\nabla \varphi(y)|^2 K(y)\, dy - \lambda \int_{\mathbb{R}^n} |\varphi(y)|^2 K(y)\, dy < a_\lambda^{\frac{p+1}{p-1}},$$

then $T^* = \infty$, so we have global (in time) existence.

Corollary 17.1.2 *Assume that $1 < p < p_{Fuj}(n)$ and the non-vanishing data φ is nonnegative in the Cauchy problem*

$$u_t - \Delta u = |u|^{p-1}u, \quad u(0, x) = \varphi(x), \quad (t, x) \in (0, \infty) \times \mathbb{R}^n.$$

Then every solution

$$u \in C\big([0, T), L^{p+1}(K)(\mathbb{R}^n) \cap H^1(K)(\mathbb{R}^n)\big) \cap C^1\big([0, T), L^2(K)(\mathbb{R}^n)\big)$$

blows up in finite time T^*.

Remark 17.1.3 In [107] it is shown that Corollary 17.1.2 remains valid for $p = p_{Fuj}(n)$. The statements of Corollary 17.1.1 explain that a nonpositive "energy" $E_{\frac{1}{p-1}}(\varphi)$ of the data φ has a deteriorating influence on global (in time) existence results while a positive small energy $E_{\frac{1}{p-1}}(\varphi)$ has an improving influence on global (in time) existence results.

Besides global existence results, the question of the long time behavior or the asymptotic profile of solutions is of interest. We shall only explain two results from [107]. The interested reader can follow the results with proofs in this paper.

On the one hand, it is shown in [107] that under the assumptions $\varphi \in H^1(K)(\mathbb{R}^n)$ and $p \in \big(p_{Fuj}(n), \frac{n+2}{n-2}\big)$ every classical solution u of

$$u_t - \Delta u = |u|^{p-1}u, \quad u(0, x) = \varphi(x), \quad (t, x) \in (0, \infty) \times \mathbb{R}^n$$

satisfies the estimate

$$\sup_{t \geq 1} t^{\frac{1}{p-1}}\big(\|u(t, \cdot)\|_{L^\infty} + \|\nabla u(t, \cdot)\|_{L^\infty} + \|\nabla^2 u(t, \cdot)\|_{L^\infty}\big) < \infty.$$

On the other hand, it is shown that there exists a close connection from the profile to a self-similar solution

$$t^{-\frac{1}{p-1}} f\Big(\frac{x}{\sqrt{t}}\Big) \text{ of } u_t - \Delta u = |u|^{p-1}u.$$

It is shown that there is a self-similar solution with $f \in H^1(K)(\mathbb{R}^n)$ and a sequence $\{t_k\}_k$ tending to infinity such that the relation

$$\lim_{t_k \to \infty} \Big\| t_k^{\frac{1}{p-1}} u(t_k, \cdot) - f\Big(\frac{\cdot}{\sqrt{t_k}}\Big)\Big\|_{L^\infty} = 0$$

is true.

In the case of a source nonlinearity we renounced, up to now, to present some $L^p - L^q$ estimates for global (in time) solutions. We expect the following result on global (in time) small data solutions to be similar to Theorem 18.1.1 under the assumption of additional regularity for the data.

Theorem 17.1.6 *Consider the Cauchy problem with source power nonlinearity*

$$u_t - \Delta u = |u|^{p-1}u, \quad u(0, x) = \varphi(x).$$

Let $p > p_{Fuj}(n) = 1 + \frac{2}{n}$ and $\varphi \in L^1(\mathbb{R}^n) \cap L^\infty(\mathbb{R}^n)$. Then, the following statement holds with a suitable constant $\epsilon_0 > 0$:

If $\|\varphi\|_{L^1 \cap L^\infty} \leq \epsilon_0$, then there exists a unique global (in time) energy solution

$$u \in C\big([0, \infty), L^2(\mathbb{R}^n) \cap L^\infty(\mathbb{R}^n)\big).$$

Moreover, there exists a constant $C > 0$ such that the solution satisfies the decay estimates (see also [152])

$$\|u(t, \cdot)\|_{L^2} \leq C(1 + t)^{-\frac{n}{4}} \|\varphi\|_{L^1 \cap L^\infty},$$

$$\|u(t, \cdot)\|_{L^\infty} \leq C(1 + t)^{-\frac{n}{2}} \|\varphi\|_{L^1 \cap L^\infty}.$$

The interested reader will find a research project related to semilinear heat models in Sect. 23.10.

17.2 Semilinear Heat Models with Absorbing Power Nonlinearity

Consider for $p > 1$ the Cauchy problem with absorbing power nonlinearity (cf. with Sect. 11.2.5)

$$u_t - \Delta u = -|u|^{p-1}u, \quad u(0, x) = \varphi(x).$$

Suppose that we choose the data $\varphi \equiv 1$. Then, there exists the time-dependent solution $u_1(t) = ((p-1)t)^{-\frac{1}{p-1}}$. Moreover, there exists the super-solution

$$u_2(t, x) = \frac{1}{(4\pi t)^{\frac{n}{2}}} \int_{\mathbb{R}^n} e^{-\frac{|x-y|^2}{4t}} \, dy,$$

that is, a solution of the differential inequality

$$u_t - \Delta u + |u|^{p-1}u \geq 0.$$

Let us compare the asymptotical behavior of both solutions for $t \to \infty$. If $p \in (1, 1 + \frac{2}{n})$, then the asymptotical behavior of u_1 prevails, but oppositely, if $p > 1 + \frac{2}{n}$, the asymptotical behavior of u_2 prevails. Both behavior coincide in the case $p = 1 + \frac{2}{n} = p_{Fuj}(n)$. Consequently, for this Cauchy problem the critical exponent with respect to the asymptotical behavior is the Fujita exponent. The results of the following sections can be found in [52] and [67] (see also the survey article [153]) and references therein.

17.2.1 Well-Posedness Results for the Cauchy Problem

We introduce the notation $L_0^\infty(\mathbb{R}^n)$ as the closure of $C_0^\infty(\mathbb{R}^n)$ with respect to the L^∞-norm. Define for $q \in [1, \infty]$ and $p > 1$ the family of operators

$$A_q : u \in D(A_q) \to A_q u := -\Delta u + |u|^{p-1} u$$

with the domain of definition

$$D(A_q) = \{u \in L^q(\mathbb{R}^n) \cap L^{pq}(\mathbb{R}^n) : \Delta u \in L^q(\mathbb{R}^n)\} \text{ for } q \in [1, \infty),$$

$$D(A_\infty) = \{u \in L_0^\infty(\mathbb{R}^n) : \Delta u \in L_0^\infty(\mathbb{R}^n)\} \text{ for } q = \infty.$$

In the following some properties of the family $\{A_q\}_q$ of elliptic operators A_q is used. Due to [15], the operator A_q is m-T-accretive in $L^q(\mathbb{R}^n)$ for $q \in [1, \infty)$ or in $L_0^\infty(\mathbb{R}^n)$ in the sense that for any $\lambda > 0$ the operator $I + \lambda A_q$ is surjective. Moreover, the estimate

$$\|(u - v)^+\|_{L^q} \leq \|((I + \lambda A_q)(u - v))^+\|_{L^q} \text{ for all } u, v \in D(A_q)$$

is valid. Here $(u)^+$ denotes the positive part of u.

To get the existence of a solution to the Cauchy problem for the semilinear heat equation with absorbing power nonlinearity a result of [28] is used. Rothe's method (semi-discretization in time) is applied. Therefore, we fix $T > 0$ and consider the implicit iteration scheme

$$\frac{u_N^{(k)} - u_N^{(k-1)}}{t_N^{(k)} - t_N^{(k-1)}} + A_q u_N^{(k)} = 0 \text{ for } k = 1, \cdots, N,$$

$$u_N^{(0)} = \varphi \in L^q(\mathbb{R}^n) \ (\in L_0^\infty(\mathbb{R}^n)),$$

where $0 = t_N^{(0)} < t_N^{(1)} < \cdots < t_N^{(N-1)} < t_N^{(N)} = T$. The property of A_q to be of m-T-accretive implies that the sequence $\{u_N^{(k)}\}_k$ is well-defined. If we call u_N the step function taking the value $u_N^{(k)}$ on $(t_N^{(k-1)}, t_N^{(k)}]$ and if

$$\lim_{N \to \infty} \max_{1 \leq k \leq N} (t_N^{(k)} - t_N^{(k-1)}) = 0,$$

then the sequence $\{u_N\}_N$ converges uniformly on $[0, T]$ to a continuous function $u = u(t)$. Introducing the notation $u(t) =: S(t)\varphi$, the family of operators $\{S(t)\}_{t \geq 0}$ defines a semigroup of order-preserving contractions on $L^q(\mathbb{R}^n)$, $L_0^\infty(\mathbb{R}^n)$, respectively. When $q \in [1, \infty)$ the function u is even a strong solution (the reader should become familiar with this notion of solutions in the theory of semigroups) of the abstract

Cauchy problem

$$u_t + A_q u = 0, \quad u(0) = \varphi.$$

Summarizing all these observations we may conclude the following result.

Theorem 17.2.1 *Suppose $q \in [1, \infty)$. Let $\varphi \in D(A_q)$ and $u = u(t, \cdot)$ be the solution to the abstract Cauchy problem*

$$u_t + A_q u = 0, \quad u(0) = \varphi.$$

Then, $u(t, \cdot) \in D(A_q)$, $u_t(t, \cdot) \in L^q(\mathbb{R}^n)$ and u satisfies a.e. on $(0, \infty) \times \mathbb{R}^n$ the Cauchy problem

$$u_t - \Delta u = -|u|^{p-1} u, \quad u(0, x) = \varphi(x).$$

Moreover, the following $L^q - L^r$ estimates can be proved (see [67]).

Theorem 17.2.2 *Consider for $p > 1$ the Cauchy problem with absorbing power nonlinearity*

$$u_t - \Delta u = -|u|^{p-1} u, \quad u(0, x) = \varphi(x).$$

Let $\varphi \in L^q(\mathbb{R}^n)$, $q \in [1, \infty)$. Then, there exists a Sobolev solution satisfying

$$u(t, \cdot) \in L^q(\mathbb{R}^n) \cap L_0^\infty(\mathbb{R}^n).$$

Moreover, we have the estimates

$$\|u(t, \cdot)\|_{L^r} \leq C(n) t^{-\frac{n}{2}(\frac{1}{q} - \frac{1}{r})} \|\varphi\|_{L^q}$$

for all $t \in (0, \infty)$ and $r \in [q, \infty]$. If $q \in [1, 2]$, then $u_t(t, \cdot) \in L^2(\mathbb{R}^n) \cap L_0^\infty(\mathbb{R}^n)$ and the following estimates hold:

$$\|u_t(t, \cdot)\|_{L^r} \leq C(n) t^{-\frac{n}{2}(\frac{1}{q} - \frac{1}{r}) - 1} \|\varphi\|_{L^q}$$

for all $t \in (0, \infty)$ and $r \in [2, \infty]$.

Remark 17.2.1 Let us compare the estimates of Theorems 17.2.2 and 17.1.6. In the case of an absorbing nonlinearity we have the global existence of large data Sobolev solutions in all dimensions for all $p > 1$ and for all data $\varphi \in L^q(\mathbb{R}^n)$, $q \in [1, \infty)$. Oppositely, in the case of a source nonlinearity we have the global existence of small data Sobolev solutions for some range of admissible $p > p_{Fuj}(n)$ only if the data is supposed to belong to $L^1(\mathbb{R}^n) \cap L^\infty(\mathbb{R}^n)$. We note the Fujita exponent in Theorem 17.1.6.

17.2.2   Influence of the Fujita Exponent on the Profile of Solutions

Now let the data φ belong to $L^1(\mathbb{R}^n)$. Then, due to Theorem 17.2.2, there exists a global (in time) solution $u(t, \cdot) \in L^1(\mathbb{R}^n) \cap L_0^\infty(\mathbb{R}^n)$ with $u_t(t, \cdot) \in L^2(\mathbb{R}^n) \cap L_0^\infty(\mathbb{R}^n)$. Using the estimates of Theorem 17.2.2 we may conclude that the solution u belongs to $C([T, \infty], L_0^\infty(\mathbb{R}^n))$ for all $T > 0$. The following theorems give some more precise statements about the asymptotical profile of solutions (see [67]). Here we feel an influence of the Fujita exponent at some threshold. For exponents larger than the Fujita exponent the profile is described by the Gauss kernel. On the contrary, for exponents below the Fujita exponent the profile is described by the aid of a self-similar solution.

Theorem 17.2.3 *Consider the Cauchy problem with absorbing power nonlinearity*

$$u_t - \Delta u = -|u|^{p-1}u, \quad u(0, x) = \varphi(x).$$

Suppose $\varphi \in L^1(\mathbb{R}^n)$ and $p > p_{Fuj}(n)$. Then, the solution $u = u(t, x)$ satisfies the relation

$$\lim_{t \to \infty} \sup_{|x| \le C\sqrt{t}} \left| t^{\frac{n}{2}} u(t, x) - \theta_0 \frac{1}{(4\pi)^{\frac{n}{2}}} \exp\left(-\frac{|x|^2}{4t}\right) \right| = 0,$$

where C is an arbitrary positive constant and

$$\theta_0 = \int_{\mathbb{R}^n} \varphi(x) \, dx - \int_{\mathbb{R}^{n+1}_+} |u|^{p-1} u \, d(t, x).$$

Proof In order to prove this result we introduce for all $k \in \mathbb{N}$ the scaling transformation

$$u_k(t, x) = k^{\frac{n}{2}} u(kt, \sqrt{k}x).$$

Then, every u_k is in $[0, \infty) \times \mathbb{R}^n$ is a solution to the Cauchy problem

$$\partial_t u_k - \Delta u_k + k^{1 - \frac{(p-1)n}{2}} |u_k|^{p-1} u_k = 0, \quad u_k(0, x) = k^{\frac{n}{2}} u_0(\sqrt{k}x).$$

Step 1:   An auxiliary result

Theorem 17.2.4 *The sequence $\{u_k\}_k$ converges uniformly on any compact set of $\mathbb{R}^{n+1}_+$ to the function*

$$W(t, x) = \theta_0 \frac{1}{(4\pi)^{\frac{n}{2}}} \exp\left(-\frac{|x|^2}{4t}\right),$$

where θ_0 is defined as in the statement of Theorem 17.2.3.

Step 2:   Proof of Theorem 17.2.3

From the statement of the last Theorem 17.2.4 it follows that $\{u_k(t', x')\}_k$ converges to $W(t', x')$ uniformly on every compact subset of $\mathbb{R}^{n+1}_+$. If we take $t' = 1$ and replace k by t, then $t^{\frac{n}{2}} u(t, \sqrt{t}x')$ converges for $t \to \infty$ to $W(1, x')$ uniformly on compact sets, in particular, on closed balls around the origin in $\mathbb{R}^n$. Introducing $x := \sqrt{t}x'$ allows us immediately to conclude the statement of Theorem 17.2.3.

Step 3:   Sketch of the proof of Theorem 17.2.4

The proof is divided into several steps.

Step 3.1:   Estimates for $\{u_k\}_k$

From Theorem 17.2.2 the following estimates hold for $k \in \mathbb{N}$ and $t > 0$:

$$\|u_k(t, \cdot)\|_{L^\infty} \le Ct^{-\frac{n}{2}} \|\varphi\|_{L^1}, \quad \|\partial_t u_k(t, \cdot)\|_{L^\infty} \le Ct^{-\frac{n}{2}-1} \|\varphi\|_{L^1},$$

$$\|u_k(0, \cdot)\|_{L^1} = \|\varphi\|_{L^1}.$$

Using the definition of u_k we have

$$\Delta u_k = \partial_t u_k + k^{1 - \frac{(p-1)n}{2}} |u_k|^{p-1} u_k.$$

Taking account of $1 - \frac{(p-1)n}{2} < 0$, the above uniform estimates for u_k and $\partial_t u_k$ yield

$$\|\Delta u_k(t, \cdot)\|_{L^\infty} \le C_\delta \text{ for all } t \in [\delta, \infty),$$

where the constant C_δ is a suitable nonnegative constant which depends on $\delta > 0$.

Step 3.2:   Convergence of a subsequence

Applying the regularity theory for solutions to the Poisson equation $\Delta u = f$, compact embeddings in Sobolev spaces and a diagonalization procedure, there exists a function $W = W(t, x) \in C^1(\mathbb{R}^{n+1}_+)$ such that a subsequence $\{u_{k_n}\}_n$ of $\{u_k\}_k$ converges to W in $C^1_{loc}(\mathbb{R}^{n+1}_+)$. Moreover, the sequences $\{\partial_t u_{k_n}\}_n$ and $\{\Delta u_{k_n}\}_n$ converge to $\partial_t W$ and ΔW, respectively, in the weak-star topology of $L^\infty((\delta, \infty) \times \mathbb{R}^n)$ for any positive δ (see also [1]). Finally, we have

$$\int_{(0,T)\times\mathbb{R}^n} |u_k(t, x)| \, d(t, x) = \int_{(0,T)\times\mathbb{R}^n} |u(kt, x)| \, d(t, x) \le T\|\varphi\|_{L^1}$$

for any positive T, which implies

$$\int_{(0,T)\times\mathbb{R}^n} |W(t, x)| \, d(t, x) \le T\|\varphi\|_{L^1}.$$

Step 3.3:   *Integral relation for the limit*

We choose $T > 0$ and a test function $\phi \in C_0^\infty([0, T) \times \mathbb{R}^n)$. so ϕ has compact support in $[0, T) \times \mathbb{R}^n$. Using the Cauchy problem for u_k we have

$$\int_{(0,T)\times\mathbb{R}^n} \partial_t u_k \phi \, d(t, x) - \int_{(0,T)\times\mathbb{R}^n} \Delta u_k \phi \, d(t, x)$$

$$+ k^{1-\frac{(p-1)n}{2}} \int_{(0,T)\times\mathbb{R}^n} |u_k|^{p-1} u_k \phi \, d(t, x) = 0.$$

The estimate for Δu_k of Step 3.1 and the compact support property of ϕ allow us to verify

$$-\int_{\mathbb{R}^n} \Delta u_k \phi \, dx = -\int_{\mathbb{R}^n} u_k \Delta \phi \, dx.$$

Integration by parts gives

$$\int_{(0,T)\times\mathbb{R}^n} \partial_t u_k \phi \, d(t, x) = -\int_{(0,T)\times\mathbb{R}^n} u_k \partial_t \phi \, d(t, x) - \int_{\mathbb{R}^n} u_k(0, x) \phi(0, x) \, dx.$$

But

$$\int_{\mathbb{R}^n} u_k(0, x) \phi(0, x) \, dx = \int_{\mathbb{R}^n} \varphi(x) \phi(0, x/\sqrt{k}) \, dx.$$

After applying Lebesgue's theorem, it follows

$$\lim_{k\to\infty} \int_{\mathbb{R}^n} u_k(0, x) \phi(0, x) \, dx = \phi(0, 0) \int_{\mathbb{R}^n} \varphi(x) \, dx.$$

In the same way we have

$$k^{1-\frac{(p-1)n}{2}} \int_{(0,T)\times\mathbb{R}^n} |u_k|^{p-1} u_k \, \phi \, d(t, x)$$

$$= \int_{(0,kT)\times\mathbb{R}^n} |u|^{p-1} u \, \phi\left(t/k, x/\sqrt{k}\right) d(t, x).$$

Due to Theorem 17.2.2 the solution u belongs to $L^p(\mathbb{R}_+^{n+1})$. Hence,

$$\lim_{k\to\infty} k^{1-\frac{(p-1)n}{2}} \int_{(0,T)\times\mathbb{R}^n} |u_k|^{p-1} u_k \, \phi \, d(t, x) = \phi(0, 0) \int_{(0,\infty)\times\mathbb{R}^n} |u|^{p-1} u \, d(t, x).$$

Finally, we derive an integral relation for W. Set $\varepsilon > 0$. Then,

$$\left| \int_{(0,T)\times\mathbb{R}^n} (u_k - W)\partial_t\phi \, d(t,x) \right| \leq \left| \int_{(\varepsilon,T)\times\mathbb{R}^n} (u_k - W)\partial_t\phi \, d(t,x) \right|$$

$$+ \left| \int_{(0,\varepsilon)\times\mathbb{R}^n} u_k \partial_t\phi \, d(t,x) \right| + \left| \int_{(0,\varepsilon)\times\mathbb{R}^n} W\partial_t\phi \, d(t,x) \right|.$$

Together with the estimates of the previous step we deduce

$$\left| \int_{(0,T)\times\mathbb{R}^n} (u_k - W)\partial_t\phi \, d(t,x) \right| \leq \left| \int_{(\varepsilon,T)\times\mathbb{R}^n} (u_k - W)\partial_t\phi \, d(t,x) \right|$$

$$+ 2\varepsilon \|\varphi\|_{L^1} \|\partial_t\phi\|_{L^\infty}.$$

If we replace u_k by u_{k_n} and choose ε arbitrarily small, then we may conclude

$$\lim_{k_n\to\infty} \int_{(0,T)\times\mathbb{R}^n} u_{k_n} \partial_t\phi \, d(t,x) = \int_{(0,T)\times\mathbb{R}^n} W\partial_t\phi \, d(t,x).$$

In the same way we obtain

$$\lim_{k_n\to\infty} \int_{(0,T)\times\mathbb{R}^n} u_{k_n} \Delta\phi \, d(t,x) = \int_{(0,T)\times\mathbb{R}^n} W\Delta\phi \, d(t,x).$$

Finally, if we replace k by k_n in

$$-\int_{(0,T)\times\mathbb{R}^n} u_k\partial_t\phi \, d(t,x) - \int_{\mathbb{R}^n} u_k(0,x)\phi(0,x) \, dx$$

$$-\int_{(0,T)\times\mathbb{R}^n} u_k\Delta\phi \, d(t,x) + k^{1-\frac{(p-1)n}{2}} \int_{(0,T)\times\mathbb{R}^n} |u_k|^{p-1} u_k \, \phi \, d(t,x) = 0$$

and using all the relations of this step, we conclude the desired integral relation for W:

$$-\int_{(0,T)\times\mathbb{R}^n} W\partial_t\phi \, d(t,x) - \theta_0 \, \phi(0,0) - \int_{(0,T)\times\mathbb{R}^n} W\Delta\phi \, d(t,x) = 0,$$

where

$$\theta_0 = \int_{\mathbb{R}^n} \varphi(x) \, dx - \int_{\mathbb{R}^{n+1}_+} |u|^{p-1} u \, d(t,x).$$

Step 3.4:   Conclusion

From the last integral relation we get that $W = W(t, x)$ is a solution of

$$\partial_t W - \Delta W = 0, \quad W(0, x) = \theta_0 \delta_0,$$

where δ_0 is Dirac's delta distribution in $x = 0$ (see Example 24.4.2). Thus,

$$W(t, x) = \theta_0 \frac{1}{(4\pi)^{\frac{n}{2}}} \exp\left(-\frac{|x|^2}{4t}\right).$$

Taking into consideration the uniqueness of solutions to our Cauchy problem with absorbing power nonlinearity our convergence result is true for the whole sequence $\{u_k\}_k$.

Remark 17.2.2 The statement of Theorem 17.2.3 shows in the supercritical case, $p > p_{Fuj}(n)$, a relation between the solution of

$$u_t - \Delta u = -|u|^{p-1} u, \quad u(0, x) = \varphi(x)$$

and the Gauss kernel as explained at the beginning of this section. If we set $p = p_{Fuj}(n) = 1 + \frac{2}{n}$, then under the assumptions of Theorem 17.2.3 we have

$$\int_{\mathbb{R}^n} \varphi(x)\, dx = \int_{\mathbb{R}^{n+1}_+} |u|^{\frac{2}{n}} u\, d(t, x),$$

that is, $\theta_0 = 0$ (see also the following result and [67]).

Theorem 17.2.5 *Consider the Cauchy problem of Theorem 17.2.3. Suppose $\varphi \in L^1(\mathbb{R}^n)$ and $p = p_{Fuj}(n)$. Then, the solution $u = u(t, x)$ satisfies the relation*

$$\lim_{t \to \infty} \sup_{|x| \le C\sqrt{t}} t^{\frac{n}{2}} |u(t, x)| = 0,$$

where C is an arbitrary positive constant.

Proof As in the proof to Theorem 17.2.3 we define a sequence $\{u_k\}_k$, where every u_k is in $[0, \infty) \times \mathbb{R}^n$ a solution to the Cauchy problem

$$\partial_t u_k - \Delta u_k + |u_k|^{\frac{2}{n}} u_k = 0, \quad u_k(0, x) = k^{\frac{n}{2}} u_0(\sqrt{k}x).$$

Because of

$$\lim_{k \to \infty} u_k(0, x) = \delta_0 \int_{\mathbb{R}^n} \varphi(x)\, dx,$$

we conclude from a result of [14] that $\{u_k = u_k(t', x')\}_k$ converges to 0 for $k \to \infty$ uniformly on any compact set of $\mathbb{R}^{n+1}_{t',x'}$. If we take $t' = 1$, $k = t$ and $\sqrt{t}x' = x$, then we deduce the desired result.

We do not know the exact decay rate in the last result. Though at least, for nonnegative solutions we have some kind of optimality.

Theorem 17.2.6 *Consider the Cauchy problem of Theorem 17.2.3. Suppose that the non-vanishing $\varphi \in L^1(\mathbb{R}^n) \cap D(A_2)$ is nonnegative and $p = p_{Fuj}(n)$. Then, the solution $u = u(t, x)$ satisfies the following estimate:*

$$\inf_{\{(t,x)\in[T,\infty)\times K\}} (t \log t)^{\frac{n}{2}} u(t, x) \geq C(T, K)$$

for all compact sets $K \in \mathbb{R}^n$ and all $T > 1$.

Proof Fix $T > 1$. We define $v(t, x) := (\log(t + T))^{\frac{n}{2}} u(t, x)$. Then v satisfies the semilinear heat equation

$$\partial_t v - \Delta v - \frac{n}{2(t + T) \log(t + T)} v + \frac{1}{\log(t + T)} v^{1+\frac{2}{n}} = 0.$$

Let $W = W(t, x)$ be the solution of the Cauchy problem for the heat equation with nonnegative data $W_0 \in L^1(\mathbb{R}^n) \cap L_0^\infty(\mathbb{R}^n)$. So,

$$W(t, x) = \frac{1}{(4\pi t)^{\frac{n}{2}}} \int_{\mathbb{R}^n} \exp\left(-\frac{|x - y|^2}{4t}\right) W_0(y) \, dy.$$

Then,

$$\partial_t(W - v) - \Delta(W - v) - \frac{n}{2(t + T) \log(t + T)} (W - v)$$

$$+ \frac{1}{\log(t + T)} \left(W^{1+\frac{2}{n}} - v^{1+\frac{2}{n}}\right) + \frac{1}{\log(t + T)} \left(\frac{n}{2(t + T)} - W^{\frac{2}{n}}\right) W = 0.$$

The above representation for W (W_0 is nonnegative) implies the estimates

$$W(t, x) \leq \frac{1}{(4\pi t)^{\frac{n}{2}}} \|W_0\|_{L^1}, \quad W(t, x) \leq \|W_0\|_{L^\infty}.$$

Now we choose the data W_0 in such a way that the following conditions are satisfied:

$$\|W_0\|_{L^\infty} \leq \left(\frac{n}{4T}\right)^{\frac{n}{2}}, \quad \|W_0\|_{L^1} \leq (\pi n)^{\frac{n}{2}},$$

$$W_0(x) \leq (\log T)^{\frac{n}{2}} \varphi(x) \quad \text{almost everywhere.}$$

Taking into consideration the first two conditions we get

$$(W(t,x))^{\frac{2}{n}} \leq \frac{n}{2(t+T)}.$$

Together with the above equation for $W - v$, it follows

$$\partial_t(W-v) - \Delta(W-v) - \frac{n}{2(t+T)\log(t+T)}(W-v)$$

$$+\frac{1}{\log(t+T)}\left(W^{1+\frac{2}{n}} - v^{1+\frac{2}{n}}\right) \leq 0.$$

In the next step we use two conditions for a given function $f \in W_{2,loc}^1((0,\infty) \times \mathbb{R}^n)$ (this implies $f \in W_{1,loc}^1((0,\infty) \times \mathbb{R}^n)$, too) (cf. with Definitions 24.22 and 24.24). If $f \in W_{2,loc}^1((0,\infty) \times \mathbb{R}^n)$, by using that $f^+f = f^2$ if $f \geq 0$ and zero otherwise, we get on the one hand for $\phi \in C_0^\infty((0,\infty) \times \mathbb{R}^n)$ the relation

$$\int_{(0,\infty)\times\mathbb{R}^n} \partial_t(f^+f)\phi\, d(t,x) = -\int_{\{(t,x):f(t,x)\geq 0\}} f^2\partial_t\phi\, d(t,x)$$

$$= -\int_{\{(t,x):f(t,x)\geq 0\}} f\big(\partial_t(\phi f) - \phi\partial_t f\big)\, d(t,x)$$

$$= \int_{\{(t,x):f(t,x)\geq 0\}} 2f(\partial_t f)\phi\, d(t,x) = \int_{(0,\infty)\times\mathbb{R}^n} 2f^+\partial_t f\phi\, d(t,x).$$

The last relation implies

$$\partial_t(f^+f) = 2f^+\partial_t f \quad \text{in the weak sense.}$$

On the other hand we have for $f \in L^\infty((0,\infty), W_{2,loc}^2(\mathbb{R}^n))$ and $\phi \in C_0^\infty((0,\infty) \times \mathbb{R}^n)$ the relations

$$\int_{(0,\infty)\times\mathbb{R}^n} \partial_{x_j}^2((f^+)^2)\phi\, d(t,x) = \int_{\{(t,x):f(t,x)\geq 0\}} \partial_{x_j}^2 f^2\phi\, d(t,x)$$

$$= -2\int_{\{(t,x):f((t,x)\geq 0\}} f\partial_{x_j} f\partial_{x_j}\phi\, d(t,x)$$

$$= -2\int_{\{(t,x):f((t,x)\geq 0\}} \big(\partial_{x_j}(f\phi) - \phi\partial_{x_j} f\big)\partial_{x_j} f\, d(t,x)$$

$$= 2\int_{\{(t,x):f(t,x)\geq 0\}} \big(f\partial_{x_j}^2 f - \partial_{x_j} f\partial_{x_j} f\big)\phi\, d(t,x)$$

$$= 2\int_{(0,\infty)\times\mathbb{R}^n} \big(f^+\partial_{x_j}^2 f - (\partial_{x_j} f^+)^2\big)\phi\, d(t,x).$$

The last relation implies

$$\partial_{x_j}^2((f^+)^2) = f^+ \partial_{x_j}^2 f - (\partial_{x_j} f^+)^2 \quad \text{in the weak sense.}$$

Multiplying the above inequality for $W - v$ by the positive part $(W - v)^+$ of $W - v$, taking account of

$$W - v \in W_{2,loc}^1((0, \infty) \times \mathbb{R}^n) \cap L^\infty((0, \infty), W_{2,loc}^2(\mathbb{R}^n)),$$

so

$$(W - v)^+ \partial_t (W - v) = \frac{1}{2} \partial_t((W - v)^+)^2,$$

$$(W - v)^+ \Delta(W - v) = \frac{1}{2} \Delta((W - v)^+)^2 - |\nabla(W - v)^+|^2,$$

and using the monotonicity of the function $r \to r|r|^{\frac{2}{n}}$ yield

$$\partial_t((W - v)^+)^2 - \Delta((W - v)^+)^2 - \frac{n}{(t + T) \log(t + T)}((W - v)^+)^2 \leq 0.$$

After putting $Z(t, x) := (\log(t + T))^n((W - v)^+(t, x))^2$ we derived

$$\partial_t Z - \Delta Z \leq 0.$$

Thanks to Theorem 17.2.5, we have that $Z(t, x)$ goes to zero for large t. Hence, by using the weak maximum principle (cf. with Remark 9.2.1) we conclude

$$\sup_{\{(t,x) \in [T,\infty) \times K\}} Z(t, x) \leq \sup_{x \in K} Z(0, x).$$

But the third condition for W_0 from above yields $(W - v)^+(0, \cdot) \equiv 0$, consequently, $(W - v)^+(t, \cdot) \equiv 0$. It follows $Z \equiv 0$. Thus $W(t, x) \leq v(t, x)$ almost everywhere in $\mathbb{R}_+^{n+1}$. Let K be a compact subset of $\mathbb{R}^n$ and $T > 1$. Then, we have for all $(t, x) \in [T, \infty) \times K$ the estimate

$$t^{\frac{n}{2}} W(t, x) \geq \frac{1}{(4\pi)^{\frac{n}{2}}} \int_{\mathbb{R}^n} \exp\left(-\frac{d^2 + |y|^2}{4T}\right) W_0(y) \, dy =: C(T, K),$$

where $d := \max\{|x| : x \in K\}$. Summarizing all estimates we conclude the desired estimate

$$(t \log t)^{\frac{n}{2}} u(t, x) \geq C(T) t^{\frac{n}{2}} v(t, x) \geq t^{\frac{n}{2}} W(t, x) \geq C(T, K)$$

for all $(t, x) \in [T, \infty) \times K$, where K is an arbitrary compact set in $\mathbb{R}^n$ and $T > 1$ is an arbitrary given constant. This gives the desired estimate. The proof is complete.

Finally, let us devote some discussion to the subcritical case $p \in (1, p_{Fuj}(n))$. Here we have no relation between the profile of solutions and the Gauss kernel anymore, but instead the profile of the solution is similar to the profile of a self-similar solution of $u_t - \Delta u + |u|^{p-1}u = 0$ (see [52]). Actually, it has been proved in [16] that for all $p \in (1, p_{Fuj}(n))$ there exists a family of positive self-similar solutions $v_b = v_b(t, x)$ such that for all $t > 0$ the limit

$$\lim_{|x| \to \infty} |x|^{\frac{2}{p-1}} v_b(t, x) =: b \geq 0$$

exists (for the case $b = 0$ see also [52]). These solutions are of interest because they describe the asymptotical behavior of solutions of the Cauchy problem for a classical heat equation with absorbing power nonlinearity under additional assumptions on the data φ for $|x| \to \infty$.

Let us assume in the following

$$\varphi \in L^1(\mathbb{R}^n), \quad \varphi \not\equiv 0, \quad \lim_{|x| \to \infty} |x|^{\frac{2}{p-1}} \varphi(x) = b \geq 0.$$

Note, that if $p \in (1, p_{Fuj}(n))$, then the decay condition of φ at infinity is compatible with the condition $\varphi \in L^1(\mathbb{R}^n)$.

Recall Sect. 17.1.2 where we introduced spherically symmetric self-similar solutions to semilinear heat models with power nonlinearity. Following those given explanations we are now interested in solutions to

$$-g'' - \left(\frac{r}{2} + \frac{n-1}{r}\right)g' + |g|^{p-1}g = \frac{1}{p-1}g,$$

$$g(0) = a, \quad g'(0) = 0.$$

For the following results we refer to [52]. To formulate the first result we define with $a^* := (p-1)^{-\frac{1}{p-1}}$ the value $a_0 := \inf\{a : a \in (0, a^*)\}$ such that there exists a positive solution $g_a = g_a(r)$ to the last Cauchy problem.

Theorem 17.2.7 *Let $p \in (1, p_{Fuj}(n))$. Then, $a_0 \in (0, a^*)$ and for any $a \in [a_0, a^*)$ there exists a positive and decreasing solution $g_a = g_a(r)$. Moreover, the limit*

$$C(a) := \lim_{r \to \infty} r^{\frac{2}{p-1}} g_a(r)$$

exists, $C(a) \in [0, \infty)$ and $C_a = 0$ if and only if $a = a_0$. The mapping $a \to C(a)$ is continuous, increasing and convex.

In the further considerations we use the notation $g_{a,C(a)}$ instead of g_a.

Theorem 17.2.8 *Let the non-vanishing data φ belong to $L^1(\mathbb{R}^n)$. Moreover, we assume*

$$\lim_{|x|\to\infty} |x|^{\frac{2}{p-1}} \varphi(x) = b \geq 0,$$

and φ is supposed to be nonnegative if $b = 0$. If $1 < p < p_{Fuj}(n)$, then the asymptotic profile as $t \to \infty$ of the solution to

$$u_t - \Delta u + |u|^{p-1}u = 0, \quad u(0,x) = \varphi(x)$$

can be described as follows:

$$\lim_{t\to\infty} t^{\frac{1}{p-1}} \|u(t,\cdot) - v_b(t,\cdot)\|_{L^\infty} = 0,$$

where v_b is a self-similar solution with the profile $g_{a,b}$, that is,

$$v_b(t,x) = (1+t)^{-\frac{1}{p-1}} g_{a,b}\left(\frac{|x|}{\sqrt{1+t}}\right).$$

The profile $g_{a,b}$ satisfies the following ordinary differential equation with additional conditions:

$$-g'' - \left(\frac{r}{2} + \frac{n-1}{r}\right)g' + |g|^{p-1}g = \frac{1}{p-1}g,$$

$$g(0) = a, \quad g'(0) = 0, \quad \lim_{r\to\infty} r^{\frac{2}{p-1}} g(r) = b.$$

Finally, the self-similar solution v_b satisfies for all $q \in [1,\infty]$ the decay estimates

$$\|v_b(t,\cdot)\|_{L^q} \leq Ct^{-\frac{1}{p-1}+\frac{n}{2q}} \text{ for } t \to \infty.$$

Proof Let us sketch the proof.

Step 1:   A useful change of variables and properties of an elliptic operator

For a given solution $u = u(t,x)$ we define a new function

$$w(s,y) := (1+t)^{\frac{1}{p-1}} u(t,x), \quad s = \log(1+t), \quad y = \frac{x}{\sqrt{1+t}}.$$

Then, w satisfies on $\mathbb{R}^{n+1}_+$ the following Cauchy problem for a parabolic equation:

$$w_s + Lw + w|w|^{p-1} = \frac{1}{p-1}w, \quad w(0,y) = \varphi(y).$$

If w is nonnegative, then it satisfies

$$w_s + Lw + w^p = \frac{1}{p-1}w, \quad w(0,y) = \varphi(y).$$

Here (cf. with notations of Sects. 17.1.3 and 17.1.4)

$$Lw := -\Delta_y w - \frac{1}{2}y \cdot \nabla_y w = -\frac{1}{K(y)}\nabla_y \cdot \left(K(y)\nabla_y w\right), \quad \text{where } K(y) = \exp\left(\frac{|y|^2}{4}\right).$$

Due to [50], the operator L is a self-adjoint operator and has a compact inverse on the weighted Lebesgue space $L^2(K)(\mathbb{R}^n)$, its domain of definition is $H^2(K)(\mathbb{R}^n)$. Finally, the eigenvalues of L are

$$\lambda_k := \frac{n+k-1}{2} \quad \text{for integers } k \geq 1.$$

Step 2:   *On the mapping $a \to C(a)$*

In this step it is shown that the mapping $a \to C(a)$ from Theorem 17.2.7 is continuous, increasing and convex for $a \in [a_0, a^*)$. Moreover, $C(a^*) := \lim_{a \to a^*} C(a) = \infty$ and $C(a) = 0$ if and only if $a = a_0$.

Step 3:   *Invariance property in L^1 spaces with asymptotical behavior at infinity*

For $a \in [a_0, a^*]$ we introduce the sets

$$\Lambda_a := \left\{f \in L^1(\mathbb{R}^n), \ f \not\equiv 0, \ \lim_{|y| \to \infty} |y|^{\frac{2}{p-1}}f(y) = C(a)\right\}.$$

Let $\varphi \in \Lambda_a$ be nonnegative. Then, for all $s > 0$ the solution $w = w(s,y)$ of the Cauchy problem

$$w_s + Lw + w^p = \frac{1}{p-1}w, \quad w(0,y) = \varphi(y)$$

is nonnegative and satisfies $w(s, \cdot) \in \Lambda_a$. Consequently, the trajectories of the solution w remain in the set Λ_a if φ belongs to Λ_a. This observation implies the same invariance property for u.

Step 4:   *Properties of solutions to the auxiliary Cauchy problem*

Consider the Cauchy problem

$$w_s + Lw + w^p = \frac{1}{p-1}w, \quad w(0,y) = \varphi(y).$$

Then the following statements are valid:

1. Let us choose with $\lambda \geq 1$ the spherically symmetric data

$$h = \lambda g_{a_0,0} + g_{a,C(a)}.$$

Then, the solution w with $w(0, y) = h(|y|)$ satisfies $w(s_2, y) \leq w(s_1, y)$ for all $s_2 > s_1$. Moreover,

$$\lim_{s \to \infty} \|w(s, \cdot) - g_{a,C(a)}(|\cdot|)\|_{L^\infty} = 0.$$

2. Let $\varphi \in \Lambda_a$ be such that $\varphi(y) \geq \delta \exp(-\gamma|y|^2)$, where δ and γ are positive constants. Then, the solution w satisfies $g_{a,C(a)}(|y|) \leq \liminf_{s \to \infty} w(s, y)$.

Step 5:   *Proof of the theorem for nonnegative data*

Additionally, as to the conditions for φ we assume the data is nonnegative. Step 2 implies the existence of a unique $a \in [a_0, a^*]$ satisfying $C(a) = b$. Then the following statements are valid:

1. It holds $\limsup_{s \to \infty} w(s, y) \leq g_{a,C(a)}(|y|)$.
2. It holds $g_{a,C(a)}(|y|) \leq \liminf_{s \to \infty} w(s, y)$. To verify the last inequality the following result from [51] is used.

   Let u be solution to

$$u_t - \Delta u + |u|^{p-1}u = 0, \quad u(0, x) = \varphi(x),$$

   where φ is nonnegative and not identical to 0. Then, there exist positive constants δ and γ such that

$$u(1, x) \geq \delta \exp(-\gamma|x|^2) \quad \text{on } \mathbb{R}^n.$$

The convergence of $w(s, y)$ to $g_{a,b}(|y|)$ takes place in $L^1(\mathbb{R}^n) \cap L^q(\mathbb{R}^n)$ for all $q < \infty$. One can show that $w(s, \cdot)$ is bounded in $H^1(\mathbb{R}^n) \cap L^\infty(\mathbb{R}^n)$. Moreover, $w(s, y)$ converges to $g_{a,b}(|y|)$ in $C^2(\mathbb{R}^n)$, too.

Step 6:   *Proof of the theorem for general data*

In this step it is shown that if φ belongs to some class Λ_a with $a > a_0$, then, after a finite time, the solution $w(s, \cdot)$ which still belongs to the same class Λ_a becomes positive. The following statement allows us to reduce the case of general data to the case of nonnegative initial data which is treated in Step 5.

1. Let w be a solution with $\varphi \in L^1(\mathbb{R}^n)$ (not necessarily with a constant sign) satisfying $\lim_{|y| \to \infty} |y|^{\frac{2}{p-1}} \varphi(y) > 0$. Then, there exists a time t_0 such that for all $s \geq t_0$ we have $w(s, y) > 0$ on $\mathbb{R}^n$.

Step 7:   *Decay estimate for the self-similar solution*

The desired decay estimate for the self-similar solution v_b follows immediately after taking into account the representation

$$v_b(t, x) = (1 + t)^{-\frac{1}{p-1}} g_{a,b}\left(\frac{|x|}{\sqrt{1+t}}\right)$$

and the properties of g_a and $g_{a,b}$ of Theorem 17.2.7.

Remark 17.2.3 The statement of Theorem 17.2.8 remains valid for any initial data $\varphi \in L^1(\mathbb{R}^n)$ for which the corresponding solution $u = u(t, x)$ enters the set Λ_a for some $t_0 > 0$ and some $a \in (a_0, a^*)$.

Choosing $q = \infty$ in the statement of Theorem 17.2.8 we get in the subcritical case the asymptotical behavior $t^{-\frac{1}{p-1}}$ of the self-similar solution which coincides with the behavior of the special solution $u_1 = u_1(t)$ from the beginning of this section.

Exercises Relating to the Considerations of Chap. 17

Exercise 1 Prove the relation

$$d_s \int_{\mathbb{R}_y^n} v(s, y)\psi(y)K(y)\, dy = \int_{\mathbb{R}_y^n} v(s, y)^p \psi(y)K(y)\, dy$$

$$+ \int_{\mathbb{R}_y^n} v(s, y)\left(\frac{1}{p-1}\psi(y) - L\psi(y)\right)K(y)\, dy$$

is used in the proof of Theorem 17.1.3.

Exercise 2 Follow the proof of Theorem 17.1.4 in the paper [107].

Exercise 3 Prove Theorem 17.1.5 (cf. with the proof of Theorem 17.1.3).

Exercise 4 Prove Corollary 17.1.2 for $p = p_{Fuj}(n)$. Why are we not able to follow the proof to Theorem 17.1.3?

Chapter 18
Semilinear Classical Damped Wave Models

The diffusion phenomenon between linear heat and linear classical damped wave models of Sect. 14.2.3 explains the parabolic character of classical damped wave models with power nonlinearities from the point of decay estimates which are discussed in this chapter. For this reason, the reader expects an influence of the Fujita exponent as well (cf. with Sect. 17.1). Section 18.1 shows that this is really so, where a general approach is proposed to prove global (in time) existence of small data energy solutions under additional regularity of the data for source power nonlinearities. One of the tools is the Gagliardo-Nirenberg inequality. The sharpness of the result (at least in dimensions $n = 1, 2$) is proved by application of the so-called test function method. This method yields sharp results for models with a parabolic like decay for solutions. In the case of an absorbing power nonlinearity we prove global existence of energy solutions and decay behavior. Moreover, the issue of a relation between the asymptotic profile and the Gauss kernel is explained. Finally, the critical exponent for other wave models, with additional mass or with scale-invariant time-dependent mass and dissipation term, is discussed in the concluding remarks. In particular, the influence of the time-dependent coefficients on the critical exponent is of interest.

18.1 Semilinear Classical Damped Wave Models with Source Nonlinearity

In this section we consider the Cauchy problem

$$u_{tt} - \Delta u + u_t = |u|^p, \quad u(0, x) = \varphi(x), \quad u_t(0, x) = \psi(x).$$

In Sect. 11.2.5 we learned that the nonlinear term $|u|^p$ is a source nonlinearity. For this reason the global existence (in time) of small data solutions is of interest. This

© Springer International Publishing AG 2018

M.R. Ebert, M. Reissig, *Methods for Partial Differential Equations*,
https://doi.org/10.1007/978-3-319-66456-9_18

means that we try to prove that the steady-state solution $u \equiv 0$ of the Cauchy problem with homogeneous data is stable in a suitable evolution space. Small perturbations of the data in suitable Banach spaces preserve the property of the Cauchy problem to have globally (in time) solutions. It turns out that there exists a critical exponent p_{crit}, a threshold between global and non-global existence of small data solutions. For the above semilinear damped wave model this critical exponent is actually the *Fujita exponent* $p_{Fuj}(n) = 1 + \frac{2}{n}$. We shall discuss this issue in the next two sections.

Due to Sect. 11.3.2, a suitable energy of solutions is the wave energy $E_W(u)(t)$.

18.1.1 Global Existence of Small Data Solutions

18.1.1.1 Main Result

To formulate the following theorem we need the abbreviation $p_{GN}(n) = \frac{n}{n-2}$ for $n \geq 3$. This number is connected with the Gagliardo-Nirenberg inequality of Proposition 24.5.4 from Sect. 24.5. The space for the data (φ, ψ) is defined as follows: $\mathcal{A}_{1,1} := (H^1(\mathbb{R}^n) \cap L^1(\mathbb{R}^n)) \times (L^2(\mathbb{R}^n) \cap L^1(\mathbb{R}^n))$.

Theorem 18.1.1 *Let $n \leq 4$ and let*

$$\begin{cases} p > p_{Fuj}(n) & \text{if } n = 1, 2, \\ 2 \leq p \leq 3 = p_{GN}(3) & \text{if } n = 3, \\ p = 2 = p_{GN}(4) & \text{if } n = 4. \end{cases}$$

Let $(\varphi, \psi) \in \mathcal{A}_{1,1}$. Then, the following statement holds with a suitable constant $\varepsilon_0 > 0$: if

$$\|(\varphi, \psi)\|_{\mathcal{A}_{1,1}} \leq \varepsilon_0,$$

then there exists a unique globally (in time) energy solution u belonging to the function space

$$C([0, \infty), H^1(\mathbb{R}^n)) \cap C^1([0, \infty), L^2(\mathbb{R}^n)).$$

Moreover, there exists a constant $C > 0$ such that the solution and its energy terms satisfy the decay estimates

$$\|u(t, \cdot)\|_{L^2} \leq C(1 + t)^{-\frac{n}{4}} \|(\varphi, \psi)\|_{\mathcal{A}_{1,1}},$$

$$\|\nabla u(t, \cdot)\|_{L^2} \leq C(1 + t)^{-\frac{n}{4} - \frac{1}{2}} \|(\varphi, \psi)\|_{\mathcal{A}_{1,1}},$$

$$\|u_t(t, \cdot)\|_{L^2} \leq C(1 + t)^{-\frac{n}{4} - 1} \|(\varphi, \psi)\|_{\mathcal{A}_{1,1}}.$$

Remark 18.1.1 We obtain the global (in time) existence of energy solutions only in low dimensions $n \leq 4$. This depends on the weak assumptions for the data, which are chosen from the energy space only with an additional regularity L^1. More restrictions on the data space or using estimates on L^p basis with $p \in [1, 2)$ allow, in general, to prove the global existence of small data solutions in higher dimensions for $p > p_{Fuj}(n)$, too (see, for example, [94, 95, 145, 205]).

18.1.1.2 Main Steps in Our Approach

We explain the main steps in our approach to prove Theorem 18.1.1. This approach can be used in studying large classes of semilinear models.

Linear Cauchy Problem Let us consider the corresponding linear Cauchy problem

$$w_{tt} - \Delta w + w_t = 0, \quad w(0, x) = \varphi(x), \quad w_t(0, x) = \psi(x).$$

Then, the solution $w = w(t, x)$ can be written in the following form

$$w(t, x) = K_0(t, 0, x) *_{(x)} \varphi(x) + K_1(t, 0, x) *_{(x)} \psi(x).$$

Here $K_0(t, 0, x) *_{(x)} \varphi(x)$ is the solution of the above Cauchy problem with second Cauchy data $\psi \equiv 0$. On the contrary, $K_1(t, 0, x) *_{(x)} \psi(x)$ is the solution of the above Cauchy problem with first Cauchy data $\varphi \equiv 0$. Now let us turn to the following classical damped wave model with source

$$v_{tt} - \Delta v + v_t = f(t, x), \quad v(0, x) = 0, \quad v_t(0, x) = 0.$$

Using Duhamel's principle we get the solution

$$v(t, x) = \int_0^t K_1(t, s, x) *_{(x)} f(s, x) \, ds.$$

The family of terms $\{K_1(t, s, x) *_{(x)} f(s, x)\}_{s \geq 0}$ is the solution of the family of parameter-dependent Cauchy problems

$$w_{tt} - \Delta w + w_t = 0, \quad w(s, x) = 0, \quad w_t(s, x) = f(s, x).$$

So, Duhamel's principle explains that we have to take account of solutions to a family of parameter-dependent Cauchy problems where the parameter appears in the description of the hyperplane $\{(t, x) \in \mathbb{R}^{n+1} : t = s\}$, where Cauchy data are posed. The classical damped wave equation has constant coefficients. Using the change of variables $t \to t - s$ in the last Cauchy problem implies the relation $K_l(t, s, x) = K_l(t - s, 0, x)$ for $l = 0, 1$.

Choice of Spaces for Solutions and Data This is a very important step. The choice of the space for the data is, in general, connected with the choice of the space for solutions. On the one hand the choice of data (the choice of l in the function spaces below) may cause some additional difficulties in the treatment. On the other hand the space of data may influence qualitative properties of solutions (e.g. compact support for all times or decay behavior for all times). We propose as space for solutions the *evolution space*

$$X(t) := C\big([0, t], H_m^l(\mathbb{R}^n)\big) \cap C^1\big([0, t], H_m^{l-1}(\mathbb{R}^n)\big)$$

for $m \in (1, 2]$, $l \in \mathbb{N}$, $l \geq 1$ and for all $t > 0$. The data are taken from the function space

$$(H_m^l(\mathbb{R}^n) \cap L^1(\mathbb{R}^n)) \times (H_m^{l-1}(\mathbb{R}^n) \cap L^1(\mathbb{R}^n)).$$

So, we assume an additional regularity L^1 for the data (φ, ψ) (see Sects. 14.2.4, 14.3.3 or Theorem 15.3.3).

Estimates for Solutions and Some of Its Partial Derivatives To fix a norm in $X(t)$ we need so-called $(L^m \cap L^1) \to L^m$ estimates for solutions and some of their partial derivatives

$$\left\|\partial_x^\alpha u(t, \cdot)\right\|_{L^m} \leq C f_{|\alpha|}(t) \|(\varphi, \psi)\|_{(H_m^l \cap L^1) \times (H_m^{l-1} \cap L^1)} \text{ for } |\alpha| \leq l,$$

$$\left\|\partial_x^\alpha u_t(t, \cdot)\right\|_{L^m} \leq C g_{|\alpha|}(t) \|(\varphi, \psi)\|_{(H_m^l \cap L^1) \times (H_m^{l-1} \cap L^1)} \text{ for } |\alpha| \leq l - 1.$$

Then, we introduce in $X(t)$ the norm

$$\|u\|_{X(t)} := \sup_{0 \leq \tau \leq t} \Big(\sum_{|\alpha| \leq l} f_{|\alpha|}(\tau)^{-1} \|\partial_x^\alpha u(\tau, \cdot)\|_{L^m} + \sum_{|\alpha| \leq l-1} g_{|\alpha|}(\tau)^{-1} \|u_\tau(\tau, \cdot)\|_{L^m} \Big).$$

Fixed Point Formulation We introduce for arbitrarily given data

$$(\varphi, \psi) \in (H_m^l(\mathbb{R}^n) \cap L^1(\mathbb{R}^n)) \times (H_m^{l-1}(\mathbb{R}^n) \cap L^1)(\mathbb{R}^n)$$

the operator

$$N : u \in X(t) \to Nu := K_0(t, 0, x) *_{(x)} \varphi(x) + K_1(t, 0, x) *_{(x)} \psi(x)$$

$$+ \int_0^t K_1(t - s, 0, x) *_{(x)} |u(s, x)|^p \, ds.$$

Then, we show that the following estimates are satisfied:

$$\|Nu\|_{X(t)} \le C_0 \|(\varphi, \psi)\|_{(H_m^l \cap L^1) \times (H_m^{l-1} \cap L^1)} + C_1(t)\|u\|_{X_0(t)}^p,$$

$$\|Nu - Nv\|_{X(t)} \le C_2(t)\|u - v\|_{X_0(t)}\left(\|u\|_{X_0(t)}^{p-1} + \|v\|_{X_0(t)}^{p-1}\right)$$

for $t \in [0, \infty)$ with nonnegative constants C_0, $C_1(t)$ and $C_2(t)$. Here we used the evolution space $X_0(t) := C([0, t], H_m^l)$ with the norm

$$\|u\|_{X_0(t)} := \sup_{0 \le \tau \le t}\left(\sum_{|\alpha| \le l} f_{|\alpha|}(\tau)^{-1}\|\partial_x^\alpha u(\tau, \cdot)\|_{L^m}\right).$$

Application of Banach's Fixed Point Theorem The estimates for the image Nu of the last step allow us to apply Banach's fixed point theorem. In this way we get simultaneously a unique solution to $Nu = u$ locally in time for large data and globally in time for small data. To prove the local (in time) existence we use $C_1(t)$, $C_2(t)$ tend to 0 for t tends to 0, while to prove the global (in time) existence we use $C_1(t) \le C_3$ and $C_2(t) \le C_3$ for all $t \in [0, \infty)$ with a suitable nonnegative constant C_3.

Let us only verify how to prove the global existence in time.

In fact, taking the recurrence sequence $u_{-1} := 0$, $u_k := N(u_{k-1})$ for $k = 0, 1, 2, \cdots$ into account, we apply the estimate for $\|Nu\|_{X(t)}$ with small norm

$$\|(\varphi, \psi)\|_{(H_m^l \cap L^1) \times (H_m^{l-1} \cap L^1)} = \varepsilon.$$

Then, we arrive at $\|u_k\|_{X(t)} \le 2C_3\varepsilon$ for any $\varepsilon \in [0, \varepsilon_0]$ with $\varepsilon_0 = \varepsilon_0(2C_3)$ sufficiently small. Once this uniform estimate is established we use the estimate for $\|Nu - Nv\|_{X(t)}$ and find

$$\|u_{k+1} - u_k\|_{X(t)} \le C_3\varepsilon^{k-1}, \quad \|u_{k+1} - u_k\|_{X(t)} \le 2^{-1}\|u_k - u_{k-1}\|_{X(t)}$$

for $\varepsilon \le \varepsilon_0$ sufficiently small. We get inductively $\|u_k - u_{k-1}\|_{X(t)} \le C_3 2^{-k}$ so that $\{u_k\}_k$ is a Cauchy sequence in the Banach space $X(t)$ converging to the unique solution of $Nu = u$ for all $t > 0$. Here we recognized that the constant C_3 appearing in the last estimates is independent of $t \in [0, \infty)$.

18.1.1.3 Proof of the Main Result

Proof Now let us prove Theorem 18.1.1 by following all the steps of the approach of the previous section.

The space for the data is $\mathcal{A}_{1,1} := (H^1(\mathbb{R}^n) \cap L^1(\mathbb{R}^n)) \times (L^2(\mathbb{R}^n) \cap L^1(\mathbb{R}^n))$. The space of energy solutions is $X(t) = C([0, t], H^1(\mathbb{R}^n)) \cap C^1([0, t], L^2(\mathbb{R}^n))$. Taking into consideration the estimates of Theorem 14.2.4 we choose

$$f_{|\alpha|}(t) = (1+t)^{-\frac{n+2|\alpha|}{4}} \quad \text{for } |\alpha| \leq 1, \quad g_0(t) = (1+t)^{-\frac{n+4}{4}}.$$

So, we introduce in $X(t)$ the norm

$$\|u\|_{X(t)} := \sup_{0 \leq \tau \leq t} \left((1+\tau)^{\frac{n}{4}} \|u(\tau,\cdot)\|_{L^2} + (1+\tau)^{\frac{n+2}{4}} \|\nabla u(\tau,\cdot)\|_{L^2} \right.$$

$$\left. + (1+\tau)^{\frac{n+4}{4}} \|u_\tau(\tau,\cdot)\|_{L^2} \right).$$

Moreover, we define the evolution space $X_0(t) = C([0,t], H^1(\mathbb{R}^n))$ with the norm

$$\|u\|_{X_0(t)} := \sup_{0 \leq \tau \leq t} \left((1+\tau)^{\frac{n}{4}} \|u(\tau,\cdot)\|_{L^2} + (1+\tau)^{\frac{n+2}{4}} \|\nabla u(\tau,\cdot)\|_{L^2} \right).$$

It remains to show the estimates

$$\|Nu\|_{X(t)} \leq C_0 \|(\varphi,\psi)\|_{(H^1 \cap L^1) \times (L^2 \cap L^1)} + C_1(t) \|u\|_{X_0(t)}^p,$$

$$\|Nu - Nv\|_{X(t)} \leq C_2(t) \|u - v\|_{X_0(t)} \left(\|u\|_{X_0(t)}^{p-1} + \|v\|_{X_0(t)}^{p-1} \right)$$

for the operator N of the previous section. These estimates follow from the next proposition in which the restriction on the power p and on the dimension n of Theorem 18.1.1 will appear.

Proposition 18.1.1 *Let u and v be elements of $X(t)$. Then, under the assumptions of Theorem 18.1.1, the following estimates hold for $j + l = 0, 1$:*

$$(1+t)^l (1+t)^{\frac{n}{4}+\frac{j}{2}} \|\nabla^j \partial_t^l Nu(t,\cdot)\|_{L^2} \leq C \|(\varphi,\psi)\|_{\mathcal{A}_{1,1}} + C \|u\|_{X_0(t)}^p,$$

$$(1+t)^l (1+t)^{\frac{n}{4}+\frac{j}{2}} \|\nabla^j \partial_t^l (Nu(t,\cdot) - Nv(t,\cdot))\|_{L^2} \leq C \|u - v\|_{X_0(t)} \left(\|u\|_{X_0(t)}^{p-1} + \|v\|_{X_0(t)}^{p-1} \right).$$

Here the nonnegative constant C is independent of $t \in [0,\infty)$.

Proof We have

$$\nabla^j \partial_t^l Nu(t,\cdot) = \nabla^j \partial_t^l K_0(t,0,x) *_{(x)} \varphi(x) + \nabla^j \partial_t^l K_1(t,0,x) *_{(x)} \psi(x)$$

$$+ \nabla^j \partial_t^l \int_0^t K_1(t-s,0,x) *_{(x)} |u(s,x)|^p \, ds.$$

The estimates of Theorem 14.2.4 imply immediately

$$\|\nabla^j \partial_t^l K_0(t,0,x) *_{(x)} \varphi(x) + \nabla^j \partial_t^l K_1(t,0,x) *_{(x)} \psi(x)\|_{L^2}$$

$$\leq C (1+t)^{-l} (1+t)^{-\frac{n}{4}-\frac{j}{2}} \|(\varphi,\psi)\|_{\mathcal{A}_{1,1}}$$

for the admissible range of j and l. So, we restrict ourselves to the integral term in the representation of $\nabla^j \partial_t^l Nu(t, \cdot)$. Using $K_1(0, 0, x) = 0$ it follows

$$\nabla^j \partial_t^l \int_0^t K_1(t - s, 0, x) *_{(x)} |u(s, x)|^p \, ds$$

$$= \int_0^t \nabla^j \partial_t^l K_1(t - s, 0, x) *_{(x)} |u(s, x)|^p \, ds.$$

We shall use different estimates of solutions to the family of parameter-dependent Cauchy problems

$$w_{tt} - \Delta w + w_t = 0, \quad w(s, x) = 0, \quad w_t(s, x) = |u(s, x)|^p.$$

On the Interval $[0, \frac{t}{2}]$ Here we use the $L^2 \cap L^1 \to L^2$ estimates of Theorem 14.2.4, so additional regularity of the data is required.

On the Interval $[\frac{t}{2}, t]$ Here we use the $L^2 \to L^2$ estimates of Theorem 14.2.2, so no additional regularity of the data is required.

Following this strategy we get

$$\left\| \int_0^t \nabla^j \partial_t^l K_1(t - s, 0, x) *_{(x)} |u(s, x)|^p \, ds \right\|_{L^2}$$

$$\leq C \int_0^{\frac{t}{2}} (1 + t - s)^{-(\frac{n}{4} + \frac{j}{2} + l)} \big\| |u(s, x)|^p \big\|_{L^2 \cap L^1} \, ds$$

$$+ C \int_{\frac{t}{2}}^t (1 + t - s)^{-\frac{j}{2} - l} \big\| |u(s, x)|^p \big\|_{L^2} \, ds.$$

We use

$$\big\| |u(s, x)|^p \big\|_{L^1 \cap L^2} \leq \|u(s, \cdot)\|_{L^p}^p + \|u(s, \cdot)\|_{L^{2p}}^p,$$

$$\big\| |u(s, x)|^p \big\|_{L^2} = \|u(s, \cdot)\|_{L^{2p}}^p.$$

Gagliardo-Nirenberg inequality (see Proposition 24.5.4) now comes into play. We may estimate

$$\|u(s, \cdot)\|_{L^p}^p \leq C \|u(s, \cdot)\|_{L^2}^{p(1 - \theta(p))} \|\nabla u(s, \cdot)\|_{L^2}^{p \theta(p)},$$

$$\|u(s, \cdot)\|_{L^{2p}}^p \leq C \|u(s, \cdot)\|_{L^2}^{p(1 - \theta(2p))} \|\nabla u(s, \cdot)\|_{L^2}^{p \theta(2p)},$$

where

$$\theta(p) = \frac{n(p-2)}{2p}, \qquad \theta(2p) = \frac{n(p-1)}{2p}.$$

We remark that the restriction $\theta(p) \geq 0$ implies that $p \geq 2$, whereas the restriction $\theta(2p) \leq 1$ implies that $p \leq p_{GN}(n)$ if $n \geq 3$. So, we use the estimates for $u(t, \cdot)$ and $\nabla u(t, \cdot)$ only. This is the main motivation for introducing the space $X_0(t)$. Taking into consideration $\theta(p) < \theta(2p)$ implies

$$\left\| |u(s,x)|^p \right\|_{L^2 \cap L^1} \leq C\|u\|^p_{X_0(s)}(1+s)^{-p(\frac{n}{4}+\frac{\theta(p)}{2})} = \|u\|^p_{X_0(s)}(1+s)^{-\frac{(p-1)n}{2}},$$

$$\left\| |u(s,x)|^p \right\|_{L^2} \leq C\|u\|^p_{X_0(s)}(1+s)^{-p(\frac{n}{4}+\frac{\theta(2p)}{2})} = \|u\|^p_{X_0(s)}(1+s)^{-\frac{(2p-1)n}{4}}.$$

After summarizing and using $\|u\|_{X_0(s)} \leq \|u\|_{X_0(t)}$ for $s \leq t$ we may conclude

$$\left\| \int_0^t \nabla^j \partial_t^l K_1(t-s,0,x) *_{(x)} |u(s,x)|^p \, ds \right\|_{L^2}$$

$$\leq C\|u\|^p_{X_0(t)} \int_0^{\frac{t}{2}} (1+t-s)^{-(\frac{n}{4}+\frac{j}{2}+l)}(1+s)^{-\frac{(p-1)n}{2}} \, ds$$

$$+ C\|u\|^p_{X_0(t)} \int_{\frac{t}{2}}^t (1+t-s)^{-\frac{j}{2}-l}(1+s)^{-\frac{(2p-1)n}{4}} \, ds.$$

The first integral is estimated by $(1+t)^{-(\frac{n}{4}+\frac{j}{2}+l)}$. Indeed, since $p > p_{Fuj}(n)$, the function $(1+t)^{-\frac{(p-1)n}{2}}$ belongs to $L^1(\mathbb{R}^1_+)$. We treat the second integral as follows:

$$\int_{\frac{t}{2}}^t (1+t-s)^{-\frac{j}{2}-l}(1+s)^{-\frac{(2p-1)n}{4}} \, ds \leq C(1+t)^{-\frac{(2p-1)n}{4}} \int_{\frac{t}{2}}^t (1+t-s)^{-\frac{j}{2}-l} ds$$

$$\leq C(1+t)^{-\frac{(2p-1)n}{4}+1-\frac{j}{2}-l}((\log(1+t))^l \leq C(1+t)^{-(\frac{n}{4}+\frac{j}{2}+l)}$$

for $j+l = 0, 1$. This completes the estimates for $\nabla^j \partial_t^l Nu(t, \cdot)$. In Exactly the same way we prove the desired estimates for $\nabla^j \partial_t^l (Nu(t, \cdot) - Nv(t, \cdot))$. The considerations are based on the following relations:

$$\nabla^j \partial_t^l \int_0^t K_1(t-s,0,x) *_{(x)} \left(|u(s,x)|^p - |v(s,x)|^p \right) ds$$

$$= \int_0^t \nabla^j \partial_t^l K_1(t-s,0,x) *_{(x)} \left(|u(s,x)|^p - |v(s,x)|^p \right) ds,$$

and

$$\big\| |u(s,x)|^p - |v(s,x)|^p \big\|_{L^1} \le C \|u(s,\cdot) - v(s,\cdot)\|_{L^p} \left(\|u(s,\cdot)\|_{L^p}^{p-1} + \|v(s,\cdot)\|_{L^p}^{p-1} \right),$$

$$\big\| |u(s,x)|^p - |v(s,x)|^p \big\|_{L^2} \le C \|u(s,\cdot) - v(s,\cdot)\|_{L^{2p}} \left(\|u(s,\cdot)\|_{L^{2p}}^{p-1} + \|v(s,\cdot)\|_{L^{2p}}^{p-1} \right).$$

We conclude by Proposition 18.1.1 the statements of Theorem 18.1.1.

18.1.2   Application of the Test Function Method

In this section we shall show that the Fujita exponent $p_{Fuj}(n)$ is really the critical exponent. Here we apply the test function method which was introduced in the paper [232]. Our main concern is the following result.

Theorem 18.1.2 *Let us consider the Cauchy problem for the classical damped wave equation with power nonlinearity*

$$u_{tt} - \Delta u + u_t = |u|^p, \quad u(0,x) = \varphi(x), \quad u_t(0,x) = \psi(x)$$

in $[0,\infty) \times \mathbb{R}^n$ with $n \ge 1$ and $p \in (1, 1 + \frac{2}{n}]$. Let $(\varphi, \psi) \in \mathcal{A}_{1,1}$ satisfy the assumption

$$\int_{\mathbb{R}^n} (\varphi(x) + \psi(x))\, dx > 0.$$

Then, there exists a locally (in time) defined energy solution

$$u \in C([0,T), H^1(\mathbb{R}^n)) \cap C^1([0,T), L^2(\mathbb{R}^n)).$$

This solution cannot be continued to the interval $[0,\infty)$ in time.

Remark 18.1.2 Following the proof to Theorem 18.1.1 we obtain a local (in time) energy solution

$$u \in C([0,T), H^1(\mathbb{R}^n)) \cap C^1([0,T), L^2(\mathbb{R}^n)).$$

For this reason we restrict ourselves to proving that this solution does not exist globally on the interval $[0,\infty)$ in time.

Proof We first introduce test functions $\eta = \eta(t)$ and $\phi = \phi(x)$ having the following properties:

1. $\eta \in C_0^\infty[0,\infty),\ 0 \le \eta(t) \le 1$,

$$\eta(t) = \begin{cases} 1 & \text{for } 0 \le t \le \frac{1}{2}, \\ 0 & \text{for } t \ge 1, \end{cases}$$

2. $\phi \in C_0^\infty(\mathbb{R}^n)$, $0 \le \phi(x) \le 1$,

$$\phi(x) = \begin{cases} 1 & \text{for } |x| \le \frac{1}{2}, \\ 0 & \text{for } |x| \ge 1, \end{cases}$$

3. $\frac{\eta'(t)^2}{\eta(t)} \le C$ for $\frac{1}{2} < t < 1$, and $\frac{|\nabla\phi(x)|^2}{\phi(x)} \le C$ for $\frac{1}{2} < |x| < 1$.

Let $R \in [0, \infty)$ be a large parameter. We define the test function

$$\chi_R(t, x) := \eta_R(t)\phi_R(x) := \eta\left(\frac{t}{R^2}\right)\phi\left(\frac{x}{R}\right).$$

We put

$$Q_R := [0, R^2] \times B_R, \quad B_R := \{x \in \mathbb{R}^n : |x| \le R\}.$$

We note that the support of χ_R is contained in the set Q_R. Moreover, $\chi_R \equiv 1$ on $[0, \frac{R^2}{2}] \times B_{\frac{R}{2}}$. We suppose that the energy solution $u = u(t, x)$ exists globally in time. We define the functional

$$I_R := \int_{Q_R} |u(t, x)|^p \chi_R(t, x)^q \, d(x, t) = \int_{Q_R} (u_{tt} - \Delta u + u_t) \chi_R(t, x)^q \, d(x, t).$$

Here q is the Sobolev conjugate of p, that is, $\frac{1}{p} + \frac{1}{q} = 1$. After integration by parts we obtain

$$I_R = -\int_{B_R} (\varphi + \psi)\phi_R^q \, dx + \int_{Q_R} u\partial_t^2(\chi_R^q) \, d(x, t)$$

$$- \int_{Q_R} u\partial_t(\chi_R^q) \, d(x, t) - \int_{Q_R} u\Delta(\chi_R^q) \, d(x, t)$$

$$:= -\int_{B_R} (\varphi + \psi)\phi_R^q \, dx + J_1 + J_2 + J_3.$$

By the assumption on the data (φ, ψ) it follows that $I_R < J_1 + J_2 + J_3$ for sufficiently large R. We shall estimate separately J_1, J_2 and J_3. Here we use the notations

$$\hat{Q}_{R,t} := \left[\frac{R^2}{2}, R^2\right] \times B_R, \quad \hat{Q}_{R,x} := [0, R^2] \times \left(B_R \setminus B_{\frac{R}{2}}\right).$$

We first estimate J_3. Noting

$$\Delta(\chi_R^q) = R^{-2}q(q-1)\eta_R^q(t)\phi_R^{q-2}(x)\left|\nabla\phi\left(\frac{x}{R}\right)\right|^2 + R^{-2}q\eta_R^q(t)\phi_R^{q-1}(x)(\Delta\phi)\left(\frac{x}{R}\right)$$

and the assumed properties for the test functions we may conclude

$$|J_3| \le CR^{-2} \int_{\hat{Q}_{R,x}} |u| \chi_R^{q-1} \, d(x,t).$$

Application of Hölder's inequality implies

$$|J_3| \le CR^{-2} \left(\int_{\hat{Q}_{R,x}} |u|^p \chi_R^q(t,x) \, d(x,t) \right)^{1/p} \left(\int_{\hat{Q}_{R,x}} 1 \, d(x,t) \right)^{1/q}$$

$$\le CR^{-2} I_{R,x}^{\frac{1}{p}} \left(\int_{\hat{Q}_{R,x}} 1 \, d(x,t) \right)^{1/q} \le C I_{R,x}^{\frac{1}{p}} R^{\frac{n+2}{q}-2},$$

where

$$I_{R,x} := \int_{\hat{Q}_{R,x}} |u|^p \chi_R^q(t,x) \, d(x,t).$$

Since $1 < p \le 1 + 2/n$, the last inequality gives $|J_3| \le C I_{R,x}^{\frac{1}{p}}$. Next, we estimate J_1. Noting

$$\partial_t^2(\chi_R^q) = \frac{1}{R^4} q(q-1) \phi_R^q(x) \eta_R^{q-2}(t) \left(\eta'\left(\frac{t}{R^2}\right) \right)^2 + \frac{1}{R^4} q \phi_R^q(x) \eta_R^{q-1}(t) \eta''\left(\frac{t}{R}\right)$$

and using the properties of the test functions again we estimate J_1 as follows:

$$|J_1| \le C \frac{1}{R^4} \left(\int_{\hat{Q}_{R,t}} |u|^p \chi_R^q(t,x) \, d(x,t) \right)^{1/p} \left(\int_{\hat{Q}_{R,t}} 1 \, d(x,t) \right)^{1/q}$$

$$= I_{R,t}^{\frac{1}{p}} \frac{1}{R^4} \left(\int_{\hat{Q}_{R,t}} 1 \, d(x,t) \right)^{1/q} \le C I_{R,t}^{\frac{1}{p}} R^{\frac{n+2}{q}-2} \le C I_{R,t}^{\frac{1}{p}},$$

where

$$I_{R,t} := \int_{\hat{Q}_{R,t}} |u|^p \chi_R^q(t,x) \, d(x,t).$$

Finally, we estimate J_2. By

$$\partial_t(\chi_R^q) = \frac{1}{R^2} q \phi_R^q(x) \eta_R^{q-1}(t) \eta'\left(\frac{t}{R^2}\right),$$

we have

$$|J_2| \le C \frac{1}{R^2} \int_{\hat{Q}_{R,t}} |u| \chi_R^{q-1} \, d(x,t)$$

$$\le C\frac{1}{R^2}\Big(\int_{\hat{Q}_{R,t}} |u|^p \chi_R^q\, d(x,t)\Big)^{1/p}\Big(\int_{\hat{Q}_{R,t}} 1\, d(x,t)\Big)^{1/q}$$

$$\le CI_{R,t}^{\frac{1}{p}}\frac{1}{R^2}R^{\frac{n}{q}}\Big(\int_{\frac{R^2}{2}}^{R^2} 1\, dt\Big)^{1/q} \le CI_{R,t}^{\frac{1}{p}}R^{\frac{n+2}{q}-2} \le CI_{R,t}^{\frac{1}{p}}.$$

Putting the estimates of J_1, J_2 and J_3 together we obtain

$$I_R \le C\big(I_{R,t}^{1/p} + I_{R,x}^{1/p}\big)$$

for large R. It is obvious that $I_{R,t}, I_{R,x} \le I_R$. Hence, we have

$$I_R \le CI_R^{1/p}.$$

This means $I_R \le C$, that is, I_R is uniformly bounded for all R. By letting $R \to +\infty$ one may conclude

$$\int_0^\infty \int_{\mathbb{R}^n} |u|^p\, dx\, dt = \lim_{R\to\infty} I_R < \infty.$$

Now we recall the inequality

$$I_R \le C\big(I_{R,t}^{1/p} + I_{R,x}^{1/p}\big).$$

By the integrability of $|u|^p$ and noting the shape of the region $\hat{Q}_{R,t}$ and $\hat{Q}_{R,x}$ we conclude

$$\lim_{R\to\infty}\big(I_{R,t}^{1/p} + I_{R,x}^{1/p}\big) = 0.$$

This implies

$$\int_0^\infty \int_{\mathbb{R}^n} |u|^p\, d(x,t) = \lim_{R\to\infty} I_R = 0.$$

Hence, $u \equiv 0$. But this is Contrary to our assumptions for the data.

Remark 18.1.3 The test function method is based on a contradiction argument. Under suitable assumptions for the data, no global in time solution exists. Here one has to explain what kind of solutions we have in mind. We formulated Theorem 18.1.2 in correspondence with Theorem 18.1.1. Both results are related to energy solutions. Following the proof of Theorem 18.1.2 we see that the same statement holds for Sobolev solutions as well. We may also exclude global in time Sobolev solutions under suitable assumptions for the data. But, we do not get any information about blow up time or life span estimates or about blow up mechanisms.

The test function method was originally developed for proving sharpness of the Fujita exponent as the critical exponent for semilinear parabolic equations. It was later recognized that this method can also be applied for classical damped wave models, which are models with a "parabolic like behavior" from the point of view of decay estimates. Attempts to apply this method to classical wave models fail in the sense that sharpness of critical exponents cannot be expected, in general.

18.2   Semilinear Classical Damped Wave Models with Absorbing Nonlinearity

In the previous section we studied the global existence (in time) or blow up behavior of small data solutions for semilinear classical damped wave models with source nonlinearity. In this section we consider instead the Cauchy problem

$$u_{tt} - \Delta u + u_t = -u|u|^{p-1}, \quad u(0,x) = \varphi(x), \quad u_t(0,x) = \psi(x)$$

with absorbing power nonlinearity $-u|u|^{p-1}$ (see Sect. 11.2.5). Our main goal is to explain the positive influence of an absorbing nonlinearity in comparison to source nonlinearity on properties of solutions to the above Cauchy problem.

18.2.1   Global Existence of Large Data Solutions

In the next sections we discuss on the one hand the global existence (in time) of energy solutions (see [194]) and on the other hand decay estimates for the solution or a suitable energy of solutions (see [108]) under additional regularity of the data.

18.2.1.1   Global Existence of Weak Solutions

Applying the steps of the proof to Theorem 20.2.1 we may derive the following result.

Theorem 18.2.1 *Consider for $p > 1$ the Cauchy problem*

$$u_{tt} - \Delta u + u_t = -u|u|^{p-1}, \quad u(0,x) = \varphi(x), \quad u_t(0,x) = \psi(x).$$

The data (φ, ψ) are supposed to belong to $(H^1(\mathbb{R}^n) \cap L^{p+1}(\mathbb{R}^n)) \times L^2(\mathbb{R}^n)$. Then there exists a global (in time) weak solution

$$u \in L^\infty((0, \infty), H^1(\mathbb{R}^n) \cap L^{p+1}(\mathbb{R}^n)) \cap W^1_\infty((0, \infty), L^2(\mathbb{R}^n)).$$

Remark 18.2.1 If $1 < p < \frac{n+2}{n-2}$ for $n \geq 3$, $1 < p < \infty$ for $n = 1, 2$, respectively, then the solution of Theorem 18.2.1 is unique and belongs to $C([0, \infty), H^1(\mathbb{R}^n)) \cap C^1([0, \infty), L^2(\mathbb{R}^n))$ (cf. with [62] and [13]).

Remark 18.2.2 We have global (in time) solutions even for large data if the power nonlinearity is absorbing which is in opposition to the results of Theorem 18.1.1, and the results of Chap. 19 which are only valid for small data if the power nonlinearity is of source type.

18.2.1.2 Decay Estimates

The power nonlinearity is supposed to be absorbing. We learned in Sects. 11.2.5 and 11.3.2 that a suitable energy is given by

$$E_W(u)(t) = \frac{1}{2}\|u_t(t, \cdot)\|_{L^2}^2 + \frac{1}{2}\|\nabla u(t, \cdot)\|_{L^2}^2 + \frac{1}{p+1}\int_{\mathbb{R}^n} |u(t, x)|^{p+1}\, dx.$$

Due to the presence of a dissipation term we might expect a decay of the energy, but what about the long time behavior of the solution itself? We turn to these items in the following. Here we refer to [108].

Theorem 18.2.2 *Consider the Cauchy problem*

$$u_{tt} - \Delta u + u_t = -u|u|^{p-1}, \quad u(0, x) = \varphi(x), \quad u_t(0, x) = \psi(x),$$

where $1 < p < \frac{n+2}{n-2}$ for $n \geq 3$, $1 < p < \infty$ for $n = 1, 2$, respectively. The data (φ, ψ) are supposed to belong to $H^1(\mathbb{R}^n) \times L^2(\mathbb{R}^n)$. Then, there exists a global (in time) weak solution

$$u \in C([0, \infty), H^1(\mathbb{R}^n)) \cap C^1([0, \infty), L^2(\mathbb{R}^n))$$

satisfying the following estimate:

$$\|u(t, \cdot)\|_{L^2} \leq C(\varphi, \psi).$$

Proof The existence and uniqueness of solutions is mentioned in Remark 18.2.1. We restrict ourselves to proving the desired estimate. We note that

$$E'_W(u)(t) = -\|u_t(t, \cdot)\|_{L^2}^2$$

which implies

$$E_W(u)(t) + \int_0^t \|u_t(s, \cdot)\|_{L^2}^2\, ds = E_W(u)(0).$$

Moreover, multiplying the semilinear partial differential equation by u and integrating over $\mathbb{R}^n$ we get

$$\frac{d}{dt}\left((u_t(t,\cdot),u(t,\cdot)) + \frac{1}{2}\|u(t,\cdot)\|_{L^2}^2\right) - \|u_t(t,\cdot)\|_{L^2}^2$$
$$+\|\nabla u(t,\cdot)\|_{L^2}^2 + \int_{\mathbb{R}^n}|u(t,x)|^{p+1}\,dx = 0,$$

where $(\cdot,\cdot) = (\cdot,\cdot)_{L^2}$ denotes the scalar product in $L^2(\mathbb{R}^n)$. After integration over $(0,t)$ and application of Cauchy-Schwarz inequality we derive

$$\frac{1}{2}\|u(t,\cdot)\|_{L^2}^2 \leq \frac{1}{2}\|\varphi\|_{L^2}^2 + (\varphi,\psi) + \|u_t(t,\cdot)\|_{L^2}\|u(t,\cdot)\|_{L^2} + \int_0^t \|u_t(s,\cdot)\|_{L^2}^2\,ds,$$

and

$$\frac{1}{4}\|u(t,\cdot)\|_{L^2}^2 \leq \frac{1}{2}\|\varphi\|_{L^2}^2 + (\varphi,\psi) + E_W(u)(0) \leq C(\varphi,\psi),$$

which is what we wanted to have.

For the damped wave model with absorbing nonlinearity, by including the nonlinearity in the energy $E_W(u)(t)$ (cf. with the beginning of this section), one may prove, without any essential changes in the proof of Proposition 4.2 of [108], the following result.

Proposition 18.2.1 *If we assume in addition to the assumptions of Theorem 18.2.2 that*

$$\|u(t,\cdot)\|_{L^2} \leq C(1+t)^{-\frac{a}{2}},\ a \geq 0,$$

then the energy for the solution of Theorem 18.2.2 satisfies

$$E_W(u)(t) \leq C(\varphi,\psi)(1+t)^{-(1+a)}.$$

Remark 18.2.3 A possible influence of additional regularity of data is already explained in Sects. 14.2.4, 14.3.3 for linear and in Sect. 18.1.1 for semilinear models. We can also derive some benefit of additional regularity of data for semilinear classical damped wave models with absorbing power nonlinearity. The following result is taken from [108].

Theorem 18.2.3 *Besides the assumptions of Theorem 18.2.2, we assume the additional regularity $(\varphi,\psi) \in L^m(\mathbb{R}^n) \times L^m(\mathbb{R}^n)$, where $m \in [1,2)$. If $p > 1 + \frac{4}{n}$ for $n \leq 3$ with $p \leq \frac{n}{n-2}$ for $n = 3$, then the solution of Theorem 18.2.2 satisfies the* *following estimates:*

$$\|u(t,\cdot)\|_{L^2} \le C(\varphi,\psi)(1+t)^{-\frac{n}{2}(\frac{1}{m}-\frac{1}{2})},$$

$$E_W(u)(t) \le C(\varphi,\psi)(1+t)^{-1-n(\frac{1}{m}-\frac{1}{2})}.$$

Proof Using the notations of Sect. 18.1 and Duhamel's principle the solution can be represented as

$$u(t,x) = K_0(t,0,x) *_{(x)} \varphi(x) + K_1(t,0,x) *_{(x)} \psi(x)$$

$$- \int_0^t K_1(t,s,x) *_{(x)} u(s,x)|u(s,x)|^{p-1}\, ds.$$

By using the derived linear estimates in Sect. 14.2.4, namely, for $m \in [1,2)$, $L^m - L^2$ estimates for low frequencies and $L^2 - L^2$ estimates for high frequencies, we get

$$\|u(t,\cdot)\|_{L^2} \le C(\varphi,\psi)(1+t)^{-\frac{n}{2}(\frac{1}{m}-\frac{1}{2})} + C\int_0^t (1+t-s)^{-\frac{n}{4}}\|u(s,\cdot)\|_{L^p}^p\, ds$$

$$+C\int_0^t e^{-c(t-s)}\|u(s,\cdot)\|_{L^{2p}}^p\, ds.$$

For $2 \le p \le \frac{n}{n-2}$ and $2 \le p < \infty$, if $n = 1,2$, we may apply Gagliardo-Nirenberg inequality to obtain

$$\|u(s,\cdot)\|_{L^p}^p \le C\|\nabla u(s,\cdot)\|_{L^2}^{\frac{n(p-2)}{2}} \|u(s,\cdot)\|_{L^2}^{p-\frac{n(p-2)}{2}},$$

and

$$\|u(s,\cdot)\|_{L^{2p}}^p \le C\|\nabla u(s,\cdot)\|_{L^2}^{\frac{n(p-1)}{2}} \|u(s,\cdot)\|_{L^2}^{p-\frac{n(p-1)}{2}}.$$

We already proved the estimates

$$\|u(t,\cdot)\|_{L^2} \le C(\varphi,\psi) \quad \text{and} \quad \|\nabla u(t,\cdot)\|_{L^2} \le C(\varphi,\psi)(1+t)^{-\frac{1}{2}}.$$

Thanks to $p - \frac{n(p-1)}{2} > 1$ we may conclude

$$\|u(t,\cdot)\|_{L^2} \le C(\varphi,\psi)(1+t)^{-\frac{n}{2}(\frac{1}{m}-\frac{1}{2})}$$

$$+C\int_0^t (1+t-s)^{-\frac{n}{4}}(1+s)^{-\frac{n(p-2)}{4}}\|u(s,\cdot)\|_{L^2}^{p-\frac{n(p-2)}{2}}\, ds$$

$$+C\int_0^t e^{-c(t-s)}(1+s)^{-\frac{n(p-1)}{4}}\|u(s,\cdot)\|_{L^2}^{p-\frac{n(p-1)}{2}}\, ds.$$

For $p > 1 + \frac{4}{n} > 2$ and $n = 1, 2, 3$ we have

$$\frac{n}{4} + \frac{n(p-2)}{4} - 1 > 0 \quad \text{and} \quad \frac{n(p-1)}{4} > \frac{n}{4} \geq \frac{n}{2}\Big(\frac{1}{m} - \frac{1}{2}\Big).$$

By applying Proposition 24.5.8 of Sect. 24.5 we conclude that

$$\|u(t, \cdot)\|_{L^2} \leq C(\varphi, \psi)(1 + t)^{-\frac{n}{2}(\frac{1}{m} - \frac{1}{2})}.$$

Moreover, applying Proposition 18.2.1 we get

$$E_W(u)(t) \leq C(\varphi, \psi)(1 + t)^{-1 - n(\frac{1}{m} - \frac{1}{2})}$$

and the proof is complete.

18.2.2 Large Time Asymptotics

We explained in Sect. 14.2.3 the diffusion phenomenon for classical damped waves. The statement of Theorem 14.2.3 describes a connection between solutions to the Cauchy problems

$$\begin{aligned} u_{tt} - \Delta u + u_t &= 0 \\ u(0, x) &= \varphi(x), \quad u_t(0, x) = \psi(x), \end{aligned} \quad \text{and} \quad \begin{aligned} w_t - \Delta w &= 0 \\ w(0, x) &= \varphi(x) + \psi(x). \end{aligned}$$

In this section, we answer the question of a connection between solutions to the Cauchy problems

$$\begin{aligned} u_{tt} - \Delta u + u_t &= -u|u|^{p-1} \\ u(0, x) &= \varphi(x), \quad u_t(0, x) = \psi(x) \end{aligned} \quad \text{and} \quad \begin{aligned} w_t - \Delta w &= -w|w|^{p-1} \\ w(0, x) &= \chi(x). \end{aligned}$$

Consider for $p > 1$ the Cauchy problem

$$u_{tt} - \Delta u + u_t = -u|u|^{p-1}, \quad u(0, x) = \varphi(x), \quad u_t(0, x) = \psi(x).$$

We introduce results which are related to the statements of Theorems 17.2.3 and 17.2.8. The first result describes the asymptotic profile of solutions under the assumptions of Theorem 18.2.3 (see [103]).

Theorem 18.2.4 *Let $n \leq 3$. Assume the assumptions of Theorem 18.2.3 with $m = 1$. Then, the asymptotic profile of the solution is described as follows:*

$$\|u(t, \cdot) - \theta_0 G(t, \cdot)\|_{L^2} = o(t^{-\frac{n}{4}}) \text{ for } t \to \infty,$$

where

$$G(t,x) := (2\pi)^{-\frac{n}{2}} \int_{\mathbb{R}^n} e^{ix\cdot\xi - t|\xi|^2} d\xi$$

and

$$\theta_0 := \int_{\mathbb{R}^n} \left(\varphi(x) + \psi(x)\right) dx - \int_0^\infty \left(\int_{\mathbb{R}^n} u|u|^{p-1} dx\right) dt.$$

Proof Thanks to the research project in Sect. 23.8, it is sufficient to prove that for $g(u) = u|u|^{p-1}$ there exists a constant $\varepsilon > 0$ such that

$$\|g(u(t,\cdot))\|_{L^1} \leq C(1+t)^{-1-\varepsilon}, \quad \|g(u(t,\cdot))\|_{L^2} \leq C(1+t)^{-\frac{n}{4}-1-\varepsilon}$$

for all $t \geq 0$ and with a nonnegative constant C which is independent of t. Applying Theorem 18.2.3 for $m = 1$ we have

$$\|u(t,\cdot)\|_{L^2} \leq C(1+t)^{-\frac{n}{4}}.$$

Then, the energy for the solution satisfies the estimate

$$E_W(u)(t) \leq C(1+t)^{-\frac{n}{2}-1}$$

for all $t \geq 0$ and with a nonnegative constant C which is independent of t. Hence, Gagliardo-Nirenberg inequality implies that there exists a constant $\varepsilon > 0$ such that

$$\|g(u(t,\cdot))\|_{L^1} = \|u(t,\cdot)\|_{L^p}^p \leq C\|\nabla u(t,\cdot)\|_{L^2}^{\frac{n(p-2)}{2}} \|u(t,\cdot)\|_{L^2}^{p-\frac{n(p-2)}{2}}$$

$$\leq C(1+t)^{-\frac{n(p-1)}{2}} = (1+t)^{-1-\varepsilon},$$

for all $t \geq 0$ provided that $p > 1 + \frac{2}{n}$. Similarly, there exists a constant $\varepsilon > 0$ such that

$$\|g(u(t,\cdot))\|_{L^2} = \|u(t,\cdot)\|_{L^{2p}}^p \leq C\|\nabla u(t,\cdot)\|_{L^2}^{\frac{n(p-1)}{2}} \|u(t,\cdot)\|_{L^2}^{p-\frac{n(p-1)}{2}}$$

$$\leq C(1+t)^{-\frac{n(2p-1)}{4}} = (1+t)^{-\frac{n}{4}-1-\varepsilon},$$

for all $t \geq 0$ provided that $p > 1 + \frac{2}{n}$.

The next result can be found in [154] and is related to Remark 18.2.1 and Theorem 17.2.8. The data have an "additional regularity", too. The space for data is described by the condition

$$I_0^2 = I_0^2(\varphi, \psi) := \int_{\mathbb{R}^n} e^{\gamma|x|^2} \left(\varphi^2(x) + |\nabla\varphi(x)|^2 + \psi^2(x)\right) dx < \infty$$

for some $\gamma > 0$.

Theorem 18.2.5 *If the data satisfy the condition $I_0^2 < \infty$ and if $1 < p < \frac{n}{n-2}$ for $n \geq 3$, $1 < p < \infty$ for $n = 1, 2$, then there exists a unique solution*

$$u \in C\big([0, \infty), H^1(\mathbb{R}^n)\big) \cap C^1\big([0, \infty), L^2(\mathbb{R}^n)\big).$$

Moreover, if $1 < p \leq 1 + \frac{4}{n}$, then the solution satisfies the estimate

$$\|u(t, \cdot)\|_{L^2} \leq C I_0 (1 + t)^{-\frac{1}{p-1} + \frac{n}{4}}.$$

Proof We omit the proof (see Exercise 4 at the end of this chapter).

Remark 18.2.4 The decay rate coincides with that of Theorem 17.2.8 with $q = 2$ for the self-similar solution

$$v_b(t, x) := t^{-\frac{1}{p-1}} f_{a,b}\left(\frac{x}{\sqrt{t}}\right)$$

to

$$w_t - \Delta w = -w|w|^{p-1},$$

where $f_{a,b}(x) =: g_{a,b}(r)$, $r = |x|$, is the unique positive solution of the Cauchy problem

$$-g'' - \left(\frac{r}{2} + \frac{n-1}{r}\right) g' + g|g|^{p-1} = \frac{1}{p-1} g,$$

$$g(0) = a, \quad g'(0) = 0, \quad \lim_{r \to \infty} r^{\frac{2}{p-1}} g(r) = b,$$

where $r \in (0, \infty)$ (cf. with Theorem 17.2.8, see [16]).

Remark 18.2.5 If $p > 1 + \frac{4}{n}$ for $n \leq 3$ and $p \leq \frac{n}{n-2}$ for $n = 3$, then the asymptotic profile of both solutions to semilinear heat and classical damped wave models is described by the Gauss kernel. If $1 < p \leq 1 + \frac{2}{n}$, then both solutions satisfy the same decay estimates. These estimates are related to those for self-similar solutions to semilinear heat models with absorbing power nonlinearity.

Remark 18.2.6 We have a gap in the previous remark. This gap was filled by the paper [96]. Roughly speaking, the authors proved among other things for $1 + \frac{2}{n} < p \leq 1 + \frac{4}{n}$ and $n \leq 3$ the relation

$$\|u(t, \cdot) - \theta_0 G_n(t, \cdot)\|_{L^q} = o\big(t^{-\frac{n}{2}(1 - \frac{1}{q})}\big) \text{ for } q \in [1, \infty] \text{ and } t \to \infty$$

under additional assumptions to the data φ, ψ as additional regularity, higher regularity and decay behavior for $|x| \to \infty$.

Remark 18.2.7 We introduced connections between solutions to semilinear heat and classical damped wave models with absorbing power nonlinearity. There might be the question of such connections in the case of source power nonlinearity and global (in time) small data solutions. If the reader follows the proof to Theorem 18.1.1, then it will be clear that the source nonlinearity is considered for small data as a small perturbation of solutions to the corresponding linear models with zero right-hand side. So, we recall the diffusion phenomenon of Sect. 14.2.3. Consequently, a relation to Gauss kernels can be expected (see [153]).

18.3   Concluding Remarks

The approach of Sect. 18.1.1 was also applied to other semilinear models in studying the global (in time) existence of small data solutions.

18.3.1   Semilinear Classical Damped Wave Models with Mass Term

The Cauchy problem for semilinear classical damped wave models with mass term

$$u_{tt} - \Delta u + m^2 u + u_t = |u|^p, \quad u(0, x) = \varphi(x), \quad u_t(0, x) = \psi(x)$$

was studied in [165]. In Sect. 14.5 we explained the exponential type decay (in time) for energy solutions from Theorem 14.5.1. This exponential type decay gives a big benefit in the treatment of the above Cauchy problem by following the approach of Sect. 18.1.1. We may expect global existence (in time) of small data solutions for all $p > 1$. That this expectation is really true shows in the following result from [165]. Here we need the abbreviation $p_{GN}(n) = \frac{n}{n-2}$ for $n \geq 3$ (see Proposition 24.5.4 from Sect. 24.5).

Theorem 18.3.1 *Let $n \geq 2$ and let*

$$\begin{cases} 1 < p & \text{if } n = 2, \\ 1 < p \leq p_{GN}(n) & \text{if } n \geq 3. \end{cases}$$

Let $(\varphi, \psi) \in H^1(\mathbb{R}^n) \times L^2(\mathbb{R}^n)$. Then, the following statement holds with a suitable constant $\varepsilon_0 > 0$: if

$$\|(\varphi, \psi)\|_{H^1 \times L^2} \leq \varepsilon_0,$$

then there exists a uniquely determined global (in time) energy solution

$$u \in C\big([0, \infty), H^1(\mathbb{R}^n)\big) \cap C^1\big([0, \infty), L^2(\mathbb{R}^n)\big).$$

Moreover, there exist positive constants C_1 and $C_2 = C_2(m)$ such that the solution and its energy terms satisfy the decay estimates

$$\|u(t, \cdot)\|_{L^2} + \|\nabla u(t, \cdot)\|_{L^2} + \|u_t(t, \cdot)\|_{L^2} \leq C_1 e^{-C_2 t} \|(\varphi, \psi)\|_{H^1 \times L^2}.$$

Remark 18.3.1 We have global existence (in time) of small data solutions for all $p > 1$. Then one could think if it were possible to weaken in Theorem 18.3.1 the asymptotic behavior of the right-hand side to $|u| f(|u|)$ with a suitable $f = f(u)$, where $|u|^p = o(|u| f(|u|))$ for $|u| \to +0$ and all $p > 1$.

18.3.2 Semilinear Damped Wave Models with Scale-Invariant Damping and Mass Term

In Sect. 14.7 the reader was introduced to the research topic "Wave models with time-dependent dissipation and mass".

Some results are explained for qualitative and quantitative properties of solutions to

$$u_{tt} - \Delta u + b(t) u_t + m(t)^2 u = 0, \quad u(0, x) = \varphi(x), \quad u_t(0, x) = \psi(x).$$

It turns out that it is reasonable to begin the study by analyzing properties of solutions to the linear model with scale-invariant dissipation and potential. The research project in Sect. 23.13 addresses this issue if we choose there $\sigma = 1$ and a homogeneous right-hand side. If we are interested in the global (in time) existence of small data solutions to the Cauchy problem

$$u_{tt} - \Delta u + \frac{\mu_1}{1+t} u_t + \frac{\mu_2^2}{(1+t)^2} u = |u|^p, \quad u(0, x) = \varphi(x), \quad u_t(0, x) = \psi(x),$$

then, due to the explanations in Sect. 23.13, we divide our study into two cases.

Case 1 $(\mu_1 - 1)^2 - 4\mu_2^2 < 1$: In this first case the mass term is dominant. In [156] a classification of mass terms is proposed for the Cauchy problem

$$u_{tt} - \Delta u + m(t)^2 u = 0, \quad u(0, x) = \varphi(x), \quad u_t(0, x) = \psi(x).$$

We omit explaining some structural assumptions for the time-dependent coefficient in effective or non-effective mass terms as well. We only hint at the asymptotical behavior of the coefficient for $t \to \infty$.

1. The mass term $m(t)^2 u$ is effective if $\lim_{t \to \infty} t m(t) = \infty$.
2. The mass term $m(t)^2 u$ is non-effective if

$$\limsup_{t \to \infty} (1 + t) \int_t^\infty m(s)^2 ds < \frac{1}{4}.$$

This classification tells us that the mass term is "almost effective", at least not non-effective, for large values of μ_2^2. Hence, there might be a relation to properties of solutions to the Cauchy problem

$$u_{tt} - \Delta u + \mu_2^2 u = |u|^p, \quad u(0, x) = \varphi(x), \quad u_t(0, x) = \psi(x).$$

The critical exponent could be $p_{crit}(n) = p_{Fuj}(n) = 1 + \frac{2}{n}$ (see [110]). But, on the other hand, we have a damping term $\frac{\mu_1}{1+t} u_t$ which has an improving influence on the critical exponent (cf. with [39] or Sect. 20.3). The following result is proved in [156], see also [157].

Theorem 18.3.2 *Consider the Cauchy problem*

$$u_{tt} - \Delta u + \frac{\mu_1}{1+t} u_t + \frac{\mu_2^2}{(1+t)^2} u = |u|^p, \quad u(0, x) = \varphi(x), \quad u_t(0, x) = \psi(x)$$

under the constrain $(\mu_1 - 1)^2 - 4\mu_2^2 < 0$. Let $n \leq 4$ and suppose that $\mu_1 > 4$. Finally, let

$$\begin{cases} p \geq 2 & \text{if } n = 1, 2, \\ 2 \leq p \leq 3 = p_{GN}(3) & \text{if } n = 3, \\ p = 2 = p_{GN}(4) & \text{if } n = 4. \end{cases}$$

Then, there exists a constant $\varepsilon_0 > 0$ such that for all

$$(\varphi, \psi) \in \mathcal{A}_{1,1} := (H^1(\mathbb{R}^n) \cap L^1(\mathbb{R}^n)) \times (L^2(\mathbb{R}^n) \cap L^1(\mathbb{R}^n))$$

with

$$\|(\varphi, \psi)\|_{\mathcal{A}_{1,1}} \leq \varepsilon_0$$

there exists a uniquely determined global (in time) energy solution belonging to

$$C([0, \infty), H^1(\mathbb{R}^n)) \cap C^1([0, \infty), L^2(\mathbb{R}^n)).$$

Moreover, there exists a nonnegative constant C such that the solution satisfies the following decay estimates:

$$\|(u_t(t, \cdot), \nabla u(t, \cdot))\|_{L^2} \leq C(1 + t)^{-\frac{\mu_1}{2}} \|(\varphi, \psi)\|_{\mathcal{A}_{1,1}},$$

$$\|u(t,\cdot)\|_{L^2} \leq C(1+t)^{-\frac{\mu_1}{2}}\widetilde{q}_{\Delta}(t)\|(\varphi,\psi)\|_{\mathcal{A}_{1,1}}$$

for all $t \geq 0$, where $\widetilde{q}_{\Delta}(t) = 1$ for $n > 1$ and $\widetilde{q}_{\Delta}(t) = (\log(e+t))^{\frac{1}{2}}$ for $n = 1$.

Remark 18.3.2 We explained in Theorem 18.3.2 only results under the assumption $\mu_1 > 2$. One can find in [156] results in the cases of small positive values for μ_1, too. But these results seem far from being optimal, although we have not determined the critical exponent in Case 1.

Case 2 $(\mu_1 - 1)^2 - 4\mu_2^2 > 1$ In this second case the dissipation term is dominant. In [222] a classification of dissipation terms is proposed for the Cauchy problem

$$u_{tt} - \Delta u + b(t)u_t = 0, \quad u(0,x) = \varphi(x), \quad u_t(0,x) = \psi(x).$$

We omit explaining some structural assumptions for the time-dependent coefficient in effective or non-effective dissipation terms as well. We only hint at the asymptotical behavior of the coefficient for $t \to \infty$.

1. The dissipation term $b(t)u_t$ is effective if $\lim_{t\to\infty} tb(t) = \infty$.
2. The dissipation term $b(t)u_t$ is non-effective if

$$\limsup_{t\to\infty} tb(t) < 1.$$

This classification tells us, that the dissipation term is "almost effective" for very large μ_1. Hence, there might be a relation to properties of solutions to the Cauchy problem

$$u_{tt} - \Delta u + u_t = |u|^p, \quad u(0,x) = \varphi(x), \quad u_t(0,x) = \psi(x).$$

For this model, the critical exponent is $p_{crit}(n) = p_{Fuj}(n) = 1 + \frac{2}{n}$ (see Theorems 18.1.1 and 18.1.2). We have the same critical exponent for the model

$$u_{tt} - \Delta u + \frac{\mu_1}{1+t}u_t = |u|^p, \quad u(0,x) = \varphi(x), \quad u_t(0,x) = \psi(x),$$

if $\mu_1 \geq n+2$, $n \geq 3$ (see [30] and [35]). But, on the other hand, we have a mass term $\frac{\mu_2^2}{(1+t)^2}u$ which might have an improving influence on the critical exponent. The following result is proved in [158].

Theorem 18.3.3 *Consider the Cauchy problem*

$$u_{tt} - \Delta u + \frac{\mu_1}{1+t}u_t + \frac{\mu_2^2}{(1+t)^2}u = |u|^p, \quad u(0,x) = \varphi(x), \quad u_t(0,x) = \psi(x)$$

under the constrain $1 + \sqrt{(\mu_1 - 1)^2 - 4\mu_2^2} > n + 2$. *Let* $n \leq 4$ *and suppose that* $\mu_1 > 1$. *Moreover, let*

$$(\varphi, \psi) \in \mathcal{A}_{1,1} := (H^1(\mathbb{R}^n) \cap L^1(\mathbb{R}^n)) \times (L^2(\mathbb{R}^n) \cap L^1(\mathbb{R}^n)),$$

and let

$$\begin{cases} p \in [2, \infty) & \text{if } n = 1, 2, \\ p \in \left[2, \frac{n}{n-2}\right] & \text{if } n = 3, 4. \end{cases}$$

Finally, let us assume

$$p > p_{Fuj}(n - \gamma) \quad \text{with} \quad \gamma = \frac{1 - \mu_1 + \sqrt{(\mu_1 - 1)^2 - 4\mu_2^2}}{2} < 0.$$

Then, the following statement holds with a suitable constant $\varepsilon_0 > 0$: *if*

$$\|(\varphi, \psi)\|_{\mathcal{A}_{1,1}} \leq \varepsilon_0,$$

then, there exists a uniquely determined global (in time) energy solution

$$u \in C([0, \infty), H^1(\mathbb{R}^n)) \cap C^1([0, \infty), L^2(\mathbb{R}^n)).$$

Moreover, there exists a nonnegative constant C *such that the solution and its energy terms satisfy the following decay estimates:*

$$\|u(t, \cdot)\|_{L^2} \leq C(1 + t)^{-\frac{n}{2} + \gamma} \|(\varphi, \psi)\|_{\mathcal{A}_{1,1}},$$

$$\|\nabla u(t, \cdot)\|_{L^2} \leq C(1 + t)^{-\frac{n}{2} - 1 + \gamma} \|(\varphi, \psi)\|_{\mathcal{A}_{1,1}},$$

$$\|u_t(t, \cdot)\|_{L^2} \leq C(1 + t)^{-\frac{n}{2} - 1 + \gamma} \|(\varphi, \psi)\|_{\mathcal{A}_{1,1}}.$$

Remark 18.3.3 We explained in Theorem 18.3.3 only results for the case of large values of $1 + \sqrt{(\mu_1 - 1)^2 - 4\mu_2^2}$. One can find in [158] results in the cases of small positive values for $\sqrt{(\mu_1 - 1)^2 - 4\mu_2^2}$, too. But these results are not optimal because our approach of Sect. 18.1 to attack semilinear models is an appropriate one only for large values of $\sqrt{(\mu_1 - 1)^2 - 4\mu_2^2}$ (see also [30] and the next case).

Case 3 The "grey zone" is formed by the above models with the condition $\sqrt{(\mu_1 - 1)^2 - 4\mu_2^2} \in [0, n + 1]$. We cannot expect results like those in Theorems 18.3.2 and 18.3.3. The critical exponent might be changed. We will discuss the case $(\mu_1 - 1)^2 - 4\mu_2^2 = 1$, only. The following blow up result is proved in [156].

Theorem 18.3.4 *Assume that $u \in C^2([0,T) \times \mathbb{R}^n)$ is a classical solution to the Cauchy problem*

$$u_{tt} - \Delta u + \frac{\mu_1}{1+t} u_t + \frac{\mu_2^2}{(1+t)^2} u = |u|^p, \quad u(0,x) = \varphi(x), \quad u_t(0,x) = \psi(x)$$

with $(\mu_1 - 1)^2 - 4\mu_2^2 = 1$. The data $(\varphi, \psi) \neq (0,0)$ are supposed to belong to $C_0^2(\mathbb{R}^n) \times C_0^1(\mathbb{R}^n)$, where φ, ψ are positive in the interior of their support. If $p \in (1, p_{\mu_1}(n)]$, then $T < \infty$, where

$$p_{\mu_1}(n) := \max\left\{ p_{Fuj}\left(n - 1 + \frac{\mu_1}{2}\right); p_0(n + \mu_1) \right\}.$$

Here $p_{Fuj}(n) = 1 + \frac{2}{n}$ is the Fujita exponent and $p_0(n)$ is the Strauss exponent, that is, the positive root of

$$(n-1)p^2 - (n+1)p - 2 = 0.$$

Remark 18.3.4 The statement of the last theorem implies that $p_{crit}(\mu_1, n) \geq p_{\mu_1}(n)$. In the paper [39] the number $p_{\mu_1}(n)$ was calculated as follows:

1. $p_{\mu_1}(1) = p_{Fuj}\left(\frac{\mu_1}{2}\right)$,

2. $p_{\mu_1}(2) = \begin{cases} p_{Fuj}\left(1 + \frac{\mu_1}{2}\right) & \text{if } \mu_1 \geq 2, \\ p_0(2 + \mu_1) & \text{if } \mu_1 \in [0,2], \end{cases}$

3. $p_{\mu_1}(n) = p_0(n + \mu_1)$ if $n \geq 3$.

If $\mu_1 = 2$ and $\mu_2 = 0$, then due to [39], the critical exponent $p_{crit}(2,n)$ coincides with $p_2(n)$. So we have, in general, a shift of 2 in the Strauss exponent ($p_2(n) = p_0(n+2)$ in a lot of cases). For general μ_1 we have at least a shift of Strauss exponent by μ_1. So, a large μ_1 requires a large μ_2, the interplay produces at least a large shift.

Exercises Relating to the Considerations of Chap. 18

Exercise 1 Prove Duhamel's principle for solutions to the Cauchy problem

$$u_{tt} - \Delta u + u_t = f(t,x), \quad u(0,x) = \varphi(x), \quad u_t(0,x) = \psi(x),$$

that is, derive the representation of solution

$$u(t,x) = K_0(t,0,x) *_{(x)} \varphi(x) + K_1(t,0,x) *_{(x)} \psi(x)$$

$$+ \int_0^t K_1(t,s,x) *_{(x)} f(s,x) \, ds.$$

The notations are chosen as in Sect. 18.1.1.

Exercise 2 Prove the statement of Theorem 18.3.1.

Exercise 3 Think on the Remark 18.3.1.

Exercise 4 Study the proof to Theorem 18.2.5 in [154].

Chapter 19
Semilinear Wave Models with a Special Structural Dissipation

The considerations of this chapter are take in conjunction with the preceding chapter. This helps to clarify and extend the approach of Chap. 18 to different classes of semilinear wave models. On the one hand a special structural damping term is included in the linear part of the model, while on the other hand different power nonlinearities of modulus of partial derivatives of the solution are allowed. First we explain $L^p - L^q$ estimates not necessarily on the conjugate line for linear structurally damped wave models. Then we discuss different treatments of the nonlinear term and application of fractional Gagliardo-Nirenberg inequality or fractional powers as well. The optimality of the results is proved by application of the test function method. That this approach can also be applied to treat semilinear viscoelastic damped wave models or semilinear structurally damped σ-evolution models is verified in the concluding remarks.

19.1 Semilinear Wave Models with a Special Structural Damping Term

In Sect. 18 we explained an approach to study the Cauchy problem for classical damped waves with power-nonlinearities. In this section we will show that the same approach can be used to study Cauchy problems for semilinear wave models with a special structural damping and other nonlinearities. The Cauchy problems we have in mind are

$$u_{tt} - \Delta u + \mu(-\Delta)^{\frac{1}{2}} u_t = ||D|^a u|^p, \quad u(0,x) = \varphi(x), \quad u_t(0,x) = \psi(x),$$

$$u_{tt} - \Delta u + \mu(-\Delta)^{\frac{1}{2}} u_t = |u_t|^p, \quad u(0,x) = \varphi(x), \quad u_t(0,x) = \psi(x),$$

© Springer International Publishing AG 2018
M.R. Ebert, M. Reissig, *Methods for Partial Differential Equations*,
https://doi.org/10.1007/978-3-319-66456-9_19

where the parameter μ is positive and $a \in [0, 1)$. Structurally damped wave models were already introduced in Sect. 11.3.3. In the following sections we will

1. derive $L^p - L^q$ estimates not necessarily on the conjugate line for solutions to the linear Cauchy problem

$$v_{tt} - \Delta v + \mu(-\Delta)^{\frac{1}{2}} v_t = 0, \quad v(0, x) = \varphi(x), \quad v_t(0, x) = \psi(x),$$

2. show how the approach of Sect. 18.1.1 is modified to prove the global (in time) existence of small data solutions to semilinear structurally damped wave models.

19.2 $L^p - L^q$ Estimates Not Necessarily on the Conjugate Line

In this section we are interested in obtaining $L^p - L^q$ estimates not necessarily on the conjugate line for solutions to the Cauchy problem

$$v_{tt} - \Delta v + \mu(-\Delta)^{\frac{1}{2}} v_t = 0, \quad v(0, x) = \varphi(x), \quad v_t(0, x) = \psi(x),$$

where $\mu > 0$. After application of partial Fourier transformation we get

$$w_{tt} + |\xi|^2 w + \mu|\xi| w_t = 0, \quad w(0, \xi) = F(\varphi)(\xi), \quad w_t(0, \xi) = F(\psi)(\xi).$$

Here we introduce the notation $w(t, \xi) := F_{x \to \xi}(v)(t, \xi)$. We apply ideas from [146] (see also Sect. 24.2.2). We divide the considerations into two sub-cases: $\mu = 2$ and $\mu \neq 2$.

Special Case $\mu = 2$ We have a double root $\lambda_{1,2}(\xi) = -|\xi|$. The solution w is

$$w(t, \xi) = e^{-|\xi|t} \big(w(0, \xi) + t(|\xi| w(0, \xi) + w_t(0, \xi)) \big).$$

Transforming back we get the following representation for $v = v(t, x)$:

$$v(t, x) = F_{\xi \to x}^{-1} \big(e^{-|\xi|t} \big(w(0, \xi) + t(|\xi| w(0, \xi) + w_t(0, \xi)) \big) \big)$$

$$= F_{\xi \to x}^{-1} \big(e^{-|\xi|t} \big) * \varphi + F_{\xi \to x}^{-1} \big(|\xi| e^{-|\xi|t} \big) * t\varphi + F_{\xi \to x}^{-1} \big(e^{-|\xi|t} \big) * t\psi,$$

where we suppose that the Fourier inversion formula holds for the data φ and ψ. In this representation there appear oscillatory integrals which can be estimated by using Corollaries 24.2.1–24.2.3 of Sect. 24.2.2. Applying Young's inequality of Proposition 24.5.2 we may conclude immediately the following statement:

Corollary 19.2.1 *The solutions to*

$$v_{tt} - \Delta v + 2(-\Delta)^{\frac{1}{2}} v_t = 0, \quad v(0, x) = \varphi(x), \quad v_t(0, x) = \psi(x),$$

satisfy the $L^p - L^q$ estimates

$$\|v(t,\cdot)\|_{L^q} \leq Ct^{-n\left(1-\frac{1}{r}\right)}\|\varphi\|_{L^p} + Ct^{1-n\left(1-\frac{1}{r}\right)}\|\psi\|_{L^p}$$

for $1 \leq p \leq q \leq \infty$, $1 + \dfrac{1}{q} = \dfrac{1}{r} + \dfrac{1}{p}$ and for general dimensions n.

In our further considerations we are not only interested in estimates for solutions, but also of their energy $E_W(u)(t)$. By choosing the values $q = r = m$ and $p = 1$ for $t \in [1, \infty)$, and then $p = q = m$ and $r = 1$ for $t \in (0, 1]$ in the representations of solutions for v, v_t and $|D|^\sigma v$, respectively, we may conclude the following statement.

Corollary 19.2.2 *Let $m \in (1, 2]$. The solutions to*

$$v_{tt} - \Delta v + 2(-\Delta)^{\frac{1}{2}}v_t = 0, \quad v(0,x) = \varphi(x), \quad v_t(0,x) = \psi(x),$$

satisfy the following $(L^1 \cap L^m) - L^m$ estimates:

$$\|v(t,\cdot)\|_{L^m} \leq C(1+t)^{-n\left(1-\frac{1}{m}\right)}\|\varphi\|_{L^m \cap L^1} + C(1+t)^{1-n\left(1-\frac{1}{m}\right)}\|\psi\|_{L^m \cap L^1},$$

$$\|(|D|v(t,\cdot), v_t(t,\cdot))\|_{L^m} \leq C(1+t)^{-1-n\left(1-\frac{1}{m}\right)}\|\varphi\|_{H_m^1 \cap L^1} + C(1+t)^{-n\left(1-\frac{1}{m}\right)}\|\psi\|_{L^m \cap L^1},$$

and the following $L^m - L^m$ estimates:

$$\|v(t,\cdot)\|_{L^m} \leq C\|\varphi\|_{L^m} + C(1+t)\|\psi\|_{L^m},$$

$$\|(|D|v(t,\cdot), v_t(t,\cdot))\|_{L^m} \leq C(1+t)^{-1}\|\varphi\|_{H_m^1} + C\|\psi\|_{L^m}$$

for all dimensions n.

Remark 19.2.1 There is a difference in the estimates from Corollaries 19.2.1 and 19.2.2. In Corollary 19.2.2 we assume additional regularity for the data which was in opposite to our assumptions in Corollary 19.2.1, where data are supposed to belong to L^q spaces only. This is the reason we can use $1 + t$ in Corollary 19.2.2 instead of t only which is what we have in the estimates from Corollary 19.2.1 (cf. with Theorem 14.2.4).

Corollary 19.2.3 *We are able to prove estimates which are similar for the solutions and their energies (even of higher order) to those in Corollary 19.2.2. Namely, for any $a \geq 0$ we have the following $(L^1 \cap L^m) - L^m$ estimates $(m \in (1, 2])$:*

$$\||D|^a v_t(t,\cdot)\|_{L^m} \leq C(1+t)^{-1-a-n\left(1-\frac{1}{m}\right)}\|\varphi\|_{H_m^{a+1} \cap L^1}$$
$$+ C(1+t)^{-a-n\left(1-\frac{1}{m}\right)}\|\psi\|_{H_m^a \cap L^1},$$

$$\||D|^a v(t, \cdot)\|_{L^m} \le C(1 + t)^{-a-n\left(1-\frac{1}{m}\right)} \|\varphi\|_{H_m^a \cap L^1}$$
$$+ C(1 + t)^{1-a-n\left(1-\frac{1}{m}\right)} \|\psi\|_{H_m^{\max(a-1,0)} \cap L^1}.$$

Moreover, we have the $L^m - L^m$ estimates ($m \in (1, 2]$)

$$\|(|D|^{a+1} v(t, \cdot), |D|^a v_t(t, \cdot))\|_{L^m} \le C(1 + t)^{-1-a} \|\varphi\|_{H_m^{a+1}} + C(1 + t)^{-a} \|\psi\|_{H_m^a}.$$

Case $\mu \in (2, \infty)$ We have the roots $\lambda_{1,2}(\xi) = \frac{1}{2}|\xi|\left(-\mu \pm \sqrt{\mu^2 - 4}\right)$.

Proposition 19.2.1 *The solutions of the Cauchy problem*

$$v_{tt} - \Delta v + \mu(-\Delta)^{\frac{1}{2}} v_t = 0, \ v(0, x) = \varphi(x), \ v_t(0, x) = \psi(x), \ \mu \in (2, \infty)$$

satisfy the following $L^p - L^q$ estimates:

$$\|v(t, \cdot)\|_{L^q} \le Ct^{-n\left(1-\frac{1}{r}\right)} \|\varphi\|_{L^p} + Ct^{1-n\left(1-\frac{1}{r}\right)} \|\psi\|_{L^p},$$

for $1 \le p \le q \le \infty$, $1 + \dfrac{1}{q} = \dfrac{1}{r} + \dfrac{1}{p}$ and for general dimensions n.

Proof We have $v = K_0 * \varphi + K_1 * \psi$, where the partial Fourier transforms $\hat{K}_0$ and $\hat{K}_1$ of the fundamental solutions K_0 and K_1 are defined as follows:

$$\hat{K}_0 := \frac{\lambda_1 e^{\lambda_2 t} - \lambda_2 e^{\lambda_1 t}}{\lambda_1 - \lambda_2}, \quad \hat{K}_1 := \frac{e^{\lambda_1 t} - e^{\lambda_2 t}}{\lambda_1 - \lambda_2} \quad \text{if } \lambda_1 \ne \lambda_2.$$

The functions $\hat{K}_j = \hat{K}_j(t, \xi)$, $j = 0, 1$, are smooth with respect to $\{\xi \in \mathbb{R}^n : |\xi| > 0\}$. Let us assume that the Fourier inversion formula is applicable for the fundamental solutions and the data. We may then conclude

$$v = F_{\xi \to x}^{-1}(\hat{K}_0) * \varphi + F_{\xi \to x}^{-1}(\hat{K}_1) * \psi.$$

On the one hand $\hat{K}_0 \sim e^{-c|\xi|t}$ with a positive $c = c(\mu)$, while on the other hand the Newton-Leibniz formula implies

$$\hat{K}_1 = t \int_0^1 e^{\frac{1}{2}\left(-\mu+(2\theta-1)\sqrt{\mu^2-4}\right)|\xi|t} d\theta.$$

By estimating

$$-\mu + (2\theta - 1)\sqrt{\mu^2 - 4} \le -\mu + \sqrt{\mu^2 - 4} \le -\frac{2}{\mu} < 0$$

and after using Corollaries 24.2.1–24.2.3 of Sect. 24.2.2 for estimating both oscillating integrals we get the desired $L^p - L^q$ estimate.

Case $\mu \in (0,2)$ This remaining case is of special interest since the characteristic roots are

$$\lambda_{1,2}(\xi) = \frac{1}{2}|\xi|\left(-\mu \pm i\sqrt{4 - \mu^2}\right).$$

The related multipliers are of trigonometric type as $e^{-c_1|\xi|t}h(t,\xi)$, where $h(t,\xi) = \cos(c_2|\xi|t)$ or $h(t,\xi) = \sin(c_2|\xi|t)$. Here $c_1 = c_1(\mu) > 0$ and $c_2 = c_2(\mu) \neq 0$, are real constants. So, we shall discuss the oscillatory integrals

$$F_{\xi \to x}^{-1}\left(e^{-c_1|\xi|t}\cos(c_2|\xi|t)\right) \text{ and } F_{\xi \to x}^{-1}\left(e^{-c_1|\xi|t}\sin(c_2|\xi|t)\right).$$

We use the following result of [146].

Lemma 19.2.1 *The following estimates hold in* $\mathbb{R}^n$ *for* $n \geq 2$:

$$\left\|F_{\xi \to x}^{-1}\left(e^{-c_1|\xi|t}\cos(c_2|\xi|t)\right)\right\|_{L^p(\mathbb{R}^n)} \leq Ct^{-n(1-\frac{1}{p})},$$

$$\left\|F_{\xi \to x}^{-1}\left(e^{-c_1|\xi|t}\sin(c_2|\xi|t)\right)\right\|_{L^p(\mathbb{R}^n)} \leq Ct^{-n(1-\frac{1}{p})}$$

for $p \in [1, \infty]$ *and* $t > 0$.

Proof Here we can follow the proofs to all theorems of Sect. 24.2.2. Let us only sketch how to obtain the estimate for

$$\left\|F_{\xi \to x}^{-1}\left(e^{-c_1|\xi|t}\cos(c_2|\xi|t)\right)\right\|_{L^p(\mathbb{R}^3)}.$$

The change of variables from the proof to Theorem 24.2.1 reduces the considerations to the Fourier multiplier

$$F_{\eta \to x}^{-1}\left(e^{-c_1|\eta|}\cos(c_2|\eta|)\right).$$

The integrand is radial symmetric with respect to η. Hence, the inverse Fourier transform is radial symmetric with respect to x, too. Modified Bessel functions of Sect. 24.2.2 are applicable and we get the representation

$$F_{\eta \to x}^{-1}\left(e^{-c_1|\eta|}\cos(c_2|\eta|)\right) = \frac{1}{|x|^2}\int_0^\infty d_r\left(e^{-c_1 r}\cos(c_2 r)r\right)\cos(r|x|)\,dr.$$

In the next step of partial integration we use $\sin(r|x|) = 0$ for $r = 0$. A fourth step of partial integration gives immediately the estimate

$$|F_{\eta \to x}^{-1}\left(e^{-c_1|\eta|}\cos(c_2|\eta|)\right)| = \left|\frac{1}{|x|^2}\int_0^\infty d_r\left(e^{-c_1 r}\cos(c_2 r)r\right)\cos(r|x|)\,dr\right|$$

$$\leq C\frac{1}{\langle x \rangle^4}.$$

This implies

$$\left\| F_{\eta \to x}^{-1} \left(e^{-c_1 |\eta|} \cos(c_2 |\eta|) \right) \right\|_{L^1(\mathbb{R}^3)} \leq C$$

(cf. with the proof to Theorem 24.2.1), which follows the desired statement. Taking into consideration Lemma 19.2.1 we conclude the following results.

Corollary 19.2.4 *The solutions of the Cauchy problem*

$$v_{tt} - \Delta v + \mu(-\Delta)^{\frac{1}{2}} v_t = 0, \quad v(0, x) = \varphi(x), \quad v_t(0, x) = \psi(x), \quad \mu \in (0, 2),$$

satisfy the following $L^p - L^q$ estimates:

$$\|v(t, \cdot)\|_{L^q} \leq C t^{-n\left(1 - \frac{1}{r}\right)} \|\varphi\|_{L^p} + C t^{1-n\left(1 - \frac{1}{r}\right)} \|\psi\|_{L^p},$$

for $1 \leq p \leq q \leq \infty$, $1 + \dfrac{1}{q} = \dfrac{1}{r} + \dfrac{1}{p}$ and for general dimensions n.

Corollary 19.2.5 *The solutions to*

$$v_{tt} - \Delta v + \mu(-\Delta)^{\frac{1}{2}} v_t = 0, \quad v(0, x) = \varphi(x), \quad v_t(0, x) = \psi(x),$$

satisfy for positive μ and $m \in (1, 2]$ the $(L^1 \cap L^m) - L^m$ estimates and $L^m - L^m$ estimates from Corollaries 19.2.2 and 19.2.3.

19.3 Structurally Damped Wave Models with Nonlinearity $||D|^a u|^p$

19.3.1 Main Result

Theorem 19.3.1 *Let us consider the Cauchy problem*

$$u_{tt} - \Delta u + \mu(-\Delta)^{\frac{1}{2}} u_t = ||D|^a u|^p, \quad u(0, x) = \varphi(x), \quad u_t(0, x) = \psi(x)$$

with $a \in [0, 1)$ and $\mu > 0$, where the data (φ, ψ) are taken from the space

$$(H_m^1(\mathbb{R}^n) \cap L^1(\mathbb{R}^n)) \times (L^m(\mathbb{R}^n) \cap L^1(\mathbb{R}^n))$$

with $m \in (1, 2]$. We assume for the dimension n the condition

$$n \leq \frac{m^2(1 - a)}{m - 1}.$$

Moreover, the exponent p satisfies the conditions

$$p \in \left[m, \frac{n}{[n + m(a - 1)]^+}\right] \quad and \quad p > 1 + \frac{2 - a}{n + a - 1}.$$

Then, there exists a constant $\varepsilon_0 > 0$ such that for any small data (φ, ψ) satisfying the condition

$$\|(\varphi, \psi)\|_{(H_m^1 \cap L^1) \times (L^m \cap L^1)} \le \varepsilon_0$$

we have a uniquely determined global (in time) small data Sobolev solution

$$u \in C\left([0, \infty), H_m^1(\mathbb{R}^n)\right) \cap C^1\left([0, \infty), L^m(\mathbb{R}^n)\right).$$

Moreover, the solution satisfies the following $L^m \cap L^1 \to L^m$ estimates:

$$\|u(t, \cdot)\|_{L^m} \le C(1 + t)^{1 - n\left(1 - \frac{1}{m}\right)} \|(\varphi, \psi)\|_{(H_m^1 \cap L^1) \times (L^m \cap L^1)},$$

$$\|(|D|u(t, \cdot), u_t(t, \cdot))\|_{L^m} \le C(1 + t)^{-n\left(1 - \frac{1}{m}\right)} \|(\varphi, \psi)\|_{(H_m^1 \cap L^1) \times (L^m \cap L^1)}.$$

Remark 19.3.1 Let us explain the conditions for p and n of the last theorem. The condition $p > 1 + \frac{2-a}{n+a-1}$ provides the same decay estimates of solutions to the semilinear model as for the solutions to the corresponding linear model with vanishing right-hand side. In some cases we, know that this lower bound is optimal (see Sect. 19.3.3). So, the nonlinearity is interpreted as a small perturbation. We apply the tool of fractional Gagliardo-Nirenberg inequality (see Corollary 24.5.1). For this reason, the other conditions for p come into play. The upper bound for n arises from the admissible range for the power p. Smaller p imply that this set is not empty. In the case $a = 0$ (we do not have a pseudodifferential action on the right-hand side) and $m = 2$ we conclude the assumptions $n \in [2, 4]$, $p \in [2, \frac{n}{n-2}]$ and $p > 1 + \frac{2}{n-1}$ (see [37]).

Example 19.3.1 Let us choose $n = m(1 - a)$. Then, the admissible p are from the interval $[m, \infty)$. Taking into account $1 + \frac{2-a}{(m-1)(1-a)} > m$, we can apply the theorem for $p \in \left(1 + \frac{2-a}{(m-1)(1-a)}, \infty\right)$.

19.3.2 Proof

We explained the philosophy for proving the existence of global (in time) small data solutions in Sect. 18.1.1.2. We can follow this approach to prove Theorem 19.3.1, too. For this reason we skip some details, sketch some steps and describe only the changes in the proof which are necessary due to the presence of the nonlinearity $\||D|^a u|^p$ with $a \in [0, 1)$.

We introduce the space of data

$$A := (H_m^1(\mathbb{R}^n) \cap L^1(\mathbb{R}^n)) \times (L^m(\mathbb{R}^n) \cap L^1(\mathbb{R}^n)).$$

Moreover, we introduce for all $t > 0$ the function spaces

$$X(t) := C\big([0, t], H_m^1(\mathbb{R}^n)\big) \cap C^1\big([0, t], L^m(\mathbb{R}^n)\big)$$

with the norm

$$\|u\|_{X(t)} := \sup_{0 \le \tau \le t} \big(f_0(\tau)^{-1} \|u(\tau, \cdot)\|_{L^m} + f_1(\tau)^{-1} \||D|u(\tau, \cdot)\|_{L^m} + g_0(\tau)^{-1} \|u_t(\tau, \cdot)\|_{L^m}\big),$$

and the space

$$X_0(t) := C\big([0, t], H_m^1(\mathbb{R}^n)\big)$$

with the norm

$$\|u\|_{X_0(t)} := \sup_{0 \le \tau \le t} \big(f_0(\tau)^{-1} \|u(\tau, \cdot)\|_{L^m} + f_1(\tau)^{-1} \||D|u(\tau, \cdot)\|_{L^m}\big),$$

where we choose from the estimates of Corollaries 19.2.5, 19.2.2 and 19.2.3 the weights

$$f_0(\tau) := (1 + \tau)^{1-n(1-\frac{1}{m})}, \quad f_1(\tau) = g_0(\tau) := (1 + \tau)^{-n(1-\frac{1}{m})}.$$

We define for any $u \in X(t)$ the operator

$$N : u \in X(t) \to Nu \in X(t) \text{ by}$$

$$Nu(t, x) = K_0(t, 0, x) *_{(x)} \varphi(x) + K_1(t, 0, x) *_{(x)} \psi(x)$$

$$+ \int_0^t K_1(t - \tau, 0, x) *_{(x)} \big||D|^a u(\tau, x)\big|^p d\tau.$$

To apply Banach's fixed point theorem we have to prove the required estimates for $\|Nu(t, \cdot)\|_{X(t)}$ and $\|Nu(t, \cdot) - Nv(t, \cdot)\|_{X(t)}$. In this way we may simultaneously conclude local (in time) large data solutions and global (in time) small data solutions as well. The decay estimates for the solution and its partial derivatives of first order follow immediately the definition of the norm in $X(t)$.

Following the strategy to estimate the integral term in the representation of Nu the estimates of Corollary 19.2.5 allows us to get

$$\|\partial_t^j |D|^k Nu(t,\cdot)\|_{L^m} \leq C(1+t)^{1-n(1-\frac{1}{m})-(k+j)} \|(\varphi,\psi)\|_{(H_m^{k+j}\cap L^1)\times(H_m^{k+j-1}\cap L^1)}$$

$$+C\int_0^{\frac{t}{2}} (1+t-\tau)^{1-n(1-\frac{1}{m})-(k+j)} \||D|^a u(\tau,\cdot)|^p\|_{L^m\cap L^1} d\tau$$

$$+C\int_{\frac{t}{2}}^t (1+t-\tau)^{1-(k+j)} \||D|^a u(\tau,\cdot)|^p\|_{L^m} d\tau,$$

where $j,k = 0,1$ and $(j,k) \neq (1,1)$. It is required to estimate $\||D|^a u(\tau,\cdot)|^p$ in $L^m(\mathbb{R}^n) \cap L^1(\mathbb{R}^n)$ and in $L^m(\mathbb{R}^n)$, so we estimate $\||D|^a u(\tau,\cdot)\|_{L^p}$ and $\||D|^a u(\tau,\cdot)\|_{L^{mp}}$. Here, fractional Gagliardo-Nirenberg inequalities come into play (see Proposition 24.5.5). The condition

$$p \in \left[m, \frac{n}{[n+m(a-1)]^+}\right]$$

allows us to apply this inequality for $q = p$ and $q = pm$. After using the estimates of Corollary 19.2.5, Proposition 24.5.5 and Remark 24.5.2, we may conclude

$$\||D|^a u(\tau,\cdot)|^p\|_{L^m\cap L^1} \leq C(1+\tau)^{p(1-n(1-\frac{1}{m})-\theta_{a,1}(p,m))} \|u(\tau,\cdot)\|_{X_0(\tau)}^p$$

$$= C(1+\tau)^{-p(n+a-1)+n} \|u(\tau,\cdot)\|_{X_0(\tau)}^p$$

because of $\theta_{a,1}(p,m) < \theta_{a,1}(pm,m)$. Here we use $\theta_{a,1}(q,m) = n(\frac{1}{m}-\frac{1}{q}+\frac{a}{n})$, whereas

$$\||D|^a u(\tau,\cdot)|^p\|_{L^m} \leq C(1+\tau)^{p(1-n(1-\frac{1}{m})-\theta_{a,1}(mp,m))} \|u(\tau,\cdot)\|_{X_0(\tau)}^p$$

$$= C(1+\tau)^{-p(n+a-1)+\frac{n}{m}} \|u(\tau,\cdot)\|_{X_0(\tau)}^p.$$

Summarizing both estimates yields

$$\|\partial_t^j |D|^k Nu(t,\cdot)\|_{L^m} \leq C(1+t)^{1-n(1-\frac{1}{m})-(k+j)} \|(\varphi,\psi)\|_{(H_m^{k+j}\cap L^1)\times(H_m^{k+j-1}\cap L^1)}$$

$$+C(1+t)^{1-n(1-\frac{1}{m})-(k+j)} \|u\|_{X_0(t)}^p \int_0^{\frac{t}{2}} (1+\tau)^{-p(n+a-1)+n} d\tau$$

$$+C(1+t)^{-p(n+a-1)+\frac{n}{m}} \|u\|_{X_0(t)}^p \int_{\frac{t}{2}}^t (1+t-\tau)^{1-(k+j)} d\tau.$$

If $p > 1 + \frac{2-a}{n+a-1}$, then the term $(1+\tau)^{-p(n+a-1)+n}$ is integrable over $\mathbb{R}^1_+$. Moreover, we have

$$(1+t)^{-p(n+a-1)+\frac{n}{m}} \int_{t/2}^t (1+t-\tau)^{1-(k+j)}\, d\tau$$

$$= (1+t)^{-p(n+a-1)+\frac{n}{m}} \int_0^{t/2} (1+\tau)^{1-(k+j)}\, d\tau \leq C(1+t)^{1-n(1-\frac{1}{m})-(k+j)}.$$

In this way we derived the desired estimate for $\|Nu(t,\cdot)\|_{X(t)}$. Analogously, the necessary estimate for $\|Nu(t,\cdot) - Nv(t,\cdot)\|_{X(t)}$ can be derived. Therefore, we need to prove suitable estimates for $\partial_t^j |D|^k \big(Nu(t,\cdot) - Nv(t,\cdot)\big)$. We arrive at the following relations:

$$\partial_t^j |D|^k \int_0^t K_1(t-\tau,0,x) *_{(x)} \big(||D|^a u(\tau,x)|^p - ||D|^a v(\tau,x)|^p\big)\, d\tau$$

$$= \int_0^t \partial_t^j |D|^k K_1(t-\tau,0,x) *_{(x)} \big(||D|^a u(\tau,x)|^p - ||D|^a v(\tau,x)|^p\big)\, d\tau,$$

and

$$\Big\| ||D|^a u(\tau,x)|^p - ||D|^a v(\tau,x)|^p \Big\|_{L^1}$$

$$\leq C \||D|^a(u(\tau,\cdot) - v(\tau,\cdot))\|_{L^p} \big(\||D|^a u(\tau,\cdot)\|_{L^p}^{p-1} + \||D|^a v(\tau,\cdot)\|_{L^p}^{p-1}\big),$$

$$\Big\| ||D|^a u(\tau,x)|^p - ||D|^a v(\tau,x)|^p \Big\|_{L^m}$$

$$\leq C \||D|^a(u(\tau,\cdot) - v(\tau,\cdot))\|_{L^{mp}} \big(\||D|^a u(\tau,\cdot)\|_{L^{mp}}^{p-1} + \||D|^a v(\tau,\cdot)\|_{L^{mp}}^{p-1}\big).$$

Repeating the above considerations we are able to estimate the terms

$$\||D|^a(u(\tau,\cdot) - v(\tau,\cdot))\|_{L^{rp}}, \quad \||D|^a u(\tau,\cdot)\|_{L^{rp}}, \quad \||D|^a v(\tau,\cdot)\|_{L^{rp}}$$

for $r = 1, m$.

Finally, we mention that the upper bound for the dimension n is determined by the range of admissible powers p. These observations complete the proof.

19.3.3 Optimality

We are able to prove the optimality of the results of Theorem 19.3.1 only in the special case $a = 0$. In this case, the condition for admissible p is $p > p_{crit} := 1 + \frac{2}{n-1}$. We are going to prove the following result from [29] (see also [37]).

Theorem 19.3.2 *Let us consider the Cauchy problem*

$$u_{tt} - \Delta u + 2(-\Delta)^{\frac{1}{2}} u_t = |u|^p, \quad u(0,x) = 0, \quad u_t(0,x) = \psi(x).$$

If the nonnegative data $\psi \in L^1_{loc}(\mathbb{R}^n)$, $n \geq 1$, *and if* $1 < p \leq p_{crit} = 1 + \frac{2}{n-1}$, *then there exists no global in time nontrivial Sobolev solution belonging to* $L^p_{loc}(\mathbb{R}^{n+1}_+)$. This result shows the optimality of the critical exponent $p_{crit} = p_{crit}(n)$. The main tool of the proof is the application of the test function method (see [232] and [229]) we introduced in Sect. 18.1.2. But now we have a new difficulty, the linear operator contains a pseudodifferential operator $(-\Delta)^{\frac{1}{2}}$. Nevertheless, we may apply the test function method due to the following observation (see for example [26] or [37]).

Lemma 19.3.1 *Let u be a local or global (in time) Sobolev solution to the Cauchy problem of Theorem 19.3.2 with nonnegative initial data* ψ *for some* $p > 1$. *Then,* $u = u(t,x)$ *is nonnegative.*
This allows us to write the above Cauchy problem in the form

$$u_{tt} - \Delta u + 2(-\Delta)^{\frac{1}{2}} u_t = u^p, \quad u(0,x) = 0, \quad u_t(0,x) = \psi(x).$$

Now let us prove Theorem 19.3.2.

Proof We assume by contradiction that $u \in L^p_{loc}(\mathbb{R}^{n+1}_+)$ is a global nontrivial Sobolev solution. Therefore, for any test function $\chi \in C_0^\infty([0,\infty) \times \mathbb{R}^n)$ it holds

$$\int_0^\infty \int_{\mathbb{R}^n} u \left(\chi_{tt} - \Delta \chi - 2(-\Delta)^{\frac{1}{2}} \chi_t \right) dx\, dt$$

$$= \int_0^\infty \int_{\mathbb{R}^n} u^p \chi\, dx\, dt + \int_{\mathbb{R}^n} \psi(x)\, \chi(0,x)\, dx.$$

Let $\phi \in C_0^\infty[0,\infty)$ be a nontrivial, non-increasing function which is compactly supported in $[0,1]$, and let $\ell > p'$, where $p' = \frac{p}{p-1}$ is the conjugate exponent to p.

First, we assume that $1 < p < 1 + \frac{2}{n-1}$. For any $R > 1$, we choose $\chi_R(t,x) :=$ $\phi\left(\frac{t}{R}\right)^\ell \phi\left(\frac{|x|}{R}\right)^\ell$ for some $\ell > p'$. Recalling that ϕ, $-\phi'$, and ψ are nonnegative and that (see [26])

$$(-\Delta)^\theta \phi\left(\frac{|x|}{R}\right)^\ell \leq \ell \phi\left(\frac{|x|}{R}\right)^{\ell-1} (-\Delta)^\theta \phi\left(\frac{|x|}{R}\right)$$

for any $\theta \in (0,1]$ and $\ell > 1$, we may derive

$$I_R := \int_0^\infty \int_{\mathbb{R}^n} u^p \chi_R\, dx\, dt \leq R^{-2} \ell \int_0^\infty \int_{\mathbb{R}^n} u \chi_R^{\frac{\ell-1}{\ell}} h\left(\frac{t}{R}, \frac{|x|}{R}\right) dx\, dt,$$

$$h(t,|x|) = \left(\phi''(t) + (\ell-1) \frac{(\phi')^2(t)}{\phi(t)} \right) \phi(|x|)$$

$$- \phi(t) \Delta \phi(|x|) - 2\ell\, \phi'(t)(-\Delta)^{\frac{1}{2}} \phi(|x|).$$

We notice that the function $\chi_R^\kappa(t,x)h\left(\frac{t}{R},\frac{|x|}{R}\right)$ is a bounded function with compact support in $[0,R]\times B_R$ for any $\kappa>0$ since h is bounded. After setting $\kappa=\frac{\ell-1}{\ell}-\frac{\ell}{p}$ by Hölder's inequality we obtain

$$I_R \le CR^{-2}I_R^{\frac{1}{p}}\left(\int_0^R\int_{B_R}\left(\chi_R^\kappa(t,x)\left|h\left(\frac{t}{R},\frac{|x|}{R}\right)\right|\right)^{p'}dx\,dt\right)^{\frac{1}{p'}}$$

$$\le CR^{-2+\frac{n+1}{p'}}I_R^{\frac{1}{p}}.$$

This gives $I_R \le CR^{-2p'+n+1}$, which vanishes as $R\to\infty$. By applying Beppo-Levi convergence theorem it follows that $u\equiv 0$.

Now let $n\ge 2$ and $p=1+\frac{2}{n-1}$, i.e. $2p'=n+1$. The previous approach gives only a uniform bound for I_R, that is, $u^p\chi_R\in L^1(\mathbb{R}^n)$. Now we fix ϕ such that it also satisfies $\phi(\rho)=1$ for any $\rho\in[0,\frac{1}{2}]$. Moreover, we define $\chi_{R,\delta}(t,x):=\phi(t)^\ell\phi\left(\frac{|x|}{\delta R}\right)^\ell$ with some $\delta>0$. Following the above reasoning, we may derive, in particular,

$$-\int_0^\infty\int_{\mathbb{R}^n}u\,\Delta\chi_{R,\delta}\,dx\,dt = -\int_0^\infty\int_{|x|>\delta\frac{R}{2}}u\,\Delta\chi_{R,\delta}\,dx\,dt$$

$$\le CJ_R^{\frac{1}{p}}\equiv\left(\int_0^\infty\int_{|x|>\delta\frac{R}{2}}u^p\chi_{R,\delta}\,dx\,dt\right)^{\frac{1}{p}}\to 0\quad\text{as }R\to\infty,$$

thanks to $u^p\chi_{R,\delta}\in L^1(\mathbb{R}^n)$ for any fixed $\delta>0$. For the other two terms we proceed as before. Taking into account the presence of $\delta>0$ in the definition of ψ we obtain

$$I_R\le CJ_R^{\frac{1}{p}}+C(\delta^{\frac{n}{p'}}+\delta^{\frac{n}{p'}-1})I_R^{\frac{1}{p}}.$$

Being δ arbitrarily small and $n=2p'-1>p'$ we conclude again $u\equiv 0$. This completes the proof.

19.4   Structurally Damped Wave Models with Nonlinearity $|u_t|^p$

19.4.1   Main Result

Theorem 19.4.1 *Let us consider for $n\ge 1$ the Cauchy problem*

$$u_{tt}-\Delta u+\mu(-\Delta)^{\frac{1}{2}}u_t=|u_t|^p,\quad u(0,x)=\varphi(x),\quad u_t(0,x)=\psi(x),$$

with $\mu>0$. The data (φ,ψ) are assumed to belong to the space

$$(H_m^s(\mathbb{R}^n)\cap L^1(\mathbb{R}^n))\times(H_m^{s-1}(\mathbb{R}^n)\cap L^1(\mathbb{R}^n))$$

with $s > 1 + \frac{n}{m}$ and $m \in (1, 2]$. *Moreover, the exponent p satisfies the condition*

$$p > \max \left\{ s; m; 1 + \frac{1}{n} \right\}.$$

Then, there exists a constant $\varepsilon_0 > 0$ such that for any small data (φ, ψ) satisfying the condition

$$\|(\varphi, \psi)\|_{(H_m^s \cap L^1) \times (H_m^{s-1} \cap L^1)} \leq \varepsilon_0$$

there exists a uniquely determined global (in time) small data Sobolev solution

$$u \in C\big([0, \infty), H_m^s(\mathbb{R}^n)\big) \cap C^1\big([0, \infty), H_m^{s-1}(\mathbb{R}^n)\big).$$

Moreover, the solution satisfies the following decay estimates:

$$\|u(t, \cdot)\|_{L^m} \leq C(1 + t)^{1 - \frac{n(m-1)}{m}} \|(\varphi, \psi)\|_{(H_m^s \cap L^1) \times (H_m^{s-1} \cap L^1)},$$

$$\|u_t(t, \cdot)\|_{L^m} \leq C(1 + t)^{-\frac{n(m-1)}{m}} \|(\varphi, \psi)\|_{(H_m^s \cap L^1) \times (H_m^{s-1} \cap L^1)},$$

$$\|(|D|^s u(t, \cdot), |D|^{s-1} u_t(t, \cdot))\|_{L^m}$$

$$\leq C(1 + t)^{1 - \frac{n(m-1) + ms}{m}} \|(\varphi, \psi)\|_{(H_m^s \cap L^1) \times (H_m^{s-1} \cap L^1)}.$$

Remark 19.4.1 Let us explain the conditions for p and s. If we want to use fractional powers (see Corollary 24.5.2), then $p > s$ is a necessary supposition. We apply the tool Gagliardo-Nirenberg inequality (see Proposition 24.5.4), but we are not interested in having a restriction above for p. For this reason we assume $s > 1 + \frac{n}{m}$. The restriction below $p \geq m$ is a straight forward condition related to this tool. The remaining condition for p implies the same decay estimates of solutions to the semilinear model as for the solutions to the corresponding linear model with vanishing right-hand side. So, the nonlinearity is interpreted as a small perturbation. In the case of $m = 2$ we assume $s > 1 + \frac{n}{2}$ and $p > \max \{s; 2\}$.

19.4.2 Proof

We explained the philosophy for proving the existence of global (in time) small data solutions in Sect. 18.1.1.2. We can follow this approach to prove Theorem 19.4.1, too. For this reason we skip some details, sketch some steps and describe only the changes in the proof which are necessary due to the presence of the nonlinearity $|u_t|^p$.

We introduce the data space

$$A := (H^s_m(\mathbb{R}^n) \cap L^1(\mathbb{R}^n)) \times (H^{s-1}_m(\mathbb{R}^n) \cap L^1(\mathbb{R}^n)).$$

The statements of Corollaries 19.2.5, 19.2.2 and 19.2.3 suggest using for $t > 0$ the solution space

$$X(t) = C([0, t], H^s_m(\mathbb{R}^n)) \cap C^1([0, t], H^{s-1}_m(\mathbb{R}^n))$$

with the norm

$$\|u(t, \cdot)\|_{X(t)} = \sup_{\tau \in [0,t]} \left((1 + \tau)^{\frac{n(m-1)-m}{m}} \|u(\tau, \cdot)\|_{L^m} \right.$$
$$+ (1 + \tau)^{\frac{n(m-1)}{m}} \|u_t(\tau, \cdot)\|_{L^m} + (1 + \tau)^{\frac{n(m-1)+m(s-1)}{m}} \||D|^{s-1} u_t(\tau, \cdot)\|_{L^m}$$
$$+ (1 + \tau)^{\frac{n(m-1)+m(s-1)}{m}} \||D|^s u(\tau, \cdot)\|_{L^m} \right).$$

As in the previous section, we define the operator N and prove for Nu the required estimates for applying Banach's fixed point argument. The decay estimates for small data solutions and some derivatives follow immediately by the definition of the norm in $X(t)$.

We begin to estimate the L^m-norm of Nu itself. We apply the $(L^1 \cap L^m) - L^m$ estimates on the interval $[0, \frac{t}{2}]$ and the $L^m - L^m$ estimates on the interval $[\frac{t}{2}, t]$ to conclude

$$\|Nu(t, \cdot)\|_{L^m} \leq C(1 + t)^{1 - \frac{n(m-1)}{m}} \|(\varphi, \psi)\|_A$$
$$+ C \int_0^{\frac{t}{2}} (1 + t - \tau)^{1 - \frac{n(m-1)}{m}} \||u_t(\tau, \cdot)|^p\|_{L^m \cap L^1} d\tau$$
$$+ C \int_{\frac{t}{2}}^t (1 + t - \tau) \||u_t(\tau, \cdot)|^p\|_{L^m} d\tau.$$

We have

$$\||u_t(\tau, \cdot)|^p\|_{L^m \cap L^1} \leq C \|u_t(\tau, \cdot)\|_{L^p}^p + \|u_t(\tau, \cdot)\|_{L^{mp}}^p.$$

To estimate the norm $\|u_t(\tau, \cdot)\|_{L^{kp}}, k = 1, m$, we apply the Gagliardo-Nirenberg inequality in the form

$$\|w(\tau, \cdot)\|_{L^q} \leq C \||D|^{s-1} w(\tau, \cdot)\|_{L^m}^{\theta_{0,s-1}(q,m)} \|w(\tau, \cdot)\|_{L^m}^{1 - \theta_{0,s-1}(q,m)}$$

with $w(\tau, \cdot) = u_t(\tau, \cdot)$, where for $q \geq m$, as in Corollary 24.5.1 and Remark 24.5.2, we have to guarantee

$$\theta_{0,s-1}(q, m) = \frac{n}{s-1} \left(\frac{1}{m} - \frac{1}{q} \right) \in [0, 1],$$

that is,

$$m \le q \le \frac{mn}{n - m(s - 1)} \quad \text{or} \quad m \le q \text{ if } \frac{n}{(s - 1)m} \le 1.$$

Since $\theta_{0,s-1}(p, m) < \theta_{0,s-1}(mp, m)$, here $m \le p \le \frac{n}{n-m(s-1)}$ or $m \le p$ if $\frac{n}{(s-1)m} \le 1$, we have on the interval $(0, \frac{t}{2})$ the estimate

$$\int_0^{\frac{t}{2}} (1 + t - \tau)^{1 - \frac{n(m-1)}{m}} \||u_t(\tau, \cdot)|^p\|_{L^m \cap L^1} \, d\tau$$

$$\le C(1 + t)^{1 - \frac{n(m-1)}{m}} \|u\|_{X(t)}^p \int_0^{\frac{t}{2}} (1 + \tau)^{-p\left(\frac{n(m-1)}{m} + \theta_{0,s-1}(p,m)(s-1)\right)} \, d\tau,$$

where

$$-p\left(\frac{n(m - 1)}{m} + \theta_{0,s-1}(p, m)(s - 1)\right) = -p\left(\frac{n(m - 1)}{m} + n\left(\frac{1}{m} - \frac{1}{p}\right)\right) < -1$$

$$\text{for } p > \frac{n + 1}{n}.$$

The last condition is satisfied due to an assumption for p. On the interval $(\frac{t}{2}, t)$ we have

$$\int_{\frac{t}{2}}^t (1 + t - \tau) \||u_t(\tau, \cdot)|^p\|_{L^m} \, d\tau$$

$$\le C(1 + t) \|u\|_{X(t)}^p \int_{\frac{t}{2}}^t (1 + \tau)^{-p\left(\frac{n(m-1)}{m} + \theta_{0,s-1}(mp,m)(s-1)\right)} \, d\tau,$$

where

$$-p\left(\frac{n(m - 1)}{m} + \theta_{0,s-1}(mp, m)(s - 1)\right)$$

$$= -p\left(\frac{n(m - 1)}{m} + n\left(\frac{1}{m} - \frac{1}{mp}\right)\right) < -\frac{n(m - 1)}{m} - 1$$

due to the assumption $p > \frac{n+1}{n}$. Combining the last two estimates we arrive at the following desired estimate:

$$(1 + \tau)^{\frac{n(m-1)-m}{m}} \|Nu(\tau, \cdot)\|_{L^m} \le C\|(\varphi, \psi)\|_A + C\|u\|_{X(t)}^p \quad \text{for all } \tau \in [0, t].$$

In the next step we consider

$$\partial_t Nu = \partial_t v(t, x) + \int_0^t \partial_t \big(K_1(t - \tau, 0, x) *_{(x)} |u_t(\tau, \cdot)|^p\big) \, d\tau,$$

where we introduce

$$v = v(t, x) := K_0(t, 0, x) *_{(x)} \varphi(x) + K_1(t, 0, x) *_{(x)} \psi(x)$$

as the solution of the corresponding linear Cauchy problem with vanishing right-hand side and with initial data (φ, ψ). Using the same approach as above we may conclude

$$(1 + \tau)^{\frac{n(m-1)}{m}} \|(\partial_t Nu)(\tau, \cdot)\|_{L^m} \leq C\|(\varphi, \psi)\|_A + C\|u\|_{X(t)}^p \quad \text{for all } \tau \in [0, t]$$

under the same assumptions for p.

Now let us turn to estimate $\|\partial_t |D|^{s-1} Nu(t, \cdot)\|_{L^m}$. We use

$$\partial_t |D|^{s-1} Nu = \partial_t |D|^{s-1} v(t, x) + \int_0^t \partial_t |D|^{s-1} \left(K_1(t - \tau, 0, x) *_{(x)} |u_t(\tau, \cdot)|^p \right) d\tau.$$

Taking into account the estimates of Corollaries 19.2.5 and 19.2.3 with $a = s - 1$ and using again the $(L^1 \cap L^m) - L^m$ estimates on the interval $(0, \frac{t}{2})$ and the $L^m - L^m$ estimates on the interval $(\frac{t}{2}, t)$, we obtain

$$\|\partial_t |D|^{s-1} (Nu)\|_{L^m} \leq C(1 + t)^{-\frac{n(m-1)+m(s-1)}{m}} \|(\varphi, \psi)\|_A$$

$$+ C \int_0^{\frac{t}{2}} (1 + t - \tau)^{-\frac{n(m-1)+m(s-1)}{m}} \left(\||u_t(\tau, \cdot)|^p\|_{L^m \cap L^1} + \||u_t(\tau, \cdot)|^p\|_{\dot{H}_m^{s-1}} \right) d\tau$$

$$+ C \int_{\frac{t}{2}}^t (1 + t - \tau)^{-\frac{n(m-1)+m(s-1)}{m}} \left(\||u_t(\tau, \cdot)|^p\|_{L^m} + \||u_t(\tau, \cdot)|^p\|_{\dot{H}_m^{s-1}} \right) d\tau.$$

The integrals with $\||u_t(\tau, \cdot)|^p\|_{L^m \cap L^1}$ or $\||u_t(\tau, \cdot)|^p\|_{L^m}$ will be handled as before if we apply as above

$$-p\left(\frac{n(m-1)}{m} + \theta_{0,s-1}(p, m)(s-1) \right) = -p\left(\frac{n(m-1)}{m} + n\left(\frac{1}{m} - \frac{1}{p} \right) \right) < -1$$

$$\text{for } p > \frac{n+1}{n}.$$

To estimate the integrals with $\||u_t(\tau, \cdot)|^p\|_{\dot{H}_m^{s-1}}$ we apply Corollary 24.5.2 with $p > s$. So, we may estimate

$$\int_0^{\frac{t}{2}} (1 + t - \tau)^{-\frac{n(m-1)+m(s-1)}{m}} \||u_t(\tau, \cdot)|^p\|_{\dot{H}_m^{s-1}} d\tau$$

$$\leq C \int_0^{\frac{t}{2}} (1 + t - \tau)^{-\frac{n(m-1)+m(s-1)}{m}} \|u_t(\tau, \cdot)\|_{\dot{H}_m^{s-1}} \|u_t(\tau, \cdot)\|_{L^\infty}^{p-1} d\tau$$

$$\leq C \int_0^{\frac{t}{2}} (1+t-\tau)^{-\frac{n(m-1)+m(s-1)}{m}} \|u_t(\tau,\cdot)\|_{\dot{H}_m^{s-1}} \|u_t(\tau,\cdot)\|_{H_m^{s_0}}^{p-1} \, d\tau$$

$$\leq C \int_0^{\frac{t}{2}} (1+t-\tau)^{-\frac{n(m-1)+m(s-1)}{m}} \|u_t(\tau,\cdot)\|_{\dot{H}_m^{s-1}}$$

$$\times \left(\|u_t(\tau,\cdot)\|_{L^m} + \|u_t(\tau,\cdot)\|_{\dot{H}_m^{s-1}} \right)^{p-1} \, d\tau.$$

Here we used $s - 1 > s_0 > \frac{n}{m}$. This implies only the restriction $p \geq m$. Using again the estimates of Corollaries 19.2.5 and 19.2.3 gives

$$\int_0^{\frac{t}{2}} (1+t-\tau)^{-\frac{n(m-1)+m(s-1)}{m}} \||u_t(\tau,\cdot)|^p\|_{\dot{H}_m^{s-1}} \, d\tau$$

$$\leq C\|u\|_{X(t)}^p \int_0^{\frac{t}{2}} (1+t-\tau)^{-\frac{n(m-1)+m(s-1)}{m}} (1+\tau)^{-(s-1)-pn\left(1-\frac{1}{m}\right)} \, d\tau$$

$$\leq C(1+t)^{-\frac{n(m-1)+m(s-1)}{m}} \|u\|_{X(t)}^p \int_0^{\frac{t}{2}} (1+\tau)^{-(s-1)-pn\left(1-\frac{1}{m}\right)} \, d\tau.$$

The assumption $s > 1 + \frac{n}{m}$ yields the condition $s > \frac{2m}{n(m-1)+m}$. This condition and the assumption $p > s$ imply $(s-1) + pn\left(1 - \frac{1}{m}\right) > 1$. In an analogous way we estimate

$$\int_{\frac{t}{2}}^t (1+t-\tau)^{-\frac{n(m-1)+m(s-1)}{m}} \||u_t(\tau,\cdot)|^p\|_{\dot{H}_m^{s-1}} \, d\tau$$

$$\leq C(1+t)^{-\frac{n(m-1)+m(s-1)}{m}} \int_{\frac{t}{2}}^t \||u_t(\tau,\cdot)|^p\|_{\dot{H}_m^{s-1}} \, d\tau$$

$$\leq C(1+t)^{-\frac{n(m-1)+m(s-1)}{m}} \|u\|_{X(t)}^p \int_{\frac{t}{2}}^t (1+\tau)^{-(s-1)-pn\left(1-\frac{1}{m}\right)} \, d\tau.$$

Summarizing, we have shown for all $\tau \in [0, t]$ the estimate

$$(1+\tau)^{\frac{n(m-1)+m(s-1)}{m}} \|\partial_t |D|^{s-1}(Nu)(\tau,\cdot)\|_{L^m} \leq C\|(\varphi, \psi)\|_A + C\|u\|_{X(t)}^p.$$

By using the same approach we arrive for all $\tau \in [0, t]$ at

$$(1+\tau)^{\frac{n(m-1)+m(s-1)}{m}} \||D|^s(Nu)(\tau,\cdot)\|_{L^m} \leq C\|(\varphi, \psi)\|_A + C\|u\|_{X(t)}^p.$$

From all these estimates we may conclude

$$\|Nu\|_{X(t)} \leq C\|(\varphi, \psi)\|_A + C\|u\|_{X(t)}^p.$$

In principle, one can use the same approach to prove the estimate

$$\|Nu - Nv\|_{X(t)} \le C\|u - v\|_{X(t)}\big(\|u\|_{X(t)}^{p-1} + \|v\|_{X(t)}^{p-1}\big).$$

There are some minor modifications however. First of all, we use the representation

$$\partial_t^j |D|^k \int_0^t K_1(t - \tau, 0, x) *_{(x)} \big(|u_t(\tau, x)|^p - |v_t(\tau, x)|^p\big) d\tau$$

$$= \int_0^t \partial_t^j |D|^k K_1(t - \tau, 0, x) *_{(x)} \big(|u_t(\tau, x)|^p - |v_t(\tau, x)|^p\big) d\tau$$

$$= p \int_0^t \partial_t^j |D|^k K_1(t - \tau, 0, x) *_{(x)} \Big(\int_0^1 |v_t + r(u_t - v_t)|^{p-2}(v_t + r(u_t - v_t)) dr\Big)$$

$$\times (u_t - v_t) d\tau.$$

If we choose $j = 1$ and $k = s - 1$, then it remains to estimate the following norms:

$$\big\| |u_t(\tau, x)|^p - |v_t(\tau, x)|^p \big\|_{L^1},$$

$$\Big\| \Big(\int_0^1 |v_t + r(u_t - v_t)|^{p-2}(v_t + r(u_t - v_t)) dr\Big)(\tau, x)(u_t - v_t)(\tau, x)\Big\|_{H_m^{s-1}}.$$

To estimate the L^1-norm and the L^m-norm we proceed as at the end of the proof to Proposition 18.1.1 if we replace there L^2 by L^m and L^{2p} by L^{mp}. To estimate the $\dot{H}_m^{s-1}$-norm we use the algebra property of $H_m^{s-1}(\mathbb{R}^n)$, the embedding of $H_m^{s-1}(\mathbb{R}^n)$ into $L^\infty(\mathbb{R}^n)$ and Proposition 24.5.6 as follows:

$$\Big\| \Big(\int_0^1 |v_t + r(u_t - v_t)|^{p-2}(v_t + r(u_t - v_t)) dr\Big)(\tau, x)(u_t - v_t)(\tau, x)\Big\|_{\dot{H}_m^{s-1}}$$

$$\le C \Big\| \Big(\int_0^1 |v_t + r(u_t - v_t)|^{p-2}(v_t + r(u_t - v_t)) dr\Big)(\tau, x)\Big\|_{H_m^{s-1}}$$

$$\times \|(u_t - v_t)(\tau, x)\|_{H_m^{s-1}}$$

$$\le C \int_0^1 \Big\| |v_t + r(u_t - v_t)|^{p-2}(v_t + r(u_t - v_t))\Big\|_{H_m^{s-1}} dr$$

$$\times \|(u_t - v_t)(\tau, x)\|_{H_m^{s-1}}$$

$$\le C \int_0^1 \|v_t + r(u_t - v_t)\|_{H_m^{s-1}} \|v_t + r(u_t - v_t)\|_{L^\infty}^{p-2} dr$$

$$\times \|(u_t - v_t)(\tau, x)\|_{H_m^{s-1}}$$

$$\le C \big(\|v_t + r(u_t - v_t)\|_{\dot{H}_m^{s-1}}^{p-1} + \|v_t + r(u_t - v_t)\|_{L^m}^{p-1}\big)$$

$$\times \big(\|(u_t - v_t)(\tau, x)\|_{\dot{H}_m^{s-1}} + \|(u_t - v_t)(\tau, x)\|_{L^m}\big).$$

Applying fractional Gagliardo-Nirenberg inequality (see Proposition 24.5.5) and the decay estimates for the corresponding linear Cauchy problem with vanishing right-hand side (see Corollary 19.2.3) we arrive at the desired inequality

$$\|Nu - Nv\|_{X(t)} \leq C\|u - v\|_{X(t)}\left(\|u\|_{X(t)}^{p-1} + \|v\|_{X(t)}^{p-1}\right).$$

All these explanations complete the proof.

19.5 Concluding Remarks

The approach of Sects. 19.3 and 19.4 was applied to other structurally damped semilinear models to study the global (in time) existence of small data Sobolev solutions.

19.5.1 Semilinear Viscoelastic Damped Wave Models

The Cauchy problem for viscoelastic damped wave models with different power nonlinearities was studied in [165]. One of the models of interest to us is

$$u_{tt} - \Delta u - \Delta u_t = \||D|^a u|^p, \quad u(0, x) = \varphi(x), \quad u_t(0, x) = \psi(x),$$

where

$$(\varphi, \psi) \in (H^s(\mathbb{R}^n) \cap L^1(\mathbb{R}^n)) \times (H^{s-2}(\mathbb{R}^n) \cap L^1(\mathbb{R}^n))$$

for a certain $s \geq 2$ and $a \in (0, 2)$ (cf. with Sects. 14.3 and 19.3).

First we suppose low regularity of data, we choose $s = 2$. This choice of regularity of data implies more restrictions on the exponent p, on the parameter a and on the spatial dimension n. Following the steps of the proof of Sect. 19.3.2, the following result is proved in [165].

Theorem 19.5.1 *Let*

$$A := (H^2(\mathbb{R}^n) \cap L^1(\mathbb{R}^n)) \times (L^2(\mathbb{R}^n) \cap L^1(\mathbb{R}^n)) \ \text{for } n \geq 2.$$

Let $a \in \left(2 - \frac{n}{2}, 2 - \frac{n}{4}\right)$ *and* $p \in \left[2, \frac{n}{n+2(a-2)}\right]$ *or* $a \in \left(0, 2 - \frac{n}{2}\right]$ *and* $p > \frac{2+n}{n+a-1}$.

Then, there exists a constant $\varepsilon_0 > 0$ such that for any small data $(\varphi, \psi) \in A$ satisfying the condition $\|(\varphi, \psi)\|_A \leq \varepsilon_0$ there exists a uniquely determined global (in time) energy solution

$$u \in C([0, \infty), H^2(\mathbb{R}^n)) \cap C^1([0, \infty), L^2(\mathbb{R}^n)).$$

Moreover, the solution, its first partial derivative in time and its partial derivatives in space of the second order satisfy the following decay estimates:

$$\|u(t,\cdot)\|_{L^2} \le C \begin{cases} (1+t)^{-\frac{n-2}{4}}\|(\varphi,\psi)\|_A & \text{if } n \ge 3, \\ \log(e+t)\|(\varphi,\psi)\|_A & \text{if } n = 2, \end{cases}$$

$$\|u_t(t,\cdot)\|_{L^2} \le C(1+t)^{-\frac{n}{4}}\|(\varphi,\psi)\|_A,$$

$$\||D|^2 u(t,\cdot)\|_{L^2} \le C(1+t)^{-\frac{n+2}{4}}\|(\varphi,\psi)\|_A.$$

In the last theorem we assume some restrictions on the parameter a and on the dimension n. If we want to avoid such restrictions, then one way is to require more regularity for the data. Combining the steps of the proofs of Sects. 19.3.2 and 19.4.2, the following result is proved in [165].

Theorem 19.5.2 *Let*

$$A := (H^s(\mathbb{R}^n) \cap L^1(\mathbb{R}^n)) \times (H^{s-2}(\mathbb{R}^n) \cap L^1(\mathbb{R}^n)).$$

Let $p > \max\{s; 2\}$ and $\max\{a + \frac{n}{2}; \frac{n+8}{n+2a}\} < s \le n$.

Then, there exists a constant $\varepsilon_0 > 0$ such that for any small data $(\varphi, \psi) \in A$ satisfying the condition $\|(\varphi, \psi)\|_A \le \varepsilon_0$ there exists a uniquely determined global (in time) energy solution

$$u \in C([0,\infty), H^s(\mathbb{R}^n)) \cap C^1([0,\infty), H^{s-2}(\mathbb{R}^n)).$$

Moreover, the solution, its first partial derivative in time and its partial derivatives which are connected with the corresponding energy of higher order satisfy the following decay estimates:

$$\|u(t,\cdot)\|_{L^2} \le C \begin{cases} (1+t)^{-\frac{n-2}{4}}\|(\varphi,\psi)\|_A & \text{if } n \ge 3, \\ \log(e+t)\|(\varphi,\psi)\|_A & \text{if } n = 2, \end{cases}$$

$$\|u_t(t,\cdot)\|_{L^2} \le C(1+t)^{-\frac{n}{4}}\|(\varphi,\psi)\|_A,$$

$$\||D|^s u(t,\cdot)\|_{L^2} \le C(1+t)^{-\frac{n+2(s-1)}{4}}\|(\varphi,\psi)\|_A,$$

$$\||D|^{s-2} u_t(t,\cdot)\|_{L^2} \le C(1+t)^{-\frac{n+2(s-2)}{4}}\|(\varphi,\psi)\|_A.$$

Finally, the model

$$u_{tt} - \Delta u - \Delta u_t = |u_t|^p, \quad u(0,x) = \varphi(x), \quad u_t(0,x) = \psi(x)$$

was discussed in [165] (cf. with Sect. 19.4). This model can be treated as the previous model assuming large regularity for the data. Consequently, the following result is similar to Theorem 19.5.2.

Theorem 19.5.3 *Let*

$$A := (H^s(\mathbb{R}^n) \cap L^1(\mathbb{R}^n)) \times (H^{s-2}(\mathbb{R}^n) \cap L^1(\mathbb{R}^n)).$$

Let $s > 2 + \frac{n}{2}$ for $n \geq 2$ and $p > s$.

Then, there exists a constant $\varepsilon_0 > 0$ such that for any small data $(\varphi, \psi) \in A$ satisfying the condition $\|(\varphi, \psi)\|_A \leq \varepsilon_0$ there exists a uniquely determined global (in time) energy solution

$$u \in C([0, \infty), H^s(\mathbb{R}^n)) \cap C^1([0, \infty), H^{s-2}(\mathbb{R}^n)).$$

Moreover, the solution, its first partial derivative in time and its partial derivatives which are connected with the corresponding energy of higher order satisfy the following decay estimates:

$$\|u(t, \cdot)\|_{L^2} \leq C \begin{cases} (1 + t)^{-\frac{n-2}{4}} \|(\varphi, \psi)\|_A & \text{if } n \geq 3, \\ \log(e + t)\|(\varphi, \psi)\|_A & \text{if } n = 2, \end{cases}$$

$$\|u_t(t, \cdot)\|_{L^2} \leq C(1 + t)^{-\frac{n}{4}} \|(\varphi, \psi)\|_A,$$

$$\||D|^s u(t, \cdot)\|_{L^2} \leq C(1 + t)^{-\frac{n+2(s-1)}{4}} \|(\varphi, \psi)\|_A,$$

$$\||D|^{s-2} u_t(t, \cdot)\|_{L^2} \leq C(1 + t)^{-\frac{n+2(s-2)}{4}} \|(\varphi, \psi)\|_A.$$

19.5.2 Semilinear Structurally Damped σ-Evolution Models

In this section we discuss the global existence of small data solutions for semilinear structurally damped σ-evolution models of the form

$$u_{tt} + (-\Delta)^\sigma u + \mu(-\Delta)^\delta u_t = f(|D|^a u, u_t), \quad u(0, x) = \varphi(x), \quad u_t(0, x) = \psi(x)$$

with $\sigma \geq 1$, $\mu > 0$ and $\delta \in [0, \sigma]$. This is a family of structurally damped σ-evolution models (in the case $\sigma = 1$ we have structurally damped wave models) interpolating between models with friction or exterior damping $\delta = 0$ and those with viscoelastic type damping $\delta = \sigma$. The function $f(|D|^a u, u_t)$ stands for different power nonlinearities. If we choose $||D|^a u|^p$ for $a \in [0, \sigma]$, then we have in mind a model power nonlinearity with pseudodifferential action. If we choose $|u_t|^p$, then we have in mind a model power nonlinearity with derivative. Our goal is to propose a Fujita type exponent dividing the admissible range of powers p into those allowing for the global existence of small data solutions (stability of zero solution) and those producing a blow up behavior, even for small data.

The beginner becomes familiar with this topic during the study of the projects in Sects. 23.9, 23.11 and 23.12. One can find more information in [34, 37] or [164].

There the authors proved $L^p - L^q$ estimates not necessarily on the conjugate line for solutions to the linear Cauchy problem

$$v_{tt} + (-\Delta)^\sigma v + \mu(-\Delta)^\delta v_t = 0, \quad v(0,x) = \varphi(x), \quad v_t(0,x) = \psi(x)$$

with $\sigma \geq 1$, $\mu > 0$ and $\delta \in (0,\sigma)$. The main tools to prove such estimates are the ones of Sect. 24.2. Applying the approach of Sects. 19.3.2 and 19.4.2, the following two results are proved in [164]:

Theorem 19.5.4 *Let us consider the Cauchy problem*

$$u_{tt} + (-\Delta)^\sigma u + \mu(-\Delta)^{\frac{\sigma}{2}} u_t = ||D|^a u|^p, \quad u(0,x) = \varphi(x), \quad u_t(0,x) = \psi(x),$$

with $\sigma \geq 1$, $a \in [0,\sigma)$ and $\mu > 0$, where the data are taken from the space

$$A := (H_m^\sigma(\mathbb{R}^n) \cap L^1(\mathbb{R}^n)) \times (L^m(\mathbb{R}^n) \cap L^1(\mathbb{R}^n)) \quad \text{with } m \in (1,2].$$

We assume for dimension n the condition

$$n \leq \frac{m^2(\sigma - a)}{m - 1}.$$

Moreover, the exponent p satisfies the conditions

$$p \in \left[m, \frac{n}{[n + m(a - \sigma)]^+} \right] \quad \text{and} \quad p > 1 + \frac{2\sigma - a}{[n + a - \sigma]^+}.$$

Then, there exists a constant $\varepsilon_0 > 0$ such that for any small data (φ, ψ) satisfying the condition $\|(\varphi, \psi)\|_A \leq \varepsilon_0$ we have a uniquely determined global (in time) small data Sobolev solution

$$u \in C([0,\infty), H_m^\sigma(\mathbb{R}^n)) \cap C^1([0,\infty), L^m(\mathbb{R}^n)).$$

The following estimates hold for the solution and its "energy terms":

$$\|u(t,\cdot)\|_{L^m} \leq C(1+t)^{1 - \frac{n}{\sigma}\left(1 - \frac{1}{m}\right)} \|(\varphi, \psi)\|_A,$$

$$\|(|D|^\sigma u(t,\cdot), u_t(t,\cdot))\|_{L^m} \leq C(1+t)^{-\frac{n}{\sigma}\left(1 - \frac{1}{m}\right)} \|(\varphi, \psi)\|_A.$$

Remark 19.5.1 Let us explain the conditions for p and n. The condition

$$p > 1 + \frac{2\sigma - a}{[n + a - \sigma]^+}$$

implies the same decay estimates as for the solutions to the corresponding linear model with vanishing right-hand side. So, the nonlinearity is interpreted as a small perturbation. We apply the tool fractional Gagliardo-Nirenberg inequality (see Proposition 24.5.5). The other conditions for p come into play for this reason. The upper bound for n arises from the set of admissible range for p. Smaller p imply that this set is not empty.

Theorem 19.5.5 *Let us consider the Cauchy problem*

$$u_{tt} + (-\Delta)^\sigma u + \mu(-\Delta)^{\frac{\sigma}{2}} u_t = |u_t|^p, \quad u(0, x) = \varphi(x), \quad u_t(0, x) = \psi(x)$$

with $\sigma \geq 1$ and $\mu > 0$. The small data (φ, ψ) are assumed to belong to the space

$$A := (H_m^s(\mathbb{R}^n) \cap L^1(\mathbb{R}^n)) \times (H_m^{s-\sigma}(\mathbb{R}^n) \cap L^1(\mathbb{R}^n))$$

with $s > \max\{\frac{2m\sigma}{n(m-1)+m}; \sigma + \frac{n}{m}\}$ and $m > 1$, that is, $\|(\varphi, \psi)\|_A \leq \varepsilon_0$. Then, for any $p > \max\{s; m; 1 + \frac{\sigma}{n}\}$ there exists a uniquely determined global (in time) small data Sobolev solution belonging to

$$C([0, \infty), H_m^s(\mathbb{R}^n)) \cap C^1([0, \infty), H_m^{s-\sigma}(\mathbb{R}^n)).$$

Moreover, the following estimates hold:

$$\|u(t, \cdot)\|_{L^m} \leq C(1+t)^{1-\frac{n(m-1)}{m\sigma}} \|(\varphi, \psi)\|_A,$$

$$\|u_t(t, \cdot)\|_{L^m} \leq C(1+t)^{-\frac{n(m-1)}{m\sigma}} \|(\varphi, \psi)\|_A,$$

$$\|(|D|^s u(t, \cdot), |D|^{s-\sigma} u_t(t, \cdot))\|_{L^m} \leq C(1+t)^{1-\frac{n(m-1)+ms}{m\sigma}} \|(\varphi, \psi)\|_A.$$

Remark 19.5.2 Let us explain the conditions for p and s. If we want to use fractional powers (see Proposition 24.5.6 and Corollary 24.5.2), then $p > s$ is a necessary Supposition. We apply the tool Gagliardo-Nirenberg inequality (see Proposition 24.5.4). But, we are not interested in having a restriction above for the admissible powers p. For this reason we assume $s > \sigma + \frac{n}{m}$. The restriction below $p \geq m$ is a straight forward condition related to this tool. The remaining conditions for s and p imply the same decay estimates as for the solutions to the corresponding linear model with vanishing right-hand side. So, the nonlinearity is interpreted as a small perturbation. In the case $\sigma = 1$ and $m = 2$ we assume $s > 1 + \frac{n}{2}$ and $p > \max\{s; 2\}$.

The loss of regularity in $L^q - L^q$ estimates, with $q \in (1, 2)$, for solutions to the linear Cauchy problem

$$v_{tt} + (-\Delta)^\sigma v + \mu(-\Delta)^\delta v_t = 0, \quad v(0, x) = \varphi(x), \quad v_t(0, x) = \psi(x)$$

with $\sigma \geq 1$, $\mu > 0$ and $2\delta \in (0, \sigma)$ is related to the wave structure of the evolution equation for high frequencies. However, the presence of the structural damping when $\delta > 0$ generates a smoothing effect for the solution which does not appear for the classical damping or friction v_t. This smoothing effect allows us to recover the additional regularity by using estimates which are singular at $t = 0$. The singularity order is proportional to $\frac{n(\sigma - 2\delta)}{2\delta}$. Hence, it vanishes at $\sigma = 2\delta$ (see Proposition 4.3 of [34]). This effect explains, roughly speaking, the possibility to employ singular $L^r - L^q$ estimates for $1 \leq r \leq q \leq \infty$ to derive the global existence of small data solutions for semilinear problems in higher space dimensions when $\sigma = 2\delta$. In Theorem 5 of [34], using estimates also on the L^∞ basis, the authors avoid the restriction on the space dimension coming from Gagliardo-Nirenberg estimates and derived the following result:

Theorem 19.5.6 *Let us consider the Cauchy problem*

$$u_{tt} + (-\Delta)^\sigma u + (-\Delta)^{\frac{\sigma}{2}} u_t = |\partial_t^k u|^p, \quad u(0, x) = \varphi(x), \quad u_t(0, x) = \psi(x),$$

with $k = 0, 1$ and $\sigma \geq 1$ is an integer. Let $n \geq 1$ and $p > p_0 = 1 + \frac{2\sigma}{(n - \sigma)_+}$ if $k = 0$, or $p > p_1 = 1 + \frac{\sigma}{n}$ if $k = 1$.
Then, there exists a sufficiently small $\varepsilon_0 > 0$ such that for any data

$$(\varphi, \psi) \in A := (W_1^\sigma(\mathbb{R}^n) \cap W_\infty^\sigma(\mathbb{R}^n)) \times (L^1(\mathbb{R}^n) \cap L^\infty(\mathbb{R}^n)) \ \text{with} \ \|(\varphi, \psi)\|_A \leq \varepsilon_0,$$

there exists a global (in time) small data Sobolev solution

$$u \in C([0, \infty), W_1^\sigma(\mathbb{R}^n) \cap W_\infty^\sigma(\mathbb{R}^n)) \cap C^1([0, \infty), L^1(\mathbb{R}^n) \cap L^\infty(\mathbb{R}^n)).$$

For any $q \in [1, \infty]$ the solution satisfies the decay estimates

$$\|(|D|^\sigma u, u_t)(t, \cdot)\|_{L^q} \leq C(1 + t)^{-\frac{n}{\sigma}\left(1 - \frac{1}{q}\right)} \|(\varphi, \psi)\|_A.$$

Moreover, the solution itself satisfies the estimate

$$\|u(t, \cdot)\|_{L^q} \leq C(1 + t)^{1 - \frac{n}{\sigma}\left(1 - \frac{1}{q}\right)} \|(\varphi, \psi)\|_A.$$

The interested reader can also find in [34] the critical exponent for global small data solution to the Cauchy problem

$$u_{tt} + (-\Delta)^\sigma u + (-\Delta)^\delta u_t = |\partial_t^k u|^p, \quad u(0, x) = \varphi(x), \quad u_t(0, x) = \psi(x),$$

with $k = 0, 1$, $\sigma, \delta \in \mathbb{N} \setminus \{0\}$ and $2\delta \in (0, \sigma)$. The nonexistence of global (in time) solutions in the critical and subcritical cases is proved by using the test function method (under suitable sign assumptions on the initial data) (see Sect. 18.1.2). Finally, life span estimates are obtained.

Exercises Relating to the Considerations of Chap. 19

Exercise 1 Prove for all $\tau \in [0, t]$ the estimate

$$(1 + \tau)^{\frac{n(m-1)}{m}} \|(\partial_t Nu)(\tau, \cdot)\|_{L^m} \leq C\|(\varphi, \psi)\|_A + C\|u\|_{X(t)}^p$$

appearing in the proof to Theorem 19.4.1.

Exercise 2 Prove for all $\tau \in [0, t]$ the estimate

$$(1 + \tau)^{\frac{n(m-1)+m(s-1)}{m}} \||D|^s(Nu)(\tau, \cdot)\|_{L^m} \leq C\|(\varphi, \psi)\|_A + C\|u\|_{X(t)}^p$$

appearing in the proof to Theorem 19.4.1.

Exercise 3 Prove the statement of Theorem 19.5.1 by following the steps of the proof of Sect. 19.3.2.

Exercise 4 Prove the statement of Theorem 19.5.2 by combining the steps of the proofs of Sects. 19.3.2 and 19.4.2.

Chapter 20
Semilinear Classical Wave Models

In the previous Chaps. 18 and 19, we investigated semilinear wave models with different types of damping mechanisms. The main concern of this chapter is to give an overview on results for semilinear wave models without any damping, a never ending story in the theory of wave models. Here we distinguish between semilinear models with source and those with absorbing power nonlinearity. First we explain the Strauss conjecture for the case of source power nonlinearity. We give an overview on results for the global (in time) existence of small data solutions and for blow up behavior of solutions as well. Moreover, an overview on life span estimates completes the discussion. We explain in detail the local (in time) existence of Sobolev solutions and show for special models how Kato's lemma is used to prove blow up of classical solutions. For wave models with absorbing power we have a more efficient energy conservation which allows for proving a well-posedness result for large data. The description of recent results on critical exponents (Strauss exponent versus Fujita exponent) for special damped wave models with power nonlinearity and on the influence of a time-dependent propagation speed in wave models on the global existence of small data weak solutions completes this chapter.

Many thanks to Sandra Lucente (Bari) and Hiroyuki Takamura (Hakodate, Hokkaido) for useful discussions on the content of this chapter.

20.1 Semilinear Classical Wave Models with Source Nonlinearity

The application of estimates for solutions to linear equations is a very useful tool on studying the global (in time) existence of small data solutions for semilinear equations. A lot of activities have been devoted to the Cauchy problem for wave equations with power nonlinearity

$$u_{tt} - \Delta u = |u|^p, \quad u(0, x) = \varphi(x), \quad u_t(0, x) = \psi(x).$$

© Springer International Publishing AG 2018

M.R. Ebert, M. Reissig, *Methods for Partial Differential Equations*,

https://doi.org/10.1007/978-3-319-66456-9_20

For $1 < p < p_K(n) = \frac{n+1}{n-1}$ ($p_K(n)$ denotes the Kato exponent) the nonexistence of global (in time) generalized solutions for data with compact support was proved in [104]. On the other hand, in [99] it was shown that $p_{crit} = 1 + \sqrt{2}$ is the critical exponent for the global existence of classical small data solutions when $n = 3$. Here classical solution means $u \in C^2([0, \infty) \times \mathbb{R}^n)$. It was later conjectured in [196] that the critical exponent $p_{crit}(n)$ ($p_{crit}(3) = 1 + \sqrt{2}$) is the positive root of the quadratic equation

$$(n - 1)p^2 - (n + 1)p - 2 = 0.$$

This critical exponent is called the Strauss exponent. In the further considerations we use the notation $p_0(n)$ for the Strauss exponent. This conjecture was verified in [66] and [65] for classical solutions when $n = 2$. For $n > 3$, the paper [186] proved the nonexistence of global (in time) solutions in $C([0, \infty), L^{\frac{2(n+1)}{n-1}}(\mathbb{R}^n))$ for suitable small data and for $1 < p < p_0(n)$. Later, the supercritical case $p > p_0(n)$ was treated in [125]. There the authors proved the existence of global weak solutions belonging to $L^\infty([0, \infty), L^q(\mathbb{R}^n, d\mu))$ with a weighted measure $d\mu$ up to $n \leq 8$ and for all n in the case of radial initial data. In [59] the authors removed the assumption of spherical symmetry. The global existence also breaks down at the critical exponent $p = p_0(n)$ as was shown in [183] for $n = 2, 3$ and in [229] and, independently, in [236] for $n \geq 4$.

Some results verifying *Strauss' conjecture* are summarized in the following table which is taken from the paper [202].

	$p < p_0(n)$	$p = p_0(n)$	$p_0(n) < p < p_{conf}(n)$
$n = 2$	Glassey [66]	Schaeffer [183]	Glassey [65]
$n = 3$	John [99]	Schaeffer [183]	John [99]
$n \geq 4$	Sideris [186]	Yordanov-Zhang [229], Zhou Yi [236]	Georgiev-Lindblad-Sogge [59]

Remark 20.1.1 The power $p_{conf}(n) = \frac{n+3}{n-1}$ is well-known as "conformal power" and one can obtain (see [124]) the global (in time) existence of small data solutions when $p \geq p_{conf}(n)$, too, under suitable regularity assumptions for the data.
In the following two sections we are going to derive a local (in time) existence result (Sect. 20.1.1) and show a blow up result in the special case $n \leq 3$ (Sect. 20.1.2). We skip describing methods of how to prove the global (in time) existence of small data solutions for $p > p_0(n)$. The reason being that due to the lack of $L^1 - L^q$ estimates (see Sect. 16.7), the necessary tools are more complicated (see for instance [191]) than the ones for the classical damped waves of Sect. 18.1.1 or for the special structurally damped waves of Sect. 19.3.

20.1.1   Local Existence (in Time) of Sobolev Solutions

According to Duhamel's principle, the Sobolev solution u of the Cauchy problem

$$u_{tt} - \Delta u = |u|^p, \quad u(0, x) = \varphi(x), \quad u_t(0, x) = \psi(x)$$

satisfies

$$u(t, x) = u_0(t, x) + \int_0^t K_1(t - s, 0, x) *_{(x)} |u(s, x)|^p \, ds,$$

where

$$u_0(t, x) = K_0(t, 0, x) *_{(x)} \varphi(x) + K_1(t, 0, x) *_{(x)} \psi(x),$$

$$\widehat{K}_0(t, 0, |\xi|) = \cos(t|\xi|) \quad \text{and} \quad \widehat{K}_1(t, 0, |\xi|) = \frac{\sin(t|\xi|)}{|\xi|}.$$

To derive a local (in time) existence result we are going to use the $L^r - L^q$ estimates from Sect. 16.7. So, in the next two results we fix $P_1 = (\frac{1}{r}, \frac{1}{q})$ with

$$q = \frac{2(n+1)}{n-1} \quad \text{and} \quad r = \frac{2(n+1)}{n+3}.$$

Lemma 20.1.1 *If $(\varphi, \psi) \in H^1(\mathbb{R}^n) \times L^2(\mathbb{R}^n)$, $n \geq 2$, with supp $\varphi, \psi \subset \{|x| \leq R\}$, then $u_0 \in C([0, T], L^q(\mathbb{R}^n))$ for all $T > 0$ with supp $u_0(t, \cdot) \subset \{|x| \leq t + R\}$.*

Proof The statement of Theorem 14.1.1 implies $u_0 \in C([0, T], H^1(\mathbb{R}^n))$ for all $n \geq 1$ and $T > 0$. The conclusion of the lemma follows by using the well-known domain of dependence property of solutions to the wave equation (see Sects. 10.1.3 and 10.5) and that $H^1(\mathbb{R}^n) \subset L^q(\mathbb{R}^n)$, thanks to Sobolev's embedding theorem. Now we shall prove the existence of a uniquely determined local (in time) Sobolev solution for compactly supported data.

Theorem 20.1.1 *Let $(\varphi, \psi) \in H^1(\mathbb{R}^n) \times L^2(\mathbb{R}^n)$, $n \geq 2$ with supp $\varphi, \psi \subset \{|x| \leq R\}$. If $1 \leq p \leq \frac{n+3}{n-1}$, then there exists a positive T and a uniquely determined local (in time) Sobolev solution*

$$u \in C([0, T], L^{\frac{2(n+1)}{n-1}}(\mathbb{R}^n)) \quad \text{with} \quad supp \; u(t, \cdot) \subset \{|x| \leq t + R\}.$$

Proof With $q = \frac{2(n+1)}{n-1}$ we define the space

$$X(T) := \{u \in C([0, T], L^q(\mathbb{R}^n)) \; : \; \text{supp } u(t, \cdot) \subset \{|x| \leq t + R\} \text{ for } t \in [0, T]\}.$$

This is a Banach space with the norm $\|u\|_{X(T)} := \max_{t \in [0, T]} \|u(t, \cdot)\|_{L^q}$. We introduce the operator

$$N : u \in X(T) \to Nu := u_0 + \int_0^t K_1(t - s, 0, x) *_{(x)} |u(s, x)|^p \, ds, \quad t \in (0, T].$$

As in Sect. 18.1.1.2, our goal is to show that for some $T = T(\varphi, \psi)$ the operator N maps $X(T)$ into itself and is Lipschitz continuous for all $(u, v) \in X(T) \times X(T)$.

In other words, we are going to prove the estimates

$$\|Nu\|_{X(T)} \le C_0(\varphi, \psi) + C_1(\varphi, \psi) T^{\frac{2}{n+1}} \|u\|_{X(T)}^{p},$$

$$\|Nu - Nv\|_{X(T)} \le C_2(\varphi, \psi) T^{\frac{2}{n+1}} \|u - v\|_{X(T)} \left(\|u\|_{X(T)}^{p-1} + \|v\|_{X(T)}^{p-1}\right).$$

Lemma 20.1.1 shows that $u_0 \in X(T)$. Now we use the $L^r - L^q$ estimate, away of the conjugate line, from Sect. 16.7 for solutions to the free wave equation. Here we take account of the fact that $(\frac{1}{r}, \frac{1}{q})$ coincides with the point P_1 from Sect. 16.7. Then we have

$$\|K_1(t, 0, x) *_{(x)} \psi(x)\|_{L^q} \le C\, t^{1 - \frac{n}{r} + \frac{n}{q}} \|\psi\|_{L^r} \quad \text{for} \quad t > 0.$$

Hence, the integral term can be estimated as follows:

$$\left\| \int_0^t K_1(t - s, 0, x) *_{(x)} |u(s, x)|^p \, ds \right\|_{L^q} \le \int_0^t \left\| K_1(t - s, 0, \cdot) *_{(x)} |u(s, \cdot)|^p \right\|_{L^q} ds$$

$$\le C \int_0^t (t - s)^{1 - \frac{n}{r} + \frac{n}{q}} \||u(s, \cdot)|^p\|_{L^r} \, ds = C \int_0^t (t - s)^{1 - \frac{n}{r} + \frac{n}{q}} \|u(s, \cdot)|\|_{L^{rp}}^p \, ds$$

$$\le C\|u\|_{X(T)}^p \int_0^t (t - s)^{1 - \frac{n}{r} + \frac{n}{q}} \, ds \le C t^{\frac{2}{n+1}} \|u\|_{X(T)}^p,$$

because of $1 - \frac{n}{r} + \frac{n}{q} > -1$ for $n \ge 2$. Here we use, on the one hand, the compact support property of $u(t, \cdot)$ and suppose, on the other hand, that $rp \le q$. The last inequality implies the condition $1 \le p \le \frac{n+3}{n-1}$. This leads to $Nu \in L^\infty\big((0, T), L^q(\mathbb{R}^n)\big)$. Moreover, using similar arguments as in the proof to Lemma 20.1.1 we conclude that N maps $X(T)$ into itself.

Thanks to Young's inequality we have

$$\left||u|^p - |v|^p\right| \le C|u - v|(|u|^{p-1} + |v|^{p-1}).$$

Using Hölder's inequality we conclude

$$\||u|^p - |v|^p\|_{L^r} \le C\|u - v\|_{L^{rp}} \left(\|u\|_{L^{rp}}^{p-1} + \|v\|_{L^{rp}}^{p-1}\right).$$

Therefore,

$$\|Nu - Nv\|_{X(T)} \le C T^{\frac{2}{n+1}} \|u - v\|_{X(T)} \left(\|u\|_{X(T)}^{p-1} + \|v\|_{X(T)}^{p-1}\right)$$

for any $u, v \in X(T)$. The term $T^{\frac{2}{n+1}}$ implies that N is a contraction mapping on $X(T)$ if T is sufficiently small. This completes the proof of the existence of a uniquely determined local (in time) Sobolev solution after applying a contraction argument for, in general, a small T.

20.1.2   Nonexistence of Global (in Time) Classical Solutions

Now our aim is to introduce tools to show that the Strauss' conjecture is really true. We restrict our attention to lower space dimensions only. For higher dimensions, besides the result in [186], we also refer the reader to [98] for a more elementary treatment.

The main idea is to consider the functional

$$F(t) = \int_{\mathbb{R}^n} u(t, x)\, dx,$$

and to verify that this functional satisfies a nonlinear ordinary differential inequality and, additionally, admits a lower bound in order to apply a version of Kato's lemma in the form of Proposition 24.5.9 given in Sect. 24.5 and taken from [186]. We apply Proposition 24.5.9 to prove that a solution can not exist beyond a certain time.

Theorem 20.1.2 *Let $n \leq 3$ and $u \in C^2([0, T) \times \mathbb{R}^n)$ be a classical solution of*

$$u_{tt} - \Delta u = |u|^p, \quad u(0, x) = \varphi(x), \quad u_t(0, x) = \psi(x).$$

Assume that $\varphi, \psi \in C_0^\infty(\mathbb{R}^n)$, $supp(\varphi, \psi) \subset \{|x| \leq R\}$, where

$$C_\varphi = \int_{\mathbb{R}^n} \varphi(x)\, dx > 0 \quad and \quad C_\psi = \int_{\mathbb{R}^n} \psi(x)\, dx > 0.$$

If $1 < p < p_0(n)$ ($p > 1$ for $n = 1$), then T is necessarily finite.

Remark 20.1.2 Under the assumptions of Theorem 20.1.2 and the hypothesis $\varphi, \psi \in C_0^\infty(\mathbb{R}^n)$ it is well-known that there exists a unique local (in time) classical solution to the semilinear wave equation with power nonlinearity for all $p > 1$. For simplicity, we prove the nonexistence of global (in time) classical solutions, but the argument does not use at all the smoothness of the solution. For more details see [186].

Proof We will not explicitly consider the case $n = 1$. The proof we will give can be easily extended to the case $n = 1$ (see Exercise 1). If $n = 1$, then no critical value of p appears because the solution does not decay uniformly to zero as $t \to \infty$. So, one can expect blow up for all $p > 1$.

Let $n = 3$. By Theorem 20.1.1 we have supp $u(t, \cdot) \subset \{|x| \leq t + R\}$. Hence, after integration with respect to the spatial variables (boundary integrals vanish), we obtain

$$F''(t) = d_t^2 \int_{\mathbb{R}^n} u(t, x)\, dx = \int_{\mathbb{R}^n} \partial_t^2 u(t, x)\, dx = \int_{\mathbb{R}^n} |u(t, x)|^p\, dx,$$

thanks to the divergence theorem. Using the compact support property of $u(t, \cdot)$ and Hölder's inequality with $q = \frac{p}{p-1}$ we get

$$\left| \int_{\mathbb{R}^n} u(t, x)\, dx \right|^p = \left| \int_{|x| \leq t+R} u(t, x)\, dx \right|^p$$

$$\leq \left(\int_{|x| \leq t+R} 1\, dx \right)^{\frac{p}{q}} \left(\int_{\mathbb{R}^n} |u(t, x)|^p\, dx \right) \leq C(t + R)^{n(p-1)} F''(t).$$

Thus, we have obtained the following differential inequality:

$$F''(t) \geq C(t + R)^{-n(p-1)} |F(t)|^p \quad \text{for all } 0 \leq t < T.$$

If $u_0 = u_0(t, x)$ is a classical solution to the Cauchy problem for the free wave equation $u_{tt} - \Delta u = 0$ with data φ and ψ, then the divergence theorem implies

$$\int_{\mathbb{R}^n} u_0(t, x)\, dx = C_\psi t + C_\varphi.$$

In three or lower dimensions the Riemann function $K(t - s, 0, \cdot)$ is non-negative. So, we may conclude $u(t, x) \geq u_0(t, x)$. Moreover, in three dimensions the Huygens' principle states that

$$\text{supp } u_0(t, \cdot) \subset \Omega = \{x \in \mathbb{R}^3 : t - R < |x| < t + R\} \quad \text{for } t > R.$$

Therefore, using Hölder's inequality we have

$$C_\psi t + C_\varphi = \int_{\mathbb{R}^n} u_0(t, x)dx = \int_{\Omega} u_0(t, x)dx$$

$$\leq \int_{\Omega} u(t, x)dx \leq (\text{vol } \Omega)^{\frac{p-1}{p}} \left(\int_{\mathbb{R}^3} |u(t, x)|^p dx \right)^{1/p}$$

$$\leq C(t + R)^{\frac{2(p-1)}{p}} \left(\int_{\mathbb{R}^3} |u(t, x)|^p dx \right)^{\frac{1}{p}} = C(t + R)^{\frac{2(p-1)}{p}} (F''(t))^{\frac{1}{p}}.$$

By our hypothesis we have $C_\psi > 0$, thus, we may conclude

$$F''(t) \geq C t^{2-p} \quad \text{for large } t.$$

Integrating twice gives, under the assumption $p < 3$, the estimate

$$F(t) \geq Ct^{4-p} \quad \text{for large } t.$$

Proposition 24.5.9 implies that $T < \infty$, provided that $1 < p < 1 + \sqrt{2}$. Indeed, for $n = 3$, the functional $F(t)$ satisfies a nonlinear ordinary differential inequality and, additionally, admits a lower bound as in Proposition 24.5.9, where we choose $q = 3(p-1)$ and $r = 4-p$. Moreover, $(p-1)r > q-2$ if and only if $p < 1 + \sqrt{2}$. Hence, the application of Proposition 24.5.9 implies $T < \infty$.

By the lack of Huygens' principle, the proof for $n = 2$ is more delicate and we refer to Glassey [66] for more details.

20.1.3 Some Remarks: Life Span Estimates

Remark 20.1.3 Let us explain the solvability behavior of the Cauchy problem

$$u_{tt} - \Delta u = |u|^p, \quad u(0,x) = \varphi(x), \quad u_t(0,x) = \psi(x)$$

in the 3d case. Due to Theorem 20.1.1, we know that for $p \in [1, 3]$ we have a local (in time) solution belonging to the evolution space $C([0,T], L^4(\mathbb{R}^3))$. Theorem 20.1.2 yields that this solution has, in general, a blow up behavior for $p \in (1, 1 + \sqrt{2})$. It is known by [183] that for $p = 1 + \sqrt{2}$ the solution may blow up in finite time for suitable small data. The existence of global (in time) classical solutions for sufficiently smooth initial data with compact support is proved in [99] for $p \in (1 + \sqrt{2}, 3]$. In [59] the authors proved the existence of global (in time) Sobolev solutions in the space $L^{p+1}(\mathbb{R}^{3+1}, d\mu)$ with a weighted measure $d\mu$ for $p \in (1 + \sqrt{2}, 3]$ and for initial data in $C_0^\infty(\mathbb{R}^3)$. More recently, under the assumption of radial initial data $\psi \in L^2(\mathbb{R}^3) \cap L^p(\mathbb{R}^3)$ (for simplicity take $\varphi \equiv 0$), in [48] the authors obtained the existence of global Sobolev solutions $u \in C([0,\infty), L^3(\mathbb{R}^3))$ without any assumption on the support of the data.

Remark 20.1.4 In Remark 18.1.3, we verified that the test function method is based on a contradiction argument. So, this method does not give any information about blow up time or life span estimate or about the blow up mechanism.

The application of Kato type lemmas (see Proposition 24.5.9) gives information that the functional

$$F(t) := \int_{\mathbb{R}^n} u(t,x)\, dx$$

may blow up in finite time. This describes a blow up mechanism. We can expect also estimates for the life span time $T(\varepsilon)$ (see Remark 20.1.5).

Remark 20.1.5 The conclusion of Theorem 20.1.2 is true even by assuming small data, let us say, $u(0, x) = \varepsilon \varphi(x)$ and $u_t(0, x) = \varepsilon \psi(x)$ with small ε. An important topic of recent research is determining the lifespan $T = T(\varepsilon)$ of solutions. Here we define $T(\varepsilon) = \sup\{t_0 > 0\}$, where the solution exists on the time interval $[0, t_0]$ for arbitrarily fixed (φ, ψ). One should pay attention to in which sense solutions do exist, as classical ones, energy solutions, Sobolev solutions or distributional solutions. In order to have a good overview about results on lower and upper bounds for the lifespan we refer to the paper [200].

The following estimates for the lifespan $T(\varepsilon)$ were conjectured for $1 < p < p_0(n)$ $(n \geq 3)$ or $2 < p < p_0(2)$ $(n = 2)$ in [200]:

$$c \varepsilon^{-\frac{2p(p-1)}{\gamma(p,n)}} \leq T(\varepsilon) \leq C \varepsilon^{-\frac{2p(p-1)}{\gamma(p,n)}}, \quad \gamma(p,n) = 2 + (n+1)p - (n-1)p^2,$$

where the positive constants c, C are independent of ε. Results verifying this conjecture are summarized in the following table from [200]:

	Lower bounds for $T(\varepsilon)$	Upper bounds for $T(\varepsilon)$
$n = 2$	Zhou [234]	Zhou [234]
$n = 3$	Lindblad [123]	Lindblad [123]
$n \geq 4$	Lai-Zhou [119]	(Rescaling argument of Sideris [186])

In [200] the author presents a simpler proof for upper bounds for $T(\varepsilon)$ by using an improved Kato type lemma without any rescaling argument.

If $p = p_0(n)$, it was conjectured that

$$\exp\left(c \varepsilon^{-p(p-1)}\right) \leq T(\varepsilon) \leq \exp\left(C \varepsilon^{-p(p-1)}\right),$$

where the positive constants c, C are independent of ε. Results verifying this conjecture are summarized in the following table from [200]:

	Lower bounds for $T(\varepsilon)$	Upper bounds for $T(\varepsilon)$
$n = 2$	Zhou [234]	Zhou [234]
$n = 3$	Zhou [233]	Zhou [233]
$n \geq 4$	Lindblad–Sogge [125] (for $n \leq 8$ or radially symmetric solutions)	Takamura–Wakasa [201]

20.2 Semilinear Classical Wave Models with Absorbing Nonlinearity

In this section we study the Cauchy problem for wave models with so-called absorbing nonlinearity

$$u_{tt} - \Delta u = -u|u|^{p-1}, \quad u(0, x) = \varphi(x), \quad u_t(0, x) = \psi(x),$$

with $p > 1$. In order to state a result on the existence of global (in time) Sobolev solutions, following Sect. 11.2.5, one can verify that the energy

$$E_W(u)(t) = \frac{1}{2}\|u_t(t, \cdot)\|_{L^2}^2 + \frac{1}{2}\|\nabla u(t, \cdot)\|_{L^2}^2 + \frac{1}{p+1}\int_{\mathbb{R}^n} |u(t, x)|^{p+1}\, dx$$

is conserved. We are going to prove the following result from [194].

Theorem 20.2.1 *Consider for $p > 1$ the Cauchy problem*

$$u_{tt} - \Delta u = -u|u|^{p-1}, \quad u(0, x) = \varphi(x), \quad u_t(0, x) = \psi(x).$$

The data (φ, ψ) are supposed to have compact support and belong to the function space $(H^1(\mathbb{R}^n) \cap L^{p+1}(\mathbb{R}^n)) \times L^2(\mathbb{R}^n)$. Then, there exists a global (in time) energy solution

$$u \in L^\infty\big((0, \infty), H^1(\mathbb{R}^n) \cap L^{p+1}(\mathbb{R}^n)\big) \cap W_\infty^1\big((0, \infty), L^2(\mathbb{R}^n)\big).$$

One important step in the proof to Theorem 20.2.1 is the following lemma (see [194]):

Lemma 20.2.1 *Let us consider the Cauchy problem*

$$u_{tt} - \Delta u + f(u) = 0, \quad u(0, x) = \varphi(x), \quad u_t(0, x) = \psi(x),$$

where the data $(\varphi, \psi) \in H^1(\mathbb{R}^n) \times L^2(\mathbb{R}^n)$ have compact support and f is globally Lipschitz continuous such that $uf(u) \geq 0$ and $f(0) = 0$. Then, there exists a uniquely determined energy solution

$$u \in C\big([0, T], H^1(\mathbb{R}^n)\big) \cap C^1\big([0, T], L^2(\mathbb{R}^n)\big).$$

Moreover, the energy equality $E_W(u)(t) = E_W(u)(0)$ is valid where

$$E_W(u)(t) = \frac{1}{2}\|u_t(t, \cdot)\|_{L^2}^2 + \frac{1}{2}\|\nabla u(t, \cdot)\|_{L^2}^2 + \int_{\mathbb{R}^n} F(u(t, \cdot))\, dx,$$

and $F(u) = \int_0^u f(s)ds$.

Proof The solution is constructed by successive approximations in the following way:

Define $u_0(t, x) \equiv 0$ and $u_k = u_k(t, x)$ inductively as the solution to the linear Cauchy problem

$$\partial_t^2 u_k - \Delta u_k = -f(u_{k-1}), \quad u_k(0, x) = \varphi(x), \quad \partial_t u_k(0, x) = \psi(x).$$

The Lipschitz condition on f implies that the operator $u(t, \cdot) \to f(u(t, \cdot))$ is Lipschitz from $L^2(\mathbb{R}^n)$ to $L^2(\mathbb{R}^n)$. Thanks to Exercise 2 in Chap. 16 we obtain for

$v_k = u_k - u_{k-1}$ the inequality

$$E_W(v_k)(t) = \frac{1}{2}\|\partial_t v_k(t, \cdot)\|_{L^2}^2 + \frac{1}{2}\|\nabla v_k(t, \cdot)\|_{L^2}^2 \leq C \int_0^t \|v_{k-1}(s, \cdot)\|_{L^2}^2 \, ds.$$

Using the domain of dependence property and Poincaré's inequality implies

$$\|v_{k-1}(s, \cdot)\|_{L^2} \leq C(1 + s)\|\nabla v_{k-1}(s, \cdot)\|_{L^2}.$$

Hence,

$$E_W(v_k)(t) \leq C \int_0^t (1 + s)^2 \|\nabla v_{k-1}(s, \cdot)\|_{L^2}^2 \, ds \leq C(1 + t)^2 \int_0^t E_W(v_{k-1})(s) \, ds.$$

Therefore,

$$E_W(v_k)(t) \leq (Ct^2)^{k-1} \frac{1}{(k-1)!} E_W(v_1)(t)$$

and the sequence $\{u_k\}_k$ converges to the uniquely determined energy solution u. The assumption for f to be globally Lipschitz continuous may be reasonable for a significant class of applied mathematical problems. It excludes the interesting case of power nonlinearities with exponent larger than one. In [194], the author considered a class of semilinear models including absorbing power nonlinearities as well that can be approximated by a sequence of globally Lipschitz continuous functions.

Lemma 20.2.2 *Let $f = f(x, u)$ be a continuous function such that $uf(x, u) \geq 0$. Then f can be approximated by a sequence of continuous function $\{f_k\}_k$ such that:*

1. *f_k tends to f uniformly on bounded sets;*
2. *there exist continuous functions $c_k = c_k(x)$ with*

$$|f_k(x, u) - f_k(x, v)| \leq c_k(x)|u - v|;$$

3. *$uf_k(x, u) \geq 0$.*

Proof For $k^{-1} \leq u \leq k$ we define

$$f_k(x, u) = \varepsilon k \Big(F\Big(x, u + \frac{\varepsilon}{k}\Big) - F(x, u) \Big),$$

where $\varepsilon = \text{sign}(u)$ and $F(x, u) = \int_0^u f(x, s) \, ds$. For $u \geq k$ and $u \leq -k$, we define $f_k(x, u)$ independent of u. For $0 \leq u \leq k^{-1}$ and $-k^{-1} \leq u \leq 0$ we define it to be linear and to vanish at $u = 0$.

Now let us turn to the proof of Theorem 20.2.1.

Proof (Theorem 20.2.1) We only sketch the proof. For further details see [194].

Step 1 By applying Lemma 20.2.2 with $f(x, u) = u|u|^{p-1}, p > 1$, there exists a sequence of global Lipschitz functions $\{f_k\}_k$ that tends uniformly to f. Let $d_u F_k(u) = f_k(u)$ with $F_k(0) = 0$. It is clear that the sequence $\{F_k\}_k$ tends to F uniformly as $k \to \infty$. Moreover, there is a subsequence of $\{F_k\}_k$ such that

$$\int_{\mathbb{R}^n} |F_k(u) - F(u)|\, dx \to 0.$$

Step 2 By Lemma 20.2.1, for each f_k given in Step 1 there is an energy solution $u_k = u_k(t, x)$ to the Cauchy problem

$$u_{tt} - \Delta u + f_k(u) = 0, \quad u(0, x) = \varphi(x), \quad u_t(0, x) = \psi(x).$$

The energy equality asserts $E_k(u_k)(t) = E_k(u_k)(0)$, where

$$E_k(u)(t) = \frac{1}{2}\|u_t(t, \cdot)\|_{L^2}^2 + \frac{1}{2}\|\nabla u(t, \cdot)\|_{L^2}^2 + \int_{\mathbb{R}^n} F_k(u(t, \cdot))\, dx.$$

Thanks to $u f_k(u) \geq 0$, this energy is well-defined. Thus each term in $E_k(u_k)(t)$ is bounded and $\{u_k\}_k$ is also bounded in $L_{loc}^{\infty}([0, \infty), L^2(\mathbb{R}^n))$. Therefore, there is a convergent subsequence $u_k \to u$ weakly in $L_{loc}^{\infty}([0, \infty), H^1(\mathbb{R}^n))$ and $\partial_t u_k \to w$ weakly in $L^{\infty}([0, \infty), L^2(\mathbb{R}^n))$.

Step 3 By compactness, the subsequence can be chosen to converge strongly and therefore $f_k(u_k) \to f(u)$ almost everywhere. Taking into consideration the inequality $u_k f_k(u_k) \geq 0$ it follows that

$$\int_0^T \int_{|x| \leq R+T} u_k f_k(u_k)\, dx\, dt$$

is bounded. Applying Theorem 1.1 of [194] we conclude that $f(u)$ is locally integrable and is the limit of $f_k(u_k)$ strongly in $L^1(\mathbb{R}^n)$ on any bounded set. In particular, a passage to the limit yields $w = u_t$ and

$$u_{tt} - \Delta u + f(u) = 0, \quad u(0, x) = \varphi(x), \quad u_t(0, x) = \psi(x).$$

In addition, by using Fatou's lemma the following energy inequality is valid:

$$E_w(u)(t) = \frac{1}{2}\|u_t(t, \cdot)\|_{L^2}^2 + \frac{1}{2}\|\nabla u(t, \cdot)\|_{L^2}^2 + \int_{\mathbb{R}^n} F(u(t, \cdot))\, dx \leq E_w(u)(0).$$

Finally, we may conclude that $u \in L^{\infty}([0, \infty), L^{p+1}(\mathbb{R}^n))$.

Remark 20.2.1 We have global (in time) solutions even for large data if the power nonlinearity is absorbing in opposition to the results with power nonlinearity of source type.

Remark 20.2.2 If $1 < p < \frac{n+2}{n-2}$ for $n \geq 3$, $1 < p < \infty$ for $n = 1, 2$, respectively, then the solution of Theorem 20.2.1 is uniquely determined and belongs to $C\big([0, \infty) \times H^1(\mathbb{R}^n)\big) \cap C^1\big([0, \infty) \times L^2(\mathbb{R}^n)\big)$ (cf. with [13] and [62]).

Remark 20.2.3 The conclusion of Theorem 20.2.1 is still true for more general semilinear models. Consider

$$u_{tt} - \Delta u = f(u), \quad u(0, x) = \varphi(x), \quad u_t(0, x) = \psi(x),$$

where the right-hand side $f(u)$ satisfies $uf(u) \leq 0$ (see [194]). Then, a conserved energy is

$$E_W(u)(t) = \frac{1}{2}\|u_t(t, \cdot)\|_{L^2}^2 + \frac{1}{2}\|\nabla u(t, \cdot)\|_{L^2}^2 + \int_{\mathbb{R}^n} F(u)\, dx,$$

where $F(u) := -\int_0^u f(\tau)d\tau$.

Remark 20.2.4 If one is interested in classical solutions, it turns out in space dimensions $n \geq 3$, that $p = 1 + \frac{4}{n-2}$ is a critical exponent as explained in [197]. This critical exponent is one less than the Sobolev number $\frac{2n}{n-2}$. Moreover, this critical exponent is important for semilinear Schrödinger equations in H^1-theory (cf. with Theorem 21.4.1).

20.3   Concluding Remarks

20.3.1   *Strauss Exponent Versus Fujita Exponent*

In Sect. 20.1 we introduced the Strauss exponent $p_0(n)$ as a critical exponent for the Cauchy problem for the wave equation with source power nonlinearity

$$u_{tt} - \Delta u = |u|^p, \quad u(0, x) = \varphi(x), \quad u_t(0, x) = \psi(x).$$

A dissipation term may have an improving influence on the critical exponent. If we are interested in the Cauchy problem for the classical damped wave equation

$$u_{tt} - \Delta u + u_t = |u|^p, \quad u(0, x) = \varphi(x), \quad u_t(0, x) = \psi(x),$$

then it is shown in Sects. 18.1.1 and 18.1.2 that the critical exponent is the Fujita exponent $p_{Fuj}(n)$. It holds $p_{Fuj}(n) < p_0(n)$. In this way we may understand the improving influence of the classical dissipation term u_t. There exists a class of damped wave models for which the critical exponent depends somehow on the Fujita exponent and the Strauss exponent as well. This class is described by scale-invariant linear damped wave operators and reads as follows:

$$u_{tt} - \Delta u + \frac{\mu}{1+t} u_t = |u|^p, \quad u(0,x) = \varphi(x), \quad u_t(0,x) = \psi(x),$$

where $\mu > 0$ is a real parameter (cf. with the research project for scale-invariant p-evolution models in Sect. 23.13). It was recently shown in [30] and [35] that $p_{Fuj}(n)$ is still the critical exponent when $\mu \geq \frac{5}{3}$ if $n = 1$, $\mu \geq 3$ if $n = 2$ and $\mu \geq n + 2$ if $n \geq 3$.

It seems to be a challenge to determine the critical exponent in the case $\mu \in (0, n + 2)$. In particular, it seems to be interesting to understand the transfer of $p_{Fuj}(n)$ to $p_0(n)$. The interested reader can find a first result in [39]. The authors consider the above model for $\mu = 2$, that is, the Cauchy problem

$$u_{tt} - \Delta u + \frac{2}{1+t} u_t = |u|^p, \quad u(0,x) = \varphi(x), \quad u_t(0,x) = \psi(x).$$

In this special case a change of variables transforms this Cauchy problem in to the Cauchy problem

$$v_{tt} - \Delta v = (1 + t)^{-(p-1)} |v|^p, \quad v(0,x) = v_0(x), \quad v_t(0,x) = v_1(x).$$

So, one can apply tools for wave models with power nonlinearity and a time-dependent coefficient. The authors prove the conjecture $p_{crit}(n) = p_0(n + 2)$, so we have a shift of the Strauss exponent by 2 in dimensions $n = 2, 3$. Later, the first two authors of [39] proved this conjecture for all odd dimensions [36]. We still feel an improving influence of the dissipation term because of $p_{Fuj}(2) = p_0(4) = 2 < p_0(2)$ for $n = 2$ and $p_{Fuj}(3) < p_0(5) < p_0(3)$ for $n = 3$.

To prove the conjecture for $n = 2, 3$ the authors use the following tools:

1. the blow up technique of Glassey (see [66]), in particular, a Kato type lemma (see Proposition 24.5.9) and the considerations in Sect. 20.1.2,
2. for $n = 2$, Klainerman's vector fields are used to derive a suitable energy estimate in Klainerman-Sobolev spaces (see [115] and [235]),
3. for $n = 3$, radial data are supposed and the existence of small data radial solutions is proved by the aid of pointwise estimates (see [2] and [118]).

20.3.2   A Special Class of Quasilinear Wave Equations with Time-Dependent Speed of Propagation

To find the influence of a time-dependent propagation speed on the global (in time) existence of weak solutions we consider the special class of Cauchy problems

$$u_{tt} - a(t)^2 \Delta u = u_t^2 - a(t)^2 |\nabla u|^2, \quad u(0,x) = \varphi(x), \quad u_t(0,x) = \psi(x),$$

where the propagation speed $a = a(t)$ is sufficiently regular and positive for $t \geq 0$. The special structure of the right-hand side allows us to apply Nirenberg's transformation (see, e.g., [114])

$$v(t, x) = 1 - e^{-u(t,x)}.$$

Under the constrain condition $v(t, x) < 1$ in $\mathbb{R}^1_+ \times \mathbb{R}^n$ the equation is transformed into the linear wave equation with time-dependent propagation speed

$$v_{tt} - a(t)^2 \Delta v = 0.$$

The constrain condition requires small data (in a suitable function space) as a natural assumption to have global (in time) Sobolev solutions.

Let us briefly discuss the influence of the time-dependent propagation speed on the global (in time) existence of solutions.

Using Floquet theory, in [226] it was proved that if a is a non-constant, 1-periodic and positive coefficient belonging to $C^\infty(\mathbb{R}^1)$, then the Cauchy problem has no global (in time) classical solution even for suitable small initial data.

On the other hand, if $a(t) = (1 + t)^\ell$ with $\ell > 0$, by applying the theory of confluent hypergeometric functions the following a priori estimate is proved in [171]:

$$\|v(t, \cdot)\|_{L^\infty} \leq C\big(\|\langle D\rangle^s v_0\|_{L^1} + \|\langle D\rangle^{s-1} v_1\|_{L^1}\big)$$

for every $t \geq 0$ and $s > 1$. Here C depends on ℓ and s, but is independent of t. This implies a global existence result for small data Sobolev solutions to the above Cauchy problem.

These examples tell us that there is an interesting interplay between the improving influence coming from a strictly increasing coefficient $a(t)$ and the deteriorating influence coming from the oscillating behavior of $a(t)$ on the global existence of small data solutions. In [45] and [220], under the assumption that $a(t)$ is sufficiently regular and has some control on its oscillations, the authors derived results about global (in time) existence of small data classical solutions to the above Cauchy problem. Also, the question for optimality of the results is discussed.

Finally, we want to mention the paper [227]. In [227] the author proved results on the existence and nonexistence of global (in time) small data solutions to the Cauchy problem for the more general wave equation

$$u_{tt} - a(t)^2 \Delta u + f(u)\big(u_t^2 - a(t)^2 |\nabla u|^2\big) = 0, \quad u(0, x) = \varphi(x), \quad u_t(0, x) = \psi(x).$$

The constant coefficient case $a \equiv 1$ was studied in [144]. There, the authors proposed necessary and sufficient conditions on $f(u)$ for which the Cauchy problem has a global (in time) smooth solution for any smooth initial data.

Exercises Relating to the Considerations of Chap. 20

Exercise 1 Prove the statements of Theorem 20.1.2 for $n = 1$.

Exercise 2 The reason that we stated Theorem 20.1.2 only for $n \leq 3$ is that for higher dimensions n the Riemann function is no longer nonnegative. Follow the proof of Theorem 20.1.2 for $n \geq 4$, verify that $F(t) \geq C_\psi t$ for large t, and conclude a blow up result for $1 < p < \frac{n+1}{n-1}$.

Exercise 3 Under the assumption of small data (in a suitable function space), prove the global existence (in time) of classical solutions to the Cauchy problem

$$u_{tt} - \Delta u = u_t^2 - |\nabla u|^2, \quad u(0, x) = \varphi(x), \quad u_t(0, x) = \psi(x).$$

Chapter 21
Semilinear Schrödinger Models

In this chapter we introduce results for semilinear Schrödinger models with power nonlinearity in the focusing and defocusing cases as well. First of all, we show how by a scaling argument a proposal for a critical exponent appears. This critical exponent heavily depends on the regularity of the data. The issue of L^2 and H^1 data is explained. As for the linear Schrödinger equation (see Sect. 11.2.3), some conserved quantities are given. Then, a global (in time) well-posedness result is proved for weak solutions in the subcritical L^2 case. This result is valid for both cases focusing and defocusing, respectively. Finally, the subcritical H^1 case is treated. Here the main concern is to show differences between both focusing and defocusing cases. A local (in time) well-posedness result is proved. This result contains, moreover, a blow up result in the focusing and a global (in time) well-posedness result in the defocusing case.

Many thanks to Vladimir Georgiev (Pisa) for useful discussions on the content of this chapter.

21.1 Examples of Semilinear Schrödinger Models

A large class of problems of the theory of waves is described by systems of semilinear differential equations of Schrödinger type

$$\partial_t u_k = i a_k \Delta u_k + f_k(u, \bar{u}), \quad x \in \mathbb{R}^n, \ t > 0, \ k = 1, \cdots, m.$$

Here $u(t, x) = (u_1(t, x), \cdots, u_m(t, x))$ is an unknown vector-function, a_k, $k = 1, \cdots, m$, are non-zero real constants. Additionally, we prescribe Cauchy conditions

$$u_k(0, x) = u_{0k}(x), \quad x \in \mathbb{R}^n, \ k = 1, \cdots, m.$$

© Springer International Publishing AG 2018
M.R. Ebert, M. Reissig, *Methods for Partial Differential Equations*,
https://doi.org/10.1007/978-3-319-66456-9_21

The question of singularities arising in the solution to this Cauchy problem might be of interest. This phenomenon is called collapse, blow up, and also nonlinear focusing in the literature. Singularities arising in solutions describe phenomenons for many physical problems such as self-focusing bundle of waves, self compression of wave packets in a nonlinear medium, self-focusing electromagnetic waves in plasma and Langmuir wave collapse (see [43]). Therefore, on the one hand blow up considerations for solutions and on the other hand global (in time) existence of solutions are of interest. We will give only a short overview beyond some of the known results.

For this reason we consider the semilinear Schrödinger model with power nonlinearity

$$iu_t + \Delta u = \pm|u|^{p-1}u, \quad u(0, x) = \varphi(x).$$

Here $\pm|u|^{p-1}u$ is an example of a focusing nonlinearity (negative sign) and a defocusing nonlinearity (positive sign).

The following treatise of this chapter is based on [61, 209] and the lecture notes "Properties of solutions to the semilinear Schrödinger equation" written by N. Tzirakis (University of Illinois).

21.2   How Do We Arrive at a Critical Exponent?

Consider the semilinear Schrödinger model with power nonlinearity

$$iu_t + \Delta u = \lambda|u|^{p-1}u, \quad u(0, x) = \varphi(x),$$

where λ is a positive constant and $p > 1$.

If $u = u(t, x)$ is, for example, a classical solution on the time interval $[0, T]$, then the family of functions $\{u_\lambda = u_\lambda(t, x)\}_{\lambda > 0}$ with

$$u_\lambda(t, x) := \lambda^{-\frac{2}{p-1}} u\left(\frac{x}{\lambda}, \frac{t}{\lambda^2}\right)$$

is a family of classical solutions to the family of Cauchy problems

$$iu_t + \Delta u = \lambda|u|^{p-1}u, \quad u_\lambda(0, x) = \varphi_{0,\lambda}(x) := \lambda^{-\frac{2}{p-1}} \varphi\left(\frac{x}{\lambda}\right).$$

The computation of the $\dot{H}^s$-norm of $\varphi_{0,\lambda}$ for $s > 0$ leads to

$$\|\varphi_{0,\lambda}\|_{\dot{H}^s} = \lambda^{s_{crit}-s} \|\varphi\|_{\dot{H}^s}, \quad \text{where } s_{crit} = \frac{n}{2} - \frac{2}{p-1}.$$

Now, let us fix $s \in [0, \frac{n}{2})$. In this way we require some regularity of the data φ. If we form $\lambda \to \infty$ in the last relation, then the following three cases can be observed:

1. The *subcritical case* $s > s_{crit}$, corresponding to $p < 1 + \frac{4}{n-2s}$: The norm $\|\varphi_{0,\lambda}\|_{\dot{H}^s}$ can be made small while at the same time the solution exists on the time interval $[0, \lambda^2 T]$. The time interval becomes larger with larger λ. This is the best possible scenario for *local well-posedness*.
2. The *supercritical case* $s < s_{crit}$, corresponding to $p > 1 + \frac{4}{n-2s}$: The norm $\|\varphi_{0,\lambda}\|_{\dot{H}^s}$ grows as the time interval is made longer. For this reason we can not expect *local well-posedness*.
3. The *critical case* $s = s_{crit}$, corresponding to $p_{crit} = p_{crit}(n, s) = 1 + \frac{4}{n-2s}$: The norm $\|\varphi_{0,\lambda}\|_{\dot{H}^s}$ remains invariant as the time interval is made longer. So, we are between both scenarios which are described by the first two cases. In general, the treatment of the critical case needs very precise tools.

Example 21.2.1 If $s = 0$, then $p_{crit} = p_{crit}(n, 0) = 1 + \frac{4}{n}$. If $s = 1$, then $p_{crit} = p_{crit}(n, 1) = 1 + \frac{4}{n-2}$.

In the following sections we discuss results of subcritical L^2 and H^1 theory.

21.3 Semilinear Models with Power Nonlinearity in the Subcritical Case with L^2 Data

In this section we discuss the Cauchy problem for the semilinear Schrödinger equation with power nonlinearity

$$iu_t + \Delta u = \lambda |u|^{p-1} u, \quad u(0, x) = \varphi(x), \quad \lambda \in \mathbb{R}^1, \quad p > 1.$$

The data φ belongs to $L^2(\mathbb{R}^n)$. We are interested in weak solutions. Therefore, we study the integral equation

$$u(t, x) = G(t, 0, x) *_{(x)} \varphi(x) - i \int_0^t G(t, s, x) *_{(x)} \lambda(|u|^{p-1} u)(s, x) \, ds,$$

where

$$v(t, x) := G(t, s, x) *_{(x)} \varphi(x)$$

is the solution to the linear Cauchy problem

$$v_t - i \Delta v = 0, \quad v(s, x) = \varphi(x).$$

We have $G(t, s, x) = G(t - s, x)$ due to the invariance of the linear Cauchy problem by the shift in time. The integral is the Bochner integral in H^{-1} (the Bochner integral generalizes the Lebesgue integral to functions with values in a Banach space, see, for example, [9]). If $u \in C([-T, T], L^2(\mathbb{R}^n))$ solves the integral equation, then we call $u = u(t, x)$ a weak solution to the given semilinear Cauchy problem. This is the integral version of the given semilinear Cauchy problem. We applied Duhamel's principle. We discuss the questions for global (in time) existence of weak solutions. One important tool is the following conservation law.

Lemma 21.3.1 *Let us assume that $u \in C([-T, T], L^2(\mathbb{R}^n))$, $T > 0$, is a weak solution to the given semilinear Cauchy problem*

$$iu_t + \Delta u = \lambda |u|^{p-1} u, \quad u(0, x) = \varphi(x), \quad \lambda \in \mathbb{R}^1, \quad p > 1.$$

Then, the L^2-norm of the solution u is conserved, that is, $\|u(t, \cdot)\|_{L^2} = \|\varphi\|_{L^2}$ for all $t \in [-T, T]$.

We are going to prove the following well-posedness result from [209].

Theorem 21.3.1 *Assume that $p \in (1, 1 + \frac{4}{n})$. Then, for any $\varphi \in L^2(\mathbb{R}^n)$ and $\lambda \in \mathbb{R}^1$ there exists a uniquely determined global (in time) weak solution $u = u(t, x)$ of*

$$iu_t + \Delta u = \lambda |u|^{p-1} u, \quad u(0, x) = \varphi(x).$$

The solution belongs to

$$C(\mathbb{R}, L^2(\mathbb{R}^n)) \cap L^r_{loc}(\mathbb{R}, L^{p+1}(\mathbb{R}^n)),$$

where $r = \frac{4(p+1)}{n(p-1)}$. Moreover, the solution belongs to (cf. with Remarks 13.1.4 and 13.1.5)

$$L^r_{loc}(\mathbb{R}, L^q(\mathbb{R}^n))$$

for every sharp admissible pair (r, q). Finally, the solution u depends continuously on the data φ in the following sense

If the sequence $\{\varphi_k\}_k$ converges to φ in $L^2(\mathbb{R}^n)$, then the sequence $\{u_k\}_k$ of the corresponding weak solutions converges to u in the following sense:

$$\lim_{k \to \infty} \|u_k(t, \cdot) - u(t, \cdot)\|_{L^2} \to 0 \ \text{for all} \ t \in \mathbb{R}^1.$$

Proof We will sketch the proof. The proof is divided into several steps.

Step 1: Some auxiliary results

In later steps of the proof the following statements will be used (the reader can find the proofs in [209]).

Proposition 21.3.1 *Let I be an open interval in $\mathbb{R}^1$. Let $r, q \in (1, \infty)$ and $a, b > 0$. We introduce the set*

$$M_{a,b} := \left\{ v \in L^\infty\left(I, L^2(\mathbb{R}^n)\right) \cap L^r\left(I, L^q(\mathbb{R}^n)\right) : \right.$$

$$\left. \|v(t, \cdot)\|_{L^\infty(I, L^2)} \le a \ \text{ and } \ \|v\|_{L^r(I, L^q)} \le b \right\}.$$

$M_{a,b}$ *is a closed set in $L^r\left(I, L^q(\mathbb{R}^n)\right)$.*

Proposition 21.3.2 *Let T_1 and T_2 be constants with $T_1 < T_2$. Assume that $u \in C\left([T_1, T_2], H^{-1}(\mathbb{R}^n)\right)$. Moreover, with some positive constant K it holds $\|u(t, \cdot)\|_{L^2} \le K$ a.e. on $[T_1, T_2]$. Then, $u \in C_w\left([T_1, T_2], L^2(\mathbb{R}^n)\right)$ and $\|u(t, \cdot)\|_{L^2} \le K$ holds for all $t \in [T_1, T_2]$.*

Here we use the notation $C_w(I, H)$ for a closed interval $I \in \mathbb{R}^1$ and a Hilbert space H, which denotes the set of all weakly continuous functions from I to H.

A function $u : t \in I \to u(t) \in H$ is called *weakly continuous on the interval I* if for all $t_0, t \in I$ and $v \in H$ it holds $\lim_{t \to t_0} v(u(t)) = v(u(t_0))$.

Step 2:   *A local (in time) existence result*

By $I_t, \bar{I}_t, t > 0$, we denote an open interval $(t_0 - t, t_0 + t)$, closed interval $[t_0 - t, t_0 + t]$, respectively. In the following we set $r = \frac{4(p+1)}{n(p-1)}$.

Proposition 21.3.3 *Assume that $p \in (1, 1 + \frac{4}{n})$. Then, for any $t_0 \in \mathbb{R}^1$ and $\varphi \in L^2(\mathbb{R}^n)$ there exists a positive constant $T = T(p, n, \lambda, \|\varphi\|_{L^2})$ such that the Cauchy problem*

$$iu_t + \Delta u = \lambda |u|^{p-1} u, \ \ u(t_0, x) = \varphi(x), \ \lambda \in \mathbb{R}^1$$

has a uniquely determined local (in time) solution u as a solution of the integral equation

$$u(t, x) = G(t, t_0, x) *_{(x)} \varphi(x) - i \int_{t_0}^t G(t, s, x) *_{(x)} \lambda(|u|^{p-1} u)(s, x) \, ds \ \text{ for } t \in \bar{I}_T,$$

where $G(t, s, x) = G(t - s, 0, x)$ and the integral is the Bochner integral in H^{-1} (the Bochner integral generalizes the Lebesgue integral to functions with values in a Banach space, see, for example, [9]). The solution u belongs to

$$C\left(\bar{I}_T, L^2(\mathbb{R}^n)\right) \cap L^r\left(I_T, L^{p+1}(\mathbb{R}^n)\right) \ \text{ and satisfies}$$

$$\|u(t, \cdot)\|_{L^2} = \|\varphi\|_{L^2} \ \text{ for } t \in \bar{I}_T.$$

Proof We sketch the main steps of the proof.

Step 2.1:   *Construction of a sequence of solutions with data from $H^1(\mathbb{R}^n)$*

First we choose $\varepsilon_k = \frac{1}{k}$ and the functions g_{ε_k} as in Sect. 24.1.1. Then we introduce the sequence of data $\{\varphi_k\}_k := \{J_{\varepsilon_k}(\varphi)\}_k$. This sequence belongs to $H^1(\mathbb{R}^n)$.

Applying Proposition 2.5 of [209] or Theorem 3.1 of [61] we get a sequence $\{u_k\}_k$ of solutions to

$$u_k(t, x) = G(t, t_0, x) *_{(x)} \varphi_k(x) - i \int_{t_0}^t G(t, s, x) *_{(x)} \lambda(|u_k|^{p-1} u_k)(s, x) \, ds$$

belonging to $C(\mathbb{R}, H^1(\mathbb{R}^n))$.

Step 2.2:   A closed set in $L^r(I_T, L^{p+1}(\mathbb{R}^n))$

Let $\rho = \|\varphi\|_{L^2}$. By $\delta = \delta(n, p)$ we denote the constant C appearing in Remark 13.1.4 with $q = p + 1$ and $r = \frac{4(p+1)}{n(p-1)}$. Then we define the set

$$M_{\rho, 2\delta\rho} := \{v \in L^\infty(I_T, L^2(\mathbb{R}^n)) \cap L^r(I_T, L^{p+1}(\mathbb{R}^n)) :$$

$$\|v(t, \cdot)\|_{L^\infty(I_T, L^2)} \leq \rho \text{ and } \|v\|_{L^r(I_T, L^{p+1})} \leq 2\delta\rho\},$$

where T is a small constant to be determined later. We note that by Proposition 21.3.1 this set $M_{\rho, 2\delta\rho}$ is closed in $L^r(I_T, L^{p+1}(\mathbb{R}^n))$. The main goal of this step is to show that the sequence $\{u_k\}_k$ belongs to $M_{\rho, 2\delta\rho}$ if T is small enough. On the one hand we have

$$\|u_k(t, \cdot)\|_{L^2} = \|\varphi_k\|_{L^2} \leq \|\varphi\|_{L^2} \text{ for all } k \text{ and } t \in \bar{I}_T.$$

To show that the second estimate for u_k belongs to $M_{\rho, 2\delta\rho}$ we define for $s \geq 0$ a new sequence of functions $\{u_k^s\}_k$ with $u_k^s = u_k$ on the interval I_s and 0 otherwise. By using the integral equation for u_k, Young's inequality, Theorem 13.1.3, Remark 13.1.4 and the convolution structure of $G(t, t_0, x) = G(t - t_0, 0, x)$ in time (which follows from the invariance of the linear Cauchy problem by shifts in time) we may conclude for all k the estimates

$$\|u_k\|_{L^r(I_s, L^{p+1})} \leq \delta\rho + C\|u_k^s\|_{L^{q_1}(\mathbb{R}^1, L^{p+1})}^p \leq \delta\rho + C\|u_k\|_{L^{q_1}(I_s, L^{p+1})}^p,$$

where $q_1 = \frac{4p(p+1)}{n+4-(n-4)p}$. The exponent q_1 belongs to the interval $(1, r)$ for $p \in (1, 1 + \frac{4}{n})$. Applying Hölder's inequality yields

$$\|u_k\|_{L^{q_1}(I_s, L^{p+1})} \leq CT^{\frac{1}{q_2}} \|u_k\|_{L^r(I_s, L^{p+1})} \text{ for all } k \text{ and } s \in (0, T],$$

where $q_2 = \frac{4p}{n+4-np}$. Summarizing both estimates leads to

$$\|u_k\|_{L^r(I_s, L^{p+1})} \leq \delta\rho + CT^{\frac{p}{q_2}} \|u_k\|_{L^r(I_s, L^{p+1})}^p \text{ for all } k \text{ and } s \in (0, T],$$

where $C = C(n, p)$ is used as a universal constant.
   Putting $X_k(s) := \|u_k\|_{L^r(I_s, L^{p+1})}$ we have

$$X_k(s) \leq \delta\rho + CT^{\frac{p}{q_2}} X_k(s)^p \text{ for all } k \text{ and } s \in (0, T], \ X_k(0) = 0.$$

Then, a sufficiently small T allows us to prove $X_k(s) \leq 2\delta\rho$ for all k and $s \in (0, T]$. Finally, from Lebesgue-Fatou Lemma (see [230]), it follows $\|u_k\|_{L^r(I_T, L^{p+1})} \leq 2\delta\rho$ for all k and a sufficiently small T. So, we arrived at $u_k \in M_{\rho, 2\delta\rho}$ for all k.

Step 2.3: Convergence properties of the sequence of solutions with data from $H^1(\mathbb{R}^n)$

The same tools as in the previous step allow us to prove

$$\|u_j - u_k\|_{L^r(I_T, L^{p+1})} \leq \delta\|\varphi_j - \varphi_k\|_{L^2} + CT^{\frac{p}{q_2}}\|u_j - u_k\|_{L^r(I_T, L^{p+1})} \text{ for all } j, k,$$

where $q_2 = \frac{4p}{n+4-np}$ and $C = C(n, p, \lambda, \delta\rho)$. After the choice of a sufficiently small T we have

$$\|u_j - u_k\|_{L^r(I_T, L^{p+1})} \leq 2\delta\|\varphi_j - \varphi_k\|_{L^2} \text{ for all } j, k.$$

Hence, $\|u_j - u_k\|_{L^r(I_T, L^{p+1})} \to 0$ for $j, k \to \infty$ for a small T. Furthermore, the definition of $\{u_k\}_k$, Sobolev embedding theorem and Hölder's inequality imply for all test functions $\phi \in H^1(\mathbb{R}^n)$ the estimates

$$|(u_j(t, \cdot) - u_k(t, \cdot), \phi)| \leq \|\varphi_j - \varphi_k\|_{L^2}\|\phi\|_{L^2}$$

$$+ C\|\phi\|_{H^1}\int_{t_0-T}^{t_0+T} \left(\|u_j(s, \cdot)\|_{L^{p+1}}^{p-1} + \|u_k(s, \cdot)\|_{L^{p+1}}^{p-1}\right)\|u_j(s, \cdot) - u_k(s, \cdot)\|_{L^{p+1}} ds$$

$$\leq \|\varphi_j - \varphi_k\|_{L^2}\|\phi\|_{L^2} + CT^{q_3}\|\phi\|_{H^1}\|u_j - u_k\|_{L^r(I_T, L^{p+1})},$$

where $q_3 = \frac{4+(n+4)p-np^2}{4(p+1)}$ and $C = C(n, p, \lambda, \delta\rho)$. We note that $q_3 > 0$ for $p \in (1, 1 + \frac{4}{n})$. Since by Proposition 21.3.2 the sequence $\{u_k\}_k$ belongs to $C(\bar{I}_T, H^{-1}(\mathbb{R}^n))$, the last inequality implies that $\{u_k\}_k$ is a Cauchy sequence in $C(\bar{I}_T, H^{-1}(\mathbb{R}^n))$. Passing to the limit $k \to \infty$ we get a limit element u which, due to Proposition 21.3.1, belongs to $M_{\rho, 2\delta\rho}$. Consequently, we have shown for this u the following properties:

1. $u \in L^\infty(I_T, L^2(\mathbb{R}^n)) \cap L^r(I_T, L^{p+1}(\mathbb{R}^n)) \cap C(\bar{I}_T, H^{-1}(\mathbb{R}^n))$,
2. $\|u(t, \cdot)\|_{L^2} \leq \|\varphi\|_{L^2}$ for almost all $t \in \bar{I}_T$.

These properties for u and the Proposition 21.3.2 imply that $u \in C_w(\bar{I}_T, L^2(\mathbb{R}^n))$.

Step 2.4: Uniqueness of the solution

It remains for us to prove the uniqueness, but to do this requires only standard considerations. Let us assume the existence of two different solutions. Then, we derive corresponding estimates to those of the previous steps to show that both solutions coincide on $\bar{I}_T$. In fact, we assume that

$$t_1 := \sup\{t \in [0, T] : u = v \text{ on } [0, t]\}.$$

Then, with a suitable $t_2 > t_1$, t_2 is close to t_1 by using the above mentioned estimate for $\|u - v\|_{L^r((t_1,t_2),L^{p+1})}$, we conclude $u = v$ on (t_1, t_2), too. So, both solutions coincide on $\bar{I}_T$. Finally, reversing the roles of the initial time t_0 and the solution u in t (we solve the backward Cauchy problem) we obtain $\|u(t, \cdot)\|_{L^2} \geq \|\varphi\|_{L^2}$ for almost all $t \in \bar{I}_T$. Summarizing, it follows $\|u(t, \cdot)\|_{L^2} = \|\varphi\|_{L^2}$ for all $t \in \bar{I}_T$. Hence, $u \in C(\bar{I}_T, L^2(\mathbb{R}^n))$. The proof is complete.

Step 3: *Continuous dependence on the data*

The unique global existence of L^2-solutions for the Cauchy problem

$$iu_t + \Delta u = \lambda |u|^{p-1}u, \quad u(0, x) = \varphi(x), \quad \lambda \in \mathbb{R}^1, \quad p > 1$$

follows directly from Proposition 21.3.3 which shows the unique local solvability in $L^2(\mathbb{R}^n)$ and the a priori bound of the L^2-norm of these solutions.

It remains only for us to prove the continuous dependence of L^2-solutions on the initial data. Let $\{\varphi_k\}_k$ be a sequence from $L^2(\mathbb{R}^n)$ converging to φ in $L^2(\mathbb{R}^n)$. Let $\{u_k\}_k$ and u be the corresponding global L^2-solutions with $u_k(0, x) = \varphi_k(x)$ and $u(0, x) = \varphi(x)$. We put $\rho = \sup_k\{\|\varphi\|_{L^2}, \|\varphi_k\|_{L^2}\}$. Then, we choose T small enough as in Steps 2.2 and 2.3. Following the approach as in the proof to Proposition 21.3.3 we have

$$u_k(t, \cdot) \to u(t, \cdot) \text{ in } C(\bar{I}_t, H^{-1}(\mathbb{R}^n)) \text{ for } k \to \infty.$$

This implies, among other things,

$$(u_k(t, \cdot), \phi) \to (u, \phi) \text{ for all } t \in \bar{I}_t \text{ and } \phi \in L^2(\mathbb{R}^n).$$

Together with the boundedness in $L^2(\mathbb{R}^n)$ of the sequence of solutions $\{u_k(t, \cdot)\}_k$ it follows

$$u_k(t, \cdot) \to u(t, \cdot) \text{ weakly in } L^2(\mathbb{R}^n) \text{ for } k \to \infty, t \in \bar{I}_t.$$

Finally, using $\|u_k(t, \cdot)\|_{L^2} \to \|u(t, \cdot)\|_{L^2}$ we may conclude

$$u_k(t, \cdot) \to u(t, \cdot) \text{ in } L^2(\mathbb{R}^n) \text{ for } k \to \infty, t \in \bar{I}_T.$$

The length of the interval $(-T, T)$ is determined only by n, p, λ and ρ. The L^2-conservation law allows us to apply the above arguments with the initial time T, $-T$, respectively, to obtain

$$u_k(t, \cdot) \to u(t, \cdot) \text{ in } L^2(\mathbb{R}^n) \text{ for } k \to \infty, t \in \bar{I}_{2T}.$$

Repeating this procedure we obtain

$$\lim_{k\to\infty} \|u_k(t,\cdot) - u(t,\cdot)\|_{L^2} \to 0 \text{ for all } t \in \mathbb{R}^1.$$

This completes the proof.

Remark 21.3.1 Using a TT^* argument (see [109]), the property

$$u \in C\big(\mathbb{R}, L^2(\mathbb{R}^n)\big) \cap L^r_{loc}\big(\mathbb{R}, L^{p+1}(\mathbb{R}^n)\big),$$

where $r = \frac{4(p+1)}{n(p-1)}$ allows us to conclude

$$u \in L^r_{loc}\big(\mathbb{R}, L^q(\mathbb{R}^n)\big)$$

for every sharp admissible pair, that is, for every admissible pair (r, q) satisfying $\frac{2}{r} + \frac{n}{q} = \frac{n}{2}$.

Remark 21.3.2 The statements of Theorem 21.3.1 imply a global well-posedness result for weak solutions in both focusing and defocusing cases as well.
What could be the next step? One possibility could be a regularity result for weak solutions. This can be expected. The following result is from [61].

Theorem 21.3.2 *Assume that $p \in (1, 1 + \frac{4}{n})$. Then, for any $\varphi \in H^1(\mathbb{R}^n)$ there exists a uniquely determined global (in time) weak solution $u = u(t, x)$ of*

$$iu_t + \Delta u = \lambda |u|^{p-1} u, \quad u(0, x) = \varphi(x), \quad \lambda \in \mathbb{R}^1.$$

The solution belongs to $C\big(\mathbb{R}^1, H^1(\mathbb{R}^n)\big)$.
But, due to Example 21.2.1, we know that for $s = 1$ the critical exponent is $p_{crit}(n, 1) = 1 + \frac{4}{n-2}$. Hence, the subcritical case covers the interval $(1, 1 + \frac{4}{n-2})$ for p instead of $(1, 1 + \frac{4}{n})$ only. So, the result of Theorem 21.3.2 is incomplete for the subcritical case for H^1 data.

21.4   Semilinear Models with Power Nonlinearity in the Subcritical Case with H^1 Data

In this section we discuss to the Cauchy problem for the semilinear Schrödinger equation with power nonlinearity

$$iu_t + \Delta u = \lambda |u|^{p-1} u, \quad u(0, x) = \varphi(x), \quad \lambda \in \mathbb{R}, \quad p > 1.$$

Now the data φ is supposed to belong to $H^1(\mathbb{R}^n)$. We are interested in weak solution, which were introduced in Sect. 21.3. Let $u \in C\big([-T, T], H^1(\mathbb{R}^n)\big)$ be a weak solution to the given semilinear Cauchy problem. Due to Lemma 21.3.1 we have one conservation law, but there exists a second one.

Lemma 21.4.1 *Let us assume that* $u \in C([-T, T], H^1(\mathbb{R}^n))$, $T > 0$, *is a weak solution to the given semilinear Cauchy problem*

$$iu_t + \Delta u = \lambda |u|^{p-1} u, \quad u(0, x) = \varphi(x), \quad \lambda \in \mathbb{R}^1, \quad p > 1.$$

Then, the functional

$$E(u)(t) := \frac{1}{2} \|\nabla u(t, \cdot)\|_{L^2}^2 + \frac{\lambda}{p+1} \int_{\mathbb{R}^n} |u(t, x)|^{p+1} \, dx$$

is conserved, that is, $E(u)(t) = E(u)(0)$ *for all* $t \in [-T, T]$.

Here we feel a difference between the defocusing and focusing case. In the defocusing case $\lambda > 0$ (it corresponds to an absorbing power nonlinearity that is explained and studied in Sects. 11.2.5, 18.2 and 20.2), the functional $E(u)(t)$ defines an energy. So, we may expect global (in time) weak solutions. In the focusing case $\lambda < 0$ (it corresponds to a source power nonlinearity that is explained and studied in Sects. 11.2.5, 18.2 and 20.2), the functional might become negative. This hints to local (in time) solutions, only. For this reason we discuss the questions for local (in time) existence of solutions and a possible blow up behavior as well. The blow up behavior for solutions is rather deeply studied. In particular, in [64] it is proved that in the focusing case for $p \geq 1 + \frac{4}{n}$ a solution with certain initial data becomes infinite in a finite time in the norm of the spaces $L^\infty(\mathbb{R}^n)$ and $H^1(\mathbb{R}^n)$.

Remark 21.4.1 The blow up behavior for solutions to the system

$$\partial_t u_1 = ia_1 \Delta u_1 + i\gamma_1 \bar{u}_2 u_3, \quad \partial_t u_2 = ia_2 \Delta u_2 + i\gamma_2 \bar{u}_1 u_3,$$

$$\partial_t u_3 = ia_3 \Delta u_3 + i\gamma_3 u_1 u_2,$$

is studied in [182]. It is proved that if $n \geq 4$, $a_k \gamma_k = A$, $k = 1, 2, 3$, $\gamma_1 + \gamma_2 = \gamma_3$, and for suitable initial data, a solution becomes infinite in a finite time in the norm of the spaces $L^\infty(\mathbb{R}^n)$ and $H^1(\mathbb{R}^n)$.

Now, let us come back to our model of interest

$$iu_t + \Delta u = \lambda |u|^{p-1} u, \quad u(0, x) = \varphi(x), \quad \lambda \in \mathbb{R}^1, \quad p > 1.$$

We are going to prove the following result from [61]. We restrict ourselves to the forward Cauchy problem.

Theorem 21.4.1 *Assume that* $n \geq 2$, $p > 1$ *and* $p \in (1, 1 + \frac{4}{n-2})$ *for* $n \geq 3$. *Then, for any* $\varphi \in H^1(\mathbb{R}^n)$ *there exists a uniquely determined local (in time) weak solution* $u = u(t, x)$ *of*

$$iu_t + \Delta u = \lambda |u|^{p-1} u, \quad u(0, x) = \varphi(x), \quad \lambda \in \mathbb{R}^1.$$

The solution belongs to $C([0, T_{max}), H^1(\mathbb{R}^n))$. Moreover, the solution belongs to (cf. with Remarks 13.1.4 and 13.1.5)

$$L^r_{loc}((0, T_{max}), W^1_q(\mathbb{R}^n))$$

for every sharp admissible pair (r, q). In addition, the following blow up behavior is only valid for $T_{max} < \infty$:

$$\lim_{t \to T_{max}-0} \|u(t, \cdot)\|_{H^1} = \infty.$$

Finally, the solution depends continuously on the data φ in the following sense: If the sequence $\{\varphi_k\}_k$ converges to φ in $H^1(\mathbb{R}^n)$, then it holds for the sequence $\{u_k\}_k$ and u of the corresponding weak solutions

$$\lim_{k \to \infty} \|u_k(t, \cdot) - u(t, \cdot)\|_{C([0,T],H^1)} \to 0 \text{ for all } T < T_{max} \text{ and } k \geq k_0(T).$$

Proof We only sketch the proof. The proof is divided into several steps.

Step 1:   Integral equation

We find for $t > 0$ a solution of the integral equation

$$u(t, x) = \Phi(u) := G(t, 0, x) *_{(x)} \varphi(x) - i \int_0^t G(t, s, x) *_{(x)} \lambda(|u|^{p-1}u)(s, x) \, ds.$$

Here $G(t, s, x) = G(t - s, 0, x)$.

Step 2:   A closed set in $L^r((0, T), W^1_{p+1}(\mathbb{R}^n))$

Let $\rho = \|\varphi\|_{H^1}$. We choose $q = p + 1$ and $r = \frac{4(p+1)}{n(p-1)}$. This follows from the admissibility condition $\frac{2}{r} + \frac{n}{q} = \frac{n}{2}$. Then we define the set

$$M_{A,A} := \{v \in L^\infty((0, T), H^1(\mathbb{R}^n)) \cap L^r((0, T), W^1_{p+1}(\mathbb{R}^n)) :$$

$$\|v(t, \cdot)\|_{L^\infty((0,T),H^1)} \leq A \text{ and } \|v\|_{L^r((0,T),W^1_{p+1})} \leq A\},$$

where A, T are constants to be determined later.

Step 3:   Some useful estimates

Now we give some useful estimates for the further reasoning. Let r' and q' be the conjugate exponents to r and q. Since $p \in (1, 1 + \frac{4}{n-2})$ we have $r > 2$ and, thus, $r > r'$, too. Notice that for $q = p + 1$ we have

$$\|(|u|^{p-1}u)(t, \cdot)\|_{L^{q'}} \leq C\|u(t, \cdot)\|_{L^q}^p.$$

After integration with respect to time it follows

$$\||u|^{p-1}u\|_{L^r((0,T),L^{q'})} \leq C\|u\|_{L^\infty((0,T),L^q)}^{p-1}\|u\|_{L^r((0,T),L^q)}.$$

By Sobolev embedding it holds

$$\|u(t,\cdot)\|_{L^{p+1}} \leq C\|u(t,\cdot)\|_{H^1}.$$

After integration with respect to time it follows

$$\||u|^{p-1}u\|_{L^r((0,T),L^{q'})} \leq C\|u\|_{L^\infty((0,T),H^1)}^{p-1}\|u\|_{L^r((0,T),L^q)}.$$

Similarly, since the nonlinearity is differentiable, we get

$$\|\nabla(|u|^{p-1}u)\|_{L^r((0,T),L^{q'})} \leq C\|u\|_{L^\infty((0,T),H^1)}^{p-1}\|\nabla u\|_{L^r((0,T),L^q)}.$$

Summarizing, we derived the estimate

$$\||u|^{p-1}u\|_{L^r((0,T),W_{q'}^1)} \leq C\|u\|_{L^\infty((0,T),H^1)}^{p-1}\|u\|_{L^r((0,T),W_q^1)}$$

$$\leq CA^{p-1}\|u\|_{L^r((0,T),W_q^1)}.$$

Finally, using Hölder's inequality in time we have

$$\||u|^{p-1}u\|_{L^{r'}((0,T),W_{q'}^1)} \leq CT^{\frac{r-r'}{rr'}}\||u|^{p-1}u\|_{L^r((0,T),W_{q'}^1)}$$

$$\leq CT^{\frac{r-r'}{rr'}}\|u\|_{L^\infty((0,T),H^1)}^{p-1}\|u\|_{L^r((0,T),W_q^1)}.$$

Step 4:   *Application of Banach's fixed point argument*

Now we have all tools to study the mapping

$$u \in M_{A,A} \to \Phi(u).$$

Applying Theorem 13.1.3, Duhamel's principle together with Strichartz estimates from Remarks 13.1.4, 13.1.5 with two admissible pairs gives the following estimates for $\Phi(u)$:

$$\|\Phi(u)\|_{L^\infty((0,T),H^1)} \leq C\|G(t,0,x) *_{(x)} \varphi(x)\|_{L^\infty((0,T),H^1)} + C\||u|^{p-1}u\|_{L^{r'}((0,T),W_{q'}^1)}$$

$$\leq C\|\varphi\|_{H^1} + CT^{\frac{r-r'}{rr'}}\|u\|_{L^\infty((0,T),H^1)}^{p-1}\|u\|_{L^r((0,T),W_q^1)},$$

and

$$\|\Phi(u)\|_{L^r((0,T),W^1_{p+1})} \le C\|G(t,0,x) *_{(x)} \varphi(x)\|_{L^r((0,T),W^1_q)} + C\||u|^{p-1}u\|_{L^{r'}((0,T),W^1_{q'})}$$

$$\le C\|\varphi\|_{H^1} + CT^{\frac{r-r'}{rr'}}\|u\|^{p-1}_{L^\infty((0,T),H^1)}\|u\|_{L^r((0,T),W^1_q)}.$$

Consequently,

$$\|\Phi(u)\|_{L^\infty((0,T),H^1)} + \|\Phi(u)\|_{L^r((0,T),W^1_{p+1})}$$

$$\le C\|\varphi\|_{H^1} + CT^{\frac{r-r'}{rr'}}A^{p-1}\|u\|_{L^r((0,T),W^1_q)}.$$

The constant C is independent of φ. We choose $A := 2C\|\varphi\|_{H^1}$. Then, we can choose $T = T(\varphi)$ small enough such that $\Phi(u) \in M_{A,A}$ for all $u \in M_{A,A}$. Thus Φ maps $M_{A,A}$ into itself. As above, we can show that a probably smaller T allows for showing that Φ is a contraction on $M_{A,A}$, that is, for arbitrary $u, v \in M_{A,A}$ we have

$$\|\Phi(u) - \Phi(v)\|_{L^\infty((0,T),H^1)} + \|\Phi(u) - \Phi(v)\|_{L^r((0,T),W^1_{p+1})}$$

$$\le c_0\big(\|u - v\|_{L^\infty((0,T),H^1)} + \|u - v\|_{L^r((0,T),W^1_{p+1})}\big),$$

where the positive constant c_0 is smaller than 1. Banach's fixed point theorem provides a uniquely determined solution in $M_{A,A}$ which is a solution of $u = \Phi(u)$. The right-hand side and the derived estimates for the solution yield $u \in C([0,T), H^1(\mathbb{R}^n))$, too.

Step 5:   *Uniqueness on the whole space and continuous dependence on the data*

To prove uniqueness in the solution space $C([0,T), H^1(\mathbb{R}^n))$ one proceeds as in the previous steps. First we show, as above, uniqueness of both solutions on a small interval $[0, \delta]$. Then, we iterate the argument finite times to cover every compact interval of $[0,T)$. The proof of continuous dependence is almost identical to the arguments to prove uniqueness. For this reason we skip it and leave it as an exercise to the reader.

Step 6:   *Discussion of blow up*

Let us define

$$T_{max} = \sup\{T > 0 : \text{there exist, a weak solution on } [0,T)\}.$$

We suppose $T_{max} < \infty$. Moreover, we assume the existence of a sequence $\{t_k\}_k \to T_{max}$ and a positive constant K such that $\|u(t_k, \cdot)\|_{H^1} \le K$. We choose, as in Step 4, the constant $A := 2CK$. Then, for all data φ satisfying $\|\varphi\|_{H^1} \le K$ there exists a uniform time $T = T(A)$ such that the solution exists in the time interval $[t_k, t_k + T)$. If we choose t_{k_0} close to T_{max}, then $t_{k_0} + T > T_{max}$ and we contradict the definition of T_{max}. Thus, $\lim_{t \to T_{max}-0} \|u(t, \cdot)\|_{H^1} = \infty$ if T_{max} is supposed to be smaller than infinity.

Finally, in the defocusing case $\lambda > 0$ the mass and energy conservation in Lemmas 21.3.1 and 21.4.1 provides an a priori bound $\sup_{t \in [0,T)} \|u(t, \cdot)\|_{H^1} \leq C(\varphi)$. By the blow up alternative we may conclude $T_{max} = \infty$ and the Cauchy problem is globally (in time) well-posed.

This completes the proof.

Remark 21.4.2 Using a TT^* argument (see [109]), then the property

$$u \in C\big([0, T), H^1(\mathbb{R}^n)\big) \cap L^r((0, T), W^1_{p+1}(\mathbb{R}^n)),$$

where $r = \frac{4(p+1)}{n(p-1)}$ allows to conclude

$$u \in L^{\tilde{r}}\big((0, T), W^1_{\tilde{q}}(\mathbb{R}^n)\big)$$

for every sharp admissible pair, that is, for every admissible pair $(\tilde{r}, \tilde{q})$ satisfying $\frac{2}{\tilde{r}} + \frac{n}{\tilde{q}} = \frac{n}{2}$.

Remark 21.4.3 Theorem 21.4.1 contains different statements. In the defocusing case we conclude $T_{max} = \infty$ after applying Lemmas 21.3.1 and 21.4.1. So, we have a global result with well-posedness in this case. In the focusing case we get a local well-posedness result together with a blow up mechanism for $t \to T_{max} - 0$.

21.5   Concluding Remarks

21.5.1   Some Remarks to Critical and Supercritical Cases

In Sects. 21.3 and 21.4 we derived results about local or global (in time) existence of solutions in the subcritical case, where the range of admissible powers p depends on the L^2 or H^1 regularity of the data. As it was pointed out in the previous chapters about semilinear heat models and wave (or damped wave) models, the assumption of small initial data in a suitable function space may have some influence on the range of the power nonlinearity for which one is able to prove the global (in time) existence of small data solutions.

The same effect can be observed for semilinear Schrödinger equations. Regarding to the L^2 theory, under the assumption that the norm of the data $\|\varphi\|_{L^2}$ is supposed to be small, the conclusion of Theorem 21.3.1 is still true in the critical case $p = 1 + \frac{4}{n}$ (see for instance Corollary 5.2 in [122]).

Related to the H^1 theory, differences between the focusing nonlinearity (negative sign) and defocusing nonlinearity (positive sign) become evident. Thanks to

Theorem 21.4.1, in the case of a defocusing nonlinearity we already concluded the existence of global (in time) solutions in the subcritical case, i.e., for $p \in (1, 1 + \frac{4}{n-2})$, even by taking large data. But, in the case of a focusing nonlinearity we have to split the analysis for $p \in (1, 1 + \frac{4}{n})$ and $p \in [1 + \frac{4}{n}, 1 + \frac{4}{n-2})$. In the first interval we have global (in time) existence of solutions even for large data, but in the second interval this is true only under the assumption that $\|\varphi\|_{H^1}$ is supposed to be small (see Section 6.2 of [18]). We recall that in [64] it is proved that in the focusing case for $p \geq 1 + \frac{4}{n}$ a solution with initial data $\varphi = k\psi \in H^1(\mathbb{R}^n)$, with $0 \neq \psi \in H^1(\mathbb{R}^n)$, becomes infinite in a finite time in the norm of the spaces $L^\infty(\mathbb{R}^n)$ and $H^1(\mathbb{R}^n)$, provided $|k|$ is large enough.

Finally, in the critical case $p = 1 + \frac{4}{n-2}$ if $\|\varphi\|_{H^1}$ is small, then the solution is global (in time) with the same regularity given by Theorem 21.4.1 (see Corollary 5.4. of [122]).

21.5.2 Some Remarks to Critical Cases

In this chapter we restricted ourselves to data belonging to $L^2(\mathbb{R}^n)$ or $H^1(\mathbb{R}^n)$. But, one can assume different regularity and remember the relation $s_{crit} = \frac{n}{2} - \frac{2}{p-1}$ introduced in Sect. 21.2. The regularity of data in $H^s(\mathbb{R}^n)$ with $s = -\frac{1}{2}$ (it corresponds to $p_{crit}(n, -\frac{1}{2}) = 1 + \frac{4}{n+1}$) is, for example, discussed in [60] in connection with cubic semilinear Schrödinger equations in the 1d case.

21.5.3 Some Remarks to the Asymptotical Profile

In Sects. 17.2.2 and 18.2.2 we explained the asymptotic profile of solutions to semilinear heat and classical damped wave models with absorbing power nonlinearity. The main part of the profile is given by the Gauss kernel of the heat operator. Let us now turn to the issue of the asymptotical profile of solutions to semilinear Schrödinger equations with defocusing power nonlinearity. Our main concern is in explaining a relation to solutions of the free Schrödinger equation. Here we follow the considerations of [210]. Due to Theorem 21.4.1 and Remark 21.4.3 we know that the Cauchy problem (here the factor $\frac{1}{2}$ has no influence on the statement of Theorem 21.4.1)

$$iu_t + \frac{1}{2}\Delta u = \lambda |u|^{p-1}u, \quad u(0, x) = \varphi(x), \quad \lambda > 0,$$

has for $p \in (1, 1 + \frac{4}{n-2})$ and $\varphi \in H^1(\mathbb{R}^n)$ a global weak solution u belonging to $C([0, \infty), H^1(\mathbb{R}^n))$. The following result is then proved in [210].

Theorem 21.5.1 *Let $p \in (1 + \frac{2}{n}, 1 + \frac{4}{n-2})$, $n \geq 2$. Moreover, let φ belong to the function space*

$$H^1_{\langle x \rangle} := \{\varphi \in H^1(\mathbb{R}^n) : \|\langle x \rangle u\|_{L^2} < \infty\}.$$

Then, there exist uniquely determined functions $\varphi_\pm = \varphi_\pm(x) \in L^2(\mathbb{R}^n)$ such that

$$\lim_{t \to \pm\infty} \|u(t, \cdot) - w_\pm(t, \cdot)\|_{L^2} = 0,$$

where $w_\pm$ are solutions to the Cauchy problem for the free Schrödinger equation

$$iw_t + \frac{1}{2}\Delta w = 0, \quad w(0, x) = \varphi_\pm(x).$$

Remark 21.5.1 It is shown by [63] and [195] that for $p \in (1, 1 + \frac{2}{n}]$ and $\varphi \in S(\mathbb{R}^n) \subset H^1_{\langle x \rangle}$ any nontrivial solution u of

$$iu_t + \frac{1}{2}\Delta u = \lambda |u|^{p-1}u, \quad u(0, x) = \varphi(x), \quad \lambda > 0,$$

does not satisfy the statement of Theorem 21.5.1. Consequently, the range of admissible p is optimal there.

Exercises Relating to the Considerations of Chap. 21

Exercise 1 Prove the statement of Lemma 21.3.1 (cf. [19]).

Exercise 2 Prove the statement of Lemma 21.4.1 (cf. [19]).

Exercise 3 Prove the property of "continuous dependence on the data" in Theorem 21.4.1.

Chapter 22
Linear Hyperbolic Systems

This chapter is devoted to aspects of linear hyperbolic systems. We have in mind mainly two classes of systems, symmetric hyperbolic and strictly hyperbolic ones. First we discuss these classes of systems with constant coefficients. Fourier analysis coupled with function-theoretical methods imply well-posedness results for different classes of solutions. Then, we treat such systems with variable coefficients. On the one hand we apply the method of characteristics introduced in Chap. 6 to derive a local existence result in time and space variables. On the other hand we discuss the issue of energy estimates and well-posedness for both cases of symmetric hyperbolic and strictly hyperbolic systems.

22.1   Plane Wave Solutions

We investigate in this chapter hyperbolic systems of linear first-order partial differential equations having the form

$$\partial_t U + \sum_{k=1}^{n} A_k(t,x)\partial_{x_k} U = 0 \ \text{ in } \ [0,\infty) \times \mathbb{R}^n$$

subject to the initial condition

$$U(0,x) = U_0(x).$$

For our further considerations we introduce the notation

$$A(t,x,\xi) := \sum_{k=1}^{n} \xi_k A_k(t,x) \ \text{ for } \ t \geq 0 \ \text{ and } \ (x,\xi) \in \mathbb{R}^{2n}.$$

© Springer International Publishing AG 2018
M.R. Ebert, M. Reissig, *Methods for Partial Differential Equations*,
https://doi.org/10.1007/978-3-319-66456-9_22

From the beginning we assume that $A(t, x, \xi) = A(\xi)$. Thus all matrices A_k are constant. Let us look for plane wave solutions, that is, we try to find a solution U having the form

$$U(t, x) = V(x \cdot \xi - \lambda t), \quad \lambda \in \mathbb{R}^1,$$

for some direction $\xi \in \mathbb{R}^n$, with velocity $\frac{|\lambda|}{|\xi|}$, and with profile V (cf. with Sect. 2.2). Plugging this ansatz into the system we compute

$$\left(-\lambda I + \sum_{k=1}^n \xi_k A_k \right) V' = 0.$$

This equality asserts that V' is an eigenvector of the matrix $A(\xi)$ corresponding to the eigenvalue λ. The hyperbolicity condition (see Definition 3.5) requires that there are m plane wave solutions to the starting system for each direction $\xi \neq 0$, where the matrices A_k, $k = 1, \cdots, n$, are supposed to be constant $m \times m$ matrices.

22.2   Symmetric Systems with Constant Coefficients

In this section we apply the partial Fourier transformation to solve the Cauchy problem for the constant coefficient system

$$\partial_t U + \sum_{k=1}^n A_k \partial_{x_k} U = 0, \quad U(0, x) = U_0(x).$$

The energy $E(U)(t)$ of a weak solution is defined by

$$E(U)(t) := \frac{1}{2} \|U(t, \cdot)\|_{L^2}^2.$$

We assume that A_k, $k = 1, \cdots, n$, are constant $m \times m$ symmetric matrices. Thanks to [219], the matrix $A(\xi) := \sum_{k=1}^n \xi_k A_k$ has m real continuous eigenvalues $\lambda_1(\xi) \leq \lambda_2(\xi) \leq \cdots \leq \lambda_m(\xi)$ and a complete set of corresponding Lebesgue measurable eigenvectors $\{e_1(\xi), \cdots, e_m(\xi)\}$ with $|e_j(\xi)| = 1$. Applying the partial Fourier transformation with respect to x, solving the auxiliary Cauchy problem for the parameter-dependent system of ordinary differential equations we obtain

$$F(U)(t, \xi) = e^{-itA(\xi)} F(U_0)(\xi).$$

Now, we may write

$$F(U_0)(\xi) = \sum_{j=1}^m F(U_{0,j})(\xi) e_j(\xi) \quad \text{with} \quad F(U_{0,j})(\xi) = F(U_0)(\xi) \cdot e_j(\xi).$$

Thanks to

$$A(\xi)e_j(\xi) = \lambda_j(\xi)e_j(\xi)$$

we derive

$$e^{-itA(\xi)}e_j(\xi) = e^{-it\lambda_j(\xi)}e_j(\xi)$$

and

$$F(U)(t,\xi) = \sum_{j=1}^{m} F(U_{0,j})(\xi)e^{-it\lambda_j(\xi)}e_j(\xi).$$

Here we used the matrix exponential series $e^B = \sum_{k=0}^{\infty} \frac{B^k}{k!}$.

Theorem 22.2.1 *Assume that $U_0 \in H^s(\mathbb{R}^n)$ with $s \geq 0$. Then, there is a uniquely determined Sobolev solution (cf. with Definition 3.11)*

$$U \in C([0,\infty), H^s(\mathbb{R}^n))$$

possessing an energy for all $t \geq 0$.

Proof Taking account of $|e_j(\xi)| = 1$ and that the eigenvalues $\lambda_j = \lambda_j(\xi)$ are real, the conclusion of the theorem follows immediately from the formula

$$F(U)(t,\xi) = \sum_{j=1}^{m} F(U_{0,j})(\xi)e^{-it\lambda_j(\xi)}e_j(\xi).$$

Example 22.2.1 Consider the Cauchy problem for the free wave equation in n spatial dimensions

$$u_{tt} - \Delta u = 0, \quad u(0,x) = \varphi(x), \quad u_t(0,x) = \psi(x).$$

Setting $U = (u_{x_1}, \cdots, u_{x_n}, u_t)^T$ and $U_0 = (\varphi_{x_1}, \cdots, \varphi_{x_n}, \psi)^T$, then

$$\partial_t U = \sum_{k=0}^{n} A_k \partial_{x_k} U, \quad U(0,x) = U_0(x),$$

where each A_k is an $(n+1) \times (n+1)$ symmetric matrix whose entries a_{ij}^k are given by

$$a_{ij}^k = \begin{cases} 1 & \text{for } k = i \text{ and } j = n+1 \text{ or } k = j \text{ and } i = n+1, \\ 0 & \text{otherwise.} \end{cases}$$

22.3   Hyperbolic Systems with Constant Coefficients

In this section we consider the Cauchy problem for the constant coefficient system

$$\partial_t U + \sum_{k=1}^{n} A_k \partial_{x_k} U = 0, \quad U(0, x) = U_0(x).$$

We assume that the $m \times m$ matrix $A(\xi) := \sum_{k=1}^{n} \xi_k A_k$ has, for each $\xi \in \mathbb{R}^n \setminus \{0\}$, m real eigenvalues $\lambda_1(\xi) \leq \lambda_2(\xi) \leq \cdots \leq \lambda_m(\xi)$. There is no hypothesis concerning the eigenvectors, so we suppose in this section instead of Definition 3.5, the fulfilment of a very weak sort of hyperbolicity, the so-called definition for *weakly hyperbolic systems*.

Theorem 22.3.1 *Assume that $U_0 \in H^s(\mathbb{R}^n)$ with $s > \frac{n}{2} + m$. Then, there is a uniquely determined classical solution $U \in C^1([0, \infty) \times \mathbb{R}^n)$.*

Proof The proof follows [53]. It is divided into several steps.

Step 1:   *Formal representation of the solution*

Applying the partial Fourier transformation with respect to x, solving the auxiliary problem and assuming the validity of Fourier inversion formula we obtain

$$U(t, x) = \frac{1}{(2\pi)^{\frac{n}{2}}} \int_{\mathbb{R}^n} e^{ix\cdot\xi} e^{-it\, A(\xi)} F(U_0)(\xi)\, d\xi.$$

First we have to check that this integral really converges. Since $U_0 \in H^s(\mathbb{R}^n)$, it holds $\langle \xi \rangle^s F(U_0) \in L^2(\mathbb{R}^n)$. So, in order to investigate the convergence we shall estimate $\|e^{-itA(\xi)}\|$. By $\|B\|$ we denote the max-norm $\|B\| = \max_{i,j} |b_{ij}|$ of the matrix B.

Step 2:   *Convergence of the integral*

For a fixed ξ, let Γ denote the path $\partial B_r(0)$ in the complex plane, traversed counterclockwise, the radius r selected so large that the eigenvalues $\lambda_1(\xi), \cdots, \lambda_m(\xi)$ are located within Γ. Then we have the formula

$$e^{-itA(\xi)} = \frac{1}{2\pi i} \int_{\Gamma} e^{-itz} (zI - A(\xi))^{-1}\, dz.$$

Indeed, let us introduce the notation

$$B(t, \xi) := \frac{1}{2\pi i} \int_{\Gamma} e^{-itz} (zI - A(\xi))^{-1}\, dz.$$

Then, for a fixed $w \in \mathbb{R}^m$ we have

$$A(\xi)B(t, \xi)w = \frac{1}{2\pi i} \int_\Gamma e^{-itz} A(\xi)(zI - A(\xi))^{-1} w \, dz$$

$$= \frac{1}{2\pi i} \int_\Gamma e^{-itz} \left(z(zI - A(\xi))^{-1} - I \right) w \, dz = i\partial_t B(t, \xi)w$$

thanks to $\int_\Gamma e^{-itz} \, dz = 0$. Hence,

$$\left(\partial_t + iA(\xi) \right) B(t, \xi) = 0.$$

Moreover, it can be proved by function-theoretical methods that $B(0, \xi)w = w$. Therefore, we may conclude $B(t, \xi) = e^{-itA(\xi)}$. Define a new path $\tilde{\Gamma}$ in the complex plane as follows: For fixed ξ draw circles $B_k = B_1(\lambda_k(\xi))$ of radius 1 centered at $\lambda_k(\xi)$, $k = 1, \cdots, m$. Then take $\tilde{\Gamma}$ to be the boundary of $\bigcup_{k=1}^m B_k$, traversed counterclockwise. Deforming the path Γ into $\tilde{\Gamma}$ we deduce that

$$e^{-itA(\xi)} = \frac{1}{2\pi i} \int_{\tilde{\Gamma}} e^{-itz}(zI - A(\xi))^{-1} dz.$$

Using that $\lambda_k(\xi)$ are real for each $\xi \in \mathbb{R}^n$ and $k = 1, \cdots, m$ we arrive at the estimate $|e^{-itz}| \le e^t$ for $z \in \tilde{\Gamma}$. Now,

$$(zI - A(\xi))^{-1} = \frac{\text{cof}\,(zI - A(\xi))^T}{\det(zI - A(\xi))}, \quad \det(zI - A(\xi)) = \prod_{k=1}^m (z - \lambda_k(\xi)),$$

where "cof" denotes the cofactor matrix. Taking into consideration the relations $|\det(zI - A(\xi))| \ge 1$ for $z \in \tilde{\Gamma}$ and $|\lambda_k(\xi)| \le C|\xi|$ for $k = 1 \cdots, m$ we get

$$\|(zI - A(\xi))^{-1}\| \le C \|\text{cof}\,(zI - A(\xi))\| \le C \langle\xi\rangle^{m-1} \quad \text{for } z \in \tilde{\Gamma}.$$

Summarizing, we derived the estimate

$$\|e^{-itA(\xi)}\| \le Ce^t \langle\xi\rangle^{m-1} \quad \text{for } \xi \in \mathbb{R}^n.$$

To show the convergence of the integral

$$\int_{\mathbb{R}^n} e^{ix\cdot\xi} e^{-it\,A(\xi)} F(U_0)(\xi) \, d\xi$$

we proceed as follows:

$$\int_{\mathbb{R}^n} |e^{ix\cdot\xi} e^{-itA(\xi)} F(U_0)(\xi)| \, d\xi \leq C \int_{\mathbb{R}^n} \|e^{-itA(\xi)}\| \, |F(U_0)(\xi)| \, d\xi$$

$$\leq C \, e^t \int_{\mathbb{R}^n} \langle\xi\rangle^{-s+m-1} \langle\xi\rangle^s |F(U_0)(\xi)| \, d\xi$$

$$\leq C \, e^t \|U_0\|_{H^s} \left(\int_{\mathbb{R}^n} \langle\xi\rangle^{-2s+2m-2} \, d\xi \right)^{1/2}.$$

Taking account of $s > \frac{n}{2} + m - 1$ the integral converges and the function

$$U = U(t,x) = \frac{1}{(2\pi)^{\frac{n}{2}}} \int_{\mathbb{R}^n} e^{ix\cdot\xi} e^{-itA(\xi)} F(U_0)(\xi) \, d\xi$$

is continuous on $[0, \infty) \times \mathbb{R}^n$.

Step 3:   Existence of a classical solution

To show that the vector U belongs to C^1 observe that for $0 < |h| \leq 1$ we have

$$\frac{U(t+h,x) - U(t,x)}{h} = \frac{1}{(2\pi)^{\frac{n}{2}} h} \int_{\mathbb{R}^n} e^{ix\cdot\xi} \left(e^{-i(t+h)A(\xi)} - e^{-itA(\xi)} \right) F(U_0)(\xi) \, d\xi.$$

Since

$$e^{-i(t+h)A(\xi)} - e^{-itA(\xi)} = -i \int_t^{t+h} A(\xi) e^{-isA(\xi)} \, ds,$$

we can estimate as in Step 2 to get

$$\left| \frac{1}{h} \left(e^{-i(t+h)A(\xi)} - e^{-itA(\xi)} \right) \right| \leq C \, e^{t+1} \langle\xi\rangle^m.$$

Therefore,

$$\left| \frac{U(t+h,x) - U(t,x)}{h} \right| \leq C \, e^{t+1} \|U_0\|_{H^s} \left(\int_{\mathbb{R}^n} \langle\xi\rangle^{-2s+2m} \, d\xi \right)^{1/2}.$$

From $s > \frac{n}{2} + m$ it follows that $\partial_t U$ exists and is continuous on $[0, \infty) \times \mathbb{R}^n$. The same statement can be concluded for $\partial_{x_k} U$. According to Lebesgue Convergence Theorem we can furthermore differentiate under the integral sign in the representation for $U = U(t,x)$ to confirm that U solves our Cauchy problem.

In general, the conclusion of the previous theorem may not be true if we consider systems with variable coefficients having an additional lower order term $A_0 U$. To verify this statement let us turn to the next example.

Example 22.3.1 Consider the example due to Qi Min-You (see [167]),

$$\partial_t^2 u - t^2 \partial_x^2 u - b \, \partial_x u = 0, \; t > 0, \; x \in \mathbb{R}^1,$$

with the initial conditions

$$u(0, x) = \varphi(x), \;\; u_t(0, x) = 0,$$

where $b = 4m + 1$, $m \geq 0$ is an integer. The uniquely determined distributional solution has the form

$$u(t, x) = \sum_{j=0}^{m} \frac{\sqrt{\pi} t^{2j}}{j!(m-j)!\Gamma(j + \frac{1}{2})} (\partial_x^j \varphi)\left(x + \frac{1}{2} t^2\right).$$

The loss of regularity increases on m. Setting $U = (u_x, u_t)^T$ and $U_0 = (\varphi', 0)^T$, then

$$\partial_t U = \begin{pmatrix} 0 & 1 \\ t^2 & 0 \end{pmatrix} \partial_x U + \begin{pmatrix} 0 & 0 \\ 4m + 1 & 0 \end{pmatrix} U, \;\; U(0, x) = U_0(x).$$

It is clear that we have real eigenvalues $\lambda_\pm(t) = \pm t$, but even by taking $U_0 \in H^s(\mathbb{R})$ with large s the solution U may not be in $C^1([0, \infty) \times \mathbb{R}^1)$ for large $m > 0$.

Remark 22.3.1 If we consider weakly hyperbolic systems with non-vanishing lower order term $A_0 U$, then so-called *Levi conditions* may be of importance (see, for example, [225]). Such conditions imply the solvability of the Cauchy problem in suitable function spaces. To avoid Levi conditions one can, on the one hand, require stronger assumptions to the system. For instance, that the system is strictly hyperbolic and, on the other hand, choose the data from special function spaces (see [140] or [178]).

22.4   Linear Strictly Hyperbolic Systems in 1d: Method of Characteristics

In this section we shall restrict ourselves to linear strictly hyperbolic systems of first order in one space variable (see Definition 3.4)

$$LU := \partial_t U + A(t, x)\partial_x U + B(t, x)U = F(t, x).$$

We assume that the $m \times m$ matrices A and B belong to $C^1(G)$ in a domain $G \subset \mathbb{R}^2$. We will show that in a neighborhood of a point $(t_0, x_0) \in G$ the system can be transformed into a very simple *canonical form* by introducing new unknowns. We make use of the eigenvalues and the eigenvectors of the matrix A in a neighborhood of $(t_0, x_0) \in \mathbb{R}^2$.

Lemma 22.1 *Let us assume that the system $LU = F$ is strictly hyperbolic in G. Then, every point $(t_0, x_0) \in G$ has an open neighborhood $G_0 \subset G$ and there exists an invertible matrix-valued function $N = N(t, x) \in C^1(G_0)$ such that the following identity is satisfied for all $(t, x) \in G_0$:*

$$N^{-1}(t, x)A(t, x)N(t, x) = \mathcal{D}(t, x),$$

where $\mathcal{D} = \mathcal{D}(t, x) := diag(\lambda_k(t, x))_{k=1}^m$ is a diagonal matrix and $\lambda_k = \lambda_k(t, x) \in C^1(G)$, $k = 1, \cdots, m$, are the distinct real eigenvalues of $A = A(t, x)$.

Proof For the proof we follow [206]. Applying the implicit function theorem we conclude that $\lambda_k = \lambda_k(t, x) \in C^1(G)$ for $k = 1, \cdots m$. Now, let us define

$$P_j = P_j(t, x) = (2\pi i)^{-1} \int_{\Gamma_j} \left(zI - A(t, x)\right)^{-1} dz,$$

where Γ_j is a circle in the complex plane on which no eigenvalues of A are located and whose interior contains one and only one eigenvalue, the j'th one, $\lambda_j = \lambda_j(t, x)$. It is clear that P_j is a $C^1(G)$ function and is the j'th spectral projector of A. The range $X_j = X_j(t, x)$ of the matrix P_j is the one-dimensional j'th eigenspace of A. For each (t, x), the vectors $X_1(t, x), \cdots, X_m(t, x)$ span the whole space $\mathbb{R}^m$. Now, let us define the diagonalizer $N = N(t, x)$. For each j we select a non-zero vector v_j^0 in $X_j(t_0, x_0)$ and we set

$$v_j(t, x) = P_j(t, x)v_j^0.$$

It is clear that $P_j(t_0, x_0)v_j^0 = v_j^0$ and that $\{v_j(t, x) : 1 \le j \le m\}$ form a basis of $\mathbb{R}^m$ in a neighborhood of (t_0, x_0). Let us write

$$v_j(t, x) = \sum_{k=1}^m \gamma_{kj}(t, x)e_k,$$

where $(e_1, \cdots, e_m)$ is the canonical basis of $\mathbb{R}^m$. Therefore $N = (\gamma_{kj})_{j,k=1}^m$. We are now in a position to prove the following result.

Theorem 22.4.1 *Let us consider the strictly hyperbolic Cauchy problem*

$$LU = F(t, x), \quad U(0, x) = U_0(x), \quad (t, x) \in G \subset \mathbb{R}^2,$$

where $A \in C^1(G)$, $F, B \in C^0(G)$ and $U_0 \in C^1(I)$ with $G = (-T, T) \times I$ is a domain in $\mathbb{R}^2$ and I is an open interval. Then, there is a classical solution $U \in C^1(G_0)$ with $G_0 = (-\delta, \delta) \times (a, b) \subset G$ is a subdomain of G.

Proof   The proof is divided into two steps.

Step 1:   Diagonalization procedure

After setting $U := N V$, where $N = N(t, x)$ is the matrix given by Lemma 22.1 when we apply it to $(0, x_0) \in G$, there exists an open neighborhood $G_0 = (-\delta, \delta) \times (a, b)$, with $x_0 \in (a, b)$, such that the Cauchy problem can be transformed into the following strictly hyperbolic Cauchy problem with diagonal principal part

$$\partial_t V + \mathcal{D}(t, x)\partial_x V + B_0(t, x)V = H(t, x), \quad V(0, x) = V_0(x),$$

where

$$F(t, x) =: N(t, x)H(t, x), \quad U_0(x) =: N(0, x)V_0(x), \quad \mathcal{D} = \mathrm{diag}\,(\lambda_k(t, x))_{k=1}^m$$

is a diagonal matrix and

$$B_0 = B_0(t, x) = N^{-1}(A\partial_x N + BN + \partial_t N).$$

The simplicity of the canonical form

$$\partial_t V + \mathcal{D}(t, x)\partial_x V + B_0(t, x)V = H(t, x)$$

becomes apparent if we write it out in component form

$$\partial_t v_k + \lambda_k(t, x)\partial_x v_k + \sum_{j=1}^m b_{kj}^0(t, x)v_j = h_k, \quad k = 1, \cdots, m,$$

where $V = (v_1, \cdots, v_m)^T$, $H = (h_1, \cdots, h_m)^T$ and $B_0 = (b_{kj}^0)_{j,k=1}^m$. The principal part of the k'th equation involves only the k'th unknown v_k.

Step 2:   Existence of a solution

To solve the transformed system in canonical form we apply Picard's successive approximation scheme. Namely, let us define for $k \geq 0$ a sequence $\{V^{(k)} = V^{(k)}(t, x)\}_k$ as follows:

$$(I\partial_t + \mathcal{D}(t, x)\partial_x)V^{(k)} = H(t, x) - B_0(t, x)V^{(k-1)},$$
$$V^{(k)}(0, x) = V_0(x), \quad V^{(-1)}(t, x) = V_0(x).$$

We shall show that for $\delta > 0$ sufficiently small, the sequence $\{V^{(k)}\}_k$ converges to a solution V uniformly in $C^1\big((-\delta, \delta) \times (a, b)\big)$. For $k = 0, 1, \cdots$ we set $W^{(k)} = V^{(k)} - V^{(k-1)}$, and derive

$$\partial_t W^{(0)} + \mathcal{D}(t, x)\partial_x W^{(0)} = H(t, x) - B_0(t, x)V_0 - \mathcal{D}(t, x)\partial_x V_0, \quad W^{(0)}(0, x) = 0,$$

and for $k > 0$

$$\partial_t W^{(k)} + \mathcal{D}(t, x)\partial_x W^{(k)} = -B_0(t, x)W^{(k-1)}, \quad W^{(k)}(0, x) = 0.$$

Let us denote by

$$w_j^{(k)}, \quad \{B_0(t, x)W^{(k-1)}\}_j \quad \text{and} \quad \{B_0(t, x)V_0 - \mathcal{D}(t, x)\partial_x V_0\}_j$$

the j'th component of

$$W^{(k)}, \quad B_0(t, x)W^{(k-1)} \quad \text{and} \quad B_0(t, x)V_0 + \mathcal{D}(t, x)\partial_x V_0,$$

respectively. Then, for $k > 0$ we have the following m scalar equations:

$$L_j w_j^{(k)} := (\partial_t + \lambda_j(t, x)\partial_x)w_j^{(k)} = -\{B_0(t, x)W^{(k-1)}\}_j, \quad w_j^{(k)}(0, x) = 0,$$

whereas for $k = 0$ we get

$$L_j w_j^{(0)} = h_j(t, x) - \{B_0(t, x)V_0\}_j - \{\mathcal{D}(t, \divideontimes)\partial_x V_0\}_j, \quad w_j^{(0)}(0, x) = 0.$$

Remember that the characteristic curve Γ_j (see Definition 6.1) is given by the solution $x = x(t, s)$ of $d_t x = \lambda_j(t, x)$, $x(0, s) = s$ (cf. with Sect. 7.1). The solution $x = x(t, s)$ is defined and C^1 is a neighborhood of $\{0\} \times (a, b)$. Along Γ_j the j'th equation of the system

$$L_j w_j^{(k)} = (\partial_t + \lambda_j(t, x)\partial_x)w_j^{(k)} = -\{B_0(t, x)W^{(k-1)}\}_j$$

is the ordinary differential equation

$$d_t w_j^{(k)} + \{B_0(t, x)W^{(k-1)}\}_j = 0.$$

The hypothesis of strict hyperbolicity implies that the characteristics do not intersect, at least for $|t| < \delta$ for a sufficiently small δ. Moreover, the set $(-\delta, \delta) \times (a, b)$ will be covered by these curves. This allows us to introduce the set $\Omega(t, x)$ as the smallest compact set containing the point (t, x) and having the following property:

If $(t', x') \in \Omega(t, x)$, then every curve $\Gamma_j(t', x'), j = 1, \cdots, m$,

passing through (t', x') is contained in $\Omega(t, x)$.

Now, we introduce the following norm for $W = W(t, x)$:

$$\|W(t,x)\|_{L^\infty(\Omega(t,x))} = \sup_{(t',x')\in\Omega(t,x)} |W(t',x')|.$$

Integrating along the characteristics of L_j we derive for $k > 0$ the estimate

$$\|W^{(k)}(t,x)\|_{L^\infty(\Omega(t,x))} \leq C|t|\,\|W^{(k-1)}(t,x)\|_{L^\infty(\Omega(t,x))}.$$

Hence, by iteration,

$$\|W^{(k)}(t,x)\|_{L^\infty(\Omega(t,x))} \leq (C|t|)^k \|W^{(0)}(t,x)\|_{L^\infty(\Omega(t,x))} \text{ for } k > 0.$$

On the other hand, $\|W^{(0)}(t,x)\|_{L^\infty(\Omega(t,x))} \leq C|t|$ and we conclude

$$\|W^{(k)}(t,x)\|_{L^\infty(\Omega(t,x))} \leq (C|t|)^{k+1} \text{ for } k = 0, 1, \cdots.$$

Therefore, for $C|t| < 1$, the series

$$\sum_{k=0}^{\infty} W^{(k)}(t,x)$$

converges uniformly to $V(t, x) - V_0(x)$, where $V = V(t, x)$ is the required solution. Similarly, using $\lambda_j \in C^1(G)$, one may also show that $V_x^{(k)}(t, x)$ and $V_t^{(k)}(t, x)$ converges uniformly on the domain G_0 to $V_x(t, x)$ and $V_t(t, x)$, respectively.

Example 22.4.1 Finally, let us explain how the method of characteristics is used to solve the following system from electrical engineering with an infinite transmission line (see Example 3.3.3):

$$\begin{aligned} L\partial_t I + \partial_x E + RI &= 0, \\ C\partial_t E + \partial_x I + GE &= 0, \\ I(0,x) = I_0(x), \quad E(0,x) &= E_0(x), \quad x \in \mathbb{R}. \end{aligned}$$

This system is strictly hyperbolic in the whole (t, x)-plane with constant eigenvalues $\lambda_1 = 1/\sqrt{LC}$ and $\lambda_2 = -1/\sqrt{LC}$. Our first step is to transform this system into its canonical form. In terms of the new variables v_1 and v_2 relating to I and E by

$$\binom{I}{E} = \begin{pmatrix} \sqrt{C} & \sqrt{C} \\ \sqrt{L} & -\sqrt{L} \end{pmatrix}\binom{v_1}{v_2}, \quad \binom{v_1}{v_2} = \frac{1}{2\sqrt{LC}}\begin{pmatrix} \sqrt{L} & \sqrt{C} \\ \sqrt{L} & -\sqrt{C} \end{pmatrix}\binom{I}{E},$$

the canonical form is

$$\frac{\partial}{\partial t}\begin{pmatrix} v_1 \\ v_2 \end{pmatrix} + \begin{pmatrix} \frac{1}{\sqrt{LC}} & 0 \\ 0 & -\frac{1}{\sqrt{LC}} \end{pmatrix} \frac{\partial}{\partial x}\begin{pmatrix} v_1 \\ v_2 \end{pmatrix}$$

$$+ \frac{1}{2LC}\begin{pmatrix} RC+LG & RC-LG \\ RC-LG & RC+LG \end{pmatrix} \begin{pmatrix} v_1 \\ v_2 \end{pmatrix} = \begin{pmatrix} 0 \\ 0 \end{pmatrix}.$$

The new unknowns satisfy for $x \in \mathbb{R}^1$ the initial conditions

$$v_1(0,x) = \varphi_1(x) := \frac{1}{2\sqrt{LC}} \left(\sqrt{L}I_0(x) + \sqrt{C}E_0(x)\right),$$

$$v_2(0,x) = \varphi_2(x) := \frac{1}{2\sqrt{LC}} \left(\sqrt{L}I_0(x) - \sqrt{C}E_0(x)\right).$$

Following the method of characteristics for solving this initial value problem, let (t,x) be an arbitrary but fixed point in the upper-half (s,x)-plane. The characteristic curves Γ_1 and Γ_2 corresponding to $\lambda_1 = 1/\sqrt{LC}$ and $\lambda_2 = -1/\sqrt{LC}$ and passing through the point (t,x) are the lines

$$\Gamma_1 : x_1 = x_1(s,t,x) = x + \frac{1}{\sqrt{LC}}(s-t),$$

$$\Gamma_2 : x_2 = x_2(s,t,x) = x - \frac{1}{\sqrt{LC}}(s-t).$$

These lines intersect the x-axis at the points (set $s = 0$)

$$x_1(0,t,x) = x - \frac{1}{\sqrt{LC}}t, \quad x_2(0,t,x) = x + \frac{1}{\sqrt{LC}}t.$$

In this case we obtain the system of Volterra integral equations

$$v_1(t,x) = \varphi_1\left(x - \frac{1}{\sqrt{LC}}t\right)$$

$$- \frac{1}{2LC}\int_0^t \left(\Theta v_1(s,x_1(s,t,x)) + \Upsilon v_2(s,x_1(s,t,x))\right)ds,$$

$$v_2(t,x) = \varphi_2\left(x + \frac{1}{\sqrt{LC}}t\right)$$

$$- \frac{1}{2LC}\int_0^t \left(\Upsilon v_1(s,x_2(s,t,x)) + \Theta v_2(s,x_2(s,t,x))\right)ds,$$

where $\Theta := RC + LG$, $\Upsilon := RC - LG$, x_1 and x_2 are given. In general, in order to proceed any further with the computation of the solution it is necessary to know the specific functional form of the initial data and apply the successive approximation scheme. However, in the *special case in which the electrical parameters of the transmission line satisfy the relation*

$$RC - LG = 0$$

it is possible to obtain a general formula for the solution. This is due to the fact that when this condition holds, the canonical form is separated in the sense that each equation involves only one of the unknowns v_1 or v_2. These unknowns restricted by the characteristics are

$$W_1(s, t, x) = v_1\left(s, x + \frac{1}{\sqrt{LC}}(s - t)\right), \quad W_2(s, t, x) = v_2\left(s, x - \frac{1}{\sqrt{LC}}(s - t)\right).$$

They satisfy the corresponding ordinary differential equations

$$\frac{dW_1}{ds} + \frac{R}{L}W_1 = 0, \quad \frac{dW_2}{ds} + \frac{R}{L}W_2 = 0,$$

and the initial conditions

$$W_1(0, t, x) = \varphi_1(x_1(0, t, x)), \quad W_2(0, t, x) = \varphi_2(x_2(0, t, x)).$$

Solving these two initial value problems we obtain

$$W_1(s, t, x) = \varphi_1(x_1(0, t, x))\, e^{-\frac{R}{L}s}, \quad W_2(s, t, x) = \varphi_2(x_2(0, t, x))\, e^{-\frac{R}{L}s}.$$

We have

$$v_1(t, x) = W_1(t, t, x), \quad v_2(t, x) = W_2(t, t, x).$$

Consequently we get, under the condition $RC - LG = 0$, the solution

$$v_1(t, x) = \varphi_1\left(x - \frac{1}{\sqrt{LC}}t\right) e^{-\frac{R}{L}t}, \quad v_2(t, x) = \varphi_2\left(x + \frac{1}{\sqrt{LC}}t\right) e^{-\frac{R}{L}t}.$$

The solution of the original problem is obtained in the form

$$I(t, x) = \left(\frac{1}{2}\left(I_0\left(x - \frac{1}{\sqrt{LC}}t\right) + I_0\left(x + \frac{1}{\sqrt{LC}}t\right)\right)\right.$$
$$\left. + \frac{1}{2}\sqrt{\frac{C}{L}}\left(E_0\left(x - \frac{1}{\sqrt{LC}}t\right) - E_0\left(x + \frac{1}{\sqrt{LC}}t\right)\right)\right) e^{-\frac{R}{L}t},$$

$$E(t,x) = \left(\frac{1}{2}\sqrt{\frac{L}{C}}\left(I_0\left(x - \frac{1}{\sqrt{LC}}t\right) - I_0\left(x + \frac{1}{\sqrt{LC}}t\right)\right)\right.$$
$$\left. + \frac{1}{2}\left(E_0\left(x - \frac{1}{\sqrt{LC}}t\right) + E_0\left(x + \frac{1}{\sqrt{LC}}t\right)\right)\right)e^{-\frac{R}{L}t}.$$

Note that this solution shows that in the case $RC = LG$ the initial data propagate along the line *without distortion* with exponential decay in time. It is for this reason that a line for which $RC = LG$ holds is called a *distortionless line*.

22.5   Energy Inequalities for Linear Symmetric Hyperbolic Systems

The main concern of this section is to derive energy inequalities for solutions to the Cauchy problem for linear symmetric hyperbolic systems. Here we follow the lecture notes "Théorie des équations d' évolution" written by J. Y. Chemin (Paris 6).

We consider the Cauchy problem for a system of partial differential equations of first order

$$\partial_t U + \sum_{k=1}^{n} A_k(t,x)\partial_{x_k} U + A_0(t,x)U = F(t,x), \quad U(0,x) = U_0(x),$$

where the matrices $A_k = A_k(t,x)$, $k = 0, 1, \cdots, n$, are real, smooth, and together with all of their derivatives with respect to the spatial variables are bounded in $[0, \infty) \times \mathbb{R}^n$. We assume that $A_k, k = 1, \cdots, n$, are $m \times m$ symmetric matrices.

Proposition 22.5.1 *Let s be a nonnegative integer. Then, any energy solution*

$$U \in C([0, T], W^{s+1}(\mathbb{R}^n)) \cap C^1([0, T], W^s(\mathbb{R}^n))$$

to the Cauchy problem

$$\partial_t U + \sum_{k=1}^{n} A_k(t,x)\partial_{x_k} U + A_0(t,x)U = F(t,x), \quad U(0,x) = U_0(x)$$

satisfies the energy inequality

$$\|U(t,\cdot)\|_{W^s}^2 \leq C(s, T)\left(\|U_0\|_{W^s}^2 + \int_0^t \|F(\tau,\cdot)\|_{W^s}^2 d\tau\right) \text{ for } t \in [0, T].$$

Proof At first we prove the desired inequality for $s = 0$, that is, we assume that

$$U \in C([0, T], W^1(\mathbb{R}^n)) \cap C^1([0, T], L^2(\mathbb{R}^n)).$$

Multiplying the system by U and integrating the space variables we derive ($(\cdot, \cdot)_{L^2}$ denotes the scalar product in L^2)

$$\frac{1}{2}\frac{d}{dt}\|U(t, \cdot)\|_{L^2}^2 = (F, U)_{L^2} - (A_0 U, U)_{L^2} - \sum_{k=1}^{n}(A_k \partial_{x_k} U, U)_{L^2}.$$

By using the fact that the matrices $A_k, k = 1, \cdots, n$, are symmetric and that $U \in C([0, T], W^1(\mathbb{R}^n)) \cap C^1([0, T], L^2(\mathbb{R}^n))$ we have

$$-(A_k \partial_{x_k} U, U)_{L^2} = \frac{1}{2}((\partial_{x_k} A_k)U, U)_{L^2}.$$

Hence, we get

$$\frac{1}{2}\frac{d}{dt}\|U(t, \cdot)\|_{L^2}^2 \leq \left(\Big\|\sum_{k=1}^{n}\partial_{x_k}A_k(t, \cdot)\Big\|_{L^\infty} + 2\|A_0(t, \cdot)\|_{L^\infty}\right)\|U(t, \cdot)\|_{L^2}^2$$

$$+2\|F(t, \cdot)\|_{L^2}\|U(t, \cdot)\|_{L^2}.$$

Applying Schwarz inequality and Gronwall's lemma we conclude

$$\|U(t, \cdot)\|_{L^2}^2 \leq C(T)\left(\|U_0\|_{L^2}^2 + \int_0^t \|F(\tau, \cdot)\|_{L^2}^2 d\tau\right) \quad \text{for } t \in [0, T],$$

where the constant $C(T)$ depends on

$$\sup_{t \in [0,T]}\left\{\Big\|\sum_{k=1}^{n}\partial_{x_k}A_k(t, \cdot)\Big\|_{L^\infty} + 2\|A_0(t, \cdot)\|_{L^\infty}\right\}.$$

We prove the energy estimates of higher order by induction on the integer s. Suppose that the energy inequality is valid for some $s = s_0 > 0$. Then, we can prove it for $s = s_0 + 1$. For this reason let us assume that

$$U \in C([0, T], W^{s_0+2}(\mathbb{R}^n)) \cap C^1([0, T], W^{s_0+1}(\mathbb{R}^n)).$$

We introduce the vector-function $\tilde{U} = \tilde{U}(t, x)$ defined by

$$\tilde{U} = (U, \partial_{x_1} U, \cdots, \partial_{x_n} U).$$

By differentiation of the original system we obtain for any $j \in \{1, \cdots, n\}$ the new system

$$\partial_t \partial_{x_j} U = -\sum_{k=1}^{n}A_k \partial_{x_k} \partial_{x_j} U - \sum_{k=1}^{n}(\partial_{x_j} A_k)\partial_{x_k} U - \partial_{x_j}(A_0 U) + \partial_{x_j} F.$$

Hence, we may write all these $n + 1$ systems (the original one and the n new ones) as the following new system:

$$\partial_t \tilde{U} + \sum_{k=1}^{n} B_k(t,x) \partial_{x_k} \tilde{U} + B_0(t,x) \tilde{U} = \tilde{F}(t,x),$$

where

$$\tilde{F} = (F, \partial_{x_1} F, \cdots, \partial_{x_n} F) \quad \text{and} \quad B_k = \begin{pmatrix} A_k & & 0 \\ & A_k & \\ 0 & & A_k \end{pmatrix}.$$

The entries of $B_0 = B_0(t,x)$ depend on $A_k, k = 0, 1, \cdots, n$, and their first order derivatives. The energy inequality for $s = s_0 + 1$ follows by applying the induction hypothesis for $s = s_0$ to the derived symmetric system for $\tilde{U}$.

22.6   Concluding Remarks

22.6.1   Well-Posedness for Linear Symmetric Hyperbolic Systems

Let us point out that the proof of the inequality of Proposition 22.5.1 done in the previous section demands exactly one additional derivative for the solution. This leads us to use a smoothing method, the so-called Friedrich's mollifiers method, which consists of smoothing both the initial data and the terms appearing in the given system. More precisely, let us consider the family of linear systems

$$\partial_t U_k + \sum_{j=1}^{n} \chi_k(A_j \partial_{x_j} U_k) + \chi_k(A_0 U_k) = \chi_k F,$$

$$\chi_k U(0,x) = \chi_k U_0(x).$$

Here the matrices $A_j = A_j(t,x), j = 0, 1, \cdots, n$, satisfy the same hypotheses as in Sect. 22.5 and $\chi_k = \chi_k(D)$ is a cut-off operator defined on $L^2(\mathbb{R}^n)$ by

$$\chi_k(D)u := F_{\xi \to x}^{-1}\big(\chi_{B_k(0)}(\xi) F_{x \to \xi}(u)\big),$$

where $\chi_{B_k(0)} = \chi_{B_k(0)}(\xi)$ denotes the characteristic function on the ball of center 0 and radius k. The operator $\chi_k(D)$ is the orthogonal projection of $L^2(\mathbb{R}^n)$ on its closed subspace $L_k^2(\mathbb{R}^n)$ having the property that the Fourier transforms of its elements are supported in the ball $B_k(0)$. Bernstein's inequality (see Proposition 24.5.7) implies that the operator ∂_{x_j} is continuous on $L_k^2(\mathbb{R}^n)$ and, thanks to the boundedness of A_j, each system of the above family of linear systems is a linear system of abstract ordinary differential equations on $L_k^2(\mathbb{R}^n)$. This implies the existence of a unique solution

$$U_k \in C\big([0, T], L^2(\mathbb{R}^n)\big).$$

Moreover, using the fact that A_j are smooth functions, we get that

$$U_k \in C^\infty\big([0, T], W^s(\mathbb{R}^n)\big)$$

for any integer s. In addition, one may prove that the functions U_k satisfy an energy inequality as derived in Proposition 22.5.1.

By taking the limit in the case of smooth enough initial data we get the following well-posedness result (for more details see the lecture notes "Théorie des équations d' évolution" written by J. Y. Chemin (Paris 6)):

Theorem 22.6.1 *Let us consider the linear symmetric system*

$$\partial_t U + \sum_{j=1}^n A_j(t, x)\partial_{x_j} U + A_0(t, x)U = F(t, x), \quad U(0, x) = U_0(x),$$

where the matrices $A_j = A_j(t, x)$, $j = 0, 1, \cdots, n$, satisfy the same hypotheses as in Sect. 22.5. If $U_0 \in W^s(\mathbb{R}^n)$ and $F \in C\big([0, T], W^s(\mathbb{R}^n)\big)$, where $s \geq 1$ is an integer, then there exists a uniquely determined energy solution

$$U \in C\big([0, T], W^s(\mathbb{R}^n)\big) \cap C^1\big([0, T], W^{s-1}(\mathbb{R}^n)\big).$$

22.6.2 Well-Posedness for Linear Strictly Hyperbolic Systems

Now we only explain some well-posedness result for solutions to the Cauchy problem for linear strictly hyperbolic systems. For this reason, we consider linear systems of first order having the form

$$LU := \partial_t U + \sum_{k=1}^n A_k(t, x)\partial_{x_k} U + B(t, x)U = F(t, x) \quad \text{in} \quad [0, \infty) \times \mathbb{R}^n,$$

where A_k, $k = 1, \cdots n$, and B are $m \times m$ matrices, subject to the initial condition

$$U(0, x) = U_0(x).$$

Recalling the definition of strictly hyperbolic systems, that is, we assume that the matrix

$$A(t, x, \xi) := \sum_{k=1}^{n} \xi_k A_k(t, x) \quad \text{for} \quad t \geq 0 \text{ and } (x, \xi) \in \mathbb{R}^{2n},$$

has m distinct real eigenvalues:

$$\lambda_1(t, x, \xi) < \lambda_2(t, x, \xi) < \cdots < \lambda_m(t, x, \xi).$$

In addition to the assumption of L being strictly hyperbolic, in order to have similar results in the case of symmetric hyperbolic systems we introduce an additional definition (see [139]).

Definition 22.1 We say that the systems of partial differential equations of first order

$$LU = F(t, x)$$

are *regularly hyperbolic* in $G = [0, T] \times \mathbb{R}^n$ if they are strictly hyperbolic and satisfy the following conditions:

1. the matrices $A_k, k = 1, \cdots, n$, are bounded in G
2. for arbitrary $(t, x) \in G, \xi \neq 0$ and $j, k = 1, \cdots, m$ we have

$$\inf_{(t,x)\in G, |\xi|=1, j\neq k} |\lambda_j(t, x, \xi) - \lambda_k(t, x, \xi)| = c > 0.$$

The meaning of regularly hyperbolic in the definition refers to the fact that the difference between distinct eigenvalues has a positive lower bound uniformly on G.

We denote by $\mathbb{B}^s(\mathbb{R}^n)$ the function space with the property that all derivatives until order s are bounded and continuous. If $p = 2$, then we denote $W_p^s(\mathbb{R}^n)$ just by $W^s(\mathbb{R}^n)$. Moreover, $W_0^s(\mathbb{R}^n)$ is the closure in this space of all infinitely differentiable functions with compact support in $\mathbb{R}^n$. The following result is proved in [139] (cf. with Theorem 6.9. in [139]).

Theorem 22.6.2 *Let us consider the Cauchy problem*

$$LU = F, \quad U(0, x) = U_0(x),$$

where L is regularly hyperbolic in $G = [0, T] \times \mathbb{R}^n$. For a fixed $s \in \mathbb{N}$, let us assume with an entire $\sigma > 0$

$$A_k \in C\big([0, T], \mathbb{B}^{\max\{1+\sigma, s\}}(\mathbb{R}^n)\big), \quad \partial_t A_k \in C\big([0, T], \mathbb{B}^s(\mathbb{R}^n)\big),$$
$$B \in C\big([0, T], \mathbb{B}^s(\mathbb{R}^n)\big).$$

If $U_0 \in W_0^s(\mathbb{R}^n)$ and $F \in C([0, T], W_0^s(\mathbb{R}^n))$, then there exists a uniquely determined energy solution

$$U \in C^0([0, T], W_0^s(\mathbb{R}^n)) \cap C^1([0, T], W_0^{s-1}(\mathbb{R}^n))$$

satisfying the following energy inequality:

$$\|U(t, \cdot)\|_{W^s}^2 \leq C(s, T)\left(\|U_0\|_{W^s}^2 + \int_0^t \|F(\tau, \cdot)\|_{W^s}^2 d\tau\right).$$

Exercises Relating to the Considerations of Chap. 22

Exercise 1   Let us consider the strictly hyperbolic Cauchy problem

$$\partial_t U + A\, \partial_x U = 0, \quad U(0, x) = U_0(x),$$

where A is a constant $m \times m$ matrix. Show that there is a uniquely determined classical solution $U \in C^1(\mathbb{R}^2)$ if U_0 is supposed to belong to $C^1(\mathbb{R}^1)$.

Exercise 2   Consider the symmetric hyperbolic system

$$\partial_t U + \sum_{k=1}^n A_k \partial_{x_k} U + A_0 U = F(t, x), \quad U(0, x) = U_0(x),$$

where the initial data $U_0(x)$ has compact support. Assuming that a solution has compact support (domain of dependence property holds), prove that there exists at most one Sobolev solution belonging to $C([0, T], L^2(\mathbb{R}^n))$.

Part V

Chapter 23
Research Projects for Beginners

In this chapter we propose research projects for beginners of PhD study. After reading chapters of this monograph these research projects serve as a preparation for ones own scientific work. This serves as a first attempt to apply mathematical techniques or basic knowledge and develop new ideas. It has been the authors that our PhD students, who were able to treat such research projects have gone on to complete their PhD theses on a certain level within a reasonable period.

23.1 Applications of the Abstract Cauchy-Kovalevskaja and Holmgren Theorems

We are interested in the study of the Cauchy problem

$$\partial_t u - a(t,x)\partial_x u - b(t,x)u = f(t,x), \quad u(0,x) = u_0(x),$$

under the following assumptions:

1. the data u_0 belongs to the function space

$$B_1 := \left\{ u \in C^\infty[h_1, h_2] : |d_x^k u(x)| \leq CR^k k! \text{ for all } k \in \mathbb{N}_0 \right\}$$

   with nonnegative constants C and R, so the data is supposed to be analytic on the interval $[h_1, h_2]$

2. the coefficients a and b and the right-hand side f belong to the function space $C([0,T], B_1)$ (see Definition 24.25).

© Springer International Publishing AG 2018
M.R. Ebert, M. Reissig, *Methods for Partial Differential Equations*,
https://doi.org/10.1007/978-3-319-66456-9_23

The norm in B_1 is defined by

$$\|u\|_1 := \sup_{k \in \mathbb{N}_0} \left\|d_x^k u(x)\right\|_{C[h_1,h_2]} \frac{1}{R^k k!}.$$

Let us introduce the scale of Banach spaces of analytic functions $\{B_s, \|\cdot\|_s\}_{s \in [0,1]}$ by

$$B_s := \left\{u \in C^\infty[h_1, h_2]\right.$$

$$\left.: \|u\|_s := \sup_{k \in \mathbb{N}_0} 2^{-k} \left\|d_x^k u(x)\right\|_{C[h_1,h_2]} \left((1+s)R\right)^k k!\right)^{-1} < \infty\right\}.$$

1. Show that $\{B_s, \|\cdot\|_s\}_{s \in [0,1]}$ is a scale of Banach spaces (see Sect. 4.2).
2. Let us define the family of linear operators

$$\left\{A(t) := a(t,x)\partial_x + b(t,x)\right\}_{t \in [0,T]}.$$

   Show that this family satisfies the assumptions of Theorem 4.2.1 (cf. with Exercise 2 of Chap. 4) in the scale $\{B_s, \|\cdot\|_s\}_{s \in [0,1]}$.
3. Apply the statement of Theorem 4.2.1. What kind of results do we conclude?
4. Try to prove that the scale $\{B_s, \|\cdot\|_s\}_{s \in [0,1]}$ is dense into itself (see Sect. 5.2).
5. If one is able to prove the density into itself, next try to understand what distributions do belong to $F_s := B'_{1-s}$. The elements of these distribution spaces are called *analytic functionals*.

Analogous to Remark 24.4.1, we may introduce the family of dual operators $A(t)'$. This family is defined by

$$\left(a(t,x)\partial_x u(t,x) + b(t,x)u(t,x)\right)\phi(x)$$

$$= u(t,x)\left(-\partial_x\left(a(t,x)\phi(x)\right) + b(t,x)\phi(x)\right)$$

for all test functions $\phi \in B_s$ and all $u \in C([0,T], F_{1-s'})$. Then we can show, that the family of dual operators

$$\left\{A(t)' := -a(t,x)\partial_x - \partial_x a(t,x) + b(t,x)\right\}_{t \in [0,T]}$$

satisfies the assumptions of Theorem 4.2.1 in the scale $\{F_{1-s}, \|\cdot\|'_{1-s}\}_{s \in [0,1]}$, too. The family of dual operators $\{A(t)'\}_{t \in [0,T]}$ satisfies the assumptions of Theorem 4.2.1 in the scale $\{B_s, \|\cdot\|_s\}_{s \in [0,1]}$. If the scale $\{B_s, \|\cdot\|_s\}_{s \in [0,1]}$ is dense into itself, then we can form the dual of $A(t)'$ and obtain $A(t)$ itself. Hence, we can apply the abstract Holmgren Theorem 5.2.1. In this way we obtain the uniqueness of solutions of the Cauchy problem

$$\partial_t u - a(t,x)\partial_x u - b(t,x)u = 0, \quad u(0,x) = 0$$

in $C([0, T], F_{1-s})$ for all $s \in [0, 1]$. The main difficulties are in proving the density into itself of the scale $\{B_s, \| \cdot \|_s\}_{s \in [0,1]}$ and understanding which distributions do belong to the dual scale $\{F_{1-s}, \| \cdot \|'_{1-s}\}_{s \in [0,1]}$.

23.2   The Robin Problem for the Heat Equation in an Interior Domain

This project addresses to the issue of mixed problems for the heat equation. Study the interior Robin problem ($G \subset \mathbb{R}^n$ is an interior domain)

$$\partial_t u - \Delta u = h(t, x) \quad \text{in } Z_T = (0, T) \times G, \quad u(0, x) = \varphi(x) \text{ for } x \in G,$$

$$\partial_\mathbf{n} u(t, x) + c(t, x)u(t, x) = g(t, x) \quad \text{on } (0, T) \times \partial G,$$

$$\text{compatibility condition } g(0, x) = \partial_\mathbf{n}\varphi(x) + c(0, x)\varphi(x) \text{ on } \{t = 0\} \times \partial G.$$

1. Fix assumptions for the regularity of the data $\varphi = \varphi(x)$, the right-hand sides $h = h(t, x)$ and $g = g(t, x)$, and the coefficient $c = c(t, x)$.
2. The assumptions for all the "data" have an influence on the choice of the space of solutions. What space of solutions seems to be reasonable?
3. Introduce the function

$$w(t, x) := \int_{Z_t} H_n(x - y, t - \tau)h(\tau, y) \, d(\tau, y) + \int_G H_n(x - y, t)\varphi(y) \, dy.$$

Verify that $w = w(t, x)$ is a solution of the auxiliary problem

$$\partial_t w - \Delta w = h(t, x) \text{ in } Z_T, \quad w(0, x) = \varphi(x) \text{ on } \{t = 0\} \times \overline{G}.$$

4. Determine regularity properties of the solution w depending on the chosen regularity for h and φ.
5. Define $v := u - w$. Derive the mixed problem for v.
6. Choose the ansatz for v in the form of a thermal single-layer potential, that is,

$$v(t, x) := \int_{S_t} H_n(t - \tau, x - y)\mu(\tau, y) \, d(\tau, \sigma_y)$$

with an unknown density μ. Study the jump condition for the normal derivative of the thermal single-layer potential (see Sect. 9.4.2.2).
7. Establish the corresponding integral equation (see Sect. 9.4.3) after setting the ansatz for v into the Robin boundary condition for v and using the jump condition in the case of an interior domain.

8. Fix a solution space for μ. The choice depends on the regularity of g and w. Solve the Volterra integral equation of second kind by successive approximation (cf. with [135]).
9. Finally, determine the regularity of u.

23.3 $L^p - L^q$ Decay Estimates for Solutions to the Heat Equation with Mass

We are interested in the Cauchy problem for heat equation with mass (cf. with Sect. 12.2)

$$u_t - \Delta u + m^2 u = 0, \quad u(0, x) = \varphi(x).$$

Our goal is to derive $L^p - L^q$ decay estimates not necessarily on the conjugate line (cf. with Sects. 12.1.2 and 12.1.3) (see also Exercise 1 from Chap. 12). As in Sect. 12.1.2, we use the corresponding representation of solution

$$u(t, x) = F_{\xi \to x}^{-1} \left(e^{-\langle \xi \rangle_m^2 t} \right) * \varphi.$$

Applying Proposition 24.5.2 implies the estimate

$$\| u(t, \cdot) \|_{L^q} \leq \left\| F_{\xi \to x}^{-1} \left(e^{-\langle \xi \rangle_m^2 t} \right) \right\|_{L^r} \| \varphi \|_{L^p}.$$

To reach our goal we have to derive L^r estimates for the oscillating integral $F_{\xi \to x}^{-1} \left(e^{-\langle \xi \rangle_m^2 t} \right)$. Here we may apply ideas and methods from Sect. 24.2.2.

23.4   The Cauchy Problem for the Free Wave Equation in Modulation Spaces

Consider the following Cauchy problem:

$$u_{tt} - \Delta u = 0, \quad u(0, x) = \varphi(x), \quad u_t(0, x) = \psi(x).$$

Suppose that the data φ, ψ are taken from the modulation spaces $M_{p,q}^s(\mathbb{R}^n)$ for $1 \leq p, q \leq \infty$ and $s \in \mathbb{R}^1$ (see Definition 24.15).

1. Show that the Cauchy problem has a classical solution in time with values in the space of tempered distributions if we assume $\varphi \in M_{p,q}^s(\mathbb{R}^n)$ and $\psi \in M_{p,q}^{s-1}(\mathbb{R}^n)$ for $1 \leq p, q \leq \infty$ and $s \in \mathbb{R}^1$.

2. What can we say about the regularity of the solution with respect to the spatial variables?

Now consider the semilinear Cauchy problem for the wave equation

$$u_{tt} - \Delta u = \lambda |u|^{2k} u, \quad u(0, x) = \varphi(x), \quad u_t(0, x) = \psi(x),$$

where $\lambda \in \mathbb{R}^1$ and $k \in \mathbb{N}$.

1. Show that this semilinear Cauchy problem is locally (in time) well-posed in the solution space

$$C\big([0, T], M_{p,1}^s(\mathbb{R}^n)\big) \cap C^1\big([0, T], M_{p,1}^{s-1}(\mathbb{R}^n)\big)$$

if we assume for the data $\varphi \in M_{p,1}^s(\mathbb{R}^n)$ and $\psi \in M_{p,1}^{s-1}(\mathbb{R}^n)$ for $1 \leq p \leq \infty$ and $s \geq 0$.
2. What do we know about the solution if we additionally assume $s \geq 2$?
3. Can we replace the condition $q = 1$ by $q \in [1, \infty]$ and obtain the same well-posedness result for the semilinear Cauchy problem as in 1? Consider the weight parameter s.

23.5   The Diffusion Phenomenon for Classical Damped Klein-Gordon Models

Prove the statement of Theorem 14.5.1 for solutions to the Cauchy problem

$$u_{tt} - \Delta u + m^2 u + u_t = 0, \quad u(0, x) = \varphi(x), \quad u_t(0, x) = \psi(x), \quad x \in \mathbb{R}^n, \quad n \geq 1.$$

Prove a similar statement to Theorem 12.2.1, but we want to be more precise. We recall the diffusion phenomenon of Sect. 14.2.3 between solutions to heat and classical damped wave models. We could prove that the difference of solutions of both models has a better decay than solutions to heat and classical damped wave models have. We want to transfer this observation to

$$
\begin{array}{ccc}
u_{tt} - \Delta u + m^2 u + u_t = 0, & & w_t - \Delta w + m^2 w = 0, \\
u(0, x) = \varphi(x), \quad u_t(0, x) = \psi(x) & \text{and} & w(0, x) = \rho(x).
\end{array}
$$

We introduce a cut-off function $\chi \in C_0^\infty(\mathbb{R}^n)$ with $\chi(s) = 1$ for $|s| \leq \frac{\varepsilon}{2} \ll 1$ and $\chi(s) = 0$ for $|s| \geq \varepsilon$. Then we pose the following question:
How should we choose the data $\rho = \rho(x)$ such that the differences

$$\left\| F_{\xi \to x}^{-1}\big(\chi(\xi) F_{x \to \xi}\big(u(t, x) - w(t, x)\big)\big) \right\|_{L^2}$$

$$\left\| F_{\xi \to x}^{-1}\big((1 - \chi(\xi)) F_{x \to \xi}\big(u(t, x) - w(t, x)\big)\big) \right\|_{L^2}$$

have a better decay behavior than

$$\left\| F_{\xi \to x}^{-1} \left(\chi(\xi) F_{x \to \xi} \big(u(t,x) \big) \right) \right\|_{L^2} \text{ and } \left\| F_{\xi \to x}^{-1} \left(\chi(\xi) F_{x \to \xi} \big(w(t,x) \big) \right) \right\|_{L^2},$$

$$\left\| F_{\xi \to x}^{-1} \left((1 - \chi(\xi)) F_{x \to \xi} \big(u(t,x) \big) \right) \right\|_{L^2} \text{ and }$$

$$\left\| F_{\xi \to x}^{-1} \left((1 - \chi(\xi)) F_{x \to \xi} \big(w(t,x) \big) \right) \right\|_{L^2}$$

have? Although all Fourier multipliers already have an exponential type decay in time, this is a very good exercise for understanding diffusion phenomena. One can find in the literature papers in which authors distinguish between diffusion phenomena for small and large frequencies in their considerations (see [145]).

23.6   The Diffusion Phenomenon for Classical Damped Plate Models

Before we study the diffusion phenomenon for plate models let us answer the following questions:

1. To which class of differential equations does the classical damped plate model belong?
2. What do we know about H^s well-posedness?
3. How do we define suitable energies (even those of higher order)?
4. What do we know about decay estimates for energies (even those of higher order)?

After these preparations we recall the main result Theorem 14.2.3 of Sect. 14.2.3. Now we are interested in a corresponding result for solutions to

$$u_{tt} + (-\Delta)^2 u + u_t = 0, \quad u(0,x) = \varphi(x), \quad u_t(0,x) = \psi(x).$$

To complete the project we proceed as follows:

1. How do we choose the related parabolic Cauchy problem? How do we choose the differential equation and the data?
2. Prove L^2 estimates for solutions to this Cauchy problem.
3. Prove L^2 estimates for solutions to the Cauchy problem for classical damped plates.
4. Pose a conjecture for the diffusion phenomenon between solutions to both Cauchy problems.
5. Prove this conjecture.
6. Interpret this diffusion phenomenon.

23.7   The Diffusion Phenomenon for Damped Wave Models with Source

Consider the Cauchy problems

$$u_{tt} - \Delta u + u_t = f(t, x),$$
$$u(0, x) = \varphi(x), \quad u_t(0, x) = \psi(x)$$

and

$$w_t - \Delta w = f(t, x),$$
$$w(0, x) = \varphi(x) + \psi(x),$$

where the data φ and ψ satisfy the following regularity properties: $\varphi, \psi \in L^1(\mathbb{R}^n)$ and $|D|\varphi, \psi \in L^2(\mathbb{R}^n)$. Moreover, we suppose that there exists a constant $\varepsilon > 0$ such that

$$\|f(t, \cdot)\|_{L^1} \leq C_2(1 + t)^{-1-\varepsilon}, \quad \|f(t, \cdot)\|_{L^2} \leq C_2(1 + t)^{-\frac{n}{4}-1-\varepsilon} \text{ for all } t \geq 0.$$

Prove that the L^2-norm of the difference of solutions to the above Cauchy problems behaves as follows:

$$\left\| u(t, \cdot) - w(t, \cdot) \right\|_{L^2} = o(t^{-\frac{n}{4}}) \text{ for } t \to \infty.$$

One may proceed as follows:

1. Following the notations of Sect. 18.1.1.2, by using Duhamel's principle the solutions are given by

$$u(t, x) = u_0(t, x) + \int_0^t K_1(t - s, 0, x) *_{(x)} f(s, x) \, ds$$

with

$$u_0(t, x) = K_0(t, 0, x) *_{(x)} \varphi(x) + K_1(t, 0, x) *_{(x)} \psi(x),$$

whereas

$$w(t, x) = G(t, 0, x) *_{(x)} (\varphi(x) + \psi(x)) + \int_0^t G(t - s, 0, x) *_{(x)} f(s, x) \, ds$$

with

$$G(t, 0, x) = (2\pi)^{-n/2} \int_{\mathbb{R}^n} e^{ix \cdot \xi - t|\xi|^2} \, d\xi.$$

2. Thanks to Exercise 8 from Chap. 14 it is sufficient to prove

$$\left\| \int_0^t \left(K_1(t-s,0,\cdot) - G(t-s,0,\cdot) \right) *_{(x)} f(s,\cdot)\, ds \right\|_{L^2} = o(t^{-\frac{n}{4}}) \quad \text{for} \quad t \to \infty.$$

To derive the last relation, it is convenient to split the integral $\int_0^t$ into $\int_0^{t/2} + \int_{t/2}^t$ and to carry out the analysis for small and large frequencies as well.

- Verify that $\chi(\xi) F_{x \to \xi}(K_1(t-s,0,x))$, where $\chi \in C_0^\infty(\mathbb{R}^n)$ is a cut-off function which localizes to small frequencies and is a bounded Fourier multiplier on L^2 satisfying

$$\left\| F_{\xi \to x}^{-1}\left(\chi(\xi) F_{x \to \xi}\left(\int_{t/2}^t K_1(t-s,0,\cdot) *_{(x)} f(s,\cdot)\, ds \right) \right) \right\|_{L^2} = o(t^{-\frac{n}{4}}) \quad \text{as} \quad t \to \infty.$$

  Similarly, show

$$\left\| F_{\xi \to x}^{-1}\left(\chi(\xi) F_{x \to \xi}\left(\int_{t/2}^t G(t-s,0,\cdot) *_{(x)} f(s,\cdot)\, ds \right) \right) \right\|_{L^2} = o(t^{-\frac{n}{4}}) \quad \text{as} \quad t \to \infty.$$

- Applying $L^1 - L^2$ estimates to

$$\left(K_1(t-s,0,x) - G(t-s,0,x) \right) *_{(x)} f(s,x)$$

  and the estimate

$$\|f(t,\cdot)\|_{L^1} \leq C_2 (1+t)^{-1-\varepsilon},$$

  one may conclude

$$\left\| F_{\xi \to x}^{-1}\left(\chi(\xi) F_{x \to \xi}\left(\int_0^{t/2} \left(K_1(t-s,0,\cdot) - G(t-s,0,\cdot) \right) *_{(x)} f(s,\cdot)\, ds \right) \right) \right\|_{L^2}$$
$$= o(t^{-\frac{n}{4}}) \quad \text{as} \quad t \to \infty.$$

- For large frequencies we no longer have an exponential decay, but one may still prove

$$\left\| F_{\xi \to x}^{-1}\left((1 - \chi(\xi)) F_{x \to \xi}\left(\int_0^t \left(K_1(t-s,0,\cdot) - G(t-s,0,\cdot) \right) *_{(x)} f(s,\cdot)\, ds \right) \right) \right\|_{L^2}$$
$$= o(t^{-\frac{n}{4}}) \quad \text{as} \quad t \to \infty.$$

23.8   Profile of Solutions to Classical Damped Waves with Source

Consider the Cauchy problem

$$u_{tt} - \Delta u + u_t = f(t, x), \quad u(0, x) = \varphi(x), \quad u_t(0, x) = \psi(x),$$

where the data φ and ψ satisfy the following regularity properties: $\varphi, \psi \in L^1(\mathbb{R}^n)$, $|D|\varphi, \psi \in L^2(\mathbb{R}^n)$. Moreover, we suppose for the source term $f = f(t, x)$ the existence of a constant $\varepsilon > 0$ such that

$$\|f(t, \cdot)\|_{L^1} \leq C_2(1 + t)^{-1-\varepsilon}, \quad \|f(t, \cdot)\|_{L^2} \leq C_2(1 + t)^{-\frac{n}{4}-1-\varepsilon} \quad \text{for all } t \geq 0.$$

Prove that the solution to the above Cauchy problem satisfies the estimate

$$\|u(t, \cdot) - \theta_0 \, G(t, 0, \cdot)\|_{L^2} = o(t^{-\frac{n}{4}}) \quad \text{for } t \to \infty,$$

where

$$G(t, 0, x) = (2\pi)^{-n/2} \int_{\mathbb{R}^n} e^{ix\cdot\xi - t|\xi|^2} \, d\xi$$

and

$$\theta_0 := \int_{\mathbb{R}^n} (\varphi(x) + \psi(x)) \, dx + \int_0^\infty \int_{\mathbb{R}^n} f(t, x) \, d(t, x).$$

One may proceed as follows:

1. Following the notations of Sect. 18.1.1.2, by using Duhamel's principle the solution is given by

$$u(t, x) = u_0(t, x) + \int_0^t K_1(t - s, 0, x) *_{(x)} f(s, x) \, ds,$$

where

$$u_0(t, x) = K_0(t, 0, x) *_{(x)} \varphi(x) + K_1(t, 0, x) *_{(x)} \psi(x).$$

By using the derived estimates in Exercise 8 of Chap. 14 for

$$(G_j(t, 0, x) - G(t, 0, x)) *_{(x)} h(x), \quad j = 0, 1,$$

conclude that

$$\left\| u_0(t, \cdot) - G(t, 0, \cdot) *_{(x)} (\varphi(\cdot) + \psi(\cdot)) \right\|_{L^2}$$
$$\leq C(1 + t)^{-\frac{n}{4} - 1} \left(\|\varphi\|_{H^1 \cap L^1} + \|\psi\|_{L^2 \cap L^1} \right).$$

2. For $h \in L^1(\mathbb{R}^n)$ verify that

$$\left\| G(t, 0, \cdot) * h(\cdot) - \left(\int_{\mathbb{R}^n} h(x) \, dx \right) G(t, 0, \cdot) \right\|_{L^2} = o(t^{-\frac{n}{4}}) \text{ as } t \to \infty$$

(for a more general version, see [44]). Then conclude that

$$\left\| u_0(t, \cdot) - \left(\int_{\mathbb{R}^n} (\varphi(\cdot) + \psi(\cdot)) \, dx \right) G(t, 0, \cdot) \right\|_{L^2} = o(t^{-\frac{n}{4}}) \text{ as } t \to \infty.$$

3. Under the assumptions imposed on f we have

$$\left\| \int_0^t K_1(t - s, 0, \cdot) *_{(x)} f(s, \cdot) \, ds - \left(\int_0^\infty \int_{\mathbb{R}^n} f(t, x) \, d(t, x) \right) G(t, 0, \cdot) \right\|_{L^2}$$
$$= o(t^{-\frac{n}{4}}) \text{ as } t \to \infty.$$

Indeed, to derive the last estimate it is convenient to split the integral $\int_0^t$ into $\int_0^{t/2} + \int_{t/2}^t$ and the analysis for small and large frequencies as well.

- Thanks to the previous research project in Sect. 23.7, we have

$$\left\| \int_0^t \left(K_1(t - s, 0, \cdot) - G(t - s, 0, \cdot) \right) *_{(x)} f(s, \cdot) \, ds \right\|_{L^2} = o(t^{-\frac{n}{4}}) \text{ for } t \to \infty.$$

- Using $\|f(t, \cdot)\|_{L^1} \leq C_2(1 + t)^{-1-\varepsilon}$ verify

$$\left\| \left(\int_{\frac{t}{2}}^\infty \int_{\mathbb{R}^n} f(t, x) \, d(t, x) \right) G(t, 0, \cdot) \right\|_{L^2} = o(t^{-\frac{n}{4}}) \text{ for } t \to \infty.$$

- Summing up we have

$$\left\| \int_0^t K_1(t - s, 0, \cdot) *_{(x)} f(s, \cdot) \, ds - \left(\int_0^\infty \int_{\mathbb{R}^n} f(t, x) \, d(t, x) \right) G(t, 0, \cdot) \right\|_{L^2}$$
$$\leq \left\| \int_0^t \left(K_1(t - s, 0, \cdot) - G(t - s, 0, \cdot) \right) *_{(x)} f(s, \cdot) \, ds \right\|_{L^2}$$
$$+ \left\| \int_0^t G(t - s, 0, \cdot) *_{(x)} f(s, \cdot) \, ds - \left(\int_0^\infty \int_{\mathbb{R}^n} f(t, x) \, d(t, x) \right) G(t, 0, \cdot) \right\|_{L^2}$$
$$\leq o(t^{-\frac{n}{4}}) + \left\| \int_0^{\frac{t}{2}} G(t - s, 0, \cdot) *_{(x)} f(s, \cdot) \, ds - \left(\int_0^{\frac{t}{2}} \int_{\mathbb{R}^n} f(t, x) \, d(t, x) \right) G(t, 0, \cdot) \right\|_{L^2}.$$

The project is completed after showing that

$$\left\| \int_0^{\frac{t}{2}} G(t-s,0,\cdot) *_{(x)} f(s,\cdot)\, ds - \left(\int_0^{\frac{t}{2}} \int_{\mathbb{R}^n} f(t,x)\, dx\, dt \right) G(t,0,\cdot) \right\|_{L^2}$$

$$= \left\| \int_0^{\frac{t}{2}} \int_{\mathbb{R}^n} \big(G(t-s,0,\cdot-y) - G(t,0,\cdot) \big) f(s,y)\, dy\, ds \right\|_{L^2} = o(t^{-\frac{n}{4}}).$$

For further details and for $L^1 - L^p$ estimates with $p \geq 2$ see [103].

23.9 $L^p - L^q$ Estimates for Solutions to Structurally Damped σ-Evolution Models

Consider the Cauchy problem for the σ-evolution equation with a structural damping term

$$u_{tt} + (-\Delta)^\sigma u + 2\mu(-\Delta)^\delta u_t = 0, \quad u(0,x) = \varphi(x), \quad u_t(0,x) = \psi(x),$$

where $\mu > 0$ is a constant and $\delta \in (0, \sigma)$. The goal of this project is to derive suitable $(L^{q_1} \cap L^{q_2}) - L^{q_2}$ estimates for the solutions with $1 \leq q_1 \leq q_2 \leq \infty$. By this notation we mean that the low frequency part of the L^{q_2}-norm of the solution or its derivatives is estimated by the L^{q_1}-norm of the data, whereas its high frequency part is estimated by the L^{q_2}-norm of the data.

For $\sigma \neq 2\delta$ it is not restrictive to assume $\mu = 1$ after a suitable change of variables. Using the Fourier transform we may write

$$\widehat{u}_{tt} + |\xi|^{2\sigma}\widehat{u} + 2|\xi|^{2\delta}\widehat{u}_t = 0, \quad \widehat{u}(0,\xi) = \widehat{\varphi}(\xi), \quad \widehat{u}_t(0,\xi) = \widehat{\psi}(\xi).$$

The roots $\lambda_\pm = \lambda_\pm(|\xi|)$ of the full symbol of the differential operator are radial symmetric, have nonpositive real parts and are given by

$$\lambda_\pm(|\xi|) = \begin{cases} \left(-1 \pm \sqrt{1 - |\xi|^{2(\sigma - 2\delta)}} \right) |\xi|^{2\delta} & \text{if } |\xi|^{\sigma - 2\delta} < 1, \\[2pt] \qquad\qquad \text{the roots are real-valued,} \\[6pt] \left(-1 \pm i\sqrt{|\xi|^{2(\sigma - 2\delta)} - 1} \right) |\xi|^{2\delta} & \text{if } |\xi|^{\sigma - 2\delta} > 1, \\[2pt] \qquad\qquad \text{the roots are complex-valued.} \end{cases}$$

Therefore, we may write

$$u(t,x) = K_0(t,0,x) *_{(x)} \varphi(x) + K_1(t,0,x) *_{(x)} \psi(x),$$

where

$$\widehat{K}_0(t,0,\xi) = \frac{\lambda_+(|\xi|)e^{\lambda_-(|\xi|)t} - \lambda_-(|\xi|)e^{\lambda_+(|\xi|)t}}{\lambda_+(|\xi|) - \lambda_-(|\xi|)},$$

$$\widehat{K}_1(t,0,\xi) = \frac{e^{\lambda_+(|\xi|)t} - e^{\lambda_-(|\xi|)t}}{\lambda_+(|\xi|) - \lambda_-(|\xi|)}.$$

Long time decay estimates for solutions have been deeply investigated in recent papers. It has been shown that a different asymptotical behavior of solutions or their derivatives appears if $\delta \in (0,\frac{\sigma}{2})$ or $\delta \in (\frac{\sigma}{2},\sigma]$. In the special case $\sigma = 2\delta$, the analysis of the asymptotical behavior depends on the size of the constant $\mu > 0$, too (see Sect. 23.11).

In the case of wave equations ($\sigma = 1$) with a structural damping term, by assuming additional L^1 regularity of the data, energy estimates and estimates for the L^2-norm of the solution have been derived in [22, 37, 92], whereas estimates for the L^1-norm of the solution have been derived in [146]. In the case $\delta \in (0,\frac{1}{2})$, the *diffusion phenomenon* proved in [103] has been extended to the general $L^p - L^q$ setting, $1 \le p \le q \le \infty$, in [32]. For wave equations with viscoelastic damping, i.e., $\sigma = 1$ and $\delta = 1$, the asymptotical behavior of solutions has been investigated in [184].

For σ-evolution equations with a time-dependent structural damping of the form $2b(t)(-\Delta)^\delta u_t$, energy estimates and $L^p - L^q$ estimates on the conjugate line have been derived in [101]. Moreover, in [33] the authors proposed a classification of the damping term, which clarifies whether the solution behaves like the solution to an anomalous diffusion problem.

In this project, the goal is to derive for $\delta \in (0,\sigma)$, $L^p - L^q$ estimates for the solutions and their partial derivatives not necessarily on the conjugate line. One may proceed as follows:

1. Study $L^p - L^q$ estimates derived in [164] for the scale invariant case $\sigma = 2\delta$.
2. Study the $L^p - L^q$ estimates derived in [34] for $2\delta \in (0,\sigma)$. For large frequencies the authors apply a multiplier theorem from [136]. Hence, they derive $L^q - L^q$ estimates only for $q \in (1,\infty)$. Try to develop new strategies to derive $L^1 - L^1$ estimates for high frequencies for $\sigma \neq 1$.
3. Consider the case $2\delta \in (\sigma, 2\sigma)$.

23.10   Semilinear Heat Models with Source Power Nonlinearities

We proposed in Theorem 17.1.6 a first result for solutions to the Cauchy problem

$$u_t - \Delta u = |u|^{p-1}u, \quad u(0,x) = \varphi(x).$$

In this research project we ask to prove well-posedness results for this Cauchy problem (see also the paper [152]).

First of all, we deal with the problem of formally transferring the result of Theorem 18.1.1 to the above Cauchy problem. It seems difficult to prove the following conjecture, why?

Conjecture 1 *Let $p \in (p_{Fuj}(n), \frac{n}{n-2}]$ and $\varphi \in L^1(\mathbb{R}^n) \cap L^2(\mathbb{R}^n)$. Then the following statement holds with a suitable constant $\varepsilon_0 > 0$: if $\|\varphi\|_{L^1 \cap L^2} \leq \varepsilon_0$, then there exists a uniquely determined global (in time) energy solution u in $C([0, \infty), L^2(\mathbb{R}^n))$. Moreover, there exists a constant $C > 0$ such that the solution satisfies the decay estimate*

$$\|u(t, \cdot)\|_{L^2} \leq C (1 + t)^{-\frac{n}{4}} \|\varphi\|_{L^1 \cap L^2}.$$

An answer to the above question depends heavily on different results on $L^p - L^q$ decay estimates for solutions to the Cauchy problem for the linear heat and classical damped wave equation (cf. with Theorems 12.1.3 and 14.2.4). A correct answer at to the special role of L^∞ in the next conjecture (see Theorem 17.1.6).

Conjecture 2 *Let $p > p_{Fuj}(n) = 1 + \frac{2}{n}$ and $\varphi \in L^1(\mathbb{R}^n) \cap L^\infty(\mathbb{R}^n)$. Then the following statement holds with a suitable constant $\varepsilon_0 > 0$: if $\|\varphi\|_{L^1 \cap L^\infty} \leq \varepsilon_0$, then there exists a uniquely determined global (in time) energy solution u in $C([0, \infty), L^2(\mathbb{R}^n) \cap L^\infty(\mathbb{R}^n))$. Moreover, there exists a constant $C > 0$ such that the solution satisfies the decay estimates*

$$\|u(t, \cdot)\|_{L^2} \leq C (1 + t)^{-\frac{n}{4}} \|\varphi\|_{L^1 \cap L^\infty},$$
$$\|u(t, \cdot)\|_{L^\infty} \leq C (1 + t)^{-\frac{n}{2}} \|\varphi\|_{L^1 \cap L^\infty}.$$

Prove this conjecture by following the ideas of the approach from Sect. 18.1.1.2. Explain differences in the approach to studying semilinear classical damped waves with power nonlinearity.

Finally, we would like to include regularity in the data φ. We suppose that φ belongs to $H^r(\mathbb{R}^n) \cap L^1(\mathbb{R}^n) \cap L^\infty(\mathbb{R}^n)$ for $r > 0$. So, we propose a third conjecture.

Conjecture 3 *Let $p > p_{Fuj}(n)$ and $\varphi \in H^r(\mathbb{R}^n) \cap L^1(\mathbb{R}^n) \cap L^\infty(\mathbb{R}^n)$ for $r \in (0, p)$. Then the following statement holds with a suitable constant $\varepsilon_0 > 0$: if $\|\varphi\|_{H^r \cap L^1 \cap L^\infty} \leq \varepsilon_0$, then there exists a uniquely determined global (in time) energy solution u in $C([0, \infty), H^r(\mathbb{R}^n) \cap L^\infty(\mathbb{R}^n))$. Moreover, there exists a constant $C > 0$ such that the solution satisfies the decay estimates*

$$\|u(t, \cdot)\|_{L^2} \leq C (1 + t)^{-\frac{n}{4}} \|\varphi\|_{H^r \cap L^1 \cap L^\infty},$$
$$\|u(t, \cdot)\|_{L^\infty} \leq C (1 + t)^{-\frac{n}{2}} \|\varphi\|_{H^r \cap L^1 \cap L^\infty},$$
$$\|u(t, \cdot)\|_{\dot{H}^r} \leq C (1 + t)^{-\frac{n}{4} - \frac{r}{2}} \|\varphi\|_{H^r \cap L^1 \cap L^\infty}.$$

Prove this conjecture by following the ideas of the approach from Sect. 18.1.1.2. Understand, how to treat the nonlinear term by using Corollary 24.5.2.
How to rewrite Conjecture 3 if $r > \frac{n}{2}$ or if $r > \frac{n}{2} + 2$?

23.11   Semilinear Structurally Damped σ-Evolution Equations

We are interested in the Cauchy problem for structurally damped σ-evolution equations ($\sigma \geq 1$)

$$u_{tt} + (-\Delta)^{\sigma} u + \mu(-\Delta)^{\frac{\sigma}{2}} u_t = |u|^p, \quad u(0,x) = \varphi(x), \quad u_t(0,x) = \psi(x),$$

where $\mu > 0$. Our goal is to determine the critical exponent $p_{crit} = p_{crit}(n)$. One can follow the approach of Sect. 18.1.1.2. We proceed as follows:

1. Derive linear $L^p - L^q$ decay estimates of solutions or their partial derivatives not necessarily on the conjugate line. Here one can apply the results and techniques of Sects. 19.2 and 24.2.2.
2. Prove the global existence of small data solutions for an admissible range of p by using the approach of Sect. 18.1.1.
3. Study the method of test functions (see Sect. 18.1.2) and prove at least in special cases for σ that the critical exponent is sharp. So, one has to prove blow up of weak solutions for, in general, (even) small data and for $p \leq p_{crit}$. One can utilize the papers [40] or [35].

One can compare their own considerations with those of [37] and [34].

23.12   Semilinear Structurally Damped Wave Equations

We are interested in the Cauchy problem for structurally damped wave equations

$$u_{tt} - \Delta u + \mu(-\Delta)^{\frac{1}{2}} u_t = ||D|u|^p, \quad u(0,x) = \varphi(x), \quad u_t(0,x) = \psi(x),$$

where $\mu > 0$. This is the model of Sect. 19.3 with $a = 1$. After a careful reading of Theorem 19.3.1 it is clear that this theorem is not applicable to the above Cauchy problem. Try to understand which steps of the proof to Theorem 19.3.1 can not be generalized to the case $a = 1$. But, if we take account of the statements of Corollary 19.2.5, in particular, that we have the same estimates for $|D|u(t, \cdot)$ and $u_t(t, \cdot)$, then we might have the idea that it is reasonable to prove a corresponding result to Theorem 19.4.1. For estimating the term $||D|u|^p$ we use Proposition 24.5.6 for fractional powers. On the other hand, we suppose data (φ, ψ) belong to the function space

$$\left(H_m^s(\mathbb{R}^n) \cap L^1(\mathbb{R}^n)\right) \times \left(H_m^{s-1}(\mathbb{R}^n) \cap L^1(\mathbb{R}^n)\right)$$

with a suitably "large" regularity s. Formulate and prove a result for global existence (in time) of small data solutions to the above Cauchy problem with sufficiently regular data.

23.13 Scale-Invariant σ-Evolution Models with Mass, Dissipation and a Power Nonlinearity

We restricted our considerations up to now to semilinear models with constant coefficients in the linear part. In this project we will discuss the following Cauchy problem:

$$u_{tt} + (-\Delta)^\sigma u + \frac{\mu_1}{1+t}u_t + \frac{\mu_2^2}{(1+t)^2}u = \||D|^a u|^r,$$

$$u(0,x) = \varphi(x), \quad u_t(0,x) = \psi(x),$$

where $a \in [0, \sigma)$. We have time-dependent coefficients making the analysis more difficult, but the linear operator of the left-hand side is scale-invariant. This means, if $u = u(t,x)$ is a solution of the partial differential equation

$$u_{tt} + (-\Delta)^\sigma u + \frac{\mu_1}{1+t}u_t + \frac{\mu_2^2}{(1+t)^2}u = 0,$$

then

$$\tilde{u}(t,x) := u\big(\lambda(1+t) - 1, \lambda^{\frac{1}{\sigma}}x\big)$$

is for all $\lambda > 0$ a solution of the same partial differential equation, too. The fact that this model is scale-invariant makes it possible for us to derive explicit representations of solutions to the corresponding linear Cauchy problem in terms of known special functions. We divide our considerations into two cases.

Case 1: $(\mu_1 - 1)^2 - 4\mu_2^2 < 1$: In this first case the mass term is dominant. Consequently, we expect that the above equation behaves like the following scale-invariant Klein-Gordon type equation:

$$v_{tt} + (-\Delta)^\sigma v + \frac{\mu}{(1+t)^2}v = 0.$$

Indeed, after a dissipative transformation we see that the solution of the linear Cauchy problem is related to a solution of the equation

$$w_{tt} + (-\Delta)^\sigma w + \frac{\tilde{\mu}}{(1+t)^2}w = 0,$$

where $\widetilde{\mu} = \frac{\mu_1}{2} - \frac{\mu_1^2}{4} + \mu_2^2$. Then, we can proceed as in [10] or [156]. Thus using confluent hypergeometric functions we are able to describe the behavior of the solutions of the previous equation and, consequently, the behavior of the solution of our given equation, but only under the assumption $\widetilde{\mu} > 0$. Let us observe that this condition is exactly the restriction given for μ_1 and μ_2 in this case. If $(\mu_1 - 1)^2 - 4\mu_2^2 < 0$, $(\mu_1 - 1)^2 - 4\mu_2^2 = 0$, $(\mu_1 - 1)^2 - 4\mu_2^2 \in (0, 1)$, then $\widetilde{\mu} > \frac{1}{4}$, $\widetilde{\mu} = \frac{1}{4}$, $\widetilde{\mu} \in (0, \frac{1}{4})$, respectively. The latter case implies a Klein-Gordon model with non-effective mass (see [10] or [156]). If $(\mu_1 - 1)^2 - 4\mu_2^2 = 1$, then $\widetilde{\mu} = 0$, so, the change of variables leads to the free σ-evolution equation

$$w_{tt} + (-\Delta)^\sigma w = 0.$$

Case 2: $(\mu_1-1)^2-4\mu_2^2 \geq 1$: In this second case the dissipation term is dominant. As in [221], we transform the study of the corresponding linear Cauchy problem to the study of a Bessel equation. More precisely, we reduce the considerations to those for the equation

$$w_{tt} + (-\Delta)^\sigma w + \frac{\mu}{1+t} w_t = 0,$$

where $\mu = 1 + \sqrt{(\mu_1 - 1)^2 - 4\mu_2^2} \geq 2$. If $(\mu_1 - 1)^2 - 4\mu_2^2 \geq 1$, then $\mu \geq 2$. As in [39], we are able to transform this equation in the case $\mu = 2$ to the free σ-evolution equation

$$w_{tt} + (-\Delta)^\sigma w = 0.$$

If we are interested in using estimates for solutions to the linear Cauchy problem to study the semilinear case with power nonlinearity, then we also need to derive estimates for solutions to the family of linear Cauchy problems depending on a parameter s:

$$v_{tt} + (-\Delta)^\sigma v + \frac{\mu_1}{1+t} v_t + \frac{\mu_2^2}{(1+t)^2} v = 0, \quad t \geq s,$$

$$v(s, x) = 0, \quad v_t(s, x) = v_1(x),$$

in order to use Duhamel's principle. Here we have to take into consideration that our scale-invariant linear σ-evolution equation is not invariant by time-translation.

Our research project is related to the case where the dissipation term is dominant (Case 2). Let us propose the following project.

1. How do you define the energy of solutions in both cases (mass is dominant or dissipation is dominant)?
2. Derive energy estimates in both cases.

3. Try to clarify, in both cases, if an additional regularity of the data (let us say the data (φ, ψ) belong to the function space

$$\left(H^\sigma(\mathbb{R}^n) \cap L^1(\mathbb{R}^n)\right) \times \left(L^2(\mathbb{R}^n) \cap L^1(\mathbb{R}^n)\right)$$

allows us to prove energy estimates with a better decay.

4. Derive estimates for solutions to the family of parameter-dependent Cauchy problems

$$v_{tt} + (-\Delta)^\sigma v + \frac{\mu}{1+t} v_t = 0, \quad t \geq s,$$

$$v(s,x) = 0, \quad v_t(s,x) = v_1(x),$$

where $\mu > 0$.

5. Prove the global (in time) existence of small data solutions to the Cauchy problem

$$u_{tt} + (-\Delta)^\sigma u + \frac{\mu_1}{1+t} u_t + \frac{\mu_2^2}{(1+t)^2} u = ||D|^a u|^r,$$

$$u(0,x) = \varphi(x), \quad u_t(0,x) = \psi(x),$$

$a \in [0, \sigma)$. Use the explanations provided in Sect. 19.3.

For more information we refer to [158] and [157].

Chapter 24
Background Material

In this chapter we gather together some background material such as "Basics of Fourier Transformation", some aspects of the "Theory of Fourier Multipliers", some "Function Spaces", "Some tools from distribution theory" and "Useful Inequalities". There is no attempt mode to present these sections in a self-contained form. Readers are encouraged to study these topics more in detail by utilizing the related literature.

24.1 Basics of Fourier Transformation

The Fourier transformation is a special integral transformation. Usually it is defined by (classical definition)

$$F(f)(\xi) := \frac{1}{(2\pi)^{\frac{n}{2}}} \int_{\mathbb{R}^n} e^{-ix\cdot\xi} f(x)\, dx$$

with $x \cdot \xi = \sum_{l=1}^{n} x_l \xi_l$. The inverse Fourier transformation is defined by (classical definition)

$$F^{-1}(g)(x) := \frac{1}{(2\pi)^{\frac{n}{2}}} \int_{\mathbb{R}^n} e^{ix\cdot\xi} g(\xi)\, d\xi.$$

© Springer International Publishing AG 2018
M.R. Ebert, M. Reissig, *Methods for Partial Differential Equations*,
https://doi.org/10.1007/978-3-319-66456-9_24

24.1.1 Application to Spaces of Infinitely Differentiable Functions

Let us choose the function space $C^\infty(\mathbb{R}^n)$ (see Definition 24.16). Functions f from this space could become unbounded for $|x| \to \infty$, for example, $f(x) = e^{|x|^4}$. Such unbounded behavior does, in general, not imply the convergence of the above integrals for $F(f)$ or for $F^{-1}(g)$. For this reason, let us choose the function space $C_0^\infty(\mathbb{R}^n)$ (see Definition 24.18). Then, elements f have a compact support. Consequently, the above integrals for $F(f)$ or for $F^{-1}(g)$ exist. We can not expect that $F(f)$ or $F^{-1}(g)$ belong to $C_0^\infty(\mathbb{R}^n)$ if nontrivial f or g are taken from this space. Why?

However, there does exist a kind of "intermediate function space" between $C_0^\infty(\mathbb{R}^n)$ and $C^\infty(\mathbb{R}^n)$ which has the property, assuming f and g to belong to this function space, that implies that $F(f)$ and $F^{-1}(g)$ belong to the same function space. This is the so-called *Schwartz space* $S(\mathbb{R}^n)$, the *space of fast decreasing functions*.

Definition 24.1 By $S(\mathbb{R}^n)$ we denote the subspace of $C^\infty(\mathbb{R}^n)$ consisting of all functions f which satisfy the conditions

$$p_{\alpha,\beta}(f) = \sup_{x \in \mathbb{R}^n} \left| x^\beta \partial_x^\alpha f(x) \right| < \infty$$

for all multi-indices α and β. The topology in $S(\mathbb{R}^n)$ is generated by the family of semi-norms $\{p_{\alpha,\beta}(f)\}_{\alpha,\beta}$.

The Schwartz space is the largest subspace of $L^1(\mathbb{R}^n)$ which is invariant with respect to the operations differentiation ∂_x^α and multiplication by x^β.

Theorem 24.1.1 *The Fourier transformation and the inverse Fourier transformation map continuously the Schwartz space into itself. The Fourier transform of $\partial_{x_k} f$ is $i\xi_k F(f)$ and the Fourier transform of $x_k f$ is $i\partial_{\xi_k} F(f)$. In this way, a differentiation in the physical space, that is in the space $\mathbb{R}_x^n$, corresponds to a multiplication by the phase space variable in the phase space, that is in the space $\mathbb{R}_\xi^n$, modulo a pure imaginary factor and conversely.*

Proof We restrict ourselves to showing that for a given f from $S(\mathbb{R}^n)$ the image $F(f)$ belongs to $S(\mathbb{R}^n)$, too. Straight forward calculations imply

$$\xi^\beta \partial_\xi^\alpha F(f)(\xi) = \frac{1}{(2\pi)^{\frac{n}{2}}} \int_{\mathbb{R}^n} e^{-ix\cdot\xi} \xi^\beta (-ix)^\alpha f(x)\, dx$$

$$= \frac{1}{(2\pi)^{\frac{n}{2}}} \int_{\mathbb{R}^n} i^{|\beta|} \partial_x^\beta (e^{-ix\cdot\xi} (-ix)^\alpha f(x)\, dx$$

$$= \frac{1}{(2\pi)^{\frac{n}{2}}} \int_{\mathbb{R}^n} e^{-ix\cdot\xi}(-i)^{|\beta|}\partial_x^\beta\left((-ix)^\alpha f(x)\right) dx$$

$$= \frac{1}{(2\pi)^{\frac{n}{2}}} \int_{\mathbb{R}^n} e^{-ix\cdot\xi}(1+|x|^2)^{-\frac{n+1}{2}}(-i)^{|\beta|}(1+|x|^2)^{\frac{n+1}{2}}\partial_x^\beta\left((-ix)^\alpha f(x)\right) dx.$$

Here the assumption $f \in S(\mathbb{R}^n)$ yields that during partial integration in the above improper integrals boundary integrals (during the process of passaging to the limit) are vanishing. Moreover, we have

$$\sup_{x\in\mathbb{R}^n}\left|(1+|x|^2)^{\frac{n+1}{2}}\partial_x^\beta\left((-ix)^\alpha f(x)\right)\right| < \infty.$$

Taking account of

$$\int_{\mathbb{R}^n}(1+|x|^2)^{-\frac{n+1}{2}}dx < \infty$$

gives in the phase space the estimate

$$p_{\alpha,\beta}(F(f)) = \sup_{\xi\in\mathbb{R}^n}\left|\xi^\beta\partial_\xi^\alpha F(f)(\xi)\right| < \infty$$

for all multi-indices α and β. In the same way we can show that $p_{\alpha,\beta}(f_k-f) \to 0$ for $k \to \infty$ implies $p_{\alpha,\beta}(F(f_k)-F(f)) \to 0$ for all multi-indices α and β. Consequently, $F: f \to F(f)$ maps the space $S(\mathbb{R}^n)$ continuously into itself. The rules

$$F(\partial_{x_k}f) = i\xi_k F(f) \quad \text{and} \quad F(x_k f) = i\partial_{\xi_k}F(f)$$

are proved in the same way.

Intuitively, we would expect the following *Fourier inversion formula*,

$$F^{-1}(F(f))(x) = f(x).$$

This inversion formula holds for all functions $f \in S(\mathbb{R}^n)$. For proving this statement we use so-called *regularizations*. For a given function

$$g \in C_0^\infty(\mathbb{R}^n), \quad g(x) \geq 0, \quad \int_{\mathbb{R}^n} g(x)\,dx = 1,$$

we define the function

$$g_\varepsilon = \varepsilon^{-n}g\left(\frac{x}{\varepsilon}\right) \quad \text{for } \varepsilon > 0.$$

Let $f \in L^p(\mathbb{R}^n)$, $p \in [1, \infty)$. Then, we define the regularization $J_\varepsilon(f)$ of f by the aid of the convolution integral

$$J_\varepsilon(f) := g_\varepsilon * f := \int_{\mathbb{R}^n} g_\varepsilon(x - y)f(y)\, dy.$$

For every $\varepsilon > 0$ the regularization $J_\varepsilon(f)$ belongs to $C^\infty(\mathbb{R}^n)$. Such a regularization satisfies the following remarkable property:

$$\lim_{\varepsilon \to +0} \|J_\varepsilon(f) - f\|_{L^p} = \lim_{\varepsilon \to +0} \|g_\varepsilon * f - f\|_{L^p} = 0.$$

Remark 24.1.1 The assumption $g \in C_0^\infty(\mathbb{R}^n)$ implies $J_\varepsilon(f) \in C^\infty(\mathbb{R}^n)$. But, $J_\varepsilon(g) \in C^\infty(\mathbb{R}^n)$ holds even for all functions $g \in L^1(\mathbb{R}^n)$.

By Remark 24.1.1 we are able to prove the Fourier inversion formula for all functions $f \in S(\mathbb{R}^n)$. We introduce a function $\chi \in C_0^\infty(\mathbb{R}^n)$ with $\chi(\eta) = 1$ for $|\eta| \leq 1$ and $\chi(\eta) = 0$ for $|\eta| \geq 2$. Then, we have

$$F^{-1}(F(f))(x) = \int_{\mathbb{R}^n} \left(\int_{\mathbb{R}^n} \frac{1}{(2\pi)^n} e^{i(x-y)\cdot\xi} f(y)\, dy \right) d\xi$$

$$= \lim_{\varepsilon \to +0} \int_{\mathbb{R}^n} \left(\int_{\mathbb{R}^n} \frac{1}{(2\pi)^n} e^{i(x-y)\cdot\xi} f(y) \chi(\varepsilon\xi)\, dy \right) d\xi$$

$$= \lim_{\varepsilon \to +0} \int_{\mathbb{R}^n} f(y)\varepsilon^{-n} \left(\int_{\mathbb{R}^n} \frac{1}{(2\pi)^n} e^{i\frac{x-y}{\varepsilon}\cdot\eta} \chi(\eta)\, d\eta \right) dy = \lim_{\varepsilon \to +0} f * g_\varepsilon(x)$$

with the function

$$g = g(x) = \frac{1}{(2\pi)^n} \int_{\mathbb{R}^n} e^{ix\cdot\eta} \chi(\eta)\, d\eta \in L^1(\mathbb{R}^n).$$

By Remark 24.1.1 we conclude the *Fourier inversion formula*

$$F^{-1}(F(f)) = f \quad \text{for all } f \in S(\mathbb{R}^n).$$

Example 24.1.1 (Important for $L^p - L^q$ estimates)

In applications appear Fourier transforms of Gauß' functions. Let $(Ax, x) = \sum_{k,l=1}^n a_{kl}x_k x_l$ be a positive definite quadratic form. Then we have

$$F\left(e^{-(Ax,x)}\right) = \frac{1}{2^{\frac{n}{2}}\sqrt{\det A}} e^{-\frac{1}{4}(\xi, A^{-1}\xi)}.$$

Let us choose $A = \frac{1}{2}I$. Then

$$F\left(e^{-\frac{|x|^2}{2}}\right) = \frac{1}{2^{\frac{n}{2}}\left(\frac{1}{2}\right)^{\frac{n}{2}}} e^{-\frac{|\xi|^2}{2}} = e^{-\frac{|\xi|^2}{2}}.$$

So, the density of the standardized normal distribution remains invariant under application of the Fourier transformation.

24.1.2   Application to L^p Spaces

This section is devoted to the Fourier transformation in L^p spaces for $p \in [1, \infty)$ (see Definition 24.6). We recall the classical definition

$$F(f)(\xi) := \frac{1}{(2\pi)^{\frac{n}{2}}} \int_{\mathbb{R}^n} e^{-ix\cdot\xi} f(x)\, dx.$$

Let $f \in L^1(\mathbb{R}^n)$. Then, $F(f)$ belongs to $L^\infty(\mathbb{R}^n)$. Moreover, $F(f)$ is continuous on $\mathbb{R}^n$ and $\lim_{\xi \to \infty} F(f)(\xi) = 0$ (see [42]). Moreover, it holds the following *convolution result* for $f, g \in L^1(\mathbb{R}^n)$:

$$F(f * g)(\xi) = (2\pi)^{\frac{n}{2}} F(f)(\xi) F(g)(\xi).$$

Here we use the property of $L^1(\mathbb{R}^n)$ to be a Banach algebra with the operation of convolution.

Let $f \in L^2(\mathbb{R}^n)$. Then, the classical definition for $F(f)$ is not applicable any more. Let us explain a suitable definition for $F(f)$ if f belongs to $L^2(\mathbb{R}^n)$. For this reason we choose functions $f, g \in S(\mathbb{R}^n)$. Then, the classical definitions for $F^{-1}(f)$ and for $F(f)$ are applicable. We obtain the relations

$$\int_{\mathbb{R}^n} F^{-1}(f)(x)\overline{g(x)}\, dx = \int_{\mathbb{R}^n} f(\xi)\overline{F(g)(\xi)}\, d\xi,$$

$$\int_{\mathbb{R}^n} F(f)(x)\overline{g(x)}\, dx = \int_{\mathbb{R}^n} f(\xi)\overline{F^{-1}(g)(\xi)}\, d\xi.$$

The first relation follows from

$$\int_{\mathbb{R}^n} F^{-1}(f)(x)\overline{g(x)}\, dx = \int_{\mathbb{R}^n} \frac{1}{(2\pi)^{\frac{n}{2}}} \left(\int_{\mathbb{R}^n} e^{ix\cdot\xi} f(\xi)\, d\xi\right)\overline{g(x)}\, dx$$

$$= \int_{\mathbb{R}^n} f(\xi)\frac{1}{(2\pi)^{\frac{n}{2}}}\left(\int_{\mathbb{R}^n} e^{ix\cdot\xi}\overline{g(x)}\, dx\right)d\xi$$

$$= \int_{\mathbb{R}^n} f(\xi)\frac{1}{(2\pi)^{\frac{n}{2}}}\left(\int_{\mathbb{R}^n} \overline{e^{-ix\cdot\xi}g(x)}\, dx\right)d\xi = \int_{\mathbb{R}^n} f(\xi)\overline{F(g)}(\xi)\, d\xi$$

if we use the classical definition for $F^{-1}(f)(x)$ and change the order of integration. So, the first relation can be written as

$$(F^{-1}(f), g)_{L^2} = (f, F(g))_{L^2} \text{ for all } f, g \in S(\mathbb{R}^n).$$

Now let us choose $f \in L^2(\mathbb{R}^n)$. Then, the scalar product $(f, F(g))_{L^2}$ is defined for all $g \in S(\mathbb{R}^n)$. Taking account of the density of $C_0^\infty(\mathbb{R}^n) \subset S(\mathbb{R}^n)$ in $L^2(\mathbb{R}^n)$, the identity

$$(w, g)_{L^2} = (f, F(g))_{L^2}$$

defines a functional on $L^2(\mathbb{R}^n)$. By Riesz theorem about representation of functionals in Hilbert spaces there exists a uniquely determined $w \in L^2(\mathbb{R}^n)$ such that the last relation is fulfilled for all $g \in S(\mathbb{R}^n)$. We define this function w as the inverse Fourier transform $F^{-1}(f)$ of $f \in L^2(\mathbb{R}^n)$. Using the second relation, by a similar reasoning we are able to define the Fourier transform $F(f) \in L^2(\mathbb{R}^n)$ for a given function $f \in L^2(\mathbb{R}^n)$. Summarizing, we explained $F(f)$ and $F^{-1}(f)$ for a given function $f \in L^2(\mathbb{R}^n)$ by the aid of the relations

$$(F^{-1}(f), g)_{L^2} = (f, F(g))_{L^2} \text{ for all } g \in S(\mathbb{R}^n),$$
$$\cdot (F(f), g)_{L^2} = (f, F^{-1}(g))_{L^2} \text{ for all } g \in S(\mathbb{R}^n).$$

Theorem 24.1.2 *The Fourier transformation is a unitary operator on $L^2(\mathbb{R}^n)$.*

Proof By the above relations and the Fourier inversion formula for $g \in S(\mathbb{R}^n)$ it holds that

$$(F^{-1}(F(f)), g)_{L^2} = (F(f), F(g))_{L^2} = (f, F^{-1}(F(g)))_{L^2} = (f, g)_{L^2}.$$

The density of $S(\mathbb{R}^n)$ in $L^2(\mathbb{R}^n)$ implies immediately the Fourier inversion formula for $f \in L^2(\mathbb{R}^n)$ because it follows $F^{-1}(F(f)) = f$ from the above described identity for functionals. Consequently, F maps $L^2(\mathbb{R}^n)$ onto itself. Moreover, F is isometric. Here we use again the above relation. It follows

$$(F(f), F(g))_{L^2} = (f, g)_{L^2}$$

for all $f \in L^2(\mathbb{R}^n)$ and all $g \in S(\mathbb{R}^n)$. Applying again the density argument gives this relation even for all $f, g \in L^2(\mathbb{R}^n)$. We obtain, in the special case $f = g$, the relation $\|F(f)\|_{L^2}^2 = \|f\|_{L^2}^2$.

Remark 24.1.2 The formula

$$(F(f), F(g))_{L^2} = (f, g)_{L^2} \text{ for } f, g \in L^2(\mathbb{R}^n)$$

is called the *formula of Parseval-Plancherel.*

24.1.2.1 Mapping Properties of the Fourier Transformation

An argument from interpolation theory:

Up to now we know the following properties of the Fourier transformation:

$$f \in L^2(\mathbb{R}^n) \Rightarrow F(f) \in L^2(\mathbb{R}^n),$$
$$f \in L^1(\mathbb{R}^n) \Rightarrow F(f) \in L^\infty(\mathbb{R}^n).$$

By Proposition 24.5.1 we conclude, for $1 < p < 2$ and (p, q) on the conjugate line $\frac{1}{p} + \frac{1}{q} = 1$ the mapping property

$$f \in L^p(\mathbb{R}^n) \Rightarrow F(f) \in L^q(\mathbb{R}^n).$$

Moreover, the following norm estimate is true:

$$\|F(f)\|_{L^q} \leq (2\pi)^{-n(\frac{1}{p} - \frac{1}{2})} \|f\|_{L^p} \text{ for } p \in [1, 2].$$

The estimate follows for $p = 2$ by Remark 24.1.2. For $p = 1$ the estimate follows by using the classical definition of the Fourier transformation.

Remark 24.1.3 If we choose $p > 2$ and $\frac{1}{p} + \frac{1}{q} = 1$, then the statement

$$f \in L^p(\mathbb{R}^n) \Rightarrow F(f) \in L^q(\mathbb{R}^n)$$

is not true.

In the following we present some basic estimates for Fourier multipliers (see [12]).

Theorem 24.1.3 *Let us suppose a given function $g = g(x)$ belongs to $L^1(\mathbb{R}^n)$. Then the following statements are true:*

1. *If $\|F^{-1}(g)\|_{L^\infty} \leq C_0$ and $v \in L^1(\mathbb{R}^n)$, then*

$$\|F^{-1}(gF(v))\|_{L^\infty} \leq C_0 \|v\|_{L^1}.$$

2. *If $\|g\|_{L^\infty} \leq C_1$ and $v \in L^2(\mathbb{R}^n)$, then*

$$\|F^{-1}(gF(v))\|_{L^2} \leq C_1 \|v\|_{L^2}.$$

3. *If $\|F^{-1}(g)\|_{L^\infty} \leq C_0$, $\|g\|_{L^\infty} \leq C_1$ and $v \in L^1(\mathbb{R}^n) \cap L^2(\mathbb{R}^n)$, then*

$$\|F^{-1}(gF(v))\|_{L^q} \leq C_0^{2\delta} C_1^{1-2\delta} \|v\|_{L^p},$$

where $p \in [1, 2]$, $\frac{1}{p} + \frac{1}{q} = 1$ and $\delta := \frac{1}{p} - \frac{1}{2}$.

The third statement follows from the first two after applying Proposition 24.5.1.

We choose a nonnegative function $\phi = \phi(\xi)$ having compact support in $\{\xi \in \mathbb{R}^n : |\xi| \in [\frac{1}{2}, 2]\}$. We set $\phi_k(\xi) := \phi(2^{-k}\xi)$ for $k \geq 1$ and $\phi_0(\xi) := 1 - \sum_{k=1}^{\infty} \phi_k(\xi)$.

Theorem 24.1.4 *Let $m \in L^{\infty}(\mathbb{R}^n)$. With $p \in (1, 2]$ we assume on the conjugate line the estimates*

$$\|F^{-1}(m\phi_k F(\varphi))\|_{L^q} \leq C\|\varphi\|_{L^p} \text{ for all } k \geq 0,$$

where C is independent of k and φ. Then, for some constant M which is independent of m we have the estimate

$$\|F^{-1}(mF(\varphi))\|_{B^0_{q,r}} \leq MC\|\varphi\|_{B^0_{p,r}}$$

for all $r \geq 1$ (see also Definition 24.10).

Remark 24.1.4 Choosing $r = 2$ in Theorem 24.1.4 and taking into consideration $L^p(\mathbb{R}^n) \subset B^0_{p,2}(\mathbb{R}^n)$ and $B^0_{q,2}(\mathbb{R}^n) \subset L^q(\mathbb{R}^n)$ on the conjugate line with $p \in (1, 2]$, then, we may conclude from Theorem 24.1.4 the estimate

$$\|F^{-1}(mF(\varphi))\|_{L^q} \leq MC\|\varphi\|_{L^p}.$$

24.1.2.2 Fourier Inversion Formulas

Up to now we learned the Fourier inversion formula for functions belonging to $S(\mathbb{R}^n)$ and $L^2(\mathbb{R}^n)$. In the following we present some inversion formulas for other function spaces. Here we restrict ourselves to the one-dimensional case.

1. Let $f \in L^1(\mathbb{R}^1)$ be of bounded variation on every compact interval $[a, b]$ and continuous there. It then holds that

$$f(x) = \frac{1}{(2\pi)^{\frac{1}{2}}} \text{ pv} \int_{\mathbb{R}} e^{ix\xi} F(f)(\xi) \, d\xi.$$

Here the integral exists in the sense of Cauchy's principal value. Without assuming the continuity we only have

$$\frac{f(x+0) + f(x-0)}{2} = \frac{1}{(2\pi)^{\frac{1}{2}}} \text{ pv} \int_{\mathbb{R}} e^{ix\xi} F(f)(\xi) \, d\xi.$$

2. Let $f \in L^p(\mathbb{R}^1)$, $p \in [1, 2]$. It then holds that

$$f(x) = \frac{1}{(2\pi)^{\frac{1}{2}}} \lim_{\varepsilon \to +0} \int_{\mathbb{R}} e^{ix\xi} F(f)(\xi) \chi(\varepsilon\xi) \, d\xi$$

with a function $\chi \in C_0^\infty(\mathbb{R}^1)$, where $\chi(\eta) = 1$ if $|\eta| \leq 1$ and $\chi(\eta) = 0$ for $|\eta| \geq 2$.

3. Let $f \in L^p(\mathbb{R}^1)$, $p \in (1, 2]$. Then

$$f(x) = \frac{1}{(2\pi)^{\frac{1}{2}}} \text{ pv} \int_\mathbb{R} e^{ix\xi} F(f)(\xi) \, d\xi.$$

24.1.3 Application to Tempered Distributions

The dual space to $S(\mathbb{R}^n)$ is denoted by $S'(\mathbb{R}^n)$. Functionals from $S'(\mathbb{R}^n)$ are called *tempered distributions* (cf. with Definition 24.28). We already applied the theory of functionals for defining the Fourier transform and inverse Fourier transform of a function from $L^2(\mathbb{R}^n)$. In $S'(\mathbb{R}^n)$ we have no scalar product. But, we can use the representation of functionals $f(g)$ for $f \in S'(\mathbb{R}^n)$ and arbitrary $g \in S(\mathbb{R}^n)$ for defining the Fourier transform and inverse Fourier transform of a tempered distribution. Here $f(g)$ denotes the action of a tempered distribution f on a Schwartz function g. We define

$$F(f)(g) = f(F^{-1}(g)) \quad \text{and} \quad F^{-1}(f)(g) = f(F(g)).$$

For tempered distributions the Fourier inversion formula $F^{-1}(F(f)) = f$ holds. This follows from the Fourier inversion formula for Schwartz functions g to get

$$F^{-1}(F(f))(g) = F(f)(F(g)) = f(F^{-1}(F(g))) = f(g)$$

and the definition when two tempered distributions coincide. The convolution theorem holds for tempered distributions $G \in S'(\mathbb{R}^n)$, $T \in E'(\mathbb{R}^n)$: $F(G * T) = (2\pi)^{\frac{n}{2}} F(G) \, F(T)$. Here we denote by $E'(\mathbb{R}^n)$ the subspace of tempered distributions having a compact support.

Example 24.1.2 (Fourier transform of Dirac's distribution δ_{x_0})
  Dirac's distribution δ_{x_0} has compact support $\{x_0\}$, so it belongs to $E'(\mathbb{R}^n)$. Actions of δ_{x_0} to arbitrary functions from $S(\mathbb{R}^n)$ are defined by $\delta_{x_0}(f) = f(x_0)$ for all functions $f \in S(\mathbb{R}^n)$. Consequently, δ_{x_0} generates a functional on $S(\mathbb{R}^n)$. To define the Fourier transform $F(\delta_{x_0})$ we use for all functions $f \in S(\mathbb{R}^n)$ the relations

$$F(\delta_{x_0})(f) = \delta_{x_0}(F^{-1}(f)) = F^{-1}(f)(x_0) = \frac{1}{(2\pi)^{\frac{n}{2}}} \int_{\mathbb{R}^n} e^{ix_0 \cdot \xi} f(x) \, dx$$

$$= \int_{\mathbb{R}^n} \frac{e^{ix_0 \cdot \xi}}{(2\pi)^{\frac{n}{2}}} \cdot f(x) \, dx = \frac{e^{ix_0 \cdot \xi}}{(2\pi)^{\frac{n}{2}}}(f).$$

Hence, we may conclude $F(\delta_{x_0}) = \frac{e^{ix_0 \cdot \xi}}{(2\pi)^{\frac{n}{2}}}$.

Example 24.1.3 Let us determine a fundamental solution to the operator $-\Delta$ in $\mathbb{R}^3$. Such a fundamental solution is a distributional solution of $-\Delta u = \delta_0$ (cf. with Definitions 24.29 and 24.30). We apply the Fourier transformation and get $|\xi|^2 F(u) = \frac{1}{(2\pi)^{\frac{3}{2}}}$. This is our auxiliary problem in the phase space and has the solution $F(u) = \frac{1}{|\xi|^2 (2\pi)^{\frac{3}{2}}}$. The main difficulty consists in determining $F^{-1}\left(\frac{1}{|\xi|^2 (2\pi)^{\frac{3}{2}}}\right)$. By using, among other things, arguments from function theory longer calculations imply $u(x) = \frac{1}{4\pi |x|}$.

24.1.3.1   Linear Operators Generated by Tempered Distributions

In this section we are interested in translation invariant operators in $L^p = L^p(\mathbb{R}^n)$ spaces (see [84]). There one can find the following explanations:

Theorem 24.1.5 *If A is a bounded translation invariant operator from L^p to L^q, then there is a unique distribution $T \in S'(\mathbb{R}^n)$ such that*

$$Af = T * f \text{ for all } f \in S(\mathbb{R}^n).$$

Definition 24.2 By $L_p^q = L_p^q(\mathbb{R}^n)$ we denote the space of tempered distributions T satisfying the estimate

$$\|T * f\|_{L^q} \le C \|f\|_{L^p}$$

for all $f \in S(\mathbb{R}^n)$ with a constant C which is independent of f.

Definition 24.3 The set of Fourier transforms $\hat{T}$ of distributions $T \in L_p^q$ is denoted by $M_p^q = M_p^q(\mathbb{R}^n)$. The elements in M_p^q are called multipliers of type (p, q).

Theorem 24.1.6 *Let f be a measurable function. Moreover, we suppose the following relation with suitable positive constants C, $b \in (1, \infty)$ and all positive l:*

$$\text{meas } \{\xi \in \mathbb{R}^n : |f(\xi)| \ge l\} \le C l^{-b}.$$

Then, $f \in M_p^q$ if $1 < p \le 2 \le q < \infty$ and $\frac{1}{p} - \frac{1}{q} = \frac{1}{b}$.

24.1.4   Application to H^s Spaces

Let us introduce the space $H^m(\mathbb{R}^n)$, $m \in \mathbb{N}$. This is the set of functions

$$H^m(\mathbb{R}^n) = \left\{ f \in S'(\mathbb{R}^n) : \|f\|_{H^m} = \left(\int_{\mathbb{R}^n} |F(f)(\xi)|^2 (1 + |\xi|^2)^m \, d\xi \right)^{1/2} < \infty \right\}.$$

This space is equivalent by norms to the Sobolev space $W_2^m(\mathbb{R}^n)$ (see Definition 24.7). Indeed, applying Parseval-Plancherel's formula and due to the rules for the Fourier transformation we have

$$\|\partial_x^\alpha f\|_{L^2} = \|\xi^\alpha F(f)\|_{L^2} \text{ for all } |\alpha| \leq m.$$

The quantities $\sum_{|\alpha| \leq m} |\xi^\alpha|^2$ and $(1 + |\xi|^2)^m$ are comparable. Therefore,

$$\int_{\mathbb{R}^n} |F(f)(\xi)|^2 (1 + |\xi|^2)^m \, d\xi < \infty$$

if and only if

$$\sum_{|\alpha| \leq m} \|\partial_x^\alpha f\|_{L^2} < \infty.$$

Defining $H^m(\mathbb{R}^n)$ by using the behavior of the Fourier transform has an advantage. It can be generalized to all real $s \in \mathbb{R}^1$ (cf. with Definition 24.9).

Definition 24.4 By $H^s(\mathbb{R}^n)$, $s \in \mathbb{R}^1$, we define the set of tempered distributions

$$H^s(\mathbb{R}^n) = \left\{ f \in S'(\mathbb{R}^n) : \|f\|_{H^s} = \left(\int_{\mathbb{R}^n} |F(f)(\xi)|^2 (1 + |\xi|^2)^s \, d\xi \right)^{1/2} < \infty \right\}.$$

Applying Sobolev's embedding theorem, the space $H^s(\mathbb{R}^n)$ is embedded in the space $C_B^2(\mathbb{R}^n)$ if $s > \frac{n}{2} + 2$. Here $C_B^2(\mathbb{R}^n)$ denotes the space of twice continuously differentiable functions with bounded derivatives. For $s \geq 0$ all elements from $H^s(\mathbb{R}^n)$ belong to $L^2(\mathbb{R}^n)$. For $s < 0$ we have spaces of distributions. To which spaces H^s does Dirac's distribution δ_0 belong?

Definition 24.5 By $L^{2,s}(\mathbb{R}^n)$, $s \in \mathbb{R}^1$, we define the set of tempered distributions

$$L^{2,s}(\mathbb{R}^n) = \left\{ f \in S'(\mathbb{R}^n) : \|f\|_{L^{2,s}} = \left(\int_{\mathbb{R}^n} |f(x)|^2 (1 + |x|^2)^s \, dx \right)^{1/2} < \infty \right\}.$$

An important tool is the Fourier inversion formula on $H^s(\mathbb{R}^n)$, $s \in \mathbb{R}^1$.

Theorem 24.1.7 *The Fourier inversion formula holds on* $H^s(\mathbb{R}^n)$, *that is,* $F^{-1}(F(f)) = f$ *for* $f \in H^s(\mathbb{R}^n)$.

Proof If $f \in H^s(\mathbb{R}^n)$, then $\langle D \rangle^s f \in L^2(\mathbb{R}^n)$. Here $\langle D \rangle^s f$ is defined by $F^{-1}(\langle \xi \rangle^s F(f))$. Then we use the Fourier inversion formula for functions belonging to $L^2(\mathbb{R}^n)$ and get

$$F^{-1}\big(F(\langle D\rangle^s f)\big) = \langle D\rangle^s f.$$

Applying $\langle D\rangle^{-s}$ to both sides of the last identity and taking into consideration standard rules of the Fourier transformation we may conclude for $s \in \mathbb{R}^1$ and $f \in H^s(\mathbb{R}^n)$ the relations

$$f = \langle D\rangle^{-s} F^{-1}\big(F(\langle D\rangle^s f)\big) = F^{-1}\big(\langle \xi\rangle^{-s} F(\langle D\rangle^s f)\big)$$
$$= F^{-1}\big(F(\langle D\rangle^{-s}\langle D\rangle^s f)\big) = F^{-1}(F(f)).$$

Using Definitions 24.4 and 24.5 we are able to prove the following result.

Theorem 24.1.8 *The Fourier transformation maps continuously the space $H^s(\mathbb{R}^n)$ onto $L^{2,s}(\mathbb{R}^n)$ for $s \in \mathbb{R}^1$.*

The inverse Fourier transformation maps continuously the space $L^{2,s}(\mathbb{R}^n)$ onto $H^s(\mathbb{R}^n)$ for $s \in \mathbb{R}^1$.

Proof The first mapping property follows by Definition 24.4. To prove the second mapping property we use the Fourier inversion formula for $H^s(\mathbb{R}^n)$, $s \in \mathbb{R}^1$.

24.2   Theory of Fourier Multipliers

24.2.1   Modified Bessel Functions

Here we summarize some rules for modified Bessel functions.

Let $J_\mu = J_\mu(s)$ be the Bessel function of order $\mu \in (-\infty, \infty)$. Then, let us define $\tilde{J}_\mu(s) := J_\mu(s)/s^\mu$ when μ is not a negative integer. These functions are called *modified Bessel functions*.

Proposition 24.2.1 *Let $f \in L^p(\mathbb{R}^n)$, $p \in [1, 2]$, be a radial function. Then, the Fourier transform $F(f)$ is also a radial function and satisfies*

$$F(f)(\xi) = c \int_0^\infty g(r) r^{n-1} \tilde{J}_{\frac{n}{2}-1}(r|\xi|)\, dr, \quad g(|x|) := f(x).$$

Proposition 24.2.2 *Assume that μ is not a negative integer. Then, the following rules hold for the scale $\{\tilde{J}_\mu\}_\mu$ of modified Bessel functions:*

(1) $s d_s \tilde{J}_\mu(s) = \tilde{J}_{\mu-1}(s) - 2\mu \tilde{J}_\mu(s),$
(2) $d_s \tilde{J}_\mu(s) = -s \tilde{J}_{\mu+1}(s),$
(3) $\tilde{J}_{-\frac{1}{2}}(s) = \sqrt{\dfrac{2}{\pi}} \cos s,$

(4) *we have for any μ the relations*

$$|\tilde{J}_\mu(s)| \le Ce^{\pi|\text{Im}\mu|} \quad if \quad |s| \le 1,$$

$$J_\mu(s) = Cs^{-\frac{1}{2}} \cos\left(s - \frac{\mu}{2}\pi - \frac{\pi}{4}\right) + O(|s|^{-\frac{3}{2}}) \quad if \quad |s| \ge 1,$$

(5) $\tilde{J}_{\mu+1}(r|x|) = -\dfrac{1}{r|x|^2}\partial_r\tilde{J}_\mu(r|x|), \quad r \ne 0, \ x \ne 0.$

Let us only verify the third relation. We have

$$\tilde{J}_{-\frac{1}{2}}(s) = s^{\frac{1}{2}}J_{-\frac{1}{2}}(s).$$

Taking into consideration the definition of $J_{-\frac{1}{2}}(s)$ we have

$$s^{\frac{1}{2}}J_{-\frac{1}{2}}(s) = s^{\frac{1}{2}}\sum_{k=0}^{\infty}\frac{(-1)^k}{k!\Gamma(k+\frac{1}{2})}\left(\frac{s}{2}\right)^{2k-\frac{1}{2}} = \sqrt{2}\sum_{k=0}^{\infty}\frac{(-1)^k s^{2k}}{k!\Gamma(k+\frac{1}{2})2^{2k}}.$$

The definition of $\Gamma(k+\frac{1}{2})$ and $\Gamma(\frac{1}{2}) = \sqrt{\pi}$ imply

$$\tilde{J}_{-\frac{1}{2}}(s) = \sqrt{\frac{2}{\pi}}\sum_{k=0}^{\infty}\frac{(-1)^k s^{2k}}{(2k)!} = \sqrt{\frac{2}{\pi}}\cos s.$$

24.2.2 L^p Estimates for Model Oscillating Integrals

In this section we derive L^p estimates for the oscillating integral

$$F^{-1}\left(e^{-c|\xi|^{2\kappa}t}\right).$$

Our main goal is to show how the theory of modified Bessel functions coupled with some new ideas can be used to prove the desired estimates. In Theorem 24.2.1 we study the 3d-case. The 2d-case is studied in Theorem 24.2.2. In Theorem 24.2.3 we will explain how the higher-dimensional case can be reduced to one of the basic cases from Theorems 24.2.1 and 24.2.2.

Theorem 24.2.1 *The following estimates hold in $\mathbb{R}^3$:*

$$\left\|F^{-1}\left(e^{-c|\xi|^{2\kappa}t}\right)\right\|_{L^p(\mathbb{R}^3)} \le Ct^{-\frac{3}{2\kappa}(1-\frac{1}{p})}$$

for $\kappa > 0$, $p \in [1, \infty]$ and $t > 0$. Here c is supposed to be a positive constant.

Proof Using the radial symmetry of $e^{-c|\xi|^{2\kappa}t}$ we have

$$F^{-1}\left(e^{-c|\xi|^{2\kappa}t}\right) = \int_0^\infty e^{-cr^{2\kappa}t} r^2 \tilde{J}_{\frac{1}{2}}(r|x|)\,dr,$$

where $\tilde{J}_{\frac{1}{2}}(r|x|)$ is a modified Bessel function of Sect. 24.2.1. For $\kappa = \frac{1}{2}$ the explicit representation of $F^{-1}(e^{-c|\xi|t})$ for $t = 1$ gives

$$F^{-1}\left(e^{-c|\xi|}\right) \sim \frac{1}{\langle x\rangle^4}.$$

Hence, our strategy is the following: First we prove

$$\left|F^{-1}\left(e^{-c|\xi|^{2\kappa}}\right)\right| \le C\frac{1}{\langle x\rangle^{3+2\kappa}} \quad \text{for all} \quad x \in \mathbb{R}^3.$$

Then, after a change of variables we derive the representation

$$F^{-1}\left(e^{-c|\xi|^{2\kappa}t}\right) = \frac{1}{t^{\frac{3}{2\kappa}}} G\left(\frac{x}{t^{\frac{1}{2\kappa}}}\right),$$

where

$$G(y) = \int_{\mathbb{R}^3} e^{iy\eta} e^{-c|\eta|^{2\kappa}}\,d\eta.$$

So, from the first step we have

$$\|G\|_{L^p(\mathbb{R}^3_y)} \le C$$

and after backward transformation

$$\left\|F^{-1}\left(e^{-c|\xi|^{2\kappa}t}\right)\right\|_{L^1(\mathbb{R}^3_x)} = \frac{1}{t^{\frac{3}{2\kappa}}}\left\|G\left(\frac{x}{t^{\frac{1}{2\kappa}}}\right)\right\|_{L^1(\mathbb{R}^3_x)} \le C,$$

$$\left\|F^{-1}\left(e^{-c|\xi|^{2\kappa}t}\right)\right\|_{L^p(\mathbb{R}^3_x)} = \frac{1}{t^{\frac{3}{2\kappa}}}\left\|G\left(\frac{x}{t^{\frac{1}{2\kappa}}}\right)\right\|_{L^p(\mathbb{R}^3_x)} \le Ct^{-\frac{3}{2\kappa}(1-\frac{1}{p})}$$

for $p \in (1, \infty]$, respectively. Let us turn now to show the basic estimate

$$\left|F^{-1}\left(e^{-c|\xi|^{2\kappa}}\right)\right| \le C\frac{1}{\langle x\rangle^{3+2\kappa}}.$$

If $|x| \leq 1$, then

$$|F^{-1}(e^{-c|\xi|^{2\kappa}})| \leq \int_{\mathbb{R}^n} e^{-c|\xi|^{2\kappa}} \, d\xi \leq C.$$

For $|x| \geq 1$ we take account of the radial symmetric representation and study for $t = 1$ the integral

$$\int_0^\infty e^{-cr^{2\kappa}t} \, r^2 \tilde{J}_{\frac{1}{2}}(r|x|) \, dr.$$

Using for the modified Bessel functions the relation

$$\tilde{J}_{\frac{1}{2}}(r|x|) = -\frac{1}{r|x|^2} \partial_r \tilde{J}_{-\frac{1}{2}}(r|x|)$$

and for $x \in \mathbb{R}^3$ the explicit representation $\tilde{J}_{-\frac{1}{2}}(r|x|) = \sqrt{\frac{2}{\pi}} \cos(r|x|)$ (for both see Proposition 24.2.2) we arrive at

$$G(x) = \sqrt{\frac{2}{\pi}} \frac{1}{|x|^2} \int_0^\infty d_r\left(e^{-cr^{2\kappa}}r\right) \cos(r|x|) \, dr$$

$$= \sqrt{\frac{2}{\pi}} \frac{1}{|x|^2} \int_0^\infty \left(1 - 2\kappa c r^{2\kappa}\right) e^{-cr^{2\kappa}} \cos(r|x|) \, dr.$$

To get the decay rate $\langle x \rangle^{-(3+2\kappa)}$ we will apply two more steps of partial integration. First

$$G(x) = -\sqrt{\frac{2}{\pi}} \frac{1}{|x|^3} \int_0^\infty d_r\left((1 - 2\kappa c r^{2\kappa})e^{-cr^{2\kappa}}\right) \sin(r|x|) \, dr$$

$$= \sqrt{\frac{2}{\pi}} \frac{1}{|x|^3} \int_0^\infty \left(c2\kappa + 4c\kappa^2 - 4c^2\kappa^2 r^{2\kappa}\right) r^{2\kappa-1} e^{-cr^{2\kappa}} \sin(r|x|) \, dr.$$

To estimate the integral

$$\int_0^\infty r^{2\kappa-1} e^{-cr^{2\kappa}} \sin(r|x|) \, dr$$

we divide it into

$$\int_0^{\frac{1}{|x|}} r^{2\kappa-1} e^{-cr^{2\kappa}} \sin(r|x|) \, dr + \int_{\frac{1}{|x|}}^\infty r^{2\kappa-1} e^{-cr^{2\kappa}} \sin(r|x|) \, dr.$$

The first integral is estimated by $\langle x \rangle^{-2\kappa}$. In the second integral we carry out one more step of partial integration and we obtain

$$\int_{\frac{1}{|x|}}^{\infty} r^{2\kappa-1} e^{-cr^{2\kappa}} \sin(r|x|)\, dr = -\frac{1}{|x|} \int_{\frac{1}{|x|}}^{\infty} r^{2\kappa-1} e^{-cr^{2\kappa}} \partial_r \cos(r|x|)\, dr$$

$$= \frac{1}{|x|} \int_{\frac{1}{|x|}}^{\infty} \left((2\kappa-1)r^{2\kappa-2} - c2\kappa r^{4\kappa-2}\right) e^{-cr^{2\kappa}} \cos(r|x|)\, dr + R(x),$$

where the term $R(x)$ can be estimated by $\langle x \rangle^{-2\kappa}$. Dividing the last integral into

$$\int_{\frac{1}{|x|}}^{\infty} \left((2\kappa-1)r^{2\kappa-2} - c2\kappa r^{4\kappa-2}\right) e^{-cr^{2\kappa}} \cos(r|x|)\, dr$$

$$= \int_{\frac{1}{|x|}}^{1} \left((2\kappa-1)r^{2\kappa-2} - c2\kappa r^{4\kappa-2}\right) e^{-cr^{2\kappa}} \cos(r|x|)\, dr$$

$$+ \int_{1}^{\infty} \left((2\kappa-1)r^{2\kappa-2} - c2\kappa r^{4\kappa-2}\right) e^{-cr^{2\kappa}} \cos(r|x|)\, dr$$

straight forward estimates lead to

$$\int_0^{\infty} r^{2\kappa-1} e^{-cr^{2\kappa}} \sin(r|x|)\, dr \le C \begin{cases} \dfrac{1}{\langle x \rangle^{2\kappa}} & \text{for } 0 < \kappa < \tfrac{1}{2}, \\[2mm] \dfrac{\log\langle x \rangle}{\langle x \rangle} & \text{for } \kappa = \tfrac{1}{2}, \\[2mm] \dfrac{1}{\langle x \rangle} & \text{for } \tfrac{1}{2} < \kappa. \end{cases}$$

Summarizing, we have shown $\|G\|_{L^p(\mathbb{R}^3)} \le C$. This completes the proof.

Corollary 24.2.1 *The following estimates hold in $\mathbb{R}^3$:*

$$\left\| F^{-1}\left(|\xi|^a e^{-c|\xi|^{2\kappa} t}\right) \right\|_{L^p(\mathbb{R}^3)} \le Ct^{-\frac{a}{2\kappa} - \frac{3}{2\kappa}(1-\frac{1}{p})}$$

for $\kappa > 0$, $p \in [1, \infty]$ and $t > 0$. Here c and a are supposed to be positive constants.

Theorem 24.2.2 *The following estimates hold in $\mathbb{R}^2$:*

$$\left\| F^{-1}\left(e^{-c|\xi|^{2\kappa} t}\right) \right\|_{L^p(\mathbb{R}^2)} \le Ct^{-\frac{2}{2\kappa}(1-\frac{1}{p})}$$

for $\kappa > 0$, $p \in [1, \infty]$ and $t > 0$. Here c is supposed to be a positive constant.

Proof As in the proof to Theorem 24.2.1 we shall, finally, study for $|x| \geq 1$ the integral

$$\int_0^\infty e^{-cr^{2\kappa}} r \tilde{J}_0(r|x|) \, dr.$$

From Proposition 24.2.2 we have the relation

$$\tilde{J}_0(s) = 2\tilde{J}_1(s) + s\frac{d}{ds}\tilde{J}_1(s).$$

Instead of the last integral we will now study

$$\int_0^\infty e^{-cr^{2\kappa}} r \left(2\tilde{J}_1(r|x|) + r\partial_r\tilde{J}_1(r|x|)\right) dr.$$

After partial integration, this integral is equal to

$$2\kappa c \int_0^\infty e^{-cr^{2\kappa}} r^{2\kappa+1} \tilde{J}_1(r|x|) \, dr.$$

We divide this integral into

$$\int_0^{\frac{1}{|x|}} e^{-cr^{2\kappa}} r^{2\kappa+1} \tilde{J}_1(r|x|) \, dr + \int_{\frac{1}{|x|}}^\infty e^{-cr^{2\kappa}} r^{2\kappa+1} \tilde{J}_1(r|x|) \, dr.$$

Using the boundedness of $\tilde{J}_1(s)$ for $s \in [0, 1]$, the first integral can be estimated by $\langle x \rangle^{-(2\kappa+2)}$. To estimate the second integral we apply the following asymptotic formula for $\tilde{J}_1(s)$ for $s \geq 1$:

$$\tilde{J}_1(s) = C_1\frac{1}{s^{\frac{3}{2}}} \cos\left(s - \frac{3}{4}\pi\right) + O\left(\frac{1}{|s|^{\frac{5}{2}}}\right).$$

Consequently, this integral can be estimated as follows:

$$\int_{\frac{1}{|x|}}^\infty e^{-cr^{2\kappa}} r^{2\kappa+1} O\left(\frac{1}{(r|x|)^{\frac{5}{2}}}\right) dr \leq C \begin{cases} \dfrac{1}{\langle x \rangle^{2\kappa+2}} & \text{for } 0 < \kappa < \frac{1}{4}, \\[2mm] \dfrac{\log\langle x \rangle}{\langle x \rangle^{\frac{5}{2}}} & \text{for } \kappa = \frac{1}{4}, \\[2mm] \dfrac{1}{\langle x \rangle^{\frac{5}{2}}} & \text{for } \frac{1}{4} < \kappa. \end{cases}$$

It remains to estimate the integrals

$$\frac{1}{|x|^{\frac{3}{2}}} \int_{\frac{1}{|x|}}^{\infty} e^{-cr^{2\kappa}} r^{2\kappa-\frac{1}{2}} \cos(r|x|)\, dr, \quad \frac{1}{|x|^{\frac{3}{2}}} \int_{\frac{1}{|x|}}^{\infty} e^{-cr^{2\kappa}} r^{2\kappa-\frac{1}{2}} \sin(r|x|)\, dr.$$

Here we proceed as in the proof to Theorem 24.2.1. We explain only the first integral, which we split into

$$\frac{1}{|x|^{\frac{3}{2}}} \int_{\frac{1}{|x|}}^{1} e^{-cr^{2\kappa}} r^{2\kappa-\frac{1}{2}} \cos(r|x|)\, dr + \frac{1}{|x|^{\frac{3}{2}}} \int_{1}^{\infty} e^{-cr^{2\kappa}} r^{2\kappa-\frac{1}{2}} \cos(r|x|)\, dr.$$

The first integral is equal to

$$\frac{1}{|x|^{\frac{5}{2}}} \int_{\frac{1}{|x|}}^{1} e^{-cr^{2\kappa}} r^{2\kappa-\frac{1}{2}} \partial_r \sin(r|x|)\, dr.$$

After partial integration, the integral limit terms behave as $\langle x \rangle^{-(2\kappa+2)}$. The new integral is estimated by

$$\frac{1}{|x|^{\frac{5}{2}}} \int_{\frac{1}{|x|}}^{1} e^{-cr^{2\kappa}} r^{2\kappa-\frac{3}{2}}\, dr \le C \begin{cases} \dfrac{1}{\langle x \rangle^{2\kappa+2}} & \text{for } 0 < \kappa < \frac{1}{4}, \\[2mm] \dfrac{\log\langle x \rangle}{\langle x \rangle^{\frac{5}{2}}} & \text{for } \kappa = \frac{1}{4}, \\[2mm] \dfrac{1}{\langle x \rangle^{\frac{5}{2}}} & \text{for } \frac{1}{4} < \kappa. \end{cases}$$

The second integral is equal to

$$\frac{1}{|x|^{\frac{5}{2}}} \int_{1}^{\infty} e^{-cr^{2\kappa}} r^{2\kappa-\frac{1}{2}} \partial_r \sin(r|x|)\, dr,$$

and can be estimated by $\langle x \rangle^{-\frac{5}{2}}$. In the same way we treat the integral

$$\frac{1}{|x|^{\frac{3}{2}}} \int_{\frac{1}{|x|}}^{\infty} e^{-cr^{2\kappa}} r^{2\kappa-\frac{1}{2}} \sin(r|x|)\, dr.$$

This completes the proof.

Corollary 24.2.2 *The following estimates hold in $\mathbb{R}^2$:*

$$\left\| F^{-1}\left(|\xi|^a e^{-c|\xi|^{2\kappa} t} \right) \right\|_{L^p(\mathbb{R}^2)} \le C t^{-\frac{a}{2\kappa} - \frac{2}{2\kappa}(1 - \frac{1}{p})}$$

for $\kappa > 0$, $p \in [1, \infty]$ and $t > 0$. Here c and a are supposed to be positive constants.

The next goal is to prove the following generalization of Theorems 24.2.1 and 24.2.2.

Theorem 24.2.3 *The following estimates hold in $\mathbb{R}^n$ for $n \geq 4$:*

$$\left\| F^{-1}\left(e^{-c|\xi|^{2\kappa}t}\right) \right\|_{L^p(\mathbb{R}^n)} \leq Ct^{-\frac{n}{2\kappa}(1-\frac{1}{p})}$$

for $\kappa > 0$, $p \in [1, \infty]$ and $t > 0$. Here c is supposed to be a positive constant.

Proof If $n \geq 4$ is odd and $|x| \geq 1$, then we carry out $\frac{n+1}{2}$ steps of partial integration. We apply in $\frac{n-1}{2}$ steps the rules of Proposition 24.2.2, in particular,

$$\tilde{J}_{\mu+1}(r|x|) = -\frac{1}{r|x|^2}\partial_r \tilde{J}_\mu(r|x|), \quad |\tilde{J}_\mu(s)| \leq C_\mu$$

for real nonnegative μ to conclude

$$F^{-1}\left(e^{-c|\xi|^{2\kappa}}\right) = \int_0^\infty e^{-cr^{2\kappa}} r^{n-1} \tilde{J}_{\frac{n}{2}-1}(r|x|)\, dr$$

$$= (-1)^{\frac{n-1}{2}} \frac{1}{|x|^{n-1}} \int_0^\infty \left(\frac{\partial}{\partial r}\frac{1}{r}\right)^{\frac{n-1}{2}} \left(e^{-cr^{2\kappa}} r^{n-1}\right) \tilde{J}_{-\frac{1}{2}}(r|x|)\, dr.$$

Among all integrals, the integrals

$$\int_0^\infty e^{-cr^{2\kappa}} \cos(r|x|)\, dr, \quad \int_0^\infty e^{-cr^{2\kappa}} r^{2\kappa} \cos(r|x|)\, dr$$

have a dominant influence. The same approach as in the proof to Theorem 24.2.1 gives immediately $\|G\|_{L^p(\mathbb{R}^3)} \leq C$. This completes the proof for odd $n \geq 4$. Let us discuss the case of even $n \geq 4$. Analogous to the odd case, we carry out $\frac{n}{2} - 1$ steps of partial integration by using again the rule

$$\tilde{J}_{\mu+1}(r|x|) = -\frac{1}{r|x|^2}\partial_r \tilde{J}_\mu(r|x|).$$

Among all integrals, the integrals

$$\int_0^\infty e^{-cr^{2\kappa}} r \tilde{J}_0(r|x|)\, dr, \quad \int_0^\infty e^{-cr^{2\kappa}} r^{1+2\kappa} \tilde{J}_0(r|x|)\, dr$$

have a dominant influence. The same approach as in the proof to Theorem 24.2.2 gives immediately $\|G\|_{L^p(\mathbb{R}^3)} \leq C$. This completes the proof.

Corollary 24.2.3 *The following estimates hold in $\mathbb{R}^n$, $n \geq 4$:*

$$\left\| F^{-1}\left(|\xi|^a e^{-c|\xi|^{2\kappa}t}\right) \right\|_{L^p(\mathbb{R}^n)} \leq Ct^{-\frac{a}{2\kappa}-\frac{n}{2\kappa}(1-\frac{1}{p})}$$

for $\kappa > 0$, $p \in [1, \infty]$ and $t > 0$. Here c and a are supposed to be positive constants.
In this section we derived several L^p estimates for the oscillating integral

$$F^{-1}\left(|\xi|^a e^{-c|\xi|^{2\kappa} t}\right)$$

under the assumptions $a \geq 0$, $\kappa > 0$ and $c > 0$. Our main goal was to show how
the theory of modified Bessel functions coupled with some new tools can be used to
prove the desired estimates. Up to now our goal was not to find some optimal range
for the parameter a under which we can prove some L^p estimates. The following
considerations are devoted to this issue.

Theorem 24.2.4 *The following estimates hold in $\mathbb{R}^n$ for $n \geq 1$:*

$$\left\|F^{-1}\left(|\xi|^a e^{-|\xi|^{2\kappa} t}\right)\right\|_{L^p(\mathbb{R}^n)} \leq C t^{-\frac{n}{2\kappa}\left(1-\frac{1}{p}\right)-\frac{a}{2\kappa}}$$

for $\kappa > 0$, $p \in [1, \infty]$ and $t > 0$ provided that

$$a + n\left(1 - \frac{1}{p}\right) > 0.$$

*In particular, if $a > 0$, then the statement is true for all $p \in [1, \infty]$. Moreover, in the
special case $a = 0$ the statement is still true for $p = 1$.*

Proof Let us put

$$G_{\kappa,a}(t, \cdot) = F^{-1}\left(|\xi|^a e^{-t|\xi|^{2\kappa}}\right),$$

where $a \in \mathbb{R}^1$. By scaling properties of the Fourier transform we get

$$\|G_{\kappa,a}(t, \cdot)\|_{L^p} = t^{-\frac{n}{2\kappa}\left(1-\frac{1}{p}\right)-\frac{a}{2\kappa}}\|G_{\kappa,a}(1, \cdot)\|_{L^p},$$

for any $p \in [1, \infty]$, so it is sufficient to study cases in which the norm on the
right-hand side is finite. It is clear that $G_{\kappa,a}(1, \cdot) \in L^\infty(\mathbb{R}^n)$ by using the Riemann-
Lebesgue theorem since $|\xi|^a e^{-|\xi|^{2\kappa}}$ belongs to $L^1(\mathbb{R}^n)$ for any $a > -n$. Hence, by
interpolation it is sufficient to prove that $G_{\kappa,a}(1, \cdot) \in L^1(\mathbb{R}^n)$ for $a \geq 0$ and, if $a < 0$,
that $G_{\kappa,a}(1, \cdot) \in L^p(\mathbb{R}^n)$ for $p \in (1, n)$ satisfying $a + n\left(1 - \frac{1}{p}\right) > 0$.

Firstly, we consider the case $a = 0$. In [8] the authors proved, after using Polya's
argument, the relation

$$\lim_{|x|\to\infty} |x|^{n+2\kappa} F^{-1}\left(e^{-|\xi|^{2\kappa}}\right)(x) = \kappa \, 4^\kappa \, \pi^{-\left(\frac{n}{2}+1\right)} \sin(\pi\kappa) \Gamma\left(\frac{n+2\kappa}{2}\right) \Gamma(\kappa)$$

for all $\kappa > 0$. Hence, using the last relation for large $|x|$ we get that $G_{\kappa,0}(1, \cdot) \in
L^1(\mathbb{R}^n)$.

After using the property

$$e^{ix\cdot\xi} = \sum_{j=1}^{n} \frac{(-ix_j)}{|x|^2} \partial_{\xi_j} e^{ix\cdot\xi}$$

and integrating by parts, we may write

$$G_{\kappa,a}(1,x) = |x|^{-l} \frac{1}{(2\pi)^{\frac{n}{2}}} \sum_{|\gamma|=l} \left(\frac{ix}{|x|}\right)^{\gamma} \int_{\mathbb{R}^n} e^{ix\cdot\xi} \partial_{\xi}^{\gamma}\left(|\xi|^a\, e^{-|\xi|^{2\kappa}}\right) d\xi$$

for any $l \in \mathbb{N}$. Moreover, we estimate

$$\left|\partial_{\xi}^{\gamma}\left(|\xi|^a\, e^{-|\xi|^{2\kappa}}\right)\right| \leq C|\xi|^{a-|\gamma|}(1 + |\xi|^{2\kappa})^{|\gamma|} e^{-|\xi|^{2\kappa}} \leq C|\xi|^{a-|\gamma|} e^{-c|\xi|^{2\kappa}}$$

with some constant $c \in (0,1)$.

Now, let us prove that $G_{\kappa,a}(1,\cdot) \in L^1(\mathbb{R}^n)$ for $a > 0$ and that $G_{\kappa,a}(1,\cdot) \in L^p(\mathbb{R}^n)$ for $a < 0$ and $p \in (1,n)$ satisfying $a + n\left(1 - \frac{1}{p}\right) > 0$. Let $|x| \geq 1$ and $\alpha \in \mathbb{N}^n$ with $|\alpha| = l$, where $l = \frac{n}{p}$ if $\frac{n}{p} \in \mathbb{N}$. Otherwise, let l be an integer satisfying $\frac{n}{p}-1 < l < \frac{n}{p}$. If $a + n > l + 1 > \frac{n}{p}$, in particular this includes the case $a > 1$ since we take $l = n$, we may trivially estimate

$$|G_{\kappa,a}(1,x)| \leq C|x|^{-(l+1)} \int_{\mathbb{R}^n} |\xi|^{a-(l+1)} e^{-c|\xi|^{2\kappa}}\, d\xi \leq C|x|^{-(l+1)}.$$

If $l + 1 \geq a + n$, then we divide the integral

$$\int_{\mathbb{R}^n} e^{ix\cdot\xi} \partial_{\xi}^{\alpha}\left(|\xi|^a e^{-|\xi|^{2\kappa}}\right) d\xi$$

into two parts

$$I_0(x) + I_1(x)$$

$$:= \int_{|\xi|\leq|x|^{-1}} e^{ix\cdot\xi} \partial_{\xi}^{\alpha}\left(|\xi|^a e^{-|\xi|^{2\kappa}}\right) d\xi + \int_{|\xi|\geq|x|^{-1}} e^{ix\cdot\xi} \partial_{\xi}^{\alpha}\left(|\xi|^a e^{-|\xi|^{2\kappa}}\right) d\xi.$$

On the one hand, we estimate

$$|I_0(x)| \leq C \int_{|\xi|\leq|x|^{-1}} |\xi|^{a-l}\, d\xi \leq C|x|^{l-(n+a)}.$$

On the other hand, we perform one additional step of integration by parts in I_1. If $l+1 > a+n$, then we obtain

$$
|I_1(x)| \le C|x|^{-1} \int_{|\xi|=|x|^{-1}} |\xi|^{a-l}\, d\sigma
$$
$$
+ |x|^{-1} \int_{|\xi|\ge |x|^{-1}} |\xi|^{a-(l+1)}\, d\xi \le C|x|^{-(n+a)},
$$

whereas, if $l+1 = a+n$, then we split each integral into two parts (for large $|x|$):

$$
\int_{\mathbb{R}^n} e^{ix\cdot\xi}\, \partial_{\xi_j}\partial_\xi^\gamma \left(|\xi|\, e^{-|\xi|^{2\kappa}}\right) d\xi = I_{1,1}(x) + I_{1,2}(x)
$$
$$
:= \int_{|x|^{-1}\le|\xi|\le 1} e^{ix\cdot\xi}\, \partial_{\xi_j}\partial_\xi^\gamma \left(|\xi|\, e^{-|\xi|^{2\kappa}}\right) d\xi + \int_{|\xi|\ge 1} e^{ix\cdot\xi}\, \partial_{\xi_j}\partial_\xi^\gamma \left(|\xi|\, e^{-|\xi|^{2\kappa}}\right) d\xi,
$$

directly estimating $I_{1,1}$, and performing one additional step of integration by parts in $I_{1,2}$. This leads to the estimates

$$
|I_{1,1}(x)| \le C\log(1+|x|), \qquad |I_{1,2}(x)| \le C.
$$

Summarizing, we proved that

$$
|G_{\kappa,a}(1,x)| \le C
\begin{cases}
|x|^{-(l+1)} & \text{for } l+1 < a+n,\\
|x|^{-(n+a)} & \text{for } l+1 > a+n,\\
|x|^{-(l+1)} \ln(1+|x|) & \text{for } l+1 = a+n.
\end{cases}
$$

Finally, using that $l = n$ for $a > 0$ and thanks to $a + n\left(1 - \frac{1}{p}\right) > 0$, there exists a positive constant ε such that $a+n = \frac{n+\varepsilon}{p}$. We may conclude

$$
|G_{\kappa,a}(1,x)| \le C
\begin{cases}
|x|^{-(n+1)} & \text{if } a > 1,\\
|x|^{-(n+1)} \log(1+|x|) & \text{if } a = 1,\\
|x|^{-(n+a)} & \text{if } a \in (0,1),\\
|x|^{-\frac{n+\varepsilon}{p}} & \text{if } a < 0 \text{ and } a + n\left(1 - \frac{1}{p}\right) > 0.
\end{cases}
$$

Recalling that $G_{\kappa,a}(1,\cdot) \in L^\infty(\mathbb{R}^n)$, we conclude $G_{\kappa,a}(1,\cdot) \in L^1(\mathbb{R}^n)$ for $a > 0$ and $G_{\kappa,a}(1,\cdot) \in L^p(\mathbb{R}^n)$ for $a < 0$ and $p \in (1,n)$ satisfying $a + n\left(1 - \frac{1}{p}\right) > 0$. This completes the proof.

24.3 Function Spaces

In this section we introduce function spaces, or spaces of distributions, which are used in this monograph (see also [179]).

First we introduce the Lebesgue spaces $L^p = L^p(\mathbb{R}^n)$ for $p \in (0, \infty]$.

Definition 24.6 Let $0 < p \leq \infty$. Then, the Lebesgue space $L^p(\mathbb{R}^n)$ is the set of all Lebesgue measurable complex-valued functions f on $\mathbb{R}^n$ such that

$$\|f\|_{L^p} = \left(\int_{\mathbb{R}^n} |f(x)|^p \, dx \right)^{\frac{1}{p}} < \infty \quad \text{for } p \in (0, \infty),$$

$$\|f\|_{L^\infty} = \operatorname*{ess\,sup}_{x \in \mathbb{R}^n} |f(x)| < \infty.$$

Now we define Sobolev spaces of integer and fractional order as well.

Definition 24.7 Let $1 \leq p \leq \infty$ and $m \in \mathbb{N}$. Then, the Sobolev spaces $W_p^m(\mathbb{R}^n)$ are defined as

$$W_p^m(\mathbb{R}^n) := \left\{ f \in L^p(\mathbb{R}^n) : \|f\|_{W_p^m} := \sum_{|\alpha| \leq m} \|\partial_x^\alpha f\|_{L^p} < \infty \right\}.$$

Here the derivatives are defined in the Sobolev sense (cf. with Definition 3.9).

Definition 24.8 Let $1 \leq p < \infty$ and $s > 0$ be a not positive integer number. By m we denote the integer part $[s]$ of s. Then, the Sobolev-Slobodeckij spaces $W_p^s(\mathbb{R}^n)$ are defined as

$$W_p^s(\mathbb{R}^n) := \Big\{ f \in W_p^m(\mathbb{R}^n) :$$

$$\|f\|_{W_p^s} := \|f\|_{W_p^m} + \sum_{|\alpha|=m} \left(\int_{\mathbb{R}_y^n} \int_{\mathbb{R}_x^n} \frac{|\partial_x^\alpha f(x) - \partial_y^\alpha f(y)|^p}{|x - y|^{n+(s-m)p}} \, dx \, dy \right)^{\frac{1}{p}} < \infty \Big\}.$$

Here the derivatives are defined in the Sobolev sense (cf. with Definition 3.9). If $p = 2$, then we also use the notation $W^s(\mathbb{R}^n)$.

Definition 24.9 Let $1 < p < \infty$ and $s \in \mathbb{R}^1$. Then, the Sobolev spaces of fractional order $H_p^s(\mathbb{R}^n)$ are defined as

$$H_p^s(\mathbb{R}^n) := \{ f \in S'(\mathbb{R}^n) : \|f\|_{H_p^s} := \left\| F^{-1}(\langle \xi \rangle^s F(f)) \right\|_{L^p} < \infty \}.$$

Here $\langle \xi \rangle$ denotes the Japanese brackets with $\langle \xi \rangle^2 := 1 + |\xi|^2$. If $p = 2$, then we also use the notation $H^s(\mathbb{R}^n)$.

Let us introduce Besov and Triebel-Lizorkin spaces. For this reason we introduce a dyadic decomposition of the phase space.

We choose a Schwartz function $\psi = \psi(\xi)$ with $\psi(\xi) \in [0,1]$, $\psi(\xi) = 1$ if $|\xi| \le 1$ and $\psi(\xi) = 0$ if $|\xi| \ge \frac{3}{2}$. Then we put $\phi_0(\xi) = \psi(\xi)$, $\phi_1(\xi) = \psi(\frac{\xi}{2}) - \psi(\xi)$ and $\phi_j(\xi) = \phi_1(2^{-j+1}\xi)$ for $j \ge 2$. These functions satisfy

$$\sum_{j=0}^{M} \phi_j(\xi) = \psi(2^{-M}\xi) \text{ and}$$

$$\operatorname{supp}\phi_j(\xi) \subset \{\xi \in \mathbb{R}^n : |\xi| \in [2^{j-1}, 3 \cdot 2^{j-1}]\} \text{ for } j \ge 1.$$

Using this dyadic decomposition we define Besov spaces.

Definition 24.10 Let $0 < p, q \le \infty$ and $s \in \mathbb{R}^1$. Then, the Besov spaces $B^s_{p,q}(\mathbb{R}^n)$ are defined as

$$B^s_{p,q}(\mathbb{R}^n) = \left\{ f \in S'(\mathbb{R}^n) : \|f\|_{B^s_{p,q}} := \left(\sum_{j=0}^{\infty} 2^{sjq} \|F^{-1}(\phi_j F(f))\|_{L^p}^q \right)^{\frac{1}{q}} < \infty \right\}.$$

Triebel-Lizorkin spaces are defined as follows:

Definition 24.11 Let $0 < q \le \infty$ and $0 < p < \infty$. The weight parameter is given by $s \in \mathbb{R}^1$. Then, the Triebel-Lizorkin spaces $F^s_{p,q}(\mathbb{R}^n)$ are defined as

$$F^s_{p,q}(\mathbb{R}^n) = \left\{ f \in S'(\mathbb{R}^n) \right.$$

$$\left. : \|f\|_{F^s_{p,q}} = \left(\int_{\mathbb{R}^n} \left(\sum_{j=0}^{\infty} 2^{sjq} |F^{-1}(\phi_j F(f))(x)|^q \right)^{\frac{p}{q}} dx \right)^{\frac{1}{p}} < \infty \right\}.$$

Remark 24.3.1 We have the following identities (in the sense of equivalent norms):

$$W^m_p(\mathbb{R}^n) = F^m_{p,2}(\mathbb{R}^n) \text{ if } 1 < p < \infty, m \in \mathbb{N},$$
$$W^s_p(\mathbb{R}^n) = F^s_{p,p}(\mathbb{R}^n) = B^s_{p,p}(\mathbb{R}^n) \text{ if } 1 \le p < \infty, 0 < s \notin \mathbb{Z},$$
$$H^s_p(\mathbb{R}^n) = F^s_{p,2}(\mathbb{R}^n) \text{ if } 1 < p < \infty.$$

Remark 24.3.2 Let $p \in (1,2]$ and $\frac{1}{p} + \frac{1}{q} = 1$. Then, the following inclusions hold (see [12] or [185]):

$$F^0_{p,2}(\mathbb{R}^n) = L^p(\mathbb{R}^n) \subset B^0_{p,2}(\mathbb{R}^n), \quad B^0_{q,2}(\mathbb{R}^n) \subset L^q(\mathbb{R}^n) = F^0_{q,2}(\mathbb{R}^n).$$

Remark 24.3.3 Let $1 \le p_1 \le p_2 \le \infty$ and $1 \le q_1 \le q_2 \le \infty$. Then, we have for any real number s the continuous embedding of $B^s_{p_1,q_1}(\mathbb{R}^n)$ in $B^{s-n(\frac{1}{p_1} - \frac{1}{p_2})}_{p_2,q_2}(\mathbb{R}^n)$ (see [3]).

Next we introduce homogeneous versions of Sobolev spaces (see also [179]). For this reason we introduce the function space

$$Z(\mathbb{R}^n) := \{f \in S(\mathbb{R}^n) : D_\xi^\alpha F(f)(\xi = 0) = 0 \text{ for all multi-indices } \alpha\}.$$

By $Z'(\mathbb{R}^n)$ we denote the topological dual to $Z(\mathbb{R}^n)$. Hence, if a distribution u belongs to $S'(\mathbb{R}^n)$, then the restriction of u to $Z(\mathbb{R}^n)$ belongs to $Z'(\mathbb{R}^n)$. Furthermore, if p is a polynomial, then the actions $(u + p)(f)$ of $u \in Z'(\mathbb{R}^n)$ on f coincide with $u(f)$. Conversely, any $u \in Z'(\mathbb{R}^n)$ can be extended linearly and continuously from $Z(\mathbb{R}^n)$ to $S(\mathbb{R}^n)$, that is, to an element of $S'(\mathbb{R}^n)$. If u_1 and u_2 are two different extensions of u, then the support of $F(u_1 - u_2)$ is equal to $\{0\}$. This implies that $u_1 - u_2$ is a polynomial. Consequently, the space of distributions $Z'(\mathbb{R}^n)$ may be identified with the factor space $S'(\mathbb{R}^n)/\{\text{all polynomials}\}$.

Definition 24.12 Let $1 < p < \infty$ and $s \in \mathbb{R}^1$. Then, the homogeneous Sobolev spaces of fractional order $\dot{H}_p^s(\mathbb{R}^n)$ are defined as

$$\dot{H}_p^s(\mathbb{R}^n) := \{f \in Z'(\mathbb{R}^n) : \|f\|_{\dot{H}_p^s} := \|F^{-1}(|\xi|^s F(f))\|_{L^p} < \infty\}.$$

To define homogeneous Besov and Triebel-Lizorkin spaces we introduce another dyadic decomposition of the phase space.

We choose a Schwartz function $\psi = \psi(\xi)$ with $\operatorname{supp}\psi(\xi) \subset \{\xi : |\xi| \in [1, 4]\}$, $\psi(\xi) = 1$ if $|\xi| \in [2, 3]$. Then, it holds

$$1 \le \sum_{j=-\infty}^{\infty} \psi(2^j \xi) \le 3.$$

We define for $j \in \mathbb{Z}$ the functions

$$\phi_j(\xi) = \frac{\psi(2^{-j+1}\xi)}{\sum\limits_{j=-\infty}^{\infty} \psi(2^j \xi)} = \phi_1(2^{-j+1}\xi).$$

These functions satisfy

$$\operatorname{supp}\phi_j(\xi) \subset \{\xi \in \mathbb{R}^n : |\xi| \in [2^{j-1}, 2^{j+1}]\} \text{ for } j \in \mathbb{Z}.$$

Using this dyadic decomposition we define homogeneous Besov spaces.

Definition 24.13 Let $0 < p, q \le \infty$ and $s \in \mathbb{R}^1$. Then, the homogeneous Besov spaces $\dot{B}_{p,q}^s(\mathbb{R}^n)$ are defined as

$$\dot{B}_{p,q}^s(\mathbb{R}^n) = \left\{f \in Z'(\mathbb{R}^n) : \|f\|_{\dot{B}_{p,q}^s} := \left(\sum_{j=-\infty}^{\infty} 2^{sjq}\|F^{-1}(\phi_j F(f))\|_{L^p}^q\right)^{\frac{1}{q}} < \infty\right\}.$$

Homogeneous Triebel-Lizorkin spaces are defined as follows:

Definition 24.14 Let $0 < q \leq \infty$ and $0 < p < \infty$. The weight parameter is given by $s \in \mathbb{R}^1$. Then, the homogeneous Triebel-Lizorkin spaces $\dot{F}_{p,q}^s(\mathbb{R}^n)$ are defined as

$$\dot{F}_{p,q}^s(\mathbb{R}^n) = \left\{ f \in Z'(\mathbb{R}^n) \right.$$

$$\left. : \|f\|_{\dot{F}_{p,q}^s} = \left(\int_{\mathbb{R}^n} \left(\sum_{j=-\infty}^{\infty} 2^{sjq} |F^{-1}(\phi_j F(f))(x)|^q \right)^{\frac{p}{q}} dx \right)^{\frac{1}{p}} < \infty \right\}.$$

Remark 24.3.4 We have the following continuous embeddings for the spaces $\dot{B}_{p,q}^s(\mathbb{R}^n)$ and $\dot{F}_{p,q}^s(\mathbb{R}^n)$:

$$Z(\mathbb{R}^n) \hookrightarrow \dot{B}_{p,q}^s(\mathbb{R}^n) \hookrightarrow Z'(\mathbb{R}^n), \quad Z(\mathbb{R}^n) \hookrightarrow \dot{F}_{p,q}^s(\mathbb{R}^n) \hookrightarrow Z'(\mathbb{R}^n).$$

Remark 24.3.5 In general, homogeneous spaces can not be compared for inclusion. Nevertheless, we have, for example, the following relations:

1. Let $s_0 \leq s \leq s_1$. Then, $\dot{H}^{s_0}(\mathbb{R}^n) \cap \dot{H}^{s_1}(\mathbb{R}^n)$ is included in $\dot{H}^s(\mathbb{R}^n)$ and we have (cf. with [3])

$$\|u\|_{\dot{H}^s} \leq \|u\|_{\dot{H}^{s_0}}^{1-\theta} \|u\|_{\dot{H}^{s_1}}^{\theta} \quad \text{with} \quad s = (1-\theta)s_0 + \theta s_1.$$

2. The inequality

$$\|u\|_{\dot{B}_{p,q}^s} \leq C \|u\|_{\dot{B}_{p_0,\infty}^{s_0}}^{1-\theta} \|u\|_{\dot{B}_{p_1,\infty}^{s_1}}^{\theta}$$

holds for all $u \in \dot{B}_{p_0,\infty}^{s_0}(\mathbb{R}^n) \cap \dot{B}_{p_1,\infty}^{s_1}(\mathbb{R}^n)$ under the assumptions of Proposition 24.5.5. For a special case see also the reference for the theory of homogeneous Besov spaces [162].

Remark 24.3.6 These homogeneous spaces have the following "property of homogeneity":

$$\frac{1}{C} \lambda^{s-\frac{n}{p}} \|f(\cdot)\|_{\dot{B}_{p,q}^s} \leq \|f(\lambda \cdot)\|_{\dot{B}_{p,q}^s} \leq C \lambda^{s-\frac{n}{p}} \|f(\cdot)\|_{\dot{B}_{p,q}^s},$$

$$\frac{1}{C} \lambda^{s-\frac{n}{p}} \|f(\cdot)\|_{\dot{F}_{p,q}^s} \leq \|f(\lambda \cdot)\|_{\dot{F}_{p,q}^s} \leq C \lambda^{s-\frac{n}{p}} \|f(\cdot)\|_{\dot{F}_{p,q}^s}$$

for all positive λ with a suitable positive constant C.

Remark 24.3.7 The property of homogeneity is an important tool for proving embeddings between spaces $\dot{H}_2^s(\mathbb{R}^n)$ and $L^p(\mathbb{R}^n)$ (see [3]). Among other things, we have the following embeddings.

1. If $s \in [0, \frac{n}{2})$, then the space $\dot{H}_2^s(\mathbb{R}^n)$ is continuously embedded in $L^q(\mathbb{R}^n)$ if and only if $q = \frac{2n}{n-2s}$. Take into consideration the following embedding for non-homogeneous spaces: if $s \in [0, \frac{n}{2})$, then the space $H_2^s(\mathbb{R}^n)$ is continuously embedded in $L^q(\mathbb{R}^n)$ if and only if $q \in [2, \frac{2n}{n-2s}]$.

2. If $p \in (1, 2]$, then $L^p(\mathbb{R}^n)$ is continuously embedded in $\dot{H}_2^s(\mathbb{R}^n)$ with $s = n(\frac{1}{2} - \frac{1}{p})$.

The above introduced function spaces are based on dyadic decompositions. There is another possibility of dividing the phase space, namely, we can use the so-called frequency-uniform decomposition. The frequency decomposition technique is explained in [69]. A special case, the so-called frequency-uniform decomposition, was independently introduced by Wang (e.g. see [215]). For that, let $\psi = \psi(\xi)$ with $\psi(\xi) \in [0, 1]$ be a Schwartz function which is compactly supported in the cube $\{\xi \in \mathbb{R}^n : \xi_k \in [-1, 1]\}$ for $k = 1, \cdots, n$. Moreover, $\psi(\xi) = 1$ if $|\xi| \leq \frac{1}{2}$. We introduce the shifted functions $\psi_j(\xi) := \psi(\xi - j)$ for $j \in \mathbb{Z}^n$. and then we put

$$\phi_j(\xi) = \frac{\psi_j(\xi)}{\sum\limits_{j \in \mathbb{Z}^n} \psi_j(\xi)} \quad \text{for } j \in \mathbb{Z}^n.$$

These functions satisfy

$$\operatorname{supp} \phi_j(\xi) \subset \{\xi \in \mathbb{R}^n : \xi_k - j_k \in [-1, 1], \ k = 1, \cdots, n\} \quad \text{for } j \in \mathbb{Z}^n,$$

$$\sum\limits_{j \in \mathbb{Z}^n} \psi_j(\xi) \equiv 1.$$

We introduce the family $\{\Box_j\}_{j \in \mathbb{Z}^n}$ of uniform decomposition operators, where

$$\Box_j := F^{-1}(\psi_j F(\cdot)).$$

By using this uniform decomposition of the phase space some scales of weighted modulation spaces are introduced, for example, in [215].

Definition 24.15 Let $1 \leq p, q \leq \infty$ and $s \in \mathbb{R}^1$. Then, the weighted modulation space $M_{p,q}^s(\mathbb{R}^n)$ is the set

$$M_{p,q}^s(\mathbb{R}^n) := \{f \in S'(\mathbb{R}^n) : \|f\|_{M_{p,q}^s} < \infty\},$$

where the norm $\|f\|_{M_{p,q}^s}$ is defined by

$$\|f\|_{M_{p,q}^s} = \left(\sum\limits_{j \in \mathbb{Z}^n} \langle j \rangle^{sq} \|\Box_j f\|_{L^p}^q \right)^{\frac{1}{q}}$$

with obvious modifications when $p = \infty$ and/or $q = \infty$.

Finally, let us introduce function spaces of differentiable (in classical or Sobolev sense) functions.

Definition 24.16 Let G be a domain in $\mathbb{R}^n$ with smooth boundary ∂G. Let $m \in \mathbb{N}$. Then, $C^m(G)$ is the space of m times continuously differentiable functions in G. By $C^\infty(G)$ we denote the space of infinitely times differentiable functions.

Definition 24.17 Let G be a domain in $\mathbb{R}^n$ with smooth boundary ∂G. Let $m \in \mathbb{N}$. Then, $C^m(\bar{G})$ is the space of m times continuously differentiable functions on the closure $\bar{G}$.

Definition 24.18 Let G be a domain in $\mathbb{R}^n$ with smooth boundary ∂G. Let $m \in \mathbb{N}$. Then, $C_0^m(G)$ is the space of m times continuously differentiable functions in G with compact support there. By $C_0^\infty(G)$ we denote the space of infinitely times differentiable functions with compact support in G.

Definition 24.19 Let G be a domain in $\mathbb{R}^n$ and $0 < p \leq \infty$. Then, the Lebesgue space $L^p(G)$ is the set of all Lebesgue measurable complex-valued functions f on G such that

$$\|f\|_{L^p} = \left(\int_G |f(x)|^p \, dx \right)^{\frac{1}{p}} < \infty \text{ for } p \in (0, \infty),$$

$$\|f\|_{L^\infty} = \operatorname*{ess\,sup}_{x \in G} |f(x)| < \infty.$$

Definition 24.20 Let G be a domain in $\mathbb{R}^n$ and $0 < p \leq \infty$. Then, the Lebesgue space $L^p_{loc}(G)$ is the set of all Lebesgue measurable complex-valued functions f on G such that ϕf belongs to $L^p(G)$ for all test functions $\phi \in C_0^\infty(G)$.

Definition 24.21 Let G be a domain in $\mathbb{R}^n$, $1 \leq p \leq \infty$ and $m \in \mathbb{N}$. Then, the Sobolev spaces $W_p^m(G)$ are defined as

$$W_p^m(G) := \left\{ f \in L^p(G) : \|f\|_{W_p^m} := \sum_{|\alpha| \leq m} \|D^\alpha f\|_{L^p} < \infty \right\}.$$

Definition 24.22 Let G be a domain in $\mathbb{R}^n$, $1 \leq p \leq \infty$ and $m \in \mathbb{N}$. Then, the Sobolev spaces $W_{p,0}^m(G)$ are defined as the closure of $C_0^\infty(G)$ with respect to the norm of $W_p^m(G)$. The Sobolev spaces $W_{p,loc}^m(G)$ are defined as the space of functions f with the property that ϕf belongs to $W_p^m(G)$ for all test functions $\phi \in C_0^\infty(G)$.

Definition 24.23 Let G be a domain in $\mathbb{R}^n$, $1 \leq p < \infty$ and $s > 0$ be a not positive integer number. By m we denote the integer part $[s]$ of s. Then, the Sobolev-Slobodeckij spaces $W_p^s(G)$ are defined as

$$W_p^s(G) := \left\{ f \in W_p^m(G) : \right.$$

$$\|f\|_{W_p^s} := \|f\|_{W_p^m} + \sum_{|\alpha|=m} \left(\int_{G_y} \int_{G_x} \frac{|\partial_x^\alpha f(x) - \partial_y^\alpha f(y)|^p}{|x-y|^{n+(s-m)p}} \, dx \, dy \right)^{\frac{1}{p}} < \infty \left. \right\}.$$

Definition 24.24 Let G be a domain in $\mathbb{R}^n$, $1 \leq p < \infty$ and $s > 0$ be a not positive integer number. Then, the Sobolev-Slobodeckij spaces $W_{p,0}^s(G)$ are defined as the closure of $C_0^\infty(G)$ with respect to the norm of $W_p^s(G)$. The Sobolev-Slobodeckij spaces $W_{p,loc}^s(G)$ are defined as the space of functions f with the property that ϕf belongs to $W_p^s(G)$ for all test functions $\phi \in C_0^\infty(G)$.
Let us define evolution spaces at the end of this section. Let B be a Banach space or a locally convex space.

Definition 24.25 By $C^k([0, T], B)$ we denote the space of all distributions f which are k times continuously differentiable in t and where all derivatives $\partial_{t}^p f$, $0 \leq p \leq k$, are continuous in t with values in the space B. By $C^k([0, \infty), B)$ we denote the projective limit of all spaces $C^k([0, T], B)$, $T > 0$.
Many thanks to Winfried Sickel (Jena) for useful discussions on the content of this section.

24.4   Some Tools from Distribution Theory

First we define the convergence in the space $C_0^\infty(G)$. Let $G \subset \mathbb{R}^n$ be a given domain. A sequence $\{K_n\}_n$ is called a sequence of regular compact sets exhausting monotonically the domain G if $G = \bigcup_n K_n$ and $K_n \subset K_m$ for $n \leq m$. A function $u \in C^m(G)$ has the property that all semi-norms

$$p_{K_n,m}(u) = \max_{|\alpha| \leq m} \|\partial_x^\alpha u\|_{C(K_n)} < \infty$$

for all K_n. If u even belongs to $C^\infty(G)$, then all $p_{K_n,m}(u)$ are finite for all K_n and all $m \in \mathbb{N}$.

Definition 24.26 A sequence $\{u_k\}_k$ tends to u in $C^m(G)$ if

$$\lim_{k \to \infty} p_{K_n,m}(u_k - u) = 0 \quad \text{for all } K_n.$$

A sequence $\{u_k\}_k$ tends to u in $C^\infty(G)$ if

$$\lim_{k \to \infty} p_{K_n,m}(u_k - u) = 0 \quad \text{for all } K_n \text{ and all } m \in \mathbb{N}.$$

This definition is independent of the chosen sequence $\{K_n\}_n$. To introduce the convergence in $C_0^m(G)$ or in $C_0^\infty(G)$ we need an additional assumption. We say that a sequence $\{\phi_k\}_k$ of functions from $C_0^m(G)$ satisfies the *support condition* if there exists a compact set $K \subset G$ which contains all supports of the functions ϕ_k.

Definition 24.27 Let $\{\phi_k\}_k$ be a sequence of functions from $C_0^m(G)$ which satisfies the support condition. Then, $\{\phi_k\}_k$ tends to ϕ in $C_0^m(G)$ if it converges to ϕ in $C^m(G)$.

Let $\{\phi_k\}_k$ be a sequence of functions from $C_0^\infty(G)$ which satisfies the support condition. Then, $\{\phi_k\}_k$ tends to ϕ in $C_0^\infty(G)$ if it converges to ϕ in $C^\infty(G)$.

Now we have all the tools to introduce the definition of a real distribution. A distribution u is characterized by its actions $u(\phi)$ on test functions ϕ.

Definition 24.28 A real distribution on a domain $G \subset \mathbb{R}^n$ is a real linear functional u on $C_0^\infty(G)$ that is continuous, i.e.,

$$u(\lambda_1\phi_1 + \lambda_2\phi_2) = \lambda_1 u(\phi_1) + \lambda_2 u(\phi_2),$$

for all real constants λ_1, λ_2 and for all test functions ϕ_1, $\phi_2 \in C_0^\infty(G)$ and if the sequence $\{\phi_k\}_k$ tends to ϕ with respect to the topology of $C_0^\infty(G)$, then the sequence $\{u(\phi_k)\}_k$ of real numbers tends to the real number $u(\phi)$ with respect to the Euclidian metric in $\mathbb{R}^1$. The set of all real distributions which are defined on a given domain G is denoted by $D'(G)$.

Due to Definition 24.28, the space $D'(G)$ is the dual space to $C_0^\infty(G)$. One can also define complex distributions, but in the following we restrict ourselves to real distributions only.

Example 24.4.1 Let $G \subset \mathbb{R}^n$ be a domain and consider any real function $u \in L_{loc}^1(G)$. Then, the linear functional

$$\Lambda_u : \phi \in C_0^\infty(G) \to \Lambda_u(\phi) \in \mathbb{R}^1$$

made by the given function u and defined by

$$\Lambda_u(\phi) = \int_G u\,\phi\,dx$$

is a real distribution.

All distributions made by functions $u \in L_{loc}^1(G)$ are called *regular distributions*. Any other distributions are called *singular distributions*.

Example 24.4.2 Let $G = \mathbb{R}^n$ and define for fixed $x_0 \in \mathbb{R}^n$ the functional

$$\delta_{x_0} : \phi \in C_0^\infty(\mathbb{R}^n) \to \delta_{x_0}(\phi) := \phi(x_0) \in \mathbb{R}^1.$$

It is clear that δ_{x_0} is linear and continuous. This distribution is the well-known Dirac's δ distribution centered at x_0. It is a singular distribution.

Now let us explain relations and operations for distributions. Let u and v be two distributions from $D'(G)$. Then,

1. $u = v$ if $u(\phi) = v(\phi)$ for all $\phi \in C_0^\infty(G)$;
2. $u + v$ is the distribution from $D'(G)$ which is defined by

$$u + v : \phi \in C_0^\infty(G) \to (u + v)(\phi) := u(\phi) + v(\phi);$$

3. λu is the distribution from $D'(G)$ which is defined by

$$\lambda u : \phi \in C_0^\infty(G) \to (\lambda u)(\phi) := \lambda u(\phi),$$

here λ is a real constant;

4. u can be multiplied by a function $a = a(x) \in C^\infty(G)$, the product au is the distribution from $D'(G)$ which is defined by

$$au : \phi \in C_0^\infty(G) \to (au)(\phi) := u(a\phi);$$

5. there exist all partial derivatives $\partial_x^\alpha u$ in the distributional sense, the partial derivative $\partial_x^\alpha u$ is the distribution from $D'(G)$ which is defined by

$$\partial_x^\alpha u : \phi \in C_0^\infty(G) \to (\partial_x^\alpha u)(\phi) := (-1)^{|\alpha|} u(\partial_x^\alpha \phi).$$

Now we have all the tools for introducing the notion of a *distributional solution*, or a *solution in the distributional sense* of a linear partial differential equation.

Consider the linear partial differential equation of order m:

$$L(x, \partial_x)u = \sum_{|\alpha| \le m} a_\alpha(x)\partial_x^\alpha u = f,$$

where the coefficients a_α are supposed to belong to $C^\infty(G)$. The source term f is supposed to belong to $D'(G)$.

Definition 24.29 A distribution $u \in D'(G)$ is called distributional solution of $L(x, \partial_x)u = f$ if the distributions $L(x, \partial_x)u$ and f are equal, that is, $L(x, \partial_x)u(\phi) = f(\phi)$ for all test functions $\phi \in C_0^\infty(G)$.

Remark 24.4.1 Applying the above rules for distributions we may conclude

$$L(x, \partial_x)u(\phi) = \sum_{|\alpha| \le m} \partial_x^\alpha u(a_\alpha(x)\phi) = \sum_{|\alpha| \le m} u\big((-1)^{|\alpha|}\partial_x^\alpha(a_\alpha(x)\phi)\big).$$

The linear partial differential operator

$$L^*(x, \partial_x)\phi := \sum_{|\alpha| \le m} (-1)^{|\alpha|}\partial_x^\alpha(a_\alpha(x)\phi)$$

is called the adjoint or dual operator to $L(x, \partial_x)$. Using Definition 24.29, a distribution $u \in D'(G)$ is a distributional solution of $L(x, \partial_x)u = f$ if and only if

$$u\big(L^*(x, \partial_x)\phi\big) = f(\phi)$$

for all test functions $\phi \in C_0^\infty(G)$.

Remark 24.4.2 In the previous remark we introduced the notion of a solution $u \in D'(G)$ of $L(x, \partial_x)u = f$ in the distributional sense, that is,

$$u\big(L^*(x, \partial_x)\phi\big) = f(\phi)$$

for all test functions $\phi \in C_0^\infty(G)$. If u and f belong to $L_{loc}^1(G)$, then both make regular distributions. So that $u\big(L^*(x, \partial_x)\phi\big)$ exists we only assume $a_\alpha \in C^m(G)$ and $\phi \in C_0^m(G)$. Then, a distributional solution of $L(x, \partial_x)u = f$ is defined by the integral relation

$$\int_G u(x)L^*(x, \partial_x)\phi(x)\, dx = \int_G f(x)\phi(x)\, dx \ \text{ for all } \ \phi \in C_0^m(G).$$

It turns out that the distributional solution u is even a Sobolev solution (cf. with Definition 3.8).

The notion of *fundamental solution* is helpful in studying the existence and regularity of solutions of differential equations.

Definition 24.30 Let

$$L(x, \partial_x) := \sum_{|\alpha| \le m} a_\alpha(x)\partial_x^\alpha$$

be a linear partial differential operator of order m with infinitely differentiable coefficients $a_\alpha \in C^\infty(\mathbb{R}^n)$. A fundamental solution for L is a distribution $K \in D'(\mathbb{R}^n)$ being a distributional solution of

$$L(x, \partial_x)K = \delta_0, \ \text{ that is, } \ K\big(L^*(x, D_x)\phi\big) = \delta_0(\phi) = \phi(0)$$

for all test functions $\phi \in C_0^\infty(G)$.

Example 24.4.3 Let us consider the operator $L = \frac{d}{dt}$ on $\mathbb{R}^1$. Then, a fundamental solution is given by the Heaviside function defined by $\theta(t) = 0$ for $t \le 0$ and $\theta(t) = 1$ for $t > 0$.

Example 24.4.4 The functions

$$K(x) = \frac{1}{2\pi} \ln \frac{1}{|x|} \ \text{ in } \mathbb{R}^2$$

$$\text{and } K(x) = \frac{1}{(n-2)\sigma_n} \frac{1}{|x|^{n-2}} \ \text{ in } \mathbb{R}^n \text{ for } n \ge 3,$$

which are defined for $x \in \mathbb{R}^n \setminus \{0\}$, where σ_n denotes the n-dimensional measure of the unit sphere, are fundamental solutions for the Laplace operator Δ.

Example 24.4.5 The function

$$K(t, x) = \frac{\theta(t)}{(4\pi t)^{n/2}} \exp\left(-\frac{|x|^2}{4t}\right),$$

where $\theta = \theta(t)$ denotes the Heaviside function, is a fundamental solution for the heat operator $\partial_t - \Delta$.

24.5   Useful Inequalities

First we remember a corollary of the Riesz-Thorin interpolation theorem (see [204]) for linear continuous operators

$$T \in L\big(L^p(\mathbb{R}^n) \to L^q(\mathbb{R}^n)\big)$$

mapping $L^p(\mathbb{R}^n)$ into $L^q(\mathbb{R}^n)$. The main concern of the Riesz-Thorin interpolation theorem is to explain that if a linear operator T is defined on both $L^{p_0}(\mathbb{R}^n)$ and $L^{p_1}(\mathbb{R}^n)$ and maps boundedly into $L^{q_0}(\mathbb{R}^n)$ and $L^{q_1}(\mathbb{R}^n)$, respectively, then the operator can be interpolated to yield a bounded operator on $L^{p_\theta}(\mathbb{R}^n)$ into $L^{q_\theta}(\mathbb{R}^n)$, where p_θ and q_θ are appropriately defined intermediate exponents.

Proposition 24.5.1 *Let* $1 \le p_0, p_1, q_0, q_1 \le \infty$. *If* T *is a linear continuous operator from*

$$L\big(L^{p_0}(\mathbb{R}^n) \to L^{q_0}(\mathbb{R}^n)\big) \cap L\big(L^{p_1}(\mathbb{R}^n) \to L^{q_1}(\mathbb{R}^n)\big),$$

then T *belongs to*

$$L\big(L^{p_\theta}(\mathbb{R}^n) \to L^{q_\theta}(\mathbb{R}^n)\big) \quad \text{for each } \theta \in (0, 1),$$

too, where

$$\frac{1}{p_\theta} = \frac{1 - \theta}{p_0} + \frac{\theta}{p_1} \quad \text{and} \quad \frac{1}{q_\theta} = \frac{1 - \theta}{q_0} + \frac{\theta}{q_1}.$$

Moreover, the following norm estimates are true:

$$\|T\|_{L(L^{p_\theta}(\mathbb{R}^n) \to L^{q_\theta}(\mathbb{R}^n))} \le \|T\|_{L(L^{p_0}(\mathbb{R}^n) \to L^{q_0}(\mathbb{R}^n))}^{1-\theta} \|T\|_{L(L^{p_1}(\mathbb{R}^n) \to L^{q_1}(\mathbb{R}^n))}^{\theta}.$$

One application of this proposition is in proving Young's inequality.

Proposition 24.5.2 (Young's Inequality) *Let* $f \in L^r(\mathbb{R}^n)$ *and* $g \in L^p(\mathbb{R}^n)$ *be two given functions. Then, the following estimates hold for the convolution* $u := f * g$:

$$\|u\|_{L^q} \leq \|f\|_{L^r} \|g\|_{L^p} \ \text{for all} \ 1 \leq p \leq q \leq \infty \ \text{and} \ 1 + \frac{1}{q} = \frac{1}{r} + \frac{1}{p}.$$

Proof First, we use

$$\|u\|_{L^1} \leq \|f\|_{L^1} \|g\|_{L^1} \ \text{and} \ \|u\|_{L^\infty} \leq \|f\|_{L^1} \|g\|_{L^\infty}.$$

Proposition 24.5.1 implies

$$\|u\|_{L^q} \leq \|f\|_{L^1} \|g\|_{L^q} \ \text{for all} \ q \in [1, \infty].$$

Finally, taking account of Hölder's inequality

$$\|u\|_{L^\infty} \leq \|f\|_{L^p} \|g\|_{L^q}, \ \frac{1}{q} + \frac{1}{p} = 1,$$

and again of Proposition 24.5.1, leads to the desired statement.
Sometimes one needs interpolation between Sobolev spaces (see Sect. 12.1.3). Here we refer to the following interpolation result from [168], Theorem A.10.

Proposition 24.5.3 *Let the linear operator* T *satisfy:*

$$T \ : \ W_1^n(\mathbb{R}^n) \rightarrow L^\infty(\mathbb{R}^n), \ \text{bounded with norm} \ M_0,$$

$$T \ : \ L^2(\mathbb{R}^n) \rightarrow L^2(\mathbb{R}^n), \ \text{bounded with norm} \ M_1.$$

Then, there exist constants $C_1 = C_1(q, n)$ *and* $C_2 = C_2(q, n)$ *such that the operator* T *satisfies the following mapping properties, too:*

$$T \ : \ W_p^{N_p}(\mathbb{R}^n) \rightarrow L^q(\mathbb{R}^n), \ \text{bounded with norm} \ M_q \leq C_1 M_0^{1-\theta} M_1^\theta,$$

$$T \ : \ H_p^{N_p}(\mathbb{R}^n) \rightarrow L^q(\mathbb{R}^n), \ \text{bounded with norm} \ M_q \leq C_2 M_0^{1-\theta} M_1^\theta$$

with $p \in (1, 2)$, $\frac{1}{p} + \frac{1}{q} = 1$, $\theta = \frac{2}{q}$ *and* $N_p > n(\frac{1}{p} - \frac{1}{q})$.
The following inequality can be found in [55], Part I, Theorem 9.3.

Proposition 24.5.4 (Classical Gagliardo-Nirenberg Inequality) *Let* $j, m \in \mathbb{N}$ *with* $j < m$, *and let* $u \in C_0^m(\mathbb{R}^n)$, *i.e.* $u \in C^m(\mathbb{R}^n)$ *with compact support. Let* $\theta \in [\frac{j}{m}, 1]$, *and let* p, q, r *in* $[1, \infty]$ *be such that*

$$j - \frac{n}{q} = \left(m - \frac{n}{r} \right)\theta - \frac{n}{p}(1 - \theta).$$

Then,

$$\|D^j u\|_{L^q} \leq C_{n,m,j,p,r,\theta} \|D^m u\|_{L^r}^{\theta} \|u\|_{L^p}^{1-\theta}$$

provided that

$$\left(m - \frac{n}{r}\right) - j \notin \mathbb{N}, \quad \text{that is,} \quad \frac{n}{r} > m - j \quad \text{or} \quad \frac{n}{r} \notin \mathbb{N}.$$

If

$$\left(m - \frac{n}{r}\right) - j \in \mathbb{N},$$

then Gagliardo-Nirenberg inequality holds provided that $\theta \in [\frac{j}{m}, 1)$.

Remark 24.5.1 Let us give some explanations. If $j = 0$, $m = 1$ and $r = p = 2$, then the Gagliardo-Nirenberg inequality reduces to the special Gagliardo-Nirenberg inequality

$$\|u\|_{L^q} \leq C \|\nabla u\|_{L^2}^{\theta(q)} \|u\|_{L^2}^{1-\theta(q)},$$

where $\theta(q)$ is given from the equation

$$-\frac{n}{q} = \left(1 - \frac{n}{2}\right)\theta(q) - \frac{n}{2}(1 - \theta(q)) = \theta(q) - \frac{n}{2}.$$

It is clear that $\theta(q) \geq 0$ if and only if $q \geq 2$. Analogously, $\theta(q) \leq 1$ if and only if either $n = 1, 2$ or $q \leq \frac{2n}{n-2}$. Applying a density argument, the above inequality holds for any $u \in H^1(\mathbb{R}^n)$. Assuming $q < \infty$, then the special Gagliardo-Nirenberg inequality holds for any finite $q \geq 2$ if $n = 1, 2$ and for any $q \in [2, \frac{2n}{n-2}]$ if $n \geq 3$. Numerous generalizations of the classical Gagliardo-Nirenberg inequality exist. As an example, we present the following fractional Gagliardo-Nirenberg type inequality from [72].

Proposition 24.5.5 *The generalized Gagliardo-Nirenberg inequality*

$$\|u\|_{\dot{B}_{p,q}^s} \leq C \|u\|_{\dot{B}_{p_0,\infty}^{s_0}}^{1-\theta} \|u\|_{\dot{B}_{p_1,\infty}^{s_1}}^{\theta}$$

holds for all $u \in \dot{B}_{p_0,\infty}^{s_0}(\mathbb{R}^n) \cap \dot{B}_{p_1,\infty}^{s_1}(\mathbb{R}^n)$ *if and only if*

$$\frac{n}{p} - s = (1 - \theta)\left(\frac{n}{p_0} - s_0\right) + \theta\left(\frac{n}{p_1} - s_1\right), \quad \frac{n}{p_0} - s_0 \neq \frac{n}{p_1} - s_1,$$

$$s \leq (1 - \theta)s_0 + \theta s_1, \quad \text{and} \quad p_0 = p_1 \quad \text{if} \quad s = (1 - \theta)s_0 + \theta s_1,$$

where $0 < q < \infty$, $0 < p, p_0, p_1 \leq \infty$, $s, s_0, s_1 \in \mathbb{R}^1$, $\theta \in (0, 1)$.

We use the following corollary from Proposition 24.5.5.

Corollary 24.5.1 *Let $a \in (0, \sigma)$. Then, we have the following inequality for $m \in (1, \infty)$:*

$$\||D|^a u\|_{L^q} \leq C \||D|^\sigma u\|_{L^m}^{\theta_{a,\sigma}(q,m)} \|u\|_{L^m}^{1-\theta_{a,\sigma}(q,m)} \quad \text{for all } u \in H_m^\sigma(\mathbb{R}^n),$$

where

$$\frac{a}{\sigma} \leq \theta_{a,\sigma}(q,m) < 1 \quad \text{and} \quad \theta_{a,\sigma}(q,m) = \frac{n}{\sigma}\left(\frac{1}{m} - \frac{1}{q} + \frac{a}{n}\right),$$

$$\text{hence, } m \leq q < \frac{mn}{[n + m(a - \sigma)]^+}.$$

Proof We use the notations from the monograph [179]. The operator $|D|^a$ generates an isomorphism from $L^p(\mathbb{R}^n)$ onto $\dot{H}_p^{-a}(\mathbb{R}^n)$ for $p \in (1, \infty)$ and $a \in \mathbb{R}^1$ (see [7] or [207]). The space $\dot{H}_p^{-a}(\mathbb{R}^n)$ coincides with $\dot{F}_{p,2}^{-a}(\mathbb{R}^n)$ for $p \in (1, \infty)$ (see [192]). The continuous embedding

$$\dot{B}_{p,\min\{p,2\}}^s(\mathbb{R}^n) \hookrightarrow \dot{F}_{p,2}^s(\mathbb{R}^n) \hookrightarrow \dot{B}_{p,\infty}^s(\mathbb{R}^n)$$

(see [207]) implies the inequality

$$\||D|^a u\|_{L^q} \leq C \|u\|_{\dot{B}_{q,\min\{q,2\}}^a}.$$

Now we apply the Gagliardo-Nirenberg inequality from Proposition 24.5.5 in the form

$$\|u\|_{\dot{B}_{q,\min\{q,2\}}^a} \leq C \|u\|_{\dot{B}_{m,\infty}^\sigma}^{\theta_{a,\sigma}(q,m)} \|u\|_{\dot{B}_{m,\infty}^0}^{1-\theta_{a,\sigma}(q,m)},$$

where all assumptions for its application are satisfied. Finally, the desired inequality follows by the chain of inequalities

$$\||D|^a u\|_{L^q} \leq C \|u\|_{\dot{B}_{m,\infty}^\sigma}^{\theta_{a,\sigma}(q,m)} \|u\|_{\dot{B}_{m,\infty}^0}^{1-\theta_{a,\sigma}(q,m)} \leq C \||D|^\sigma u\|_{\dot{F}_{m,2}^0}^{\theta_{a,\sigma}(q,m)} \|u\|_{\dot{F}_{m,2}^0}^{1-\theta_{a,\sigma}(q,m)}$$

$$\leq C \||D|^\sigma u\|_{L^m}^{\theta_{a,\sigma}(q,m)} \|u\|_{L^m}^{1-\theta_{a,\sigma}(q,m)}.$$

This completes the proof.

Remark 24.5.2 The statement of Corollary 24.5.1 remains true for

$$\frac{a}{\sigma} \leq \theta_{a,\sigma}(q,m) \leq 1, \quad \text{hence, } m \leq q \leq \frac{mn}{n + m(a - \sigma)}$$

(see [159]).

Sometimes the following result from [179] for fractional powers is very helpful.

Proposition 24.5.6 *Let* $p > 1$ *and* $v \in H_m^s(\mathbb{R}^n)$, *where* $s \in \left(\dfrac{n}{m}, p\right)$. *The following estimates then hold:*

$$\||v|^p\|_{H_m^s} \le C\|v\|_{H_m^s}\|v\|_{L^\infty}^{p-1},$$

$$\|v|v|^{p-1}\|_{H_m^s} \le C\|v\|_{H_m^s}\|v\|_{L^\infty}^{p-1}.$$

We derive the following corollary from Proposition 24.5.6.

Corollary 24.5.2 *The following estimates hold under the assumptions of Proposition 24.5.6:*

$$\||v|^p\|_{\dot{H}_m^s} \le C\|v\|_{\dot{H}_m^s}\|v\|_{L^\infty}^{p-1},$$

$$\|v|v|^{p-1}\|_{\dot{H}_m^s} \le C\|v\|_{\dot{H}_m^s}\|v\|_{L^\infty}^{p-1}.$$

Proof We only prove the first inequality. For this reason we write the estimate from Proposition 24.5.6 in the form

$$\||v|^p\|_{\dot{H}_m^s} + \||v|^p\|_{L^m} \le C\big(\|v\|_{\dot{H}_m^s} + \|v\|_{L^m}\big)\|v\|_{L^\infty}^{p-1}.$$

Using instead of v the dilation $v_\lambda(\cdot) := v(\lambda \cdot)$ in the last inequality, we obtain with

$$\|u_\lambda\|_{\dot{H}_m^s} = \lambda^{s-\frac{n}{m}}\|u\|_{\dot{H}_m^s} \quad \text{and} \quad \|u_\lambda\|_{L^m} = \lambda^{-\frac{n}{m}}\|u\|_{L^m}$$

and with λ to infinity the desired inequality. The other inequality can be proved in the same way.

The Littlewood-Paley decomposition is a localization procedure in the frequency space for tempered distributions (cf. with Definitions 24.10 and 24.11). One of the main motivations for introducing such a localization when dealing with nonlinear partial differential equations is because the derivatives act almost as homotheties on distributions with Fourier transform supported in a ball or an annulus. More precisely, we have the following proposition.

Proposition 24.5.7 (Bernstein's Inequalities) *Let D be an annulus and B a ball. Then, there exists a constant C such that for any nonnegative integer k, any couple of real (p, q) such that $q \ge p \ge 1$ and for any function $u \in L^p(\mathbb{R}^n)$ with supp $(F(u)) \subset \lambda B$ for some $\lambda > 0$, we have*

$$\sup_{|\alpha|=k} \|\partial_x^\alpha u\|_{L^q} \le C^{k+1}\lambda^{k+n(\frac{1}{p}-\frac{1}{q})}\|u\|_{L^p}.$$

On the other hand, if supp $(F(u)) \subset \lambda D$ for some $\lambda > 0$, then

$$C^{-k-1}\lambda^k\|u\|_{L^p} \leq \sup_{|\alpha|=k} \|\partial_x^\alpha u\|_{L^p} \leq C^{k+1}\lambda^k\|u\|_{L^p}.$$

The proof of decay estimates or blow up behavior of solutions to nonlinear Cauchy problems often relies on ordinary differential inequalities.

Proposition 24.5.8 *Let $y = y(t)$ be a bounded nonnegative function on the interval $[0, T)$, $T > 0$, satisfying the integral inequality*

$$y(t) \leq k_0(1 + t)^{-\alpha} + k_1 \int_0^t (1 + t - s)^{-\beta}(1 + s)^{-\gamma} y(s)^\mu \, ds$$

for some constants $k_0, k_1 > 0$, $\alpha, \beta, \gamma \geq 0$ and $0 \leq \mu < 1$. Then, we have the estimate

$$y(t) \leq C(1 + t)^{-\theta}$$

for some constant $C > 0$ and

$$\theta = \min\left\{\alpha; \beta; \frac{\gamma}{1-\mu}; \frac{\beta + \gamma - 1}{1 - \mu}\right\}$$

with an exception given in the case of $\alpha \geq \tilde{\theta}$ and

$$\tilde{\theta} := \min\left\{\beta; \frac{\gamma}{1-\mu}\right\} = \frac{\beta + \gamma - 1}{1 - \mu} \leq 1,$$

whereas

$$y(t) \leq C(1 + t)^{-\tilde{\theta}}(\log(2 + t))^{\frac{1}{1-\mu}}.$$

Remark 24.5.3 The conclusion of Proposition 24.5.8 is also true for the case $\mu = 1$. In particular, if $\gamma > 0$ and $\beta + \gamma - 1 > 0$, we may take $\theta = \min\{\alpha; \beta\}$.

Proof First we consider the case $\mu = 0$. Let us divide the interval $[0, t]$ into two subintervals $[0, \frac{t}{2}]$ and $[\frac{t}{2}, t]$. It holds

$$\frac{1}{2}(1 + t) \leq (1 + t - s) \leq 1 + t \text{ for any } s \in \left[0, \frac{t}{2}\right],$$

$$\frac{1}{2}(1 + t) \leq (1 + s) \leq 1 + t \text{ for any } s \in \left[\frac{t}{2}, t\right].$$

Hence, using the change of variables $\tau = t - s$, if needed, we get

$$I(t) := \int_0^t (1 + t - s)^{-\beta} (1 + s)^{-\gamma} \, ds$$

$$\leq (1 + t)^{-\beta} \int_0^{\frac{t}{2}} (1 + s)^{-\gamma} \, ds + (1 + t)^{-\gamma} \int_{\frac{t}{2}}^t (1 + t - s)^{-\beta} \, ds$$

$$= (1 + t)^{-\beta} \int_0^{\frac{t}{2}} (1 + s)^{-\gamma} \, ds + (1 + t)^{-\gamma} \int_0^{\frac{t}{2}} (1 + \tau)^{-\beta} \, d\tau$$

$$\approx (1 + t)^{-\min\{\beta;\gamma\}} \int_0^{\frac{t}{2}} (1 + s)^{-\max\{\beta;\gamma\}} \, ds.$$

Therefore,

$$I(t) \leq C \begin{cases} (1 + t)^{-\min\{\beta;\gamma\}} & \text{if } \max\{\beta;\gamma\} > 1, \\ (1 + t)^{-\min\{\beta;\gamma\}} \log(2 + t) & \text{if } \max\{\beta;\gamma\} = 1, \\ (1 + t)^{1-\beta-\gamma} & \text{if } \max\{\beta;\gamma\} < 1. \end{cases}$$

The proof of the desired estimate follows immediately for $\mu = 0$. If $0 < \mu < 1$, then we define

$$M(t) := \sup_{0 \leq s \leq t} (1 + s)^{\theta} y(s).$$

So, we may write

$$y(t) \leq k_0 (1 + t)^{-\alpha} + k_1 \int_0^t (1 + t - s)^{-\beta} (1 + s)^{-\gamma - \mu\theta} \, ds \, M(t)^{\mu}.$$

If $\max\{\beta; \gamma + \mu\theta\} \neq 1$, following the ideas to estimate $I(t)$, we get

$$y(t) \leq k_0 (1 + t)^{-\alpha} + C(1 + t)^{-\theta^{\sharp}} M(t)^{\mu},$$

with $\theta^{\sharp} = \min\{\beta; \gamma + \mu\theta; \beta + \gamma + \mu\theta - 1\}$. One may verify that $\min\{\alpha; \theta^{\sharp}\} = \theta$. Hence,

$$(1 + t)^{\theta} y(t) \leq k_0 + CM(t)^{\mu}.$$

Thanks to $0 < \mu < 1$, this inequality implies $M(t) \leq C$ and the proof is concluded. The exceptional case $\max\{\beta; \gamma + \mu\theta\} = 1$ can be treated in a similar way.
In Sect. 20.1.2 we applied the following version of Kato's lemma to prove a blow up behavior of solutions to the Cauchy problem for semilinear wave equations.

Proposition 24.5.9 *Suppose $F \in C^2[a, b)$ and assume that for $a \leq t < b$ we have*

$$F(t) \geq C_0(k + t)^r, \quad F''(t) \geq C_1(k + t)^{-q}F(t)^p,$$

for some positive constants C_0, C_1 and k. If $p > 1$, $r \geq 1$ and $(p-1)r > q-2$, then b must be finite.

Proof By the hypotheses of the lemma we get

$$F''(t) \geq C_1(k + t)^{-q}C_0^p(k + t)^{pr} \geq C(k + t)^{pr-q}.$$

After integration one has

$$F'(t) - F'(a) \geq C \int_a^t (k + s)^{pr-q} \, ds.$$

Taking into consideration $pr - q \geq -1$, the last inequality implies that unless b is finite, $F'(t)$ must be positive for t sufficiently large. Thus, one may assume that there exists an a_0 such that $a < a_0 < b$ and

$$F'(t) > 0 \quad \text{for all } t \in [a_0, b).$$

It follows from the assumptions on p, q and r that

$$\frac{1}{p} < 1 - \frac{q-2}{pr}.$$

Hence, there is a $\theta \in (0, 1)$ such that

$$\frac{1}{p} < \theta < 1 - \frac{q-2}{pr}.$$

By interpolating between the assumed inequalities, one has

$$F''(t) \geq C_1(k + t)^{-q} F(t)^{\theta p+(1-\theta)p} \geq C(k + t)^{rp(1-\theta)-q} F(t)^{\theta p}.$$

Our choice of θ implies $\alpha = \theta p > 1$ and $\beta = q - rp(1 - \theta) < 2$. Without loss of generality one can set $\beta \geq 0$. This leads to

$$F''(t) F'(t) \geq C(k + t)^{-\beta} F(t)^{\alpha} F'(t).$$

Integration of the last inequality yields

$$\frac{1}{2}\big(F'(t)^2 - F'(a_0)^2\big) \geq C \int_{a_0}^{t} (k+s)^{-\beta} F(s)^{\alpha} F'(s)\, ds$$

$$\geq C_2 (k+t)^{-\beta}\big(F(t)^{1+\alpha} - F(a_0)^{1+\alpha}\big).$$

Note that we can choose the constant C_2 so small such that

$$F'(a_0)^2 \geq 2C_2(k+a_0)^{-\beta} F(a_0)^{1+\alpha}.$$

Here we take account of $F'(a_0) > 0$. It follows that

$$F'(t)^2 \geq 2C_2(k+t)^{-\beta} F(t)^{1+\alpha},$$

and, therefore,

$$F(t)^{-\frac{1+\alpha}{2}} F'(t) \geq C(k+t)^{-\frac{\beta}{2}}$$

for all $a_0 < t < b$. One final integration yields ($\alpha > 1$)

$$F(a_0)^{\frac{1-\alpha}{2}} - F(t)^{\frac{1-\alpha}{2}} \geq C\big((k+t)^{1-\frac{\beta}{2}} - (k+a_0)^{1-\frac{\beta}{2}}\big).$$

Since $\beta < 2$, it is clear that the time variable t can not be arbitrarily large.

Many thanks to Winfried Sickel (Jena) for useful discussions on the content of this section.

References

1. R. Adams, *Sobolev Spaces* (Academic, New York, 1975)
2. F. Asakura, Existence of a global solution to a semilinear wave equation with slowly decreasing initial data in three space dimensions. Commun. Partial Differ. Equ. **11**, 1459–1487 (1986)
3. H. Bahouri, J.Y. Chemin, R. Danchin, *Fourier Analysis and Nonlinear Partial Differential Equations*. Grundlehren der Mathematischen Wissenschaften, vol. 343 (Springer, Berlin-Heidelberg, 2011)
4. M.B. Balk, *Polyanalytic Functions*. Mathematical Research, vol. 63 (Akademie, Berlin, 1991)
5. M.S. Baouendi, C. Goulaouic, Cauchy problems with characteristic initial hypersurface. Commun. Pure Appl. Math. **26**, 455–475 (1973)
6. M. Ben-Artzi, S. Klainerman, Decay and regularity for the Schrödinger equation. J. Anal. Math. **58**, 25–37 (1992)
7. J. Bergh, J. Löfström, *Interpolation Spaces. An Introduction*. Grundlagen der Mathematischen Wissenschaften, vol. 223 (Springer, Berlin-New York, 1976)
8. R.M. Blumenthal, R.K. Getoor, Some theorems on stable processes. Trans. Am. Math. Soc. **95**, 263–273 (1960)
9. S. Bochner, Integration von Funktionen, deren Werte die Elemente eines Vektoraumes sind. Fund. Math. **20**, 262–276 (1933)
10. C. Böhme, M. Reissig, A scale-invariant Klein-Gordon model with time-dependent potential. Ann. Univ. Ferrara Sez. VII Sci. Mat. **58**, 229–250 (2012)
11. C. Böhme, M. Reissig, Energy bounds for Klein-Gordon equations with time-dependent potential. Ann. Univ. Ferrara Sez. VII Sci. Mat. **59**, 31–55 (2013)
12. P. Brenner, On $L_p - L_{p'}$ estimates for the wave equation. Math. Z. **145**, 251–254 (1975)
13. P. Brenner, On space-time means and strong global solutions of nonlinear hyperbolic equations. Math. Z. **201**, 45–55 (1989)
14. H. Brezis, A. Friedman, Nonlinear parabolic equations involving measures as initial conditions. J. Math. Pures Appl. **62**(1), 73–97 (1983)
15. H. Brezis, W.A. Strauss, Semilinear second order elliptic equation in L^1. J. Math. Soc. Jpn. **25**, 565–590 (1973)
16. H. Brezis, L.A. Peletier, D. Terman, A very singular solution of the heat equation with absorption. Arch. Ration. Mech. Anal. **95**(3), 185–209 (1986)
17. T.B.N. Bui, M. Reissig, The interplay between time-dependent speed of propagation and dissipation in wave models, in *Fourier Analysis*, ed. by M. Ruzhansky, V. Turunen. Trends in Mathematics (Birkhäuser, Basel, 2014), pp. 9–45

© Springer International Publishing AG 2018
M.R. Ebert, M. Reissig, *Methods for Partial Differential Equations*,
https://doi.org/10.1007/978-3-319-66456-9

18. T. Cazenave, *An Introduction to Nonlinear Schrödinger Equations*. Textos de Métodos Matemáticos, vol. 26 (Universidade Federal de Rio de Janeiro, Rio de Janeiro, 1996)

19. T. Cazenave, *Semilinear Schrödinger Equations*. Courant Lecture Notes in Mathematics, vol. 10 (American Mathematical Society, Providence, RI, 2003)

20. T. Cazanave, A. Haraux, *An Introduction to Semilinear Evolution Equations*. Oxford Lecture Series in Mathematics and its Applications, vol. 13 (Oxford Science Publishers, Oxford, 1998)

21. T. Cazanave, F. Dickstein, F.B. Weissler, Multi-scale multi-profile global solutions of parabolic equations in $\mathbb{R}^n$. Discrete Contin. Dyn. Syst. S **5**(3), 449–472 (2012)

22. R.C. Charão, C.R. da Luz, R. Ikehata, Sharp decay rates for wave equations with a fractional damping via new method in the Fourier space. J. Math. Anal. Appl. **408**, 247–255 (2013)

23. H. Chihara, Smoothing effects of dispersive pseudodifferential equations. Commun. Partial Differ. Equ. **27**, 1953–2005 (2002)

24. F. Colombini, Energy estimates at infinity for hyperbolic equations with oscillating coefficients. J. Differ. Equ. **231**(2), 598–610 (2006)

25. P. Constantin, J.C. Saut, Local smoothing properties of dispersive equations. J. Am. Math. Soc. **1**, 413–439 (1988)

26. A. Córdoba, D. Córdoba, A maximum principle applied to quasi-geostrophic equations. Commun. Math. Phys. **249**, 511–528 (2004)

27. R. Courant, D. Hilbert, *Methods of Mathematical Physics*, vol. 2 (Wiley-Interscience, New-York, London, 1962)

28. M.G. Crandall, T.M. Liggett, Generation of semigroups of nonlinear transformations on general Banach spaces. Am. J. Math. **93**, 265–298 (1971)

29. M. D'Abbicco, A benefit from the L^∞ smallness of initial data for the semilinear wave equation with structural damping, in *Current Trends in Analysis and its Applications*, ed. by V. Mityushev, M. Ruzhansky. Trends in Mathematics. Proceedings of the 9th ISAAC Congress, Krakow (Birkhäuser, Basel, 2014), pp. 209–216

30. M. D'Abbicco, The threshold of effective damping for semilinear wave equations. Math. Meth. Appl. Sci. **38**, 1032–1045 (2015)

31. M. D'Abbicco, M.R. Ebert, A class of dissipative wave equations with time-dependent speed and damping. J. Math. Anal. Appl. **399**, 315–332 (2013)

32. M. D'Abbicco, M.R. Ebert, Diffusion phenomena for the wave equation with structural damping in the $L^p - L^q$ framework. J. Differ. Equ. **256**, 2307–2336 (2014)

33. M. D'Abbicco, M.R. Ebert, A classification of structural dissipations for evolution operators. Math. Meth. Appl. Sci. **39**, 2558–2582 (2016)

34. M. D'Abbicco, M.R. Ebert, A new phenomenon in the critical exponent for structurally damped semi-linear evolution equations. Nonlinear Anal. **149**, 1–40 (2017)

35. M. D'Abbicco, S. Lucente, A modified test function method for damped wave equations. Adv. Nonlinear Stud. **13**, 867–892 (2013)

36. M. D'Abbicco, S. Lucente, NLWE with a special scale-invariant damping in odd space dimension, in *Discrete and Continuous Dynamical Systems*. AIMS Proceedings (2015), pp. 312–319

37. M. D'Abbicco, M. Reissig, Semi-linear structural damped waves. Math. Meth. Appl. Sci. **37**, 1570–1592 (2014)

38. M. D'Abbicco, S. Lucente, M. Reissig, Semi-linear wave equations with effective damping. Chin. Ann. Math. Ser. B **34**(3), 345–380 (2013)

39. M. D'Abbicco, S. Lucente, M. Reissig, A shift in the Strauss exponent for semilinear wave equations with a not effective damping. J. Differ. Equ. **259**, 5040–5073 (2015)

40. L. D'Ambrosio, S. Lucente, Nonlinear Liouville theorems for Grushin and Tricomi operators. J. Differ. Equ. **193**, 511–541 (2003)

41. W. Dan, Y. Shibata, On a local energy decay of solutions of a dissipative wave equation. Funkcial. Ekvac. **38**, 545–568 (1995)

42. L. Debnath, D. Bhatta, *Integral Transforms and Their Applications*, 2nd edn. (Chapman & Hall, Boca Raton-London-New York, 2007)

43. A. Domarkas, On the blowing up of solutions of a system of nonlinear Schrödinger equations. Lith. Math. J. **35**, 144–150 (1995)
44. J. Duoandikoetxea, E. Zuazua, Moments, masses de Dirac et décomposition de fonctions. C. R. Acad. Sci. Paris Sér. I Math. **315**, 693–698 (1992)
45. M.R. Ebert, M. Reissig, The influence of oscillations on global existence for a class of semilinear wave equations. Math. Meth. Appl. Sci. **34**, 1289–1307 (2011)
46. M.R. Ebert, M. Reissig, Theory of damped wave models with integrable and decaying in time speed of propagation. J. Hyperbol. Differ. Equ. **13**(2), 417–439 (2016)
47. M.R. Ebert, R.A. Kapp, W.N. Nascimento, M. Reissig, Klein-Gordon type wave models with non-effective time-dependent potential, in *Analytic Methods of Analysis and Differential Equations*, ed. by M.V. Dubatovskaya, S.V. Rogosin. AMADE 2012 (Cambridge Scientific Publishers, Cambridge, 2014), pp. 143–161
48. M.R. Ebert, R.A. Kapp, T. Picon, $L^1 - L^p$ estimates for radial solutions of the wave equation and application. Ann. Mat. Pura Appl. **195**, 1081–1091 (2016)
49. Yu. Egorov, On an example of a linear hyperbolic equation without solutions. C. R. Acad. Sci. Paris **317**, 1149–1153 (1993)
50. M. Escobedo, O. Kavian, Variational problems related to self-similar solutions of the heat equation. Nonlinear Anal. **11**(10), 1103–1133 (1987)
51. M. Escobedo, O. Kavian, Asymptotic behavior of positive solutions of a non-linear heat equation. Houst. J. Math. **14**, 39–50 (1988)
52. M. Escobedo, O. Kavian, H. Matano, Large time behavior of solutions of a dissipative semilinear heat equation. Commun. Partial Differ. Equ. **20**(7–8), 1427–1452 (1995)
53. L.C. Evans, *Partial Differential Equations*, 2nd edn. Graduate studies in Mathematics, vol. 19 (American Mathematical Society, Providence, RI, 2010)
54. A. Friedman, A strong maximum principle for weakly subparabolic functions. Pac. J. Math. **11**, 175–184 (1961)
55. A. Friedman, *Partial Differential Equations*. Corrected reprint of the original edition. (Robert E. Krieger Publishing Co., Huntington, New York, 1976)
56. H. Fujita, On the blowing-up of solutions of the Cauchy problem for $u_t = \Delta u + u^{1+\alpha}$. J. Fac. Sci. Univ. of Tokyo, Sect. 1 **13**, 109–124 (1966)
57. G.P. Galdi, An introduction to the Navier-Stokes initial-boundary value problem, in *Fundamental Directions in Mathematical Fluid Mechanics*. Advances in Mathematical Fluid Mechanics (Birkhäuser, Basel, 2000), pp. 1–70
58. G.P. Galdi, *An Introduction to the Mathematical Theory of the Navier-Stokes Equations*. Springer Monographs in Mathematics (Springer, New York, 2011)
59. V. Georgiev, H. Lindblad, C.D. Sogge, Weighted Strichartz estimates and global existence for semilinear wave equations. Am. J. Math. **119**, 1291–1319 (1997)
60. V. Georgiev, N. Tzvetkov, N. Visciglia, On the regularity of the flow map associated with the 1d cubic periodic half-wave equation. Differ. Integral Equ. **29**(1/2), 183–200 (2016)
61. J. Ginibre, G. Velo, On a class of nonlinear Schrödinger equations. I. The Cauchy problem, general case. J. Funct. Anal. **32**, 1–32 (1979)
62. J. Ginibre, G. Velo, The global Cauchy problem for the nonlinear Klein-Gordon equation. Math. Z. **189**, 487–505 (1985)
63. R.T. Glassey, On the asymptotic behavior of nonlinear wave equations. Trans. Am. Math. Soc. **182**, 187–200 (1973)
64. R.T. Glassey, On the blowing up of solutions to the Cauchy problem for nonlinear Schrödinger equation. J. Math. Phys. **18**, 1794–1797 (1977)
65. R.T. Glassey, Existence in the large for $\Box u = F(u)$ in two space dimensions. Math. Z. **178**, 233–261 (1981)
66. R.T. Glassey, Finite-time blow-up for solutions of nonlinear wave equations. Math. Z. **177**, 323–340 (1981)
67. A. Gmira, L. Veron, Large time behavior of the solutions of a semilinear parabolic equation in $\mathbb{R}^n$. J. Differ. Equ. **53**, 258–276 (1984)

68. M. Goldberg, L. Vega, N. Visciglia, Counterexamples of Strichartz inequalities for Schrödinger equations with repulsive potentials. Int. Math. Res. Not. **2006**, 1–16 (2006)
69. K. Gröchenig, *Foundations of Time-Frequency Analysis* (Birkhäuser, Boston, 2001)
70. R.B. Guenther, J.W. Lee, *Partial Differential Equations of Mathematical Physics and Integral Equations* (Dover Publications, Prentice Hall, New-York, 1988)
71. B. Gustaffson, Applications of variational inequalities to a moving boundary value problem for Hele-Shaw flows. SIAM J. Math. Anal. **16**, 279–300 (1985)
72. H. Hajaiej, L. Molinet, T. Ozawa, B. Wang, Necessary and sufficient conditions for the fractional Gagliardo-Nirenberg inequalities and applications to Navier-Stokes and generalized boson equations, in *Harmonic Analysis and Nonlinear Partial Differential Equations*. RIMS Kokyuroku Bessatsu, B26 (Research Institute for Mathematical Sciences, Kyoto, 2011), pp. 159–175
73. Y. Hasegawa, On the initial-value problems with data on a double characteristic. J. Math. Kyoto Univ. **11**, 357–372 (1971)
74. A. Haraux, F.B. Weissler, Non-uniqueness for a semilinear initial value problem. Indiana Univ. Math. J. **31**(2), 167–189 (1982)
75. K. Hayakawa, On nonexistence of global solutions of some semi-linear parabolic differential equations. Proc. Jpn. Acad. **49**, 503–505 (1973)
76. H. Helmholtz, Über Integrale der hydrodynamischen Gleichungen, welcher der Wirbelbewegungen entsprechen. J. Reine Angew. Math. **55**, 25–55 (1858)
77. F. Hirosawa, On the asymptotic behavior of the energy for the wave equations with time depending coefficients. Math. Ann. **339**, 819–838 (2007)
78. F. Hirosawa, Energy estimates for wave equations with time dependent propagation speeds in the Gevrey class. J. Differ. Equ. **248**(12), 2972–2993 (2010)
79. F. Hirosawa, J. Wirth, C^m-theory of damped wave equations with stabilisation. J. Math. Anal. Appl. **343**(2), 1022–1035 (2008)
80. F. Hirosawa, J. Wirth, Generalised energy conservation law for wave equations with variable propagation speed. J. Math. Anal. Appl. **358**(1), 56–74 (2009)
81. Y. Hohlov, M. Reissig, On classical solvability for Hele-Shaw moving boundary problems with kinetic undercooling regularization. Eur. J. Appl. Math. **6**, 421–439 (1995)
82. E. Holmgren, Über Systeme von linearen partiellen Differentialgleichungen. Ofversigt af kongl. Vetenskapakad. Förhandlinger **58**, 91–103 (1901)
83. T. Hoshiro, On weighted L^2-estimates of solutions to wave equations. J. Anal. Math. **72**, 127–140 (1997)
84. L. Hörmander, Estimates for translation invariant operators in L^p spaces. Acta Math. **104**, 93–140 (1960)
85. L. Hörmander, *Linear Partial Differential Operators* (Springer, Berlin, 1963)
86. L. Hörmander, *The Analysis of Linear Partial Differential Operators II. Differential Operators with Constant Coefficients* (Springer, Berlin-Heidelberg-New York, 1983)
87. L. Hörmander, *Lectures on Nonlinear Hyperbolic Differential Equations* (Springer, Berlin, 1997)
88. L. Hsiao, T. Liu, Convergence to nonlinear diffusion waves for solutions of a system of hyperbolic conservations with damping. Commun. Math. Phys. **143**, 599–605 (1992)
89. R. Ikehata, Decay estimates of solutions for the wave equations with strong damping terms in unbounded domains. Math. Meth. Appl. Sci. **24**, 659–670 (2001)
90. R. Ikehata, T. Matsuyama, Remarks on the behaviour of solutions to the linear wave equations in unbounded domains. Proc. Schl. Sci. Tokai Univ. **36**, 1–13 (2001)
91. R. Ikehata, T. Matsuyama, L^2 behaviour of solutions to the linear heat and wave equations in exterior domains. Sci. Math. Jpn. **55**, 33–42 (2002)
92. R. Ikehata, M. Natsume, Energy decay estimates for wave equations with a fractional damping. Differ. Integral Equ. **25**, 939–956 (2012)
93. R. Ikehata, K. Nishihara, Diffusion phenomenon for second order linear evolution equations. Stud. Math. **158**(2), 153–161 (2003)

94. R. Ikehata, M. Ohta, Critical exponents for semilinear dissipative wave equations in $\mathbb{R}^N$. J. Math. Anal. Appl. **269**, 87–97 (2002)

95. R. Ikehata, K. Tanizawa, Global existence of solutions for semilinear damped wave equations in $\mathbb{R}^N$ with noncompactly supported initial data. Nonlinear Anal. **61**(7), 1189–1208 (2005)

96. R. Ikehata, K. Nishihara, H. Zhao, Global asymptotics of solutions to the Cauchy problem for the damped wave equation with absorption. J. Differ. Equ. **226**, 1–29 (2006)

97. H. Iwashita, L_q-L_r estimates for solutions to the nonstationary Stokes equations in an exterior domain and the Navier-Stokes initial value problems in L_q spaces. Math. Ann. **285**, 265–288 (1989)

98. H. Jiao, Z. Zhou, An elementary proof of the blow-up for semilinear wave equation in high space dimensions. J. Differ. Equ. **189**, 355–365 (2003)

99. F. John, Blow-up of solutions of nonlinear wave equations in three space dimensions. Manuscr. Math. **28**, 235–268 (1979)

100. F. John, *Partial Differential Equations*. Applied Mathematical Sciences, vol. 1, 4th edn. (Springer, New York, 1982)

101. M. Kainane, Structural damped σ-evolution operators. Ph.D. thesis, Technical University Bergakademie Freiberg, 164 pp. (2014)

102. B.V. Kapitonov, Decrease in the solution of the exterior boundary value problem for a system of elasticity theory (in Russian). Differ. Uravn. **22**, 452–458 (1986)

103. G. Karch, Selfsimilar profiles in large time asymptotics of solutions to damped wave equations. Stud. Math. **143**(2), 175–197 (2000)

104. T. Kato, Blow-up of solutions of some nonlinear hyperbolic equations. Commun. Pure Appl. Math. **33**, 501–505 (1980)

105. T. Kato, Abstract differential equations and nonlinear mixed problems. Academia Nazionale dei Lincei, Scuola Normale Superior, Lezioni Fermiane, Pisa (1985)

106. T. Kato, K. Yajima, Some examples of smooth operators and the associated smoothing effect. Rev. Math. Phys. **1**, 481–496 (1989)

107. O. Kavian, Remarks on the large time behaviour of a nonlinear diffusion equation. Annales de l' I.H.P. Sect. C **4**, 423–452 (1987)

108. S. Kawashima, M. Nakao, K. Ono, On the decay property of solutions to the Cauchy problem of the semilinear wave equation with a dissipative term. J. Math. Soc. Jpn. **47**(4), 617–653 (1995)

109. M. Keel, T. Tao, Endpoint Strichartz estimates. Am. J. Math. **120**, 955–980 (1998)

110. M. Keel, T. Tao, Small data blow-up for semilinear Klein-Gordon equations. Am. J. Math. **121**, 629–669 (1999)

111. C.E. Kenig, G. Ponce, L. Vega, Oscillatory integrals and regularity of dispersive equations. Indiana Univ. Math. J. **40**, 33–69 (1991)

112. C.E. Kenig, G. Ponce, L. Vega, Small solutions to nonlinear Schrödinger equations. Ann. Inst. H. Poincare **10**, 255–288 (1993)

113. G. Kirchhoff, Zur Theorie der Lichtstrahlen. Ann. Phys. Chem. **18**, 663–695 (1883)

114. S. Klainerman, Global existence for nonlinear wave equations. Commun. Pure Appl. Math. **33**, 43–101 (1980)

115. S. Klainerman, Uniform decay estimates and Lorentz invariance of the classical wave equation. Commun. Pure Appl. Math. **38**, 321–332 (1985)

116. K. Kobayashi, T. Sirao, H. Tanaka, On the growing up problem for semi-linear heat equations. J. Math. Soc. Jpn. **29**, 407–424 (1977)

117. S.V. Kowalevsky, Zur Theorie der partiellen Differentialgleichungen. J. Reine Angew. Math. **80**, 1–32 (1875)

118. H. Kubo, Slowly decaying solutions for semilinear wave equations in odd space dimensions. Nonlinear Anal. **28**, 327–357 (1997)

119. N. Lai, Y. Zhou, An elementary proof of Strauss conjecture. J. Funct. Anal. **267**, 1364–1381 (2014)

120. L.D. Landau, E.M. Lifshitz, *Fluid Mechanics* (Pergamon Press, Elmsford, New York, 1959)

121. H. Lewy, An example of a smooth linear partial differential equation without solution. Ann. Math. **66**, 155–158 (1957)
122. F. Linares, G. Ponce, *Introduction to Nonlinear Dispersive Equations*. Universitext (Springer, New York, 2009)
123. H. Lindblad, Blow-up for solutions of $\Box u = |u|^p$ with small initial data. Commun. Partial Differ. Equ. **15**, 757–821 (1990)
124. H. Lindblad, C. Sogge, On existence and scattering with minimal regularity for semilinear wave equations. J. Funct. Anal. **130**, 357–426 (1995)
125. H. Lindblad, C. Sogge, Long-time existence for small amplitude semilinear wave equations. Am. J. Math. **118**, 1047–1135 (1996)
126. W. Littman, Fourier transformations of surface-carried measures and differentiability of surface averages. Bull. Am. Math. Soc. **69**, 766–770 (1963)
127. T. Mandai, Characteristic Cauchy problems for some non-Fuchsian partial differential operators. J. Math. Soc. Jpn. **45**, 511–545 (1993)
128. P. Marcati, K. Nishihara, The L^p-L^q estimates of solutions to one-dimensional damped wave equations and their application to the compressible flow through porous media. J. Differ. Equ. **191**, 445–469 (2003)
129. B. Marshall, W. Strauss, S. Wainger, $L^p - L^q$ estimates for the Klein-Gordon equation. J. Math. Pures Appl. **59**, 417–440 (1980)
130. M. Mascarello, L. Rodino, *Partial Differential Equations with Multiple Characteristics* (Akademie, Berlin, 1997)
131. A. Matsumura, On the asymptotic behavior of solutions of semi-linear wave equations. Publ. RIMS **12**, 169–189 (1976)
132. T. Matsuyama, M. Reissig, Stabilization and L^p-L^q decay estimates. Asymptot. Anal. **50**, 239–268 (2006)
133. G. Métivier, Uniqueness and approximation of solutions of first order nonlinear equations. Invent. Math. **82**, 263–282 (1985)
134. G. Métivier, Counterexamples of Hölmgren's uniqueness for analytic nonlinear Cauchy problems. Invent. Math. **112**, 217–222 (1993)
135. S.G. Michlin, *Partielle Differentialgleichungen in der Mathematischen Physik* (Akademie, Berlin, 1978)
136. A. Miyachi, On some Fourier multipliers for $H^p(\mathbb{R}^n)$. J. Fac. Sci. Univ. Tokyo Sect. IA Math. **27**, 157–179 (1980)
137. A. Miyachi, On some estimates for the wave equation in L^p and H^p. J. Fac. Sci. Univ. Tokyo Sect. IA Math. **27**, 331–354 (1980)
138. S. Mizohata, Solutions nulles et solutions non analytiques. J. Math. Kyoto Univ. **1**, 271–302 (1962)
139. S. Mizohata, *The Theory of Partial Differential Equations* (University Press, Cambridge, 1973)
140. S. Mizohata, *On the Cauchy Problem* (Academic, New York, 1985)
141. K. Mochizuki, M. Nakao, Total energy decay for the wave equation in exterior domains with a dissipation near infinity. J. Math. Anal. Appl. **326**, 582–588 (2007)
142. C.S. Morawetz, Decay for solutions of the exterior problem for the wave equation. Commun. Pure Appl. Math. **28**, 229–264 (1975)
143. M. Morse, The calculus of variations in the large. American Mathematical Society Colloquium Publications, vol. 18 (American Mathematical Society, Providence, RI, 1996)
144. K. Nakanishi, M. Ohta, On global existence of solutions to nonlinear wave equations of wave map type. Nonlinear Analysis **42**(7), 1231–1252 (2000)
145. T. Narazaki, $L^p - L^q$ estimates for damped wave equations and their applications to semi-linear problem. J. Math. Soc. Jpn. **56**, 585–626 (2004)
146. T. Narazaki, M. Reissig, L^1 estimates for oscillating integrals related to structural damped wave models, in *Studies in Phase Space Analysis with Applications to PDEs*, ed. by M. Cicognani, F. Colombini, D. Del Santo. Progress in Nonlinear Differential Equations (Birkhäuser, Basel, 2013), pp. 215–258

147. L. Nirenberg, A strong maximum principle for parabolic equations. Commun. Pure Appl. Math. **6**, 167–177 (1953)
148. L. Nirenberg, An abstract form of the nonlinear Cauchy-Kowalevskaja-theorem. J. Diff. Geom. **6**, 561–576 (1972)
149. T. Nishida, A note on a theorem of Nirenberg. J. Diff. Geom. **12**, 629–633 (1977)
150. K. Nishihara, Asymptotic behavior of solutions of quasilinear hyperbolic equations with linear damping. J. Differ. Equ. **137**, 384–395 (1997)
151. K. Nishihara, $L^p - L^q$ estimates of solutions to the damped wave equations in 3-dimensional space and their applications. Math. Z. **244**, 631–649 (2003)
152. K. Nishihara, Asymptotic behavior of solutions for a system of semilinear heat equations and the corresponding damped wave system. Osaka J. Math. **49**, 331–348 (2012)
153. K. Nishihara, Diffusion phenomena of solutions to the Cauchy problem for the damped wave equations. Sugaku Expositions **26**(1), 29–47 (2013)
154. K. Nishihara, H. Zhao, Decay properties of solutions to the Cauchy problem for the damped wave equation with absorption. J. Math. Anal. Appl. **313**, 598–610 (2006)
155. R.S.O. Nunes, W.D. Bastos, Energy decay for the linear Klein-Gordon equation and boundary control. J. Math. Anal. Appl. **414**, 934–944 (2014)
156. W. Nunes do Nascimento, Klein-Gordon models with non-effective potential. Ph.D. thesis, Universidade Federal de São Carlos/Technical University Bergakademie Freiberg, (2016), 183pp.
157. W. Nunes do Nascimento, A. Palmieri, M. Reissig, Semi-linear wave models with power non-linearity and scale-invariant time-dependent mass and dissipation. Math. Nachr. **290**, 1779–1805 (2017)
158. A. Palmieri, Linear and non-linear sigma-evolution equations. Master thesis, University of Bari (2015), 117pp.
159. A. Palmieri, M. Reissig, Semi-linear wave models with power non-linearity and scale-invariant time-dependent mass and dissipation. II, to appear in Math. Nachr. 33pp.
160. H. Pecher, L^p-Abschätzungen und klassische Lösungen für nichtlineare Wellengleichungen. I. Math. Z. **150**, 159–183 (1976)
161. J. Peral, L^p estimates for the wave equation. J. Funct. Anal. **36**, 114–145 (1980)
162. J. Petree, *New Thoughts on Besov Spaces*. Duke University Mathematical Series I (Duke University Press, Durham, 1976)
163. I.G. Petrowski, *Vorlesungen über partielle Differentialgleichungen* (Teubner, Leipzig, 1955)
164. D.T. Pham, M. Kainane, M. Reissig, Global existence for semi-linear structurally damped σ-evolution models. J. Math. Anal. Appl. **431**, 569–596 (2015)
165. F. Pizichillo, Linear and non-linear damped wave equations. Master thesis, University of Bari, (2014), 62pp.
166. N.B. Pleshchinskii, M. Reissig, Hele-Shaw flows with nonlinear kinetic undercooling regularization. Nonlinear Anal. **50**, 191–203 (2002)
167. M.-Y. Qi, On the Cauchy problem for a class of hyperbolic equations with initial data on the parabolic degenerating line. Acta Math. Sin. **8**, 521–529 (1958)
168. R. Racke, *Lectures on Nonlinear Evolution Equations. Initial Value Problems*. Aspects of Mathematics, vol. E19 (Vieweg, Braunschweig/Wiesbaden, 1992)
169. J. Ralston, Solutions of the wave equation with localized energy. Commun. Pure Appl. Math. **22**, 807–823 (1969)
170. M. Reissig, The existence and uniqueness of analytic solutions for a moving boundary problem for Hele-Shaw flows in the plane. Nonlinear Anal. **23**(5), 565–576 (1994)
171. M. Reissig, On $L_p - L_q$ estimates for solutions of a special weakly hyperbolic equation, in *Nonlinear Evolution Equations and Infinite-Dimensional Dynamical Systems*, ed. by L. Ta-Tsien. (World Scientific, River Edge, NJ, 1997), pp. 153–164
172. M. Reissig, C. Reuther, $L^p - L^q$ decay estimates for Klein-Gordon models with effective mass. Int. J. Dyn. Syst. Differ. Equ. **4**, 323–362 (2012)

173. M. Reissig, S. Rogosin, Analytical and numerical treatment of a complex model for Hele-Shaw moving boundary value problems with kinetic undercooling regularization. Eur. J. Appl. Math. **10**, 561–579 (1999)

174. M. Reissig, J. Smith, $L^p - L^q$ estimate for wave equation with bounded time-dependent coefficient. Hokkaido Math. J. **34**, 541–586 (2005)

175. M. Reissig, L. v. Wolfersdorf, A simplified proof for a moving boundary problem for Hele-Shaw flows in the plane. Ark. Mat. **31**(1), 101–116 (1993)

176. M. Reissig, K. Yagdjian, About the influence of oscillations on Strichartz-type decay estimates. Rend. Sem. Mat. Univ. Politec. Torino **58**, 375–388 (2000)

177. M. Reissig, K. Yagdjian, $L_p - L_q$ decay estimates for the solutions of strictly hyperbolic equations of second order with increasing in time coefficients. Math. Nachr. **214**, 71–104 (2000)

178. L. Rodino, *Linear Partial Differential Operators in Gevrey Spaces* (World Scientific, Singapore, 1993)

179. T. Runst, W. Sickel, *Sobolev Spaces of Fractional Order, Nemytskij Operators, and Nonlinear Partial Differential Equations*. De Gruyter Series in Nonlinear Analysis and Applications (de Gruyter, Berlin, 1996)

180. M. Ruzhansky, *Regularity Theory of Fourier Integral Operators with Complex Phases and Singularities of Affine Fibrations*. CWI Tract, vol. 131 (Stichting Mathematisch Centrum, Amsterdam, 2001)

181. M. Ruzhansky, M. Sugimoto, A new proof of global smoothing estimates for dispersive equations. Oper. Theory Adv. Appl. **155**, 65–75 (2004)

182. K. Rypdal, J.J. Rasmussen, Blow-up in nonlinear Schrödinger equation, I, II. Phys. Scripta **33**, 481–504 (1986)

183. J. Schaeffer, The equation $u_{tt} - \Delta u = |u|^p$ for the critical value of p. Proc. R. Soc. Edinb. Sect. A **101**, 31–44 (1985)

184. Y. Shibata, On the rate of decay of solutions to linear viscoelastic equation. Math. Meth. Appl. Sci. **23**, 203–226 (2000)

185. W. Sickel, H. Triebel, Hölder inequalities and sharp embeddings in function spaces of $B_{p,q}^s$ and $F_{p,q}^s$ type. Z. Anal. Anwendungen **14**(1), 105–140 (1995)

186. T.C. Sideris, Nonexistence of global solutions to semilinear wave equations in high dimensions. J. Differ. Equ. **52**, 378–406 (1984)

187. P. Sjölin, Regularity of solutions to the Schrödinger equation. Duke Math. J. **55**, 699–715 (1987)

188. S. Sjöstrand, On the Riesz means of the solution of the Schrödinger equation. Ann. Scull Norm. Sup. Pisa **24**, 331–348 (1970)

189. S.L. Sobolev, *Partial Differential Equations of Mathematical Physics* (Pergamon Press, Elmsford, New York, 1965)

190. Ch.D. Sogge, *Fourier Integrals in Classical Analysis*. Cambridge Tracts in Mathematics, vol. 105 (Cambridge University Press, Cambridge, 1993).

191. Ch.D. Sogge, *Lectures on Nonlinear Wave Equations*. Monographs in Analysis, vol. II (International Press, Boston, 1995)

192. E.M. Stein, *Singular Integrals and Differentiability Properties of Functions* (Princeton University Press, Princeton, NJ, 1970)

193. E.M. Stein, *Harmonic Analysis: Real-Variable Methods, Orthogonality, and Oscillatory Integrals*. Princeton Mathematical Series, vol. 43 (Princeton University Press, Princeton, NJ, 1993)

194. W.A. Strauss, On weak solutions of semi-linear hyperbolic equations. An. Acad. Brasil. Cienc. **42**, 645–651 (1970)

195. W.A. Strauss, *Nonlinear Scattering Theory*. Scattering Theory in Mathematical Physics (Reidel, Dordrecht, 1974), pp. 53–78

196. W.A. Strauss, Nonlinear scattering theory at low energy. J. Funct. Anal. **41**, 110–133 (1981)

197. W.A. Strauss, *Nonlinear Wave Equations*. CBMS Series, vol. 73 (American Mathematical Society, Providence, RI, 1989)

198. R. Strichartz, Convolutions with kernels having singularities on a sphere. Trans. Am. Math. Soc. **148**, 461–471 (1970)
199. M. Sugimoto, Global smoothing properties of generalized Schrödinger equations. J. Anal. Math. **76**, 191–204 (1998)
200. H. Takamura, Improved Kato's lemma on ordinary differential inequality and its application to semilinear wave. Nonlinear Anal. **125**, 227–240 (2015)
201. H. Takamura, K. Wakasa, The sharp upper bound of the lifespan of solutions to critical semilinear wave equations in high dimensions. J. Differ. Equ. **251**, 1157–1171 (2011)
202. H. Takamura, K. Wakasa, Almost global solutions of semilinear wave equations with the critical exponent in high dimensions. Nonlinear Anal. **109**, 187–229 (2014)
203. R. Temam, *Navier-Stokes Equations. Theory and Numerical Analysis*. Studies in Mathematics and its Applications, vol. 2 (North-Holland, Amsterdam-New York-Oxford, 1977)
204. G.O. Thorin, Convexity theorems generalizing those of M. Riesz and Hadamard with some applications. Comm. Sem. Math. Univ. Lund. **9**, 1–58 (1948)
205. G. Todorova, B. Yordanov, Critical exponent for a nonlinear wave equation with damping. J. Differ. Equ. **174**, 464–489 (2001)
206. F. Treves, *Basic Linear Partial Differential Equations* (Academic, New York-San Francisco-London, 1975)
207. H. Triebel, *Theory of Function Spaces*. Monographs in Mathematics, vol. 78 (Birkhäuser, Basel, 1983)
208. Y. Tsutsumi, Local energy decay of solutions to the free Schrödinger equation in exterior domains. Fac. Sci. Univ. Tokyo Sect. IA Math. **31**(1), 97–108 (1984)
209. Y. Tsutsumi, L^2-Solutions for nonlinear Schrödinger equations and nonlinear groups. Funkcialaj Ekvacioj **30**, 115–125 (1987)
210. Y. Tsutsumi, K. Yajima, The asymptotic behavior of non-linear Schrödinger equations. Bull. Am. Math. Soc. **11**(1), 186–188 (1984)
211. W. Tutschke, *Solution of Initial Value Problems in Classes of Generalized Analytic Functions* (Teubner, Leipzig, 1989)
212. M.C. Vilela, Regularity of solutions to the free Schrödinger equation with radial initial data. Ill. J. Math. **45**, 361–370 (2001)
213. B.G. Walther, A sharp weighted L^2-estimate for the solution to the time-dependent Schrödinger equation. Ark. Mat. **37**, 381–393 (1999)
214. W.v. Wahl, L^p Decay rates for homogeneous wave equations. Math. Z. **120**, 93–106 (1971)
215. B. Wang, H. Hudzik, The global Cauchy problem for the NLS and NLKG with small rough data. J. Differ. Equ. **232**, 36–73 (2007)
216. K. Watanabe, Smooth perturbations of the selfadjoint operator $|\Delta|^{\frac{\alpha}{2}}$. Tokyo J. Math. **14**, 239–250 (1991)
217. F.B. Weissler, Existence and nonexistence of global solutions for a semi-linear heat equation. Isr. J. Math. **38**(1–2), 29–40 (1981)
218. M. Wiegner, The Navier-Stokes equations - a neverending challenge? Jber. d. Dt. Math.-Verein. **101**, 1–25 (1999)
219. C.H. Wilcox, Measurable eigenvectors for Hermitian matrix-valued polynomials. J. Math. Anal. Appl. **40**, 12–19 (1972)
220. J. Wirth, About the solvability behaviour for special classes of nonlinear hyperbolic equations. Nonl. Anal. **52**(2), 421–431 (2003)
221. J. Wirth, Solution representations for a wave equation with weak dissipation. Math. Meth. Appl. Sci. **27**, 101–124 (2004)
222. J. Wirth, Asymptotic properties of solutions to wave equations with time-dependent dissipation. Ph.D. thesis, Technical University Bergakademie Freiberg (2005), 146pp.
223. J. Wirth, Wave equations with time-dependent dissipation I. Non-effective dissipation. J. Differ. Equ. **222**, 487–514 (2006)
224. J. Wirth, Wave equations with time-dependent dissipation II. Effective dissipation. J. Differ. Equ. **232**, 74–103 (2007)

225. K. Yagdjian, *The Cauchy Problem for Hyperbolic Operators. Multiple Characteristics. Micro-Local Approach.* Mathematical Topics, vol. 12 (Akademie, Berlin, 1997)
226. K. Yagdjian, Parametric resonance and nonexistence of global solution to nonlinear wave equations. J. Math. Anal. Appl. **260**(1), 251–268 (2001)
227. K. Yagdjian, *Global Existence in the Cauchy Problem for Nonlinear Wave Equations with Variable Speed of Propagation.* Operator Theory: Advances and Applications, vol. 159 (Birkhäuser, Basel, 2005), pp. 301–385
228. H. Yang, A. Milani, On the diffusion phenomenon of quasilinear hyperbolic waves. Bull. Sci. Math. **124**, 415–433 (2000)
229. B.T. Yordanov, Q.S. Zhang, Finite time blow up for critical wave equations in high dimensions. J. Funct. Anal. **231**, 361–374 (2006)
230. K. Yosida, *Functional Analysis.* Classics in Mathematics (Springer, Berlin, 1995). Reprint of the sixth edition (1980)
231. E.C. Zachmanoglou, D.W. Thoe, *Introduction to Partial Differential Equations with Applications* (Dover, New York, 1986)
232. Q.S. Zhang, A blow-up result for a nonlinear wave equation with damping: the critical case. C. R. Acad. Sci. Paris Ser. I Math. **333**, 109–114 (2001)
233. Y. Zhou, Blow up of classical solutions to $\Box u = |u|^{1+\alpha}$ in three space dimensions. J. Partial Differ. Equ. **5**, 21–32 (1992)
234. Y. Zhou, Life span of classical solutions to $\Box u = |u|^p$ in two space dimensions. Chin. Ann. Math. Ser. B **14**, 225–236 (1993)
235. Y. Zhou, Cauchy problem for semilinear wave equations in four space dimensions with small initial data. J. Partial Differ. Equ. **8**(2), 135–144 (1995)
236. Y. Zhou, Blow up of solutions to semilinear wave equations with critical exponent in high dimensions. Chin. Ann. Math. Ser. B **28**, 205–212 (2007)

Notations

List of Symbols

$\lvert \cdot \rvert$	Absolute value or the norm of a vector
$[x]^+$	$[x]^+ = \max\{x; 0\}$
$\lceil \cdot \rceil$	The ceiling function, i.e., $\lceil x \rceil = \min\{m \in \mathbb{Z} : x \leq m\}$
$[\cdot]$	The integer part, i.e., $[x] = \max\{m \in \mathbb{Z} : x \geq m\}$
$\langle \cdot \rangle$	Which stands for $\langle x \rangle = \sqrt{1 + \lvert x \rvert^2}$
$\mathfrak{R}$	Real part
$\mathbf{n}$	Normal vector
$f \approx g$	If there exist constants $C_1, C_2 > 0$ such that $C_1 g \leq f \leq C_2 g$
$f \sim g$	If $\lim_{t \to \infty} \frac{f(t)}{g(t)} = 1$, i.e., f and g have the same asymptotic behavior for $t \to \infty$
$f(t) = O(g(t))$	If there exist constants $C > 0$ and $t_0 > 0$ such that $\lvert f(t) \rvert \leq C \lvert g(t) \rvert$ for all $t \geq t_0$
$f(t) = o(g(t))$	If for every positive constant ϵ there exists a constant $t_0 > 0$ such that $\lvert f(t) \rvert \leq \epsilon \lvert g(t) \rvert$ for all $t \geq t_0$
$supp\ u$	Support of u
$(\cdot, \cdot)$	Inner product in $\mathbb{R}^n$
$(\cdot, \cdot)_H$	Inner product in a Hilbert space H
$\lVert \cdot \rVert_B$	Norm in a Banach space B
$\hookrightarrow$	Continuous imbedding
D_t	$D_t = \frac{1}{i} \partial_t$
$\partial_{\mathbf{n}}$	Normal derivative
∂_x^α	Partial derivatives $\partial_{x_1}^{\alpha_1} \partial_{x_2}^{\alpha_2} \cdots \partial_{x_n}^{\alpha_n}$ with a multi-index $\alpha = (\alpha_1, \alpha_2, \cdots, \alpha_n)$, where α_i is nonnegative for all $i = 1, 2, \cdots, n$
$\langle D_x \rangle^s$	Pseudodifferential operator with symbol $\langle \cdot \rangle^s$
L^*	Adjoint or dual operator to L
Δ	Laplace operator in $\mathbb{R}^n$, i.e., $\Delta = \partial_{x_1}^2 + \partial_{x_2}^2 + \cdots + \partial_{x_n}^2$
$F(f)$ or $\hat{f}$	Fourier transform

© Springer International Publishing AG 2018

M.R. Ebert, M. Reissig, *Methods for Partial Differential Equations*,
https://doi.org/10.1007/978-3-319-66456-9

$F_{x \to \xi}$ or $\hat{u}(t, \xi)$ Partial Fourier transform in x
$F^{-1}(f)$ Inverse Fourier transform
$F^{-1}_{\xi \to x}$ Partial inverse Fourier transform in x
$f * g$ Convolution

Spaces of Functions and Distributions

$C^k(\mathbb{R}^n)$ Space of k−times continuously differentiable functions
$C^\infty(\mathbb{R}^n)$ Space of infinitely continuously differentiable functions
$C^k(I, B)$ Space of k−times continuously differentiable functions from an interval $I \subset \mathbb{R}$ into a Banach space B
$C^\infty_0(\mathbb{R}^n)$ Space of infinitely continuously differentiable functions with compact support
$S(\mathbb{R}^n)$ Schwartz space of rapidly decreasing functions
$L^p(\mathbb{R}^n)$ Lebesgue spaces
$W^m_p(\mathbb{R}^n)$ Sobolev spaces
$W^s_p(\mathbb{R}^n)$ Sobolev-Slobodeckij spaces
$H^s(\mathbb{R}^n)$ Sobolev spaces of fractional order based on $L^2(\mathbb{R}^n)$
$H^s_p(\mathbb{R}^n)$ Sobolev spaces of fractional order or Bessel potential space
$\dot{H}^s_p(\mathbb{R}^n)$ Homogeneous Sobolev spaces of fractional order
$B^s_{p,q}(\mathbb{R}^n)$ Besov spaces
$\dot{B}^s_{p,q}(\mathbb{R}^n)$ Homogeneous Besov spaces
$F^s_{p,q}(\mathbb{R}^n)$ Triebel-Lizorkin spaces
$\dot{F}^s_{p,q}(\mathbb{R}^n)$ Homogeneous Triebel-Lizorkin spaces
$S'(\mathbb{R}^n)$ Space of tempered distributions
$D'(\mathbb{R}^n)$ Space of distributions
$M^q_p(\mathbb{R}^n)$ Spaces of Multipliers inducing bounded translation invariant operators from $L(L^p \to L^q)$

Index

© Springer International Publishing AG 2018
M.R. Ebert, M. Reissig, *Methods for Partial Differential Equations*,
https://doi.org/10.1007/978-3-319-66456-9

Printed in the United States
By Bookmasters